Teodor Bulboacă, Santosh B. Joshi, and
Pranay Goswami

Complex Analysis

Theory and Applications

DE GRUYTER

Mathematics Subject Classification 2010
30-XX

Authors
Prof. Dr. Habil. Teodor Bulboacă
Babeş-Bolyai University
Faculty of Mathematics and Computer Science
Str. Mihail Kogălniceanu Nr. 1
400084 Cluj-Napoca
Romania
bulboaca@math.ubbcluj.ro

Prof. Dr. Santosh B. Joshi
Walchand College of Engineering
Department of Mathematics
416415 Maharastra
India
joshisb@hotmail.com

Prof. Dr. Pranay Goswami
Ambedkar University Delhi
Kashmere Gate Campus
Lothian Road, Kashmere Gate
110006 Delhi
India
pranaygoswami83@gmail.com

ISBN 978-3-11-065782-1
e-ISBN (PDF) 978-3-11-065786-9
e-ISBN (EPUB) 978-3-11-065803-3

Library of Congress Control Number: 2019938423

Bibliographic information published by the Deutsche Nationalbibliothek
The Deutsche Nationalbibliothek lists this publication in the Deutsche Nationalbibliografie; detailed bibliographic data are available on the Internet at http://dnb.dnb.de.

Cover image: Lokal_Profil (Grafic »conformal_grid«) – Michael Bernhart (surface)
Typesetting: VTeX UAB, Lithuania
Printing and binding: CPI books GmbH, Leck

MIX
Papier aus verantwortungsvollen Quellen
FSC www.fsc.org FSC® C083411

www.degruyter.com

Contents

Dedicated to the memory of Professor Petru T. Mocanu (1931–2016)

Preface

Complex analysis, as we know it now, is the culmination of over 500 years of mathematical development that has had tremendous influence in mathematics, physics and engineering. The numbers we now know as "complex" (a most unfortunate name due to C. F. Gauss) made their first appearance through the use of square roots of negative numbers in methods for the solution of cubic and quartic equations[1] in the works of N. F. Tartaglia and G. Cardano and proved their value in "predicting" the correct values of roots. R. Descartes, who also coined the concept of a real number, named such square roots of negative numbers imaginary, since they could "only" be imagined. Leonhard Euler made extensive use of complex numbers and also introduced these[2] in his textbooks. However, Euler also had only a vague geometric notion of complex numbers and the complex plane. The geometrization of the complex numbers had to wait until the end of the eighteenth century by the separate insights of J-R. Argand, C. Wessel and Gauss. Much of the development of classical complex analysis, the topic of this book, occurred in the nineteenth century in the hands of A-L. Cauchy and B. Riemann followed by the even more geometrical work of F. Klein, H. Poincare and many others.

This book is an in-depth and modern presentation of important classical results in complex analysis and is suitable for a first course in this topic. The organization of the material is as one would find it in a typical syllabus for an undergraduate course in complex analysis. This book is an out-growth of, and has been tested during the courses we taught at Babeş-Bolyai University, Romania, Walchand College of Engineering, India and Ambedkar University Delhi, India. The level of difficulty of the material increases gradually from chapter to chapter. Each chapter contains exercises with solutions and applications of the results. We have strived for diversity in solution methods.

Chapter 1 introduces the concept of complex numbers, and their arithmetic and geometric properties. We use stereographic projection to introduce the one-point compactification of the complex plane, the Riemann sphere.

Chapter 2 studies complex valued function, and various notions of differentiability of such functions and culminates with the concept of a holomorphic function. Our presentation of their basic properties ends with the Cauchy–Riemann equations. This chapter ends with a thorough overview of elementary entire functions and Möbius transformations that will be needed in the remaining chapters.

Chapter 3 starts with the definitions of paths and complex integrals and is followed by Cauchy's theorem and its consequences: the fundamental theorem of alge-

---

1 See, for example, Nahin, Paul J. (1998), An Imaginary Tale: The Story of $\sqrt{-1}$, Princeton University Press.

2 and the notation i for $\sqrt{-1}$ to reduce confusion in the applications of rules such as $\sqrt{ab} = \sqrt{a}\sqrt{b}$.

https://doi.org/10.1515/9783110657869-201

bra, the Cauchy integral formula for holomorphic functions defined on the disc and Morera's theorem establishing sufficient conditions for holomorphy. The chapter ends with many applications including the theory of multivalent functions.

Chapter 4 characterizes holomorphic function by their local analytic properties via power series expansions. We present important theorems on the zeroes of holomorphic functions, the uniqueness of holomorphic functions, the maximum modulus principle, the Schwarz lemma, Laurent series expansions, isolated singular points and some basic results on meromorphic functions.

Chapter 5 develops the theory of residues and its principal applications: the computation of a variety of trigonometric and improper integrals. We also apply the theory of residues to the study of zeros and poles of meromorphic functions, the principle of the argument and Rouché's theorem. This chapter ends with the open mapping theorem for nonconstant holomorphic functions and its topological consequences.

Chapter 6 starts with the fundamental theorems of Montel, Vitali and Hurwitz and is then devoted to the topic of conformal mappings, univalent functions and the Riemann mapping theorem.

Chapter 7 contains the solutions of all exercises that appeared at the end of the previous chapters.

This book evolved from lectures delivered at universities in Romania and India, hence we would especially thank to students since it was their reaction for the course of "Complex Analysis" that made us to write this book. We would like to thank to our family members for their encouragement, timely help, and patience.

We consider that this book could be used by the students that need a basic (and not only) preparation in complex analysis, and for all that are interested in this field, and any remarks and comments for improving the content of the book are welcome.

1 Complex numbers

1.1 The field of the complex numbers

Let $\mathbb{R}$ denote the set of all real numbers, where we defined the usual addition and multiplication, such that this set together with these two operations will be a field.

Definition 1.1.1.

1. Let $\mathbb{R}^2 = \mathbb{R} \times \mathbb{R}$ denote the set of all real ordered couples, i. e., $(x, y) \in \mathbb{R}^2 \Leftrightarrow x, y \in \mathbb{R}$.
2. Define in the $\mathbb{R}^2$ set the addition, and respectively, multiplication binary operations as follows:

$$+ : \mathbb{R}^2 \times \mathbb{R}^2 \to \mathbb{R}^2, \quad (x_1, y_1) + (x_2, y_2) = (x_1 + x_2, y_1 + y_2);$$
$$\cdot : \mathbb{R}^2 \times \mathbb{R}^2 \to \mathbb{R}^2, \quad (x_1, y_1) \cdot (x_2, y_2) = (x_1x_2 - y_1y_2, x_1y_2 + y_1x_2).$$

3. The above defined algebraic structure $(\mathbb{R}^2, +, \cdot)$ is called the set of complex numbers, briefly denoted by the symbol $\mathbb{C}$. We will use the notation $\mathbb{C}^* = \mathbb{C} \setminus \{(0, 0)\}$.

Theorem 1.1.1.

1. $(\mathbb{C}, +)$ *is an Abelian group; the zero element is* $(0, 0)$.
2. $(\mathbb{C}^*, \cdot)$ *is an Abelian group; the unit element is* $(1, 0)$.
3. *The multiplication "·" is distributive with respect to the addition "+".*

Proof. The points 1 and 3 can be proved directly.

2. Using simple computations, we may check that the operation "·" is associative and commutative. Supposing that there exists a unitary element, it follows easily that it will be $(1, 0)$.

We will prove that every nonzero element has an inverse. Let $(x_1, y_1) \in \mathbb{C}^*$ be an arbitrary. Then $(x_1, y_1) \neq (0, 0) \Leftrightarrow x_1^2 + y_1^2 \neq 0$, hence

$$\exists (x, y) \in \mathbb{C}^* \quad \text{such that } (x_1, y_1) \cdot (x, y) = (1, 0)$$
$$\Leftrightarrow \exists (x, y) \in \mathbb{C}^* \quad \text{such that } (x_1x - y_1y, x_1y + y_1x) = (1, 0) \Leftrightarrow \begin{cases} x_1x - y_1y = 1 \\ y_1x + x_1y = 0. \end{cases}$$

Since $\det A = x_1^2 + y_1^2 \neq 0$, because $(x_1, y_1) \neq (0, 0)$, the above system is compatible (where x and y are unknown) and has a unique solution.

This will be $(x, y) = (\frac{x_1}{x_1^2+y_1^2}, \frac{-y_1}{x_1^2+y_1^2})$, hence for all elements of the form $(x_1, y_1) \neq (0, 0)$ there exists an inverse given by the above relation, so $\mathbb{C}$ is a commutative field. □

The proof of the following result is straightforward, so it is left to the reader.

https://doi.org/10.1515/9783110657869-001

Theorem 1.1.2. *The subset* $\mathbb{R} \times \{0\} = \{(x,0) : x \in \mathbb{R}\} \subset \mathbb{C}$ *is a subfield of the* $\mathbb{C}$ *field with respect to the above operations.*

The function $\varphi : \mathbb{R} \to \mathbb{R} \times \{0\} \subset \mathbb{C}$ *defined by* $\varphi(x) = (x,0)$ *is an isomorphism between the above field.*

Conventions.
1. According to the Theorem 1.1.2, we will denote the subfield $\mathbb{R} \times \{0\}$ of the $\mathbb{C}$ field by $\mathbb{R}$, so we will simplify our notation.
2. We will introduce the notation $i = (0,1) \in \mathbb{C}$.
3. Using the above convention, we get $\mathbb{C} = \mathbb{R} + i\mathbb{R}$.

Theorem 1.1.3. *Any arbitrary complex number* $z = (x,y)$ *may be written in the form* $x + iy$, *where* $i^2 = -1$.

Proof. A simple calculation shows that $(0,1) \cdot (0,1) = (-1,0) \Rightarrow i^2 = -1$, and hence

$$z = (x,y) = (x,0) + (0,1) \cdot (y,0) = x + iy.$$

□

Definition 1.1.2.
1. The form $z = x + iy$ of the complex number $z = (x,y)$, is called **the algebraic form** of the given complex number.
2. The numbers x, y, $x - iy$, and $\sqrt{x^2+y^2}$ are called **the real part, the imaginary part, the conjugate** and **the module,** respectively, of the number $z \in \mathbb{C}$.
3. If $z \in \mathbb{C}^*$, then the inverse of it, with respect to the multiplication, is denoted by $\frac{1}{z}$.

Notation. $x = \operatorname{Re} z$, $y = \operatorname{Im} z$, $x - iy = \overline{z}$, $\sqrt{x^2+y^2} = |z|$.

Theorem 1.1.4. *For any arbitrary numbers* $z, z_1, z_2 \in \mathbb{C}$, *the next properties hold:*
1. $\operatorname{Re} z = \frac{1}{2}(z + \overline{z})$, $\operatorname{Im} z = \frac{1}{2i}(z - \overline{z})$;
2. $\overline{z_1 + z_2} = \overline{z_1} + \overline{z_2}$, $\overline{z_1 z_2} = \overline{z_1}\,\overline{z_2}$, $\overline{\overline{z}} = z$;
3. $-|z| \le \operatorname{Re} z \le |z|$, $-|z| \le \operatorname{Im} z \le |z|$, $|\overline{z}| = |z|$;
4. $|z|^2 = z\overline{z}$, $\frac{1}{z} = \frac{\overline{z}}{|z|^2}$ *if* $z \neq 0$;
5. $|z| = 0 \Leftrightarrow z = 0$, $|z_1 z_2| = |z_1||z_2|$, $|z_1 + z_2| \le |z_1| + |z_2|$.

Proof. The proof of the first four points is easy and straightforward, and we will provide the proof of the last point only. We can see that

$$\begin{aligned}|z_1 + z_2|^2 &= (z_1 + z_2)(\overline{z_1 + z_2}) = (z_1 + z_2)(\overline{z_1} + \overline{z_2}) \\ &= z_1\overline{z_1} + z_1\overline{z_2} + z_2\overline{z_1} + z_2\overline{z_2} = |z_1|^2 + |z_2|^2 + 2\operatorname{Re}(z_1\overline{z_2}).\end{aligned}$$

Since

$$2\operatorname{Re}(z_1\overline{z_2}) \le 2|z_1\overline{z_2}| = 2|z_1||\overline{z_2}| = 2|z_1||z_2|$$
$$\Rightarrow |z_1+z_2|^2 \le |z_1|^2 + |z_2|^2 + 2|z_1||z_2| = (|z_1|+|z_2|)^2 \Rightarrow |z_1+z_2| \le |z_1|+|z_2|. \qquad \square$$

1.2 The complex plane

Let xOy an orthogonal axes system on a plane. We will identify the set $\mathbb{R}^2$ with the plane xOy. Since $\mathbb{R}^2$ and $\mathbb{C}$ are identical, it follows that we could identify the complex set $\mathbb{C}$ with the plane xy, such that $x = \mathbb{R}$ and $y = i\mathbb{R}$. Thus, the axis x will be called **the real axis**, and y will be called **the imaginary axis.**

To every point from xy, it corresponds one and only one complex number of $\mathbb{C}$, and vice versa. Thus, there exists a bijective correspondence between the sets $\mathbb{R}^2$ and $\mathbb{C}$; i. e., $(x,y) \longleftrightarrow x+iy$.

Letting $z \in \mathbb{C}\setminus\{0\} = \mathbb{C}^*$, $z = x+iy$, from the relation

$$\frac{z}{|z|} = \frac{x}{\sqrt{x^2+y^2}} + i\frac{y}{\sqrt{x^2+y^2}}$$

combined with elementary knowledge of trigonometry, it follows that the system

$$\begin{cases} \cos\theta = \frac{x}{\sqrt{x^2+y^2}} \\ \sin\theta = \frac{y}{\sqrt{x^2+y^2}} \end{cases}$$

is compatible, with respect to the θ unknown. This fact allows us to give the following definition.

Definition 1.2.1.

1. Let $z \in \mathbb{C}^*$. Every solution θ of the equation

$$\cos\theta + i\sin\theta = \frac{z}{|z|}$$

   is called **the argument of the complex number** z.
2. The equation has a unique solution in the interval $(-\pi,\pi]$, and it is called **the principal argument of the complex number** z, and is denoted by $\arg z$. If we denote by $\operatorname{Arg} z$ the set of all the arguments of the number z, then

$$\operatorname{Arg} z = \{\arg z + 2k\pi : k \in \mathbb{Z}\}.$$

3. The correspondence $z \longmapsto \operatorname{Arg} z$ is **a multivalued function** from $\mathbb{C}^*$ to $\mathbb{R}$.

Theorem 1.2.1 (Properties of the argument). *For any arbitrary numbers $z_1, z_2 \in \mathbb{C}^*$, we have:*

1. $\operatorname{Arg}(z_1z_2) = \operatorname{Arg} z_1 + \operatorname{Arg} z_2$;
2. $\operatorname{Arg} \frac{z_1}{z_2} = \operatorname{Arg} z_1 - \operatorname{Arg} z_2$.

Proof. Letting $\theta_j \in \operatorname{Arg} z_j$, $j = 1, 2$, then

$$\cos\theta_j + i\sin\theta_j = \frac{z_j}{|z_j|}, \quad j = 1, 2.$$

Using simple trigonometric properties, we obtain that

$$\frac{z_1}{|z_1|}\frac{z_2}{|z_2|} = (\cos\theta_1 + i\sin\theta_1)(\cos\theta_2 + i\sin\theta_2) = \cos(\theta_1+\theta_2) + i\sin(\theta_1+\theta_2)$$

and

$$\frac{z_1}{|z_1|}\frac{z_2}{|z_2|} = \frac{z_1z_2}{|z_1z_2|} \Rightarrow \theta_1 + \theta_2 \in \operatorname{Arg} z_1z_2,$$

hence

$$\operatorname{Arg} z_1 + \operatorname{Arg} z_2 \subset \operatorname{Arg}(z_1z_2). \tag{1.1}$$

Similarly, we deduce that

$$\begin{aligned}\frac{\frac{z_1}{z_2}}{|\frac{z_1}{z_2}|} &= \frac{\frac{z_1}{|z_1|}}{\frac{z_2}{|z_2|}} = \frac{\cos\theta_1 + i\sin\theta_1}{\cos\theta_2 + i\sin\theta_2} = \frac{(\cos\theta_1 + i\sin\theta_1)(\cos\theta_2 - i\sin\theta_2)}{\cos^2\theta_2 + \sin^2\theta_2}\\ &= \cos(\theta_1-\theta_2) + i\sin(\theta_1-\theta_2),\end{aligned}$$

hence

$$\theta_1 - \theta_2 \in \operatorname{Arg}\left(\frac{z_1}{z_2}\right),$$

or

$$\operatorname{Arg} z_1 - \operatorname{Arg} z_2 \subset \operatorname{Arg}\left(\frac{z_1}{z_2}\right). \tag{1.2}$$

Let $\theta \in \operatorname{Arg}(z_1z_2)$ be arbitrary, and consider an arbitrary angle $\theta_1 \in \operatorname{Arg} z_1$. From the relation (1.2), we get

$$\theta - \theta_1 \in \operatorname{Arg}\left(\frac{z_1z_2}{z_1}\right) = \operatorname{Arg} z_2,$$

and thus

$$\theta \in \operatorname{Arg} z_2 + \theta_1 \subset \operatorname{Arg} z_1 + \operatorname{Arg} z_2. \tag{1.3}$$

In conclusion, from the relations (1.1) and (1.3) it follows the equality of the first point.

Since

$$\operatorname{Arg} z_1 = \operatorname{Arg}\left(\frac{z_1}{z_2} z_2\right) \overset{(1.1)}{=} \operatorname{Arg}\frac{z_1}{z_2} + \operatorname{Arg} z_2,$$

we have

$$\forall \theta \in \operatorname{Arg}\frac{z_1}{z_2},\ \exists \theta_1 \in \operatorname{Arg} z_1,\ \exists \theta_2 \in \operatorname{Arg} z_2 \quad \text{such that } \theta_1 = \theta + \theta_2.$$

hence

$$\theta = \theta_1 - \theta_2 \in \operatorname{Arg} z_1 - \operatorname{Arg} z_2,$$

and thus

$$\operatorname{Arg}\frac{z_1}{z_2} \subset \operatorname{Arg} z_1 - \operatorname{Arg} z_2. \tag{1.4}$$

From the relations (1.2) and (1.4), it follows the equality of the second point. □

Definition 1.2.2. If $z \in \mathbb{C}^*$, then

$$z = |z|(\cos\theta + i\sin\theta), \quad \text{where } \theta \in \operatorname{Arg} z \tag{1.5}$$

is called **the trigonometric form** of the complex number z.

Remarks 1.2.1.

1. The set $\{z \in \mathbb{C} : z = \operatorname{Re} z\}$ is **the real axis.**
2. The set $\{z \in \mathbb{C} : z = i \operatorname{Im} z\}$ is **the imaginary axis.**
3. The set of those points $z \in \mathbb{C}$, such that $\operatorname{Re} z > 0$, $\operatorname{Re} z < 0$, $\operatorname{Im} z > 0$, $\operatorname{Im} z < 0$ is called **the right, the left, the upper,** and respectively **the lower open half-plane.**
4. The equation of **a line** of the complex plane is

   $$z = a + bt, \quad \text{where } a \in \mathbb{C},\ b \in \mathbb{C}^* \text{ and } t \in \mathbb{R}\ (t \text{ runs all over the } \mathbb{R}).$$

   This line contains the point a and it is parallel with the vector b.
5. The above equation of the line can be written in the form

   $$\operatorname{Im}\frac{z-a}{b} = 0.$$

6. The set $\{z \in \mathbb{C} : \operatorname{Im}\frac{z-a}{b} > 0\}$ is called **the right half-plane bounded by the above line.**
7. The set $\{z \in \mathbb{C} : \operatorname{Im}\frac{z-a}{b} < 0\}$ is called **the left half-plane bounded by the above line.**
8. **The angle of the lines** $z = a_1 + b_1 t$ and $z = a_2 + b_2 t$ is given by $\operatorname{Arg}\frac{b_2}{b_1}$. It depends on the considered order of the lines, and has a countable number of elements.
9. The set $S_{z_0}(\alpha_1, \alpha_2) = \{z \in \mathbb{C} : \alpha_1 < \arg(z - z_0) < \alpha_2\}$ is an **angular sector** with the apex in z_0.

1.3 The topological and metric structure of the complex plane

Let us define the function $\mathrm{d} : \mathbb{C} \times \mathbb{C} \to \mathbb{R}$ by the following formula:

$$\mathrm{d}(z_1, z_2) = |z_1 - z_2|.$$

Theorem 1.3.1. *The pair* $(\mathbb{C}, \mathrm{d})$ *is a metric space, i. e.,*

$$\begin{aligned}
&(m_1) \quad \mathrm{d}(z_1, z_2) \geq 0, \quad \forall z_1, z_2 \in \mathbb{C} \quad \textit{and} \quad \mathrm{d}(z_1, z_2) = 0 \Leftrightarrow z_1 = z_2;\\
&(m_2) \quad \mathrm{d}(z_1, z_2) = \mathrm{d}(z_2, z_1), \quad \forall z_1, z_2 \in \mathbb{C};\\
&(m_3) \quad \mathrm{d}(z_1, z_3) \leq \mathrm{d}(z_1, z_2) + \mathrm{d}(z_2, z_3), \quad \forall z_1, z_2, z_3 \in \mathbb{C}.
\end{aligned}$$

Proof. The (m_1) and (m_2) properties are obvious, and it follows from the fact that $|z_1 - z_2| = 0 \Leftrightarrow z_1 - z_2 = 0$. For (m_3), it is well known that $\forall w_1, w_2 \in \mathbb{C}$, we have

$$|w_1 + w_2| \leq |w_1| + |w_2|. \tag{1.6}$$

Let $w_1 = z_1 - z_2$, $w_2 = z_2 - z_3$. Hence $w_1 + w_2 = z_1 - z_3$, and replacing these numbers in the inequality (1.6), we get

$$|z_1 - z_3| \leq |z_1 - z_2| + |z_2 - z_3|,$$

which is in fact (m_3). □

Remark 1.3.1. If we replace the complex numbers that appeared in the definition of the metric d by the corresponding couples of real numbers, we see that this metric represents the Euclidian distance of the Euclidian plane. Hence, the $(\mathbb{C}, \mathrm{d})$ metric space can be identified with the couple consisting of the plane together with the Euclidian distance.

1.3.1 Basic definitions and notation

1. If $z_0 \in \mathbb{C}$ and $r > 0$, then

   $$U(z_0; r) = \{z \in \mathbb{C} : |z - z_0| < r\}$$

   is **the (open) disc with the center** z_0 **and radius** r.
2. We denote by $\dot{U}(z_0; r) = U(z_0; r) \setminus \{z_0\}$.
3. If $z_0 \in \mathbb{C}$ and $0 \leq r_1 < r_2$, then

   $$U(z_0; r_1, r_2) = \{z \in \mathbb{C} : r_1 < |z - z_0| < r_2\}$$

   is **the (open) ring with the center** z_0 **and radius** r_1 **and** r_2.
4. The set $G \subset \mathbb{C}$ is called **an open set,** if $G = \emptyset$ or if $\forall z \in G$, $\exists r_z > 0$ such that $U(z; r_z) \subset G$.

5. The set $F \subset \mathbb{C}$ is called **a closed set,** if $\mathbb{C} \setminus F$ is an open set.
6. If $z_1, z_2 \in \mathbb{C}$, then

$$[z_1, z_2] = \{z \in \mathbb{C} : z = z_1 + (z_2 - z_1)t, t \in [0,1]\}$$

is called **the closed segment that connects** z_1 **and** z_2.
7. For all $A \subset \mathbb{C}$, $A \neq \emptyset$, the couple (A, d) is a metric space.
8. In the (A, d) metric space, **the neighborhood system of a point** b is given by

$$\mathcal{V}(b) = \{V \subset A : \exists G \subset A, G \text{ is an open set in } A, b \in G \subset V\}.$$

9. The set $B \subset A$ is said to be **a closed set in** A (in the (A, d) metric space), if

$$\forall (b_n)_{n\in\mathbb{N}} \subset B \quad \text{and} \quad \exists \lim_{n\to\infty} b_n = b \in A \Rightarrow b \in B.$$

10. The set A is closed and open in the (A, d) metric space.
11. The set A is called to be **connected,** if in the (A, d) metric space the set A is the only one set, excepted the empty set, that is closed and open.

Example 1.3.1. The set $A = \{z_1, z_2\}$, with $z_1 \neq z_2$ is not connected; also, the set $A = U(2i; 1) \cup \mathbb{R}$, etc. The set $A = U(i; 1) \cup \mathbb{R}$ is connected.

12. The subset $M \subset \mathbb{C}$ is said to be **a domain,** if M is an open and connected subset of $\mathbb{C}$.
13. The subset $D \subset \mathbb{C}$ is said to be a **starlike set with respect to the point** $z_0 \in D$, if $[z_0, z] \subset D, \forall z \in D$.
14. If the subset $D \subset \mathbb{C}$ is starlike with respect to any arbitrary point $z \in D$, then D is called a **convex set.**
15. Let $A \subset \mathbb{C}$. We denote by A^- **the closure set of** A, defined by

$$A^- = \{z \in \mathbb{C} : U(z; r) \cap A \neq \emptyset, \forall r > 0\}.$$

16. Let $A \subset \mathbb{C}$. We denote by A' **the set of accumulation points of** A, defined by

$$A' = \{z \in \mathbb{C} : \dot{U}(z; r) \cap A \neq \emptyset, \forall r > 0\}.$$

It is easy to prove that $A^- = A \cup A'$.
17. We denote by ∂A **the boundary of the set** A, defined by

$$\partial A = A^- \cap (\mathbb{C} \setminus A)^-.$$

Exercise 1.3.1.
1. The subset $A \subset \mathbb{C}$ is closed $\Leftrightarrow A = A^-$.
2. If $G \subset \mathbb{C}$, then $\partial(G^-) \subset \partial G$.
Give an example of an open set $G \subset \mathbb{C}$, such that $\partial(G^-) \neq \partial G$.

Definition 1.3.1. Let $A, B \subset \mathbb{C}$ be two nonempty sets ($A, B \neq \emptyset$). Then **the distance between the sets** A **and** B is given by

$$d(A,B) = \inf\{d(a,b) : a \in A, b \in B\}.$$

Lemma 1.3.1.
1. $z \in A^{-} \Leftrightarrow d(z,A) = 0$.
2. *The subset $A \subset \mathbb{C}$ is closed, if and only if*

$$d(z,A) = 0 \Leftrightarrow z \in A.$$

Definition 1.3.2. The subset $B \subset \mathbb{C}$ is said **compact set,** if

$$\forall (z_n)_{n\in\mathbb{N}} \subset B,\ \exists (z_{n_j})_{j\in\mathbb{N}} \subset (z_n)_{n\in\mathbb{N}} \quad \text{such that } \exists z^* = \lim_{j\to\infty} z_{n_j} \text{ and } z^* \in B.$$

Lemma 1.3.2. *Every compact set is a closed set.*

Example 1.3.2. Let us consider the set

$$A = \mathbb{R} \subset \mathbb{C}, \quad A = \{(x,0) : x \in \mathbb{R}\}$$

and

$$B = \{(x,e^x) : x \in \mathbb{R}\} = \{x + ie^x : x \in \mathbb{R}\} \subset \mathbb{C}.$$

Thus, $A \cap B = \emptyset$, A and B are closed sets, but $d(A,B) = 0$.

Exercise 1.3.2. If $G \subset \mathbb{C}$ is an open set, then $\partial G = G^{-} \setminus G$.

Theorem 1.3.2. *If $A \subset \mathbb{C}$ is a compact set, $B \subset \mathbb{C}$ is a closed set, and $A \cap B = \emptyset$, then* $d(A,B) > 0$.

Proof. Using the above notation, suppose that $d(A,B) = 0$. Then

$$\forall n \in \mathbb{N}^*,\ \exists a_n \in A,\ \exists b_n \in B, \quad \text{such that } d(a_n,b_n) < \frac{1}{n}.$$

Since A is compact,

$$\exists (a_{n_j})_{j\in\mathbb{N}} \subset (a_n)_{n\in\mathbb{N}^*}, \quad \text{such that } \exists a_0 = \lim_{j\to\infty} a_{n_j}, \text{ and } a_0 \in A.$$

Since

$$d(a_{n_j},b_{n_j}) < \frac{1}{n_j} \Rightarrow d(a_0,b_{n_j}) \le d(a_0,a_{n_j}) + d(a_{n_j},b_{n_j}) < d(a_0,a_{n_j}) + \frac{1}{n_j}.$$

But $a_{n_j} \to a_0$, if $j \to \infty$, and $n_j \to \infty$, if $j \to \infty$, and thus

$$d(a_0,a_{n_j}) \to 0 \quad \text{and} \quad \frac{1}{n_j} \to 0, \quad \text{if } j \to \infty.$$

Hence $d(a_0, b_{n_j}) \to 0$, if $j \to \infty \Rightarrow d(a_0, B) = 0 \Rightarrow a_0 \in B^- = B$, because B is a closed set. It follows that $a_0 \in B$ and $a_0 \in A$, and thus $a_0 \in A \cap B \neq \emptyset$, which contradicts the assumption. □

Corollary 1.3.1. *If $K \subset G$ and K is a compact set, $K \neq \emptyset$, and $G \subset \mathbb{C}$ is an open set, then* $d(K, \partial G) > 0$.

Proof. We will prove that $\partial G \cap G = \emptyset$. If this is not true, then there exists $z \in \partial G \cap G$, and

$$z \in \partial G \cap G \Rightarrow z \in G \Rightarrow \exists r > 0 \quad \text{such that } U(z; r) \subset G \Rightarrow U(z; r) \cap (\mathbb{C} \setminus G) = \emptyset$$
$$\Rightarrow z \notin (\mathbb{C} \setminus G)^- \Rightarrow z \notin \partial G, \quad \text{which is a contradiction.}$$

Since $K \subset G$ and $\partial G \cap G = \emptyset \Rightarrow K \cap \partial G = \emptyset$, where K is a compact and ∂G is closed. According to Theorem 1.3.2, we conclude that $d(K, \partial G) > 0$. □

Corollary 1.3.2. *If $K \subset G$ and K is a compact set, $K \neq \emptyset$, and $G \subset \mathbb{C}$ is an open set, then $\exists r > 0$ (where r depends only of K and G), such that $\forall z \in K$, $U(z; r) \subset G$.*

1.4 Complex function, limits, continuity

First, we will consider real variable complex functions. If $[a, b] \subset \mathbb{R}$, then a function $f : [a, b] \to \mathbb{C}$ has the form

$$f(t) = \alpha(t) + i\beta(t) \quad \text{where } \alpha, \beta : [a, b] \to \mathbb{R} \text{ are real valued functions.}$$

The limits and the continuity of the function f reduces to the limits and respectively to the continuity of the real-valued functions α and β.

If $A \subset \mathbb{C}$, then the function $f : A \to \mathbb{C}$ may be written in the form

$$f(z) = u(z) + iv(z), \qquad \text{where } z \longmapsto u(z),\ z \longmapsto v(z)$$

are real valued functions defined on the complex subset A.

Using the notation $z = x + iy \in \mathbb{C}$, with $x, y \in \mathbb{R}$, then

$$w = f(z) = f(x + iy) = u(x, y) + iv(x, y),$$

where u and v are two variables real functions.

Definition 1.4.1. The function $f : A \to \mathbb{C}$ has **the limit** $l \in \mathbb{C}$ **at the point** $a \in A'$ (where A' represents the set of all the accumulation points of A), written $\lim_{z \to a} f(z) = l$, if

$$\forall \varepsilon > 0,\ \exists \eta = \eta(\varepsilon) > 0 \quad \text{such that } \forall z \in A \text{ and } 0 < |z - a| < \eta \Rightarrow |f(z) - l| < \varepsilon.$$

We may easily prove that $\lim_{z\to a} f(z) = l$ is equivalent to each of the following relations:

(i) $\lim\limits_{z\to a} \overline{f(z)} = \bar{l}$;

(ii) $\lim\limits_{z\to a} \operatorname{Re} f(z) = \operatorname{Re} l$ and $\lim\limits_{z\to a} \operatorname{Im} f(z) = \operatorname{Im} l$.

Let $f : A \to \mathbb{C}$ and $z_0 \in A \cap A'$. Then, the function f is called to be **continuous at the point** z_0, if $\lim_{z\to z_0} f(z) = f(z_0)$, i. e., if

$$\forall \varepsilon > 0,\ \exists \eta = \eta(\varepsilon) > 0 \quad \text{such that}$$
$$\forall z \in A \quad \text{and} \quad 0 < |z - z_0| < \eta \Rightarrow |f(z) - f(z_0)| < \varepsilon.$$

If the function f is continuous at every point of the set A, then we say that the function f is **continuous on the set** A.

From the *triangle inequality*,

$$\big||f(z)| - |f(z_0)|\big| \leq |f(z) - f(z_0)|$$

it follows, that if f is continuous at the point z_0, then $|f|$ is also continuous at the point z_0, where $|f|(z) = |f(z)|$. The reverse implication is not true!

1.5 The compactified complex plane

As we will see in the following chapters, it is useful to introduce the notion of the ∞ point, which will "complete" the complex plane $\mathbb{C}$ with a unique "point" denoted by ∞, that is situated to an "infinite large distance" from the origin (and from any other point of $\mathbb{C}$), and satisfies the properties:

1. $\widehat{\mathbb{C}} = \mathbb{C} \cup \{\infty\}$;
2. $a + \infty = \infty + a = \infty$, $\forall a \in \mathbb{C}$;
3. $a \cdot \infty = \infty \cdot a = \infty$, $\forall a \in \widehat{\mathbb{C}} \setminus \{0\}$, (hence $-\infty = \infty$);
4. $\frac{a}{0} = \infty$, $\forall a \in \widehat{\mathbb{C}} \setminus \{0\}$;
5. $\frac{a}{\infty} = 0$, $\forall a \in \mathbb{C}$.

The next forms have no sense (are not well-defined): $\infty - \infty$, $0 \cdot \infty$, $\frac{0}{0}$, $\frac{\infty}{\infty}$.

Also, we have $|\infty| = +\infty$ (the real $+\infty$).

Definition 1.5.1. The above defined $\widehat{\mathbb{C}}$ set, together with the addition operation "+" and the multiplication operation "·" is called the **compactified complex plane.**

Conventions.

1. The ∞ is an element of every line of $\widehat{\mathbb{C}}$ (i. e., lines of $\mathbb{C}$ together with ∞).
2. The ∞ is not element of any circle.

Let us denote by $S^2 \subset \mathbb{R}^3$, $S^2 = \{(x,y,u) \in \mathbb{R}^3 : x^2+y^2+u^2 = 1\}$ the **two dimensional sphere (ball)** (Figure 1.1).

Theorem 1.5.1. *Let us define the function $\Phi : S^2 \to \widehat{\mathbb{C}}$ by*

$$z = \Phi(x,y,u) = \begin{cases} \frac{x+iy}{1-u}, & \text{if } (x,y,u) \neq (0,0,1); \\ \infty, & \text{if } (x,y,u) = (0,0,1). \end{cases}$$

Let $N = (0,0,1) \in S^2$, and denote by φ the restriction of the function Φ to $S^2 \setminus \{N\}$, i. e., $\varphi = \Phi|_{S^2\setminus\{N\}}$.

Then the function φ is a continuous bijection between $S^2 \setminus \{N\}$ and $\mathbb{C}$, and φ^{-1} is also continuous.

Proof. 1. The function φ is a bijection.

Thus,

$$z = \frac{x+iy}{1-u} \Rightarrow |z|^2 = z\overline{z} = \frac{x+iy}{1-u}\frac{x-iy}{1-u} = \frac{x^2+y^2}{(1-u)^2} = \frac{1-u^2}{(1-u)^2} = \frac{1+u}{1-u}.$$

From the above relation, we get

$$u = \frac{|z|^2-1}{|z|^2+1} \quad \text{and} \quad 1-u = \frac{2}{|z|^2+1}.$$

Since

$$z+\overline{z} = \frac{2x}{1-u}, \qquad z-\overline{z} = \frac{2iy}{1-u}$$

it follows that

$$\varphi^{-1}(z) = (x,y,u) = \left(\frac{z+\overline{z}}{1+|z|^2}, \frac{z-\overline{z}}{i(1+|z|^2)}, \frac{|z|^2-1}{|z|^2+1}\right), \quad \forall z \in \mathbb{C}.$$

2. The function $\varphi(x,y,u) = \frac{x+iy}{1-u}$ is continuous on the set $S^2 \setminus \{N\}$.

3. The inverse function

$$\varphi^{-1}(z) = (x,y,u) = \left(\frac{z+\overline{z}}{1+|z|^2}, \frac{z-\overline{z}}{i(1+|z|^2)}, \frac{|z|^2-1}{|z|^2+1}\right)$$

is continuous on $\mathbb{C}$. □

Theorem 1.5.2. *Let $N' = (0,0,-1) \in S^2$, and define the function $\Phi_1 : S^2 \to \widehat{\mathbb{C}}$ by*

$$z = \Phi_1(x,y,u) = \begin{cases} \frac{x+iy}{1+u}, & \text{if } (x,y,u) \neq (0,0,-1); \\ \infty, & \text{if } (x,y,u) = (0,0,-1). \end{cases}$$

Let us denote by φ_1 the restriction of the function Φ_1 to $S^2 \setminus \{N'\}$, i. e., $\varphi_1 = \Phi_1|_{S^2\setminus\{N'\}}$.

Then the function φ_1 is a continuous bijection between $S^2 \setminus \{N'\}$ and $\mathbb{C}$, and φ_1^{-1} is also continuous.

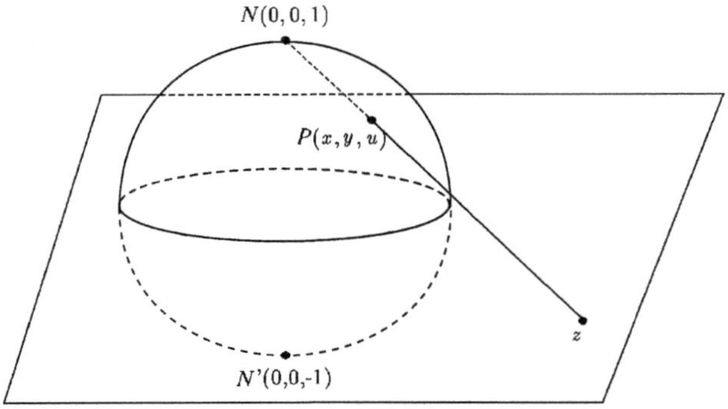

Figure 1.1: The two dimensional sphere (ball).

1.5.1 The geometric interpretation of the φ function

If $\overline{NP}(x, y, u-1)$, let $(NP) : \frac{X}{x} = \frac{Y}{y} = \frac{U-1}{u-1}$ (Figure 1.1).

Let it intersects the line (NP) with the plane $U = 0$, and let $(x_1, y_1, 0)$ be the coordinate of this intersection. Then

$$\frac{x_1}{x} = \frac{y_1}{y} = \frac{-1}{u-1} \Rightarrow x_1 = \frac{x}{1-u}, \quad y_1 = \frac{y}{1-u}.$$

Identifying the plane $U = 0$ with the set of the complex numbers $\mathbb{C}$, we get

$$z = x_1 + iy_1 = \frac{x+iy}{1-u} = \varphi(x, y, u).$$

It follows that the function φ maps each point $P \in S^2 \setminus \{N\}$ into the point $z = x_1 + iy_1 = \frac{x+iy}{1-u} = \varphi(x, y, u)$ of the plane $U = 0$, which is called **the stereographic projection** of the point P to the complex plane $\mathbb{C}$.

1.5.2 The topological structure of $\widehat{\mathbb{C}}$

The neighborhood system of the ∞ point is given by all the sets V that satisfy the next property:

$$\exists r > 0 \quad \text{such that } W_r = \{\infty\} \cup \{z \in \mathbb{C} : |z| > r\} \subset V.$$

We will denote it by $\mathcal{B}(\infty) = \{W_r : r > 0\}$ the **base of the neighborhood system of the** ∞.

The images of the W_r sets by the function Φ^{-1} are open spherical caps with the apexes in the "North pole" N.

Theorem 1.5.3. *The function Φ is a bijection between S^2 and $\widehat{\mathbb{C}}$, and Φ and Φ^{-1} are continuous functions.*

Proof. The function $\Phi|_{S^2\setminus\{N\}} = \varphi$ is a bijection between $S^2 \setminus \{N\}$ and $\mathbb{C}$; the functions φ and φ^{-1} are continuous. From Theorem 1.5.1, the function Φ is a bijection.

Now we will prove that the function Φ is continuous at the point N, while the function Φ^{-1} is continuous at the point ∞.

Since

$$|\Phi(x,y,u)| > r \Leftrightarrow \left|\frac{x+iy}{1-u}\right|^2 = \frac{1+u}{1-u} > r^2 \Leftrightarrow u > \frac{r^2-1}{r^2+1},$$

it follows that

$$\Phi(x,y,u) \to \infty \Leftrightarrow u \to 1 \Leftrightarrow (x,y,u) \to N.$$

Using these inequalities, we have

$$\Phi^{-1}(\widehat{\mathbb{C}} \setminus \overline{U}(0;r)) = \left\{(x,y,u) \in S^2 : u > \frac{r^2-1}{r^2+1}\right\},$$

hence $\Phi^{-1}(\widehat{\mathbb{C}} \setminus \overline{U}(0;r))$ is a neighborhood of the point $(0,0,1)$ in S^2.

Using the fact that

$$\lim_{z\to\infty} \Phi^{-1}(z) = \lim_{z\to\infty}\left(\frac{z+\overline{z}}{1+|z|^2}, \frac{z-\overline{z}}{i(1+|z|^2)}, \frac{|z|^2-1}{1+|z|^2}\right) = (0,0,1) = \Phi^{-1}(\infty),$$

the function Φ^{-1} is continuous at the point $z = \infty$. □

Remarks 1.5.1. The two-dimensional sphere (ball) S^2 together with the functions Φ and Φ^{-1} is called **the Riemann sphere (ball).**

Remarks 1.5.2.
1. It is not difficult to prove that the stereographic projection maps each circle of S^2, that does not contain the N point, into a circle of $\mathbb{C}$, and vice versa.
2. Moreover, the stereographic projection maps each circle of S^2 that contains the N point, into a line of $\mathbb{C}$.

Remark 1.5.3. In the set $\widehat{\mathbb{C}}$, we could determine the distance between the points z and z' by using the Euclidian distance of $\mathbb{R}^3$ between their images $\Phi^{-1}(z)$ and $\Phi^{-1}(z')$. This new distance will be denoted by $\mathrm{d}_c(z,z')$.

Since $\Phi^{-1}(z), \Phi^{-1}(z') \in S^2$, then

$$\mathrm{d}_c : \widehat{\mathbb{C}} \times \widehat{\mathbb{C}} \to [0,2],$$
$$\mathrm{d}_c(z,z') = \sqrt{(x-x')^2 + (y-y')^2 + (u-u')^2} = \sqrt{2-2(xx'+yy'+uu')},$$

where $z = x_1 + iy_1 \longmapsto P(x,y,u) \in S^2$, and $z' = x_1' + iy_1' \longmapsto P'(x',y',u') \in S^2$.

Definition 1.5.2. The distance $\mathrm{d}_c(z,z')$ between the points $z,z' \in \widehat{\mathbb{C}}$ is called **the chordal distance**, where

$$\mathrm{d}_c(z,z') = \frac{2|z-z'|}{\sqrt{(1+|z|^2)(1+|z'|^2)}} \quad \text{and} \quad \mathrm{d}_c(z,\infty) = \frac{2}{\sqrt{1+|z|^2}}.$$

Theorem 1.5.4. *The couple* $(\widehat{\mathbb{C}}, \mathrm{d}_c)$ *is a metric space, and the topology of this metric space coincides with the already defined topology of* $\widehat{\mathbb{C}}$.

Corollary 1.5.1. *The* $(\widehat{\mathbb{C}}, \mathrm{d}_c)$ *metric space is compact.*

Proof. The set $S^2 \subset \mathbb{R}^3$ is compact (since it is bounded and closed in $\mathbb{R}^3$). Since $\Phi : S^2 \to \widehat{\mathbb{C}}$ is a continuous bijection, it follows that $\widehat{\mathbb{C}}$ is compact. □

Remarks 1.5.4.
1. $\mathrm{d}_c(\frac{1}{z}, \frac{1}{z'}) = \mathrm{d}_c(z, z')$ and $\mathrm{d}_c(\frac{1}{z}, \infty) = \frac{2|z|}{\sqrt{1+|z|^2}}$, $\forall z, z' \in \hat{\mathbb{C}}$.
2. If $A \subset \mathbb{C}$ is a bounded set, i. e., $\exists R > 0$ such that $A \subset U(0; R)$, then

$$\frac{2|z - z'|}{1 + R^2} \le \mathrm{d}_c(z, z') \le 2|z - z'|, \quad \forall z, z' \in A.$$

3. The behavior of the function f in a neighborhood of the ∞ point can be easily studied by considering the new function $g(z) = f(\frac{1}{z})$, and studying the behavior of the function g in a neighborhood of the point $z = 0$.

1.6 Exercises

Exercise 1.6.1. Represent in the complex plane the following sets:
1. $\{z \in \mathbb{C} : \operatorname{Im}(z + i) = 0\}$;
2. $\{z \in \mathbb{C} : \arg(z - 2i) = \frac{\pi}{4}\}$;
3. $\{z \in \mathbb{C} : |z - \alpha| = |z - \beta|\}$, where $\alpha = -i$ and $\beta = 4 + i$;
4. $\{z \in \mathbb{C} : |z + i| = 2\}$;
5. $\{z \in \mathbb{C} : |z - 2i| + |z + 4i| = 10\}$;
6. $\{z \in \mathbb{C} : \arg \frac{z-1}{z+1} = \frac{\pi}{3}\}$;
7. $\{z \in \mathbb{C} : -\frac{\pi}{4} < \arg(z - i) < \frac{\pi}{6}\}$;
8. $\{z \in \mathbb{C} : |z| < 1 - \frac{1}{2i}(z - \overline{z})\}$;
9. $\{z \in \mathbb{C} : |\frac{z}{z+3i}| < 1\}$;
10. $\{z \in \mathbb{C} : |\frac{1-z}{1+z}| > 3\}$;
11. $\{z \in \mathbb{C} : |z - 4| - |z + 6| - 9 > 0\}$.

Exercise 1.6.2. If $a \in \mathbb{C}$, prove that:
1. $z \neq \overline{a}, |\frac{z-a}{z-\overline{a}}| < 1 \Leftrightarrow \operatorname{Im} a \operatorname{Im} z > 0$;
2. $z \neq \overline{a}, |\frac{z-a}{z-\overline{a}}| > 1 \Leftrightarrow \operatorname{Im} a \operatorname{Im} z < 0$;
3. $z \neq \overline{a}, |\frac{z-a}{z-\overline{a}}| = 1 \Leftrightarrow \operatorname{Im} a \operatorname{Im} z = 0$.

Exercise 1.6.3. If $a \in \mathbb{C}$, prove that:
1. $z \neq -\bar{a}, |\frac{z-a}{z+\bar{a}}| < 1 \Leftrightarrow \operatorname{Re} a \operatorname{Re} z > 0$;
2. $z \neq -\bar{a}, |\frac{z-a}{z+\bar{a}}| > 1 \Leftrightarrow \operatorname{Re} a \operatorname{Re} z < 0$;
3. $z \neq -\bar{a}, |\frac{z-a}{z+\bar{a}}| = 1 \Leftrightarrow \operatorname{Re} a \operatorname{Re} z = 0$.

Exercise 1.6.4. Prove that the following functions are bijections:
1. $f : \{z \in \mathbb{C} : \operatorname{Im} z > 0\} \to U(0;1), f(z) = \frac{z-a}{z-\bar{a}}$, where $\operatorname{Im} a > 0$;
2. $f : \{z \in \mathbb{C} : \operatorname{Re} z > 0\} \to U(0;1), f(z) = \frac{z-a}{z+\bar{a}}$, where $\operatorname{Re} a > 0$.

Exercise 1.6.5. Let $z_1, z_2 \in \mathbb{C}$, $z_1 \neq z_2$ be two distinct complex numbers, and let $A_1(z_1)$, $A_2(z_2)$ be their corresponding points in the complex plane, and A is the line segment between points A_1 and A_2. Prove that

$$A(z) \in A_1A_2 \Leftrightarrow \frac{z - z_1}{z_2 - z_1} = \overline{\frac{z - z_1}{z_2 - z_1}}.$$

Exercise 1.6.6. Let $z_1, z_2, z_3 \in \mathbb{C}$ be three distinct complex numbers, and let $A_k(z_k)$, $k \in \{1, 2, 3\}$ be their corresponding points in the complex plane. Prove that

$$A_1, A_2 \text{ and } A_3 \quad \text{are collinear} \Leftrightarrow \frac{z_3 - z_1}{z_2 - z_1} \in \mathbb{R}.$$

Exercise 1.6.7. Let $z_1, z_2, z_3 \in \mathbb{C}$ be three distinct complex numbers, and let $A_k(z_k)$, $k \in \{1, 2, 3\}$ be their corresponding points in the complex plane. Prove that

$$A_3 \in [A_1A_2] \Leftrightarrow \frac{z_3 - z_1}{z_2 - z_1} \in [0, 1] \Leftrightarrow \exists t \in [0, 1] : z_3 = (1 - t)z_1 + tz_2,$$

where $[A_1A_2]$ denotes the closed segment A_1A_2.

Exercise 1.6.8. Let $A_k(z_k)$, $k \in \{1, 2, 3, 4\}$ be the corresponding images in the complex plane of the four distinct complex numbers $z_1, z_2, z_3, z_4 \in \mathbb{C}$. Prove that:
1. $A_1A_2 \perp A_3A_4 \Leftrightarrow \frac{z_1 - z_2}{z_3 - z_4} \in i\mathbb{R}$;
2. $A_1A_2 \| A_3A_4 \Leftrightarrow \frac{z_1 - z_2}{z_3 - z_4} \in \mathbb{R}$.

Exercise 1.6.9. Denote by $C(a; r) = \partial U(a; r)$ the circle with the center in $a \in \mathbb{C}$, and the radius $r > 0$. Prove that

$$A(z) \in C(a; r) \Leftrightarrow z\bar{z} - \bar{a}z - a\bar{z} + a\bar{a} = r^2.$$

Exercise 1.6.10. Let $z_1, z_2, z_3 \in \mathbb{C}$ be three complex numbers, such that their corresponding images in the complex plane $A_k(z_k)$, $k \in \{1, 2, 3\}$, are not collinear points. Prove that

$$A(z) \in C(z_1, z_2, z_3) \Leftrightarrow \begin{vmatrix} z\bar{z} & z & \bar{z} & 1 \\ z_1\bar{z}_1 & z_1 & \bar{z}_1 & 1 \\ z_2\bar{z}_2 & z_2 & \bar{z}_2 & 1 \\ z_3\bar{z}_3 & z_3 & \bar{z}_3 & 1 \end{vmatrix} = 0,$$

where $C(z_1, z_2, z_3)$ represents the circle that contains the points $A_1(z_1)$, $A_2(z_2)$ and $A_3(z_3)$.

Exercise 1.6.11. Let $z_1, z_2, z_3, z_4 \in \mathbb{C}$ be four complex numbers, and let $A_k(z_k)$, $k \in \{1, 2, 3, 4\}$ be their corresponding images in the complex plane. Prove that

$$A_1, A_2, A_3 \text{ and } A_4 \quad \text{belong to the same circle} \Leftrightarrow \frac{z_2 - z_1}{z_3 - z_1} : \frac{z_2 - z_4}{z_3 - z_4} \in \mathbb{R}.$$

Exercise 1.6.12. Suppose that the continuous function $f : \mathbb{C} \to \mathbb{C}$ satisfies the conditions:

1. $f(z_1 + z_2) = f(z_1) + f(z_2)$
   and
2. $f(z_1 z_2) = f(z_1) f(z_2)$, $\forall z_1, z_2 \in \mathbb{C}$.

Prove that the function f has one of the following form:

$$f(z) = 0, \quad \forall z \in \mathbb{C},$$
or
$$f(z) = z, \quad \forall z \in \mathbb{C},$$
or
$$f(z) = \overline{z}, \quad \forall z \in \mathbb{C}.$$

2 Holomorphic functions

2.1 The derivative of the real valued complex functions

Definition 2.1.1. The function $f : [a,b] \to \mathbb{C}$ is said to be **differentiable** at the point $t_0 \in [a,b]$, if there exist the limit

$$\lim_{t\to t_0} \frac{f(t)-f(t_0)}{t-t_0} = f'(t_0) \in \mathbb{C}.$$

Using the fact that

$$\frac{f(t)-f(t_0)}{t-t_0} = \frac{\alpha(t)-\alpha(t_0)}{t-t_0} + i\frac{\beta(t)-\beta(t_0)}{t-t_0},$$

the function f is differentiable at $t_0 \in [a,b]$ if and only if there exists the limits

$$\lim_{t\to t_0} \frac{\alpha(t)-\alpha(t_0)}{t-t_0} = \alpha'(t_0) \quad \text{and} \quad \lim_{t\to t_0} \frac{\beta(t)-\beta(t_0)}{t-t_0} = \beta'(t_0),$$

i. e., the real valued functions α and β are differentiable at the point t_0. In such a case, we have

$$f'(t_0) = \alpha'(t_0) + i\beta'(t_0).$$

Theorem 2.1.1. *If the function $f : [a,b] \to \mathbb{C}$ is differentiable on the interval $[a,b]$ if $f'(t) = 0$, $\forall t \in [a,b]$, then f is constant on the interval $[a,b]$.*

Proof. We have $f'(t) = 0 \Rightarrow \alpha'(t) = \beta'(t) = 0$. But α and β are real valued real functions, and according to the Lagrange mean-value theorem we deduce that $\alpha(t) = \alpha$, $\beta(t) = \beta$, $\forall t \in [a,b]$, i. e., $f(t) = \alpha + i\beta$, $\forall t \in [a,b]$ is a constant function on $[a,b]$. □

In the proof of the above theorem, we used the Lagrange's mean-value theorem[1] for the real valued real functions. But this theorem does not hold for the complex valued real functions, as we may see in the following example.

Example 2.1.1. Let $f(t) = t^2 + it^3$, $t \in [a,b]$, $a < b$, and suppose that there exists $c \in (a,b)$, such that

$$f(b) - f(a) = f'(c)(b-a), \quad \text{i. e.,} \quad b^2 + ib^3 - a^2 - ia^3 = (2c + 3ic^2)(b-a),$$

1 Lagrange mean-value theorem states that a function is continuous in the closed interval $[a,b]$ and differentiable in an open interval (a,b). Then there is at least one point $c \in (a,b)$ such that $f'(c) = \frac{f(b)-f(a)}{b-a}$.

https://doi.org/10.1515/9783110657869-002

or

$$\begin{cases} b^3 - a^3 = 3c^2(b-a) \\ b^2 - a^2 = 2c(b-a) \end{cases} \Leftrightarrow \begin{cases} b^2 + ab + a^2 = 3c^2 \\ b + a = 2c \end{cases}$$

$$\Rightarrow b^2 + ab + a^2 = 3\left(\frac{a+b}{2}\right)^2$$

$$\Rightarrow 4b^2 + 4ba + 4a^2 = 3a^2 + 6ab + 3b^2 \Leftrightarrow b^2 - 2ab + a^2 = 0 \Leftrightarrow (b-a)^2 = 0 \Leftrightarrow a = b,$$

which contradicts the assumption. □

The next weaker version of the Lagrange theorem holds for these functions.

Theorem 2.1.2 (The Lagrange theorem for complex valued real functions). *If the function $f : [a,b] \to \mathbb{C}$ is continuous on $[a,b]$, and it is differentiable on (a,b), then there exist the numbers $c \in (a,b)$ and $\sigma \in \mathbb{C}$, with $|\sigma| \leq 1$, such that:*
1. $|f(b) - f(a)| \leq |f'(c)|(b-a)$;
2. $f(b) - f(a) = \sigma f'(c)(b-a)$.

Proof. Letting $f(t) = \alpha(t) + i\beta(t)$, then $f'(t) = \alpha'(t) + i\beta'(t)$, $t \in (a,b)$. We will define the function

$$\varphi : [a,b] \to \mathbb{R}, \quad \varphi(t) = \big(\alpha(b) - \alpha(a)\big)\alpha(t) + \big(\beta(b) - \beta(a)\big)\beta(t).$$

The function φ is continuous on $[a,b]$, and it is differentiable on (a,b), then according to the Lagrange theorem for the real valued real functions, we deduce that

$$\exists c \in (a,b) \quad \text{such that } \varphi(b) - \varphi(a) = \varphi'(c)(b-a).$$

On the other hand,

$$\varphi(b) - \varphi(a) = \big(\alpha(b) - \alpha(a)\big)^2 + \big(\beta(b) - \beta(a)\big)^2 = |f(b) - f(a)|^2$$

and

$$\varphi'(c) = \big(\alpha(b) - \alpha(a)\big)\alpha'(c) + \big(\beta(b) - \beta(a)\big)\beta'(c).$$

From the Cauchy inequality, we have

$$\begin{aligned} |\varphi'(c)| &\leq \sqrt{\big(\alpha(b) - \alpha(a)\big)^2 + \big(\beta(b) - \beta(a)\big)^2}\sqrt{\alpha'^2(c) + \beta'^2(c)} \\ &= |f(b) - f(a)||f'(c)|, \end{aligned}$$

and combining with the above relations it follows that

$$|f(b) - f(a)|^2 = \varphi(b) - \varphi(a) = \varphi'(c)(b-a) \leq (b-a)|f(b) - f(a)||f'(c)|.$$

If $f(b) \neq f(a)$, the previous inequality implies that

$$|f(b) - f(a)| \leq |f'(c)|(b - a).$$

It is obvious that this inequality holds for the case when $f(b) = f(a)$, which completes the proof of the first point.

Supposing that $f(b) = f(a)$, then we can choose $\sigma = 0$, and for this value the second conclusion holds.

Supposing that $f(b) \neq f(a)$, from the inequality of the first point it follows that $f'(c) \neq 0$, and thus $f'(c)(b - a) \neq 0$. Hence we may denote

$$\frac{f(b) - f(a)}{f'(c)(b - a)} = \sigma \in \mathbb{C},$$

and from the first point we have $|\sigma| \leq 1$. Multiplying with $f'(c)(b - a)$, we obtain the second relation of the theorem. □

2.2 The differentiability of a complex function

From the basic knowledge of mathematical analysis, it is well known that if $G \subset \mathbb{R}^2$ is an open set, the function

$$f : G \to \mathbb{R}^2$$

is said to be **(Fréchet) differentiable at the point** $(x_0, y_0) \in G$, if there exists a linear operator $A : \mathbb{R}^2 \to \mathbb{R}^2$, such that

$$\frac{\|f(x,y) - f(x_0,y_0) - A(x - x_0, y - y_0)\|}{\sqrt{(x - x_0)^2 + (y - y_0)^2}} \to 0, \quad \text{if } (x,y) \to (x_0,y_0). \tag{2.1}$$

In the $\mathbb{R}^2$ set, with an orthogonal coordinates axis, the operator A can be associated with a matrix $\begin{pmatrix} a_{11} & a_{12} \\ a_{21} & a_{22} \end{pmatrix}$, and with this notation we have

$$A(x - x_0, y - y_0) = (a_{11}(x - x_0) + a_{12}(y - y_0), a_{21}(x - x_0) + a_{22}(y - y_0)).$$

Replacing this form of the operator A in the limit formula (2.1), and denoting by u and v the components of the vector function f, it follows that

$$\begin{aligned} &\frac{1}{\sqrt{(x - x_0)^2 + (y - y_0)^2}} \|(u(x,y) - u(x_0,y_0), v(x,y) - v(x_0,y_0)) \\ &\quad -(a_{11}(x - x_0) + a_{12}(y - y_0), a_{21}(x - x_0) + a_{22}(y - y_0))\| \to 0, \end{aligned} \tag{2.2}$$

if $(x,y) \to (x_0,y_0)$.

From here, taking $y = y_0$, and letting $x \to x_0$ we get

$$a_{11} = \frac{\partial u(x_0, y_0)}{\partial x}, \quad a_{21} = \frac{\partial v(x_0, y_0)}{\partial x}.$$

Similarly, taking $x = x_0$, and letting $y \to y_0$ we have

$$a_{12} = \frac{\partial u(x_0, y_0)}{\partial y}, \quad a_{22} = \frac{\partial v(x_0, y_0)}{\partial y},$$

hence we deduce that

$$A = \begin{pmatrix} \frac{\partial u(x_0,y_0)}{\partial x} & \frac{\partial u(x_0,y_0)}{\partial y} \\ \frac{\partial v(x_0,y_0)}{\partial x} & \frac{\partial v(x_0,y_0)}{\partial y} \end{pmatrix}.$$

If we identify $\mathbb{R}^2$ with $\mathbb{C}$, then the functions f and A will become complex functions of the forms:

$$\begin{aligned} f(z) &= f(x,y) = u(x,y) + iv(x,y); \\ f(z_0) &= f(x_0,y_0) = u(x_0,y_0) + iv(x_0,y_0), \quad \text{where} \\ z &= x + iy, \ z_0 = x_0 + iy_0; \\ A(x - x_0, y - y_0) &= (a_{11} + ia_{21})(x - x_0) + (a_{12} + ia_{22})(y - y_0). \end{aligned}$$

Letting $A_1 = a_{11} + ia_{21}$ and $A_2 = a_{12} + ia_{22}$, then the limit formula (2.2) can be written as

$$\frac{|f(z) - f(z_0) - A_1(x - x_0) - A_2(y - y_0)|}{|z - z_0|} \to 0, \quad \text{if } z \to z_0. \tag{2.3}$$

Definition 2.2.1. Let $G \subset \mathbb{C}$ be an open set, and let $f : G \to \mathbb{C}$. If there exist the numbers $A_1, A_2 \in \mathbb{C}$ such that the limit formula (2.3) holds, then the function f is said to be **differentiable at the point** $z_0 \in G$.

Remarks 2.2.1.

1. Writing the complex function in this vectorial form, we see that the concept of Differentiability is identical with the concept of the (Fréchet) differentiability.
2. The next equivalent form is the following:
   The function $f : G \to \mathbb{C}$ ($G \subset \mathbb{C}$ is an open set) is differentiable at the point $z_0 \in G$, if and only if there exist the complex numbers $A_1, A_2 \in \mathbb{C}$, and there exists the function $g : G \setminus \{z_0\} \to \mathbb{C}$ with $\lim_{z \to z_0} g(z) = 0$, such that

$$f(z) = f(z_0) + A_1(x - x_0) + A_2(y - y_0) + g(z)|z - z_0|, \quad \forall z \in G \setminus \{z_0\}. \tag{2.4}$$

In fact, if (2.3) holds, letting

$$g(z) = \frac{f(z) - f(z_0) - A_1(x - x_0) - A_2(y - y_0)}{|z - z_0|},$$

then g satisfies the required condition.

Conversely, if there exists the function g with the above property, then the fraction that appeared in the formula (2.3) is $|g(z)|$, and thus $\lim_{z\to z_0} |g(z)| = 0$.

If we let $f = u + iv$, and we suppose that the function f is differentiable at $z_0 = x_0 + iy_0 \in G$, we obtain

$$A_1 = a_{11} + ia_{21} = \frac{\partial u}{\partial x} + i\frac{\partial v}{\partial x} = \frac{\partial f}{\partial x},$$
$$A_2 = a_{12} + ia_{22} = \frac{\partial u}{\partial y} + i\frac{\partial v}{\partial y} = \frac{\partial f}{\partial y}.$$

If we rewrite the relation (2.4) using the fact that

$$x - x_0 = \frac{z - z_0 + \overline{(z - z_0)}}{2}, \quad y - y_0 = \frac{z - z_0 - \overline{(z - z_0)}}{2i},$$

then we deduce that

$$\begin{aligned} f(z) &= f(z_0) + \frac{\partial f}{\partial x}\frac{z - z_0 + \overline{(z - z_0)}}{2} + \frac{\partial f}{\partial y}\frac{z - z_0 - \overline{(z - z_0)}}{2i} + g(z)|z - z_0| \\ &= f(z_0) + \frac{1}{2}\left(\frac{\partial f(z_0)}{\partial x} - i\frac{\partial f(z_0)}{\partial y}\right)(z - z_0) \\ &\quad + \frac{1}{2}\left(\frac{\partial f(z_0)}{\partial x} + i\frac{\partial f(z_0)}{\partial y}\right)\overline{(z - z_0)} + g(z)|z - z_0|. \end{aligned}$$

Remarks 2.2.2.

1. We will use the following important notation:

$$\frac{\partial f(z_0)}{\partial z} = \frac{1}{2}\left(\frac{\partial f(z_0)}{\partial x} - i\frac{\partial f(z_0)}{\partial y}\right), \quad \frac{\partial f(z_0)}{\partial \overline{z}} = \frac{1}{2}\left(\frac{\partial f(z_0)}{\partial x} + i\frac{\partial f(z_0)}{\partial y}\right).$$

2. With the above notation, the relation (2.4) becomes

$$f(z) = f(z_0) + \frac{\partial f(z_0)}{\partial z}(z - z_0) + \frac{\partial f(z_0)}{\partial \overline{z}}\overline{(z - z_0)} + g(z)|z - z_0|, \quad \forall z \in G \setminus \{z_0\}. \quad (2.5)$$

3. From the mathematical analysis, it is well known that the function $f : \mathbb{R}^2 \to \mathbb{R}^2$ is differentiable, if and only if its co-ordinates u and v are differentiable real valued functions. It is also known that a real valued function is differentiable, if it has continuous first-order partial derivatives.

Definition 2.2.2. Let $G \subset \mathbb{C}$ be an open set. We say that the function $f : G \to \mathbb{C}$ **has first-order continuous partial derivatives on** G (denoted by $f \in C^1(G)$), if both real and imaginary parts of the function f have first-order continuous partial derivatives on G.

Theorem 2.2.1. *If $f \in C^1(G)$, where $G \subset \mathbb{C}$ is an open set, then the function f is differentiable at all the points of G.*

Proof. Suppose that the function $f : G \to \mathbb{C}$ is differentiable at the point $z_0 \in G$. Let $\theta \in (-\pi, \pi]$ be an arbitrary angle, and let us compute the directional derivative of the function f in the point z_0. In fact, this will be the derivative of a complex valued real variable function.

For this, suppose that $z = z_0 + h$, where $\arg h = \theta$ and $h \to 0$. For the computation of the directional derivative denoted by $f'_\theta(z_0)$, we will use the formula (2.5), i. e.,

$$f(z_0 + h) - f(z_0) = \frac{\partial f(z_0)}{\partial z} h + \frac{\partial f(z_0)}{\partial \bar{z}} \bar{h} + g(z_0 + h)|h|. \tag{2.6}$$

We conclude that

$$f'_\theta(z_0) = \lim_{\substack{h \to 0 \\ \arg h = \theta}} \frac{f(z_0 + h) - f(z_0)}{h},$$

and the existence of this last limit follows from the relation (2.6), because $\frac{\bar{h}}{h}$ is constant whenever $\arg h = \theta$. □

Theorem 2.2.2 (Kasner theorem). *Let $G \subset \mathbb{C}$ be an open set, let $z_0 \in G$, and suppose that the function f is Differentiable at the point z_0. If the angle θ runs once the whole $(-\pi, \pi]$ interval, then the directional derivative $f'_\theta(z_0)$ runs twice the whole circle with the center $\frac{\partial f(z_0)}{\partial z}$ and radius $|\frac{\partial f(z_0)}{\partial \bar{z}}|$.*

Proof. Using the notation $\cos\theta + i\sin\theta = e^{i\theta}$, then $h = |h|e^{i\theta}$, $\bar{h} = |h|e^{-i\theta}$, and hence $\frac{\bar{h}}{h} = e^{-2i\theta}$. If we divide by h the equality (2.6), we get

$$\frac{f(z_0 + h) - f(z_0)}{h} = \frac{\partial f(z_0)}{\partial z} + \frac{\partial f(z_0)}{\partial \bar{z}} e^{-2i\theta} + g(z_0 + h)\frac{|h|}{h},$$

hence

$$f'_\theta(z_0) = \lim_{\substack{h \to 0 \\ \arg h = \theta}} \frac{f(z_0 + h) - f(z_0)}{h} = \frac{\partial f(z_0)}{\partial z} + \frac{\partial f(z_0)}{\partial \bar{z}} e^{-2i\theta}, \tag{2.7}$$

i. e.,

$$\left| f'_\theta(z_0) - \frac{\partial f(z_0)}{\partial z} \right| = \left| \frac{\partial f(z_0)}{\partial \bar{z}} \right|.$$

The proof follows immediate from these last two relations. □

Remarks 2.2.3.

1. The circle that appeared in Theorem 2.2.2 is called **the Kasner circle.**
2. The relation (2.7) shows that if the angle θ runs once in positive orientation the complete circle, then the directional derivative $f'_\theta(z_0)$ runs twice the Kasner circle in negative direction.

2.3 The derivative of a complex function

Definition 2.3.1. Let $G \subset \mathbb{C}$ be an open set, and let $z_0 \in G$. We say that the function $f : G \to \mathbb{C}$ is **differentiable** at the point z_0, if

$$\exists \lim_{z \to z_0} \frac{f(z) - f(z_0)}{z - z_0} = f'(z_0) \in \mathbb{C}.$$

The above limit is denoted by $f'(z_0)$, and it is called the derivative of the function f at the point z_0.

Definition 2.3.2. Let $G \subset \mathbb{C}$ be an open set. The function $f : G \to \mathbb{C}$ is said to be **differentiable at the point** $z_0 \in G$, if there exists the complex number $\alpha \in \mathbb{C}$, and there exists the function $g : G \setminus \{z_0\} \to \mathbb{C}$ with $\lim_{z \to z_0} g(z) = 0$, such that

$$f(z) = f(z_0) + \alpha(z - z_0) + g(z)(z - z_0), \quad z \in G \setminus \{z_0\}. \tag{2.8}$$

Theorem 2.3.1 (The Cauchy–Riemann theorem). *The function $f = u + iv$, $f : G \to \mathbb{C}$ (where $G \subset \mathbb{C}$ is an open set) is differentiable at the point $z_0 \in G$, if and only if it is differentiable at the point z_0, and the partial derivatives of u and v satisfy the relations*

$$\frac{\partial u(z_0)}{\partial x} = \frac{\partial v(z_0)}{\partial y} \quad \text{and} \quad \frac{\partial u(z_0)}{\partial y} = -\frac{\partial v(z_0)}{\partial x}$$

(called Cauchy–Riemann conditions).

Proof. Suppose that $\exists f'(z_0)$. From the previous theorem $\exists \alpha \in \mathbb{C}$, and $\exists g : G \setminus \{z_0\} \to \mathbb{C}$, with $\lim_{z \to z_0} g(z) = 0$, such that

$$f(z) = f(z_0) + \alpha(z - z_0) + g(z)(z - z_0).$$

Letting $A_1 = \alpha$ and $A_2 = i\alpha$, the above relation becomes

$$f(z) = f(z_0) + A_1(x - x_0) + A_2(y - y_0) + g(z)\frac{z - z_0}{|z - z_0|}|z - z_0|.$$

This represents the differentiability of the function f, where the condition (2.4) is written in another equivalent form, because $g(z)\frac{z-z_0}{|z-z_0|} \to 0$, if $z \to z_0$.

Since f is differentiable at z_0, and since there exists $f'(z_0)$ it follows that $f'_\theta(z_0) = f'(z_0)$, $\forall \theta \in (-\pi, \pi]$. This last relation holds if and only if the coefficient of $e^{-2i\theta}$ from the right-hand side of (2.7) vanishes (it is zero), i. e.,

$$\frac{\partial f(z_0)}{\partial \bar{z}} = \frac{1}{2}\left(\frac{\partial f(z_0)}{\partial x} + i\frac{\partial f(z_0)}{\partial y}\right) = 0.$$

From this equality, we have

$$\frac{\partial u(z_0)}{\partial x} + i\frac{\partial v(z_0)}{\partial x} + i\frac{\partial u(z_0)}{\partial y} - \frac{\partial v(z_0)}{\partial y} = 0,$$

hence

$$\frac{\partial u(z_0)}{\partial x} = \frac{\partial v(z_0)}{\partial y} \quad \text{and} \quad \frac{\partial u(z_0)}{\partial y} = -\frac{\partial v(z_0)}{\partial x},$$

which represents the Cauchy–Riemann conditions.

Suppose now that the function f is differentiable at the point z_0, and the functions u and v satisfy the Cauchy–Riemann conditions.

From the differentiation of the function f at the point z_0, it follows that u and v are two differentiable real functions at $z_0 = x_0 + iy_0$. Then

$$u(z) = u(z_0) + \frac{\partial u(z_0)}{\partial x}(x - x_0) + \frac{\partial u(z_0)}{\partial y}(y - y_0) + u_1(z)|z - z_0|, \tag{2.9}$$

$$v(z) = v(z_0) + \frac{\partial v(z_0)}{\partial x}(x - x_0) + \frac{\partial v(z_0)}{\partial y}(y - y_0) + v_1(z)|z - z_0|, \tag{2.10}$$

where $u_1(z) \to 0$ and $v_1(z) \to 0$, if $z \to z_0$.

We will multiply the both sides of (2.10) by i, and add equation (2.9) to it. Using the Cauchy–Riemann conditions, we deduce that

$$f(z) = u(z) + iv(z) = f(z_0) + \left(\frac{\partial u(z_0)}{\partial x} + i\frac{\partial v(z_0)}{\partial x}\right)(z - z_0) + g(z)(z - z_0), \tag{2.11}$$

where

$$g(z) = \frac{u_1(z) + iv_1(z)}{z - z_0}|z - z_0| \to 0, \quad \text{if } z \to z_0.$$

From the relation (2.11), we conclude that the function f is differentiable at the point z_0, and using Theorem 2.2.1 it follows the existence of the derivative $f'(z_0)$. □

Remarks 2.3.1.

1. Let $G \subset \mathbb{C}$ be an open set. The function $f : G \to \mathbb{C}$ is derivable at the point $z_0 \in G$, if and only if it is differentiable at this point, and $\frac{\partial f(z_0)}{\partial \bar{z}} = 0$.
2. The relation (2.11) shows the next very important formula:

$$f'(z_0) = \frac{\partial u(z_0)}{\partial x} + i\frac{\partial v(z_0)}{\partial x}.$$

Using the Cauchy–Riemann conditions, the derivative of the function f at the point $z_0 \in G$ can be written in the following forms:

$$f'(z_0) = \frac{\partial u(z_0)}{\partial x} - i\frac{\partial u(z_0)}{\partial y} = \frac{\partial v(z_0)}{\partial y} + i\frac{\partial v(z_0)}{\partial x} = \frac{\partial v(z_0)}{\partial y} - i\frac{\partial u(z_0)}{\partial y},$$

hence

$$f'(z_0) = \frac{\partial f(z_0)}{\partial x} = -i\frac{\partial f(z_0)}{\partial y}.$$

3. It is easy to see that

$$|f'(z_0)|^2 = \left(\frac{\partial u(z_0)}{\partial x}\right)^2 + \left(\frac{\partial v(z_0)}{\partial x}\right)^2 = \left(\frac{\partial u(z_0)}{\partial x}\right)^2 + \left(\frac{\partial u(z_0)}{\partial y}\right)^2$$
$$= \left(\frac{\partial v(z_0)}{\partial y}\right)^2 + \left(\frac{\partial u(z_0)}{\partial y}\right)^2 = \left(\frac{\partial v(z_0)}{\partial y}\right)^2 + \left(\frac{\partial v(z_0)}{\partial x}\right)^2$$
$$= \frac{\partial u(z_0)}{\partial x}\frac{\partial v(z_0)}{\partial y} - \frac{\partial u(z_0)}{\partial y}\frac{\partial v(z_0)}{\partial x}.$$

Consequently,

$$|f'(z_0)|^2 = \begin{vmatrix} \frac{\partial u(z_0)}{\partial x} & \frac{\partial u(z_0)}{\partial y} \\ \frac{\partial v(z_0)}{\partial x} & \frac{\partial v(z_0)}{\partial y} \end{vmatrix} = J(u,v)(z_0) = \frac{D(u,v)(z_0)}{D(x,y)}.$$

2.3.1 The properties of the derivative

Property 2.3.1.
1. Let $G \subset \mathbb{C}$ be an open set. Every differentiable function at the point $z_0 \in G$ is continuous at that point.
2. $(f(z)+g(z))' = f'(z)+g'(z)$; $(f(z)g(z))' = f'(z)g(z)+f(z)g'(z)$; $\left(\frac{f(z)}{g(z)}\right)' = \frac{f'(z)g(z)-f(z)g'(z)}{g^2(z)}$.
3. Let $G_1, G_2 \subset \mathbb{C}$ be two open sets, and let $f_1 : G_1 \to G_2$, $f_2 : G_2 \to \mathbb{C}$, $z_1 \in G_1$, $z_2 \in f_1(G_1)$. If the functions f_k are differentiable in G_k, $k \in \{1,2\}$, then $f_2 \circ f_1 : G_1 \to \mathbb{C}$ is differentiable at $z_1 \in G_1$, and
$$(f_2 \circ f_1)'(z_1) = f_2'(f_1(z_1))f_1'(z_1).$$
4. If the function $h : (a,b) \to \mathbb{C}$ is differentiable on (a,b), $h((a,b)) \subset G$, where $G \subset \mathbb{C}$ is an open set, and the function $f : G \to \mathbb{C}$ is differentiable at the point $h(t_0)$, then
$$(f \circ h)'(t_0) = f'(h(t_0))h'(t_0).$$

Proof. 3. Let $z \in G_1$. Since the function f_1 is differentiable at z_1, we have

$$f_1(z) = f_1(z_1) + f_1'(z_1)(z-z_1) + g_1(z)(z-z_1),$$

and since f_2 is differentiable at $z_2 \in G_2$ we have that for all $z \in G_2$,

$$f_2(z) = f_2(z_2) + f_2'(z_2)(z-z_2) + g_2(z)(z-z_2),$$

where $g_j : G_j \setminus \{z_j\} \to \mathbb{C}$ with $\lim_{z\to z_j} g_j(z) = 0$, $j = 1, 2$. Replacing in the last relation z by $f_1(z)$, and $z_2 = f_1(z_1)$, we obtain

$$\begin{aligned}(f_2 \circ f_1)(z) &= f_2(f_1(z_1))) + f_2'(f_1(z_1))(f_1(z) - f_1(z_1)) + g_2(f_1(z))(f_1(z) - f_1(z_1)) \\ &= f_2(f_1(z_1)) + f_2'(f_1(z_1))(f_1'(z_1)(z - z_1) + g_1(z)(z - z_1)) + g_2(f_1(z))(f_1(z) - f_1(z_1)) \\ &= f_2(f_1(z_1)) + f_2'(f_1(z_1))f_1'(z_1)(z - z_1) + g(z)(z - z_1),\end{aligned}$$

where

$$g(z) = f_2'(f_1(z_1))g_1(z) + g_2(f_1(z))\frac{f_1(z) - f_1(z_1)}{z - z_1} \quad \text{and} \quad \lim_{z\to z_1} g(z) = 0,$$

because $\lim_{z\to z_1} g_1(z) = 0$, $\lim_{z\to z_1} f_1(z) = z_2$, hence $\lim_{z\to z_1} g_2(f_1(z)) = 0$, and because there exists the limit $\lim_{z\to z_1} \frac{f_1(z)-f_1(z_1)}{z-z_1} = f_1'(z_1) \in \mathbb{C}$.

Consequently,

$$(f_2 \circ f_1)'(z_1) = f_2'(f_1(z_1))f_1'(z_1).$$

□

Definition 2.3.3.

1. Let $G \subset \mathbb{C}$ be an open set, $G \neq \emptyset$, and let $f : G \to \mathbb{C}$. The function f is said to be a **holomorphic function on** G, if it is differentiable in all the points of G.
2. The set of all the holomorphic functions on the open set $G \subset \mathbb{C}$ is denoted by $H(G)$; the set $H(G)$ is a complex vectorial space.
3. Those functions that are holomorphic on $\mathbb{C}$ are called **entire functions.**

Theorem 2.3.2. *Let $f = u + iv$ a complex valued function, defined on the open disc $U(z_0; r)$, with $r > 0$. If the functions u and v have partial derivatives in a neighborhood of the point z_0, that are continuous at z_0, and satisfy the Cauchy–Riemann conditions in z_0, then f is differentiable at z_0.*

Proof. From the assumption, the functions u and v are differentiable at the point z_0. Hence the function f is differentiable at z_0. Since the Cauchy–Riemann conditions are satisfied, according to the Cauchy–Riemann theorem, we get that there exists the derivative $f'(z_0)$. □

Theorem 2.3.3. *If $G \subset \mathbb{C}$ is an open set, $f = u + iv \in C^1(G)$, and the functions u and v satisfy the Cauchy–Riemann conditions in every point of the set G, then $f \in H(G)$.*

Let denote by $R(G)$ the set of all the differentiable functions on the open set G, and let denote by $F(G)$ the set of all the complex valued functions defined on G. Denoting by

$$\frac{\partial f}{\partial \bar{z}} = \frac{1}{2}\left(\frac{\partial f}{\partial x} + i\frac{\partial f}{\partial y}\right),$$

then

$$\frac{\partial}{\partial \overline{z}} : R(G) \to F(G).$$

From the Cauchy–Riemann theorem, we have $H(G) \subset R(G)$, and from the point 1 of Remark 2.3.1 we deduce that

$$H(G) = \ker \frac{\partial}{\partial \overline{z}} \Leftrightarrow \frac{\partial}{\partial \overline{z}} \equiv 0.$$

Remarks 2.3.2.
1. There are "many" functions $f \in C^1(\mathbb{C})$, that are not differentiable (derivable) in any point. The most simple example is given by the function $f(z) = \overline{z}$, that is not differentiable in any point, since it does not satisfy the Cauchy–Riemann conditions in any point of $\mathbb{C}$.
2. The Cauchy–Riemann conditions are necessary conditions for the differentiability, but they are not sufficient.

For $z = x + iy$, let define the function

$$f(z) = \begin{cases} \frac{xy}{\sqrt{x^2+y^2}} + i\frac{xy}{\sqrt{x^2+y^2}}, & \text{if } z \neq 0 \\ 0, & \text{if } z = 0. \end{cases}$$

Thus, $u = \operatorname{Re} f = \frac{xy}{\sqrt{x^2+y^2}} = \operatorname{Im} f = v$. A simple computation shows that

$$\frac{\partial u(0)}{\partial x} = \frac{\partial v(0)}{\partial x} = \frac{\partial u(0)}{\partial y} = \frac{\partial v(0)}{\partial y} = 0.$$

Hence the functions u and v have partial derivatives at the point $z = 0$, and satisfy the Cauchy–Riemann conditions.

We will prove that the function f is continuous at $z = 0$. Denoting $x = r\cos\theta$, $y = r\sin\theta$, then

$$f(z) = \begin{cases} r(\sin\theta\cos\theta + i\sin\theta\cos\theta), & r \neq 0 \\ 0, & r = 0 \end{cases}$$

$$\Rightarrow |f(z)| \le r|\sin\theta\cos\theta||1+i| \le r\left|\frac{\sin 2\theta}{2}\right||1+i| \le r\frac{\sqrt{2}}{2} < r \Rightarrow \lim_{z\to 0} f(z) = 0.$$

On the other hand

$$\frac{f(z)-f(0)}{z-0} = \frac{r\cos\theta\sin\theta(1+i)}{r(\cos\theta + i\sin\theta)} = \frac{\sin\theta(1-i)}{1+i\tan\theta}.$$

Taking $z \in \mathbb{R}^+$, $z \to 0$, then $\theta = 0$, $\sin\theta = 0$, and thus $\frac{f(z)-f(0)}{z-0} \to 0$.

Taking $\theta = \arg z = \frac{\pi}{4}$, $z \to 0$, then $\sin\theta = \frac{\sqrt{2}}{2}$, $\tan\theta = 1$, and thus $\frac{f(z)-f(0)}{z-0} \to \frac{\frac{\sqrt{2}}{2}(1+i)}{1+i} = \frac{\sqrt{2}}{2}$.

From the above two limits, it follows that $\nexists \lim_{z\to 0} \frac{f(z)-f(0)}{z-0}$.

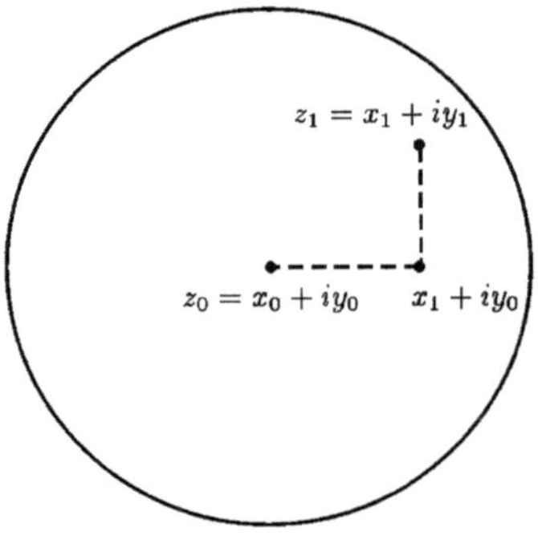

Figure 2.1: Proof of Theorem 2.3.4.

Theorem 2.3.4. *Let $D \subset \mathbb{C}$ be a domain, and let f be a holomorphic function on D, i. e., $f \in H(D)$. Then the function f is constant on D if and only if $f'(z) = 0$, $\forall z \in D$.*

Proof. If f is a constant function on D, it follows immediately that $f \in H(D)$ and $f'(z) = 0$, $\forall z \in D$.

Suppose that $f'(z) = 0$, $\forall z \in D$. Let $z_0 \in D$ be an arbitrary point. Since D is an open set, there exists $r > 0$, such that $U(z_0; r) \subset D$. For all $z \in U(z_0; r)$, we have $f'(z) = \frac{\partial u(z)}{\partial x} + i\frac{\partial v(z)}{\partial x} = 0$.

Let $z_1 \in U(z_0; r)$ be a fixed point for the moment (Figure 2.1). The real valued function

$$\varphi : [x_0, x_1] \to \mathbb{R}, \quad \varphi(x) = u(x, y_0)$$

is derivable, and $\varphi'(x) = \frac{\partial u(x,y_0)}{\partial x} = 0$. Hence φ is a constant function, thus $u(x, y_0) = u(x_0, y_0) = u(x_1, y_0)$, $\forall x \in [x_0, x_1]$.

From the similar reasons for the function v, we deduce that $v(x, y_0) = v(x_0, y_0) = v(x_1, y_0)$.

Since $f'(z) = \frac{\partial v(z)}{\partial y} - i\frac{\partial u(z)}{\partial y} = 0$, differentiating the function $\psi(y) = u(x_1, y)$ we get

$$\psi'(y) = \frac{\partial u(x_1, y)}{\partial y} = 0, \quad \forall y \in [y_0, y_1] \Rightarrow u(x_1, y_0) = u(x_1, y_1).$$

Similarly, we have $v(x_1, y_0) = v(x_1, y_1)$.

Combining all of the above results, we deduce that $f(z_1) = f(z_0)$. Since $z_1 \in U(z_0; r)$ was an arbitrary point, we obtain that $f(z) = f(z_0)$, $\forall z \in U(z_0; r)$.

Let $A = \{z \in D : f(z) = f(z_0)\}$. Using the same proof as to the above, it is easy to show that for all arbitrary $z_1 \in A$, there exists $r_1 > 0$, such that $U(z_1; r_1) \subset D$ and $f(z) = f(z_1)(= f(z_0))$, $\forall z \in U(z_1; r_1)$. Thus, the set A is an open set in D.

Let $z_n \in A$, $\forall n \in \mathbb{N}^*$, with $\lim_{n\to\infty} z_n = z \in D$. Since f is a continuous function on D, we have $\lim_{n\to\infty} f(z_n) = f(z)$. But $f(z_n) = f(z_0)$, $\forall n \in \mathbb{N}^*$, because of the fact that $z_n \in A$, $\forall n \in \mathbb{N}^*$. Thus $f(z) = f(z_0)$, so the set A is closed in D.

Since D is a connected set, from the above two results we conclude that $D = A$. □

Theorem 2.3.5. *Let $D \subset \mathbb{C}$ be a domain, and let $f \in H(D)$. If at least one of the following real valued functions*

$$|f|, \ \mathrm{Re} f, \ \mathrm{Im} f, \ \arg f$$

is constant on the domain D, then the function f is constant on D.

Proof. For example, let, $u(x,y) = \mathrm{Re} f(z) = c$, $\forall z \in D$. Then $\frac{\partial u}{\partial x} = \frac{\partial u}{\partial y} = 0$ on D, and from the Cauchy–Riemann conditions we have $\frac{\partial v}{\partial x} = \frac{\partial v}{\partial y} = 0$ on the domain D. Then $f'(z) = 0$, $\forall z \in D$, and according to the Theorem 2.3.4, the function f is constant on the domain D.

The proof is similar to the case that $\mathrm{Im} f$ is constant on D.

If $|f(z)|^2 = u^2(x,y) + v^2(x,y) = c$, $\forall z \in D$, then

$$\begin{cases} u\frac{\partial u}{\partial x} + v\frac{\partial v}{\partial x} = 0 \\ u\frac{\partial u}{\partial y} + v\frac{\partial v}{\partial y} = 0. \end{cases}$$

From the Cauchy–Riemann conditions, this system is equivalent to

$$\begin{cases} u\frac{\partial u}{\partial x} + v\frac{\partial v}{\partial x} = 0 \\ v\frac{\partial u}{\partial x} - u\frac{\partial v}{\partial x} = 0, \end{cases}$$

with the unknown $\frac{\partial u}{\partial x}$ and $\frac{\partial v}{\partial x}$, so the determinant of the system will be $-(u^2 + v^2)$.

If $u^2 + v^2 = c = 0$, then $f(z) = 0$, $\forall z \in D$.

If $u^2 + v^2 = c \neq 0$, then the system has in every point $z \in D$ only the trivial solution, i. e., $\frac{\partial u}{\partial x} = \frac{\partial v}{\partial x} = 0$, $\forall z \in D$, and thus $f'(z) = 0$, $\forall z \in D$. According to Theorem 2.3.4, the function f will be constant on the domain D.

We have $\arg f = \arctan \frac{v}{u} = k_0 \Rightarrow \frac{v}{u} = k$, $\forall z \in D$.

If $u = 0 \Rightarrow u = \mathrm{Re} f$ is constant $\Rightarrow f$ is constant.

If $u \neq 0 \Rightarrow v - ku = 0 \Rightarrow \mathrm{Re}(-k - i)f = \mathrm{Re}(-k - i)(u + iv) = \mathrm{Re}(-ku + v + i(-u - kv)) = -ku + v = 0$. Hence the function $(-k - i)f$ is constant, then f is constant on the domain D. □

2.4 The geometric interpretation of the derivative

Definition 2.4.1. The real variable continuous function

$$\gamma : [0,1] \to \mathbb{C}, \quad \gamma(t) = \alpha(t) + i\beta(t)$$

is called **a path of curve in $\mathbb{C}$**.

Then the system

$$\begin{cases} x = \alpha(t) \\ y = \beta(t), \quad t \in [0,1] \end{cases}$$

represents the parametric equation of a continuous curve in $\mathbb{R}^2$.

The point $\gamma(0)$ is called **the starting point**, while the point $\gamma(1)$ is called **the end point** of the path γ.

Definition 2.4.2.
1. The set $\{\gamma\} = \gamma([0,1])$ is called **the image of the path** γ.
2. The paths γ_1 and γ_2 are called **equivalent paths**, denoted by $\gamma_1 \sim \gamma_2$, if $\exists h : [0,1] \to [0,1]$ a continuous bijection, with h^{-1} continuous, and $h(0) = 0$, $h(1) = 1$, such that $\gamma_1 = \gamma_2 \circ h$.
3. If $\gamma \in C^1[0,1]$, with $\gamma'(t) \neq 0$, $\forall t \in [0,1]$, then γ is called a **smooth path.**

Remarks 2.4.1.
1. If $\gamma \in C^1[0,1]$, with $\gamma'(t_0) = \alpha'(t_0) + i\beta'(t_0) \neq 0$, then $\alpha'(t_0)$ and $\beta'(t_0)$ are called the **directional parameters** of the tangent to the path $\{\gamma\}$ into the point $\gamma(t_0)$.
2. If $\{\gamma\}$ is a smooth path, then there exists a tangent to it in every point on $\{\gamma\}$. If θ is the angle between the Ox axis and this tangent, then

$$\theta = \arg \gamma'(t_0) = \arctan \frac{\beta'(t_0)}{\alpha'(t_0)}.$$

3. If γ_1 and γ_2 are two smooth paths, with $\gamma_1(t_0) = \gamma_2(t_0) = z_0$, then **the angle between γ_1 and γ_2 at the point** z_0 is the angle between the tangents to both of these paths into the point z_0.

Definition 2.4.3.
1. Let $G \subset \mathbb{C}$ be an open set and let $f \in C^1(G)$. The function f is said to be a **first-order conformal map at the point** $z_0 \in G$, if the angle at $z_0 \in G$ between any arbitrary smooth paths γ_1 and γ_2, that start from z_0, is the same with the angle at $f(z_0)$ between their images $f \circ \gamma_1$ and $f \circ \gamma_2$.
2. The function f is called **direct conformal mapping** or **inverse conformal mapping**, if the directions of the angles is preserved, or respectively is changed.
3. The function f is said to be a **second-order conformal map at the point** $z_0 \in G$, if the length of any rectifiable smooth path that starts from $z_0 = \gamma(0)$ is modified "in the same way" by the function f, i. e.,

$$|(f \circ \gamma)'(0)| = k|\gamma'(0)|, \quad \forall \gamma \text{ smooth path, } \gamma(0) = z_0,$$

and the constant k is independent of γ and it is called **the linear deformation (distortion) coefficient of the function f into the point** z_0.

Theorem 2.4.1. *If the function $f : G \to \mathbb{C}$ is differentiable at the point $z_0 \in G$, where $G \subset \mathbb{C}$ is an open set, and $f'(z_0) \neq 0$, then:*
1. *The function f is first order direct conformal mapping at the point z_0.*
2. *The function f is second order conformal mapping at the point z_0.*

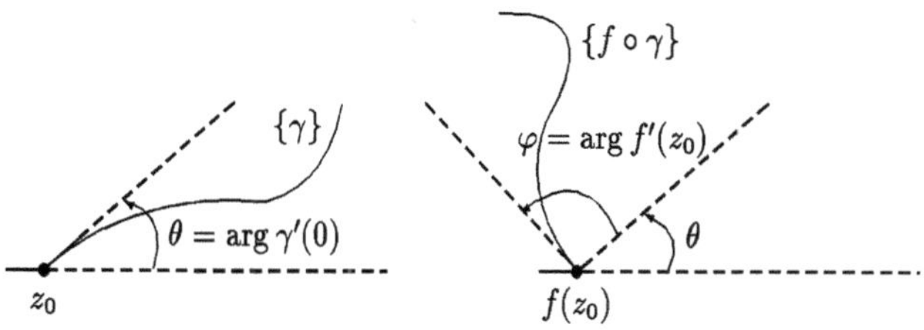

Figure 2.2: Proof of Theorem 2.4.1.

Proof. Let γ be an arbitrary smooth path that starts from z_0, i. e., $\gamma(0) = z_0$, and let $\gamma_1 = f \circ \gamma$ (Figure 2.2). Since

$$(f \circ \gamma)'(0) = f'(\gamma(0))\gamma'(0),$$

we have $\gamma_1'(0) = f'(z_0)\gamma'(0)$, so it follows that

$$\arg \gamma_1'(0) = \arg f'(z_0) + \arg \gamma'(0).$$

It means that the tangent to the image path γ_1 into the point $f(z_0)$ is obtained by rotating the tangent to the path γ into the point z_0 with the angle $\arg f'(z_0)$ (which is independent of γ), i. e.,

$$\arg \gamma_1'(0) - \arg \gamma'(0) = \arg f'(z_0).$$

This implies that the angle and the direction of the angle between any two arbitrary smooth paths that starts from z_0 will be preserved by the function f, which proves the first part of the theorem.

For any arbitrary smooth path that starts from z_0, we have

$$|(f \circ \gamma)'(0)| = |f'(z_0)||\gamma'(0)|,$$

hence f is second-order conformal mapping at the point z_0, and the linear distortion coefficient of the function f in z_0 will be $|f'(z_0)|$. □

Remarks 2.4.2.
1. If the function f has derivative at the point $z_0 \in G$, where $G \subset \mathbb{C}$ is an open set, and if $f'(z_0) \neq 0$, then $\arg f'(z_0)$ is the rotational angle of the tangent in z_0 to any smooth path that starts from z_0.
2. The condition $f'(z_0) \neq 0$ is necessary. For example, the function $f(z) = z^2$ doubles all the angles that starts from $z_0 = 0$.

Theorem 2.4.2. *Let $f \in C^1(G)$, where $G \subset \mathbb{C}$ is an open set, and let $z_0 \in G$.*
1. *If the function f is first-order direct conformal mapping at the point z_0, then f has derivative at this point.*

2. *If the function f is second-order conformal mapping at the point z_0, then f or $\overline{f}$ have derivatives at this point.*

Proof. 1. Let $\gamma(t) = (1-t)z_0 + tz_1$, where $\arg(z_1 - z_0) = \theta$, and let $\gamma_1(t) = f(\gamma(t))$, $t \in [0,1]$. Then

$$\gamma_1'(0) = f_\theta'(\gamma(0))(z_1 - z_0) = f_\theta'(z_0)\gamma'(0),$$

hence

$$\arg \gamma_1'(0) - \arg \gamma'(0) = \arg f_\theta'(z_0). \tag{2.12}$$

For the left-hand side to be independent of θ, it is necessary that $\arg f_\theta'(z_0)$ is independent of θ.

During the proof of Kasner theorem, we obtained that

$$f_\theta'(z_0) = \frac{\partial f(z_0)}{\partial z} + \frac{\partial f(z_0)}{\partial \overline{z}} e^{-2i\theta}.$$

Using this relation, it follows that if $\frac{\partial f(z_0)}{\partial \overline{z}} \neq 0$, then

$$\arg\left(f_\theta'(z_0) - \frac{\partial f(z_0)}{\partial z} \right) = \arg \frac{\partial f(z_0)}{\partial \overline{z}} - 2\theta.$$

Since $\frac{\partial f(z_0)}{\partial z}$ and $\frac{\partial f(z_0)}{\partial \overline{z}}$ are constants, then $\arg(f_\theta'(z_0) - \frac{\partial f(z_0)}{\partial z})$ cannot be constant. Hence, the vector with the starting point in $\frac{\partial f(z_0)}{\partial z}$ and the end point in $f_\theta'(z_0)$ runs an angle of 4π whenever θ runs an interval of the length of 2π. Thus, the right-hand side of the equality (2.12) is constant only if $\frac{\partial f(z_0)}{\partial \overline{z}} = 0$. Since this last equality is equivalent to the Cauchy–Riemann conditions, it follows that the function f has derivative at the point z_0.

2. Let γ be the same path as to the point 1 of the proof. From the assumption

$$|(f \circ \gamma)'(0)| = k|\gamma'(0)|,$$

where the constant k is independent of θ.

On the other hand,

$$(f \circ \gamma)'(0) = f_\theta'(z_0)\gamma'(0),$$

hence $|f_\theta'(z_0)| = k$, $\forall \theta \in (-\pi, \pi]$.

According to the Kasner theorem, are possible the next two situations:

(i) $\frac{\partial f(z_0)}{\partial \overline{z}} = 0$, and then the function f has derivative at the point z_0;

or

(ii) $\frac{\partial f(z_0)}{\partial z} = 0$, and this represents the Cauchy–Riemann conditions for the function $\overline{f}$. In this case, we have that the function $\overline{f}$ has derivative at the point z_0. □

2.5 Entire functions

2.5.1 The polynomial function

Definition 2.5.1. The function $p : \mathbb{C} \to \mathbb{C}$, $p(z) = c_0 + c_1 z + \cdots + c_n z^n$, where $c_j \in \mathbb{C}$, $j \in \{0, 1, \ldots, n\}$ and $c_n \neq 0$, is called **nth order polynomial function.**

Remarks 2.5.1.

1. Any nonzero constant polynomial is a 0th order polynomial.
2. The zero constant polynomial is a $-\infty$-th order polynomial.
3. The function $f(z) = z^n$ is an entire function, and its derivative is $f'(z) = nz^{n-1}$.
4. From the previous remark, any polynomial function in an entire function, as a sum of entire functions.

2.5.2 The exponential function

Let

$$z = x + iy \longrightarrow e^z = e^x(\cos y + i \sin y).$$

Since $\operatorname{Re} e^z = e^x \cos y$ and $\operatorname{Im} e^z = e^x \sin y$ have continuous first-order partial derivatives on $\mathbb{C}$, and satisfy the Cauchy–Riemann conditions on $\mathbb{C}$, it follows that the function e^z is holomorphic on the complex plane $\mathbb{C}$. The function, denoted by e^z, $\exp z$ or $\exp(z)$, is called **the exponential function.**

From

$$(e^z)' = \frac{\partial}{\partial x} e^z = e^z,$$

we have $(e^z)^{(n)} = e^z$, $\forall n \in \mathbb{N}$.

If $x = 0$, then

$$\begin{aligned} e^{iy} &= \cos y + i \sin y, \\ e^{-iy} &= \cos y - i \sin y \\ &\Rightarrow \cos y = \frac{e^{iy} + e^{-iy}}{2}, \quad \sin y = \frac{e^{iy} - e^{-iy}}{2i}. \end{aligned}$$

The above relations are called **Euler formulas.**

1. $e^{z+2\pi i} = e^z$, $\forall z \in \mathbb{C}$, hence the exponential function is a periodic function, with the period equal with $2\pi i$, so it is not an injective function on $\mathbb{C}$.
2. $\operatorname{Re} e^z = e^x \cos y$, $\operatorname{Im} e^z = e^x \sin y$, $|e^z| = e^x$, where $z = x + iy$.
3. $\overline{e^z} = e^{\bar{z}}$, $\forall z \in \mathbb{C}$, and $\arg e^z = y \pmod{2\pi}$, where $z = x + iy$.
4. $e^z \neq 0$, $\forall z \in \mathbb{C}$, because $|e^z| = e^x \neq 0$, $\forall x \in \mathbb{R}$.
5. $z = re^{i\theta}$, where $r = |z|$, $\theta = \arg z$, $\forall z \in \mathbb{C}^*$.

Remark 2.5.2. Let us define the following B_k strips, parallel with the Ox axis, by

$$B_k = \{z = x + iy \in \mathbb{C} : 2k\pi \le y < 2(k+1)\pi\}, \quad k \in \mathbb{Z}.$$

Then $\mathbb{C} = \bigcup_{k\in\mathbb{Z}} B_k$, and the exponential function is injective on the every B_k strip, $k \in \mathbb{Z}$. Since the exponential is holomorphic and injective in each B_k strip, we say that it is a **univalent function** on B_k.

2.5.3 Complex trigonometric functions

Starting with the Euler formulas, we will define the complex valued trigonometric functions as follows:

$$\cos z = \frac{1}{2}(e^{iz} + e^{-iz}); \quad \sin z = \frac{1}{2i}(e^{iz} - e^{-iz}).$$

Thus,

$$\cos ix = \cosh x, \quad \sin ix = i \sinh x.$$

It is easy to prove the next formulas:

$$\cos(z + w) = \cos z \cos w - \sin z \sin w,$$
$$\sin(z + w) = \sin z \cos w + \cos z \sin w,$$

which imply

$$\cos z = \cos(x + iy) = \cos x \cos iy - \sin x \sin iy = \cos x \cosh y - i \sin x \sinh y.$$

Exercise 2.5.1.
1. Prove that $|\sinh y| \le |\cos z| \le \cosh y$, if $z = x + iy$.
2. Prove that $|\cos z|^2 = \frac{1}{2}(\cosh 2y + \cos 2x)$, if $z = x + iy$.
3. The function sin and cos are unbounded functions, because

$$\lim_{z=x_0+iy\to\infty} \sin z = \infty \quad \text{and} \quad \lim_{z=x_0+iy\to\infty} \cos z = \infty.$$

Solution. It is well known that

$$\cosh^2 y - \sinh^2 y = 1, \tag{2.13}$$

$$\begin{aligned} |\cos z|^2 &= \cos^2 x \cosh^2 y + \sin^2 x \sinh^2 y = \cos^2 x(1 + \sinh^2 y) + \sin^2 x \sinh^2 y \\ &= \cos^2 x + (\cos^2 x + \sin^2 x)\sinh^2 y = \cos^2 x + \sinh^2 y \\ &\Rightarrow \sinh^2 y \le |\cos z|^2, \end{aligned} \tag{2.14}$$

$$
\begin{aligned}
|\cos z|^2 &= \cos^2 x \cosh^2 y + \sin^2 x \sinh^2 y = \cos^2 x \cosh^2 y + \sin^2 x(\cosh^2 y - 1) \\
&= \cosh^2 y - \sin^2 x \le \cosh^2 y \\
&\Rightarrow |\cos z|^2 \le \cosh^2 y. \qquad (2.15)
\end{aligned}
$$

From the relations (2.13) and (2.15), we obtain

$$|\cos z|^2 = \cosh^2 y - \sin^2 x, \quad \cosh^2 y = \frac{1}{2}(\cosh 2y + 1) \quad \text{and} \quad \sin^2 x = \frac{1}{2}(1 - \cos 2x),$$

respectively. Hence

$$|\cos z|^2 = \frac{1}{2}(\cosh 2y + 1) - \frac{1}{2}(1 - \cos 2x) = \frac{1}{2}(\cosh 2y + \cos 2x).$$

Since

$$|\cos z|^2 \ge \frac{1}{2}(\cosh 2y - 1) \Rightarrow \lim_{z=x_0+iy\to\infty} \cos z = \infty,$$

and

$$
\begin{aligned}
|\sin z|^2 &= \frac{1}{2}(\cosh 2y - \cos 2x) \Rightarrow |\sin z|^2 \ge \frac{1}{2}(\cosh 2y - 1) \\
&\Rightarrow \lim_{z=x_0+iy\to\infty} \sin z = \infty.
\end{aligned}
$$

□

The above defined functions $\cos z$ and $\sin z$, and the function $\tan z = \frac{\sin z}{\cos z}$, are called (complex) **trigonometric functions**.

2.5.4 Complex hyperbolic functions

Similar to the definition of the complex trigonometric functions, we will define the complex **hyperbolic functions** as follows:

$$\cosh z = \frac{1}{2}(e^z + e^{-z}); \quad \sinh z = \frac{1}{2}(e^z - e^{-z}).$$

Thus, the functions cosh and sinh are entire periodic functions, with the period equal with $2\pi i$.

2.6 Bilinear transforms

Let the function T be defined as

$$z \longmapsto Tz = \frac{az + b}{cz + d}, \quad a, b, c, d \in \mathbb{C}.$$

Then T is a constant function, if and only if $ad - bc = 0$. (For example, if $ac \neq 0$, then $ad - bc = 0 \Leftrightarrow \frac{b}{a} = \frac{d}{c}$, hence $Tz = \frac{a}{c}\frac{z+\frac{b}{a}}{z+\frac{d}{c}} = \frac{a}{c}$.)

Definition 2.6.1. The function T is called **bilinear transform**, if $ad - bc \neq 0$ and

$$T : \widehat{\mathbb{C}} \to \widehat{\mathbb{C}},$$

$$Tz = \begin{cases} \frac{az+b}{cz+d}, & \text{if } z \in \mathbb{C} \setminus \{-\frac{d}{c}\} \\ \infty, & \text{if } z = -\frac{d}{c} \\ \frac{a}{c}, & \text{if } z = \infty, \end{cases} \qquad \text{if } c \neq 0$$

and

$$Tz = \begin{cases} \frac{a}{d}z + \frac{b}{d}, & \text{if } z \in \mathbb{C} \\ \infty, & \text{if } z = \infty, \end{cases} \qquad \text{if } c = 0,$$

where $a, b, c, d \in \mathbb{C}$.

Theorem 2.6.1. *A bilinear transform is an injective single-valued continuous function of* $\widehat{\mathbb{C}}$ *onto* $\widehat{\mathbb{C}}$.

Proof. When $c = 0$, $Tz = \frac{a}{d}z + \frac{b}{d}$, where $\frac{a}{d} \neq 0$. Hence $\forall w \in \mathbb{C}, \exists! z \in \mathbb{C}: Tz = w$. Further, T and T^{-1} are continuous in $\mathbb{C}$, $\lim_{z\to\infty} Tz = \infty$ and $\lim_{w\to\infty} T^{-1}w = \infty$, so it follows that T and T^{-1} are continuous at the ∞ point.

If $c \neq 0$ and $z_0 \neq -\frac{d}{c}$, then $az + b$ and $cz + d$ are continuous in a neighborhood of z_0 and $cz_0 + d \neq 0$, hence T is continuous at the z_0 point.

If $z \to -\frac{d}{c} \Rightarrow cz + d \to 0$, $az + b \to a(-\frac{d}{c}) + b \neq 0 \Rightarrow Tz \to \infty \Rightarrow T$ is continuous at the $-\frac{d}{c}$ point.

If $z \neq -\frac{d}{c}$, then

$$w = \frac{az+b}{cz+d} \Rightarrow z = \frac{-dw+b}{cw-a} \Rightarrow T|_{\mathbb{C}\setminus\{-\frac{d}{c}\}} \quad \text{is injective.}$$

The value of the function T cannot be $\frac{a}{c}$, because

$$\frac{a}{c} = \frac{az+b}{cz+d} \Leftrightarrow ad = bc,$$

hence it is impossible.

Hence, the function T is a bijection between $\mathbb{C} \setminus \{-\frac{d}{c}\}$ and $\mathbb{C} \setminus \{\frac{a}{c}\}$, $T(-\frac{d}{c}) = \infty$ and $T(\infty) = \frac{a}{c}$. $Tz \to \infty$, if $z \to -\frac{d}{c}$ and $T^{-1}w \to \infty$, if $w \to \frac{a}{c}$, so it follows that T is continuous at the $-\frac{d}{c}$ point, and T^{-1} is continuous at the $\frac{a}{c}$ point. □

2.6.1 Decompositions in elementary functions

If $c \neq 0$, then

$$Tz = \frac{az+b}{c(z+\frac{d}{c})} = \frac{az + a\frac{d}{c} - a\frac{d}{c} + b}{c(z+\frac{d}{c})} = \frac{a}{c} + \frac{-a\frac{d}{c} + b}{c(z+\frac{d}{c})} = \frac{a}{c} + \frac{bc-ad}{c^2(z+\frac{d}{c})}.$$

If $c \neq 0$, then $Tz = \frac{a}{c} + \frac{bc-ad}{c^2(z+\frac{d}{c})}$.
It is well known that:

(1)	$z_1 = z + \frac{d}{c}$	translation
(2)	$z_2 = \frac{1}{z_1}$	symmetry-inversion
(3)	$z_3 = kz_2 \quad (k = \frac{bc-ad}{c^2})$	complex scaling
(4)	$w = z_3 + \frac{a}{c}$	translation.

If $c = 0$, then $Tz = \frac{a}{d}z + \frac{b}{d}$ is a first degree polynomial.

Hence, in the case $c \neq 0$, the function T can be written by composing (1), (2), (3) and (4) transformations: (1) and (4) translations, and (3) complex homotheticity:

$$z_3 = |k|\frac{k}{|k|}z_2,$$

where

(i) $z_3' = \frac{k}{|k|}z_2$ rotation (i. e., the product between z_2 and a complex number with unitary module)

(ii) $z_3 = |k|z_3'$ real homotheticity (i. e., dilatation or shrinking).

The (2) number transform is called symmetry-inversion and

$$z_2 = \frac{1}{z_1} \Rightarrow |z_1||z_2| = 1.$$

If $\zeta = \frac{1}{|z_1|}e^{i\arg z_1}$, then ζ and z_1 are inverse points with respect to the unit circle, because

(i) ζ and z_1 belongs on a half-line that starts from the origin O,

(ii) $|\zeta||z_1| = 1$.

On the other hand,

$$e^{i\arg z_1} = \frac{z_1}{|z_1|}, \quad \overline{\zeta} = \frac{1}{|z_1|}e^{-i\arg z_1} = \frac{1}{|z_1|e^{i\arg z_1}} = \frac{1}{z_1} \Rightarrow z_2 = \overline{\zeta}.$$

Hence, the point z_2-t can be obtained from z_1 as follows:

(i) First, we find the inverse of z_1 with respect to the unit circle, which will be ζ.

(ii) Second, we find the symmetric of ζ with respect to the real axis Ox, which will be $\overline{\zeta} = z_2$.

Theorem 2.6.2. *A bilinear transform maps the circles from $\widehat{\mathbb{C}}$ onto $\widehat{\mathbb{C}}$'s circles. (The lines are also circles which contain the ∞ point.)*

Proof. The general equation of a circle is $A(x^2 + y^2) + Bx + Cy + D = 0$, where $|A| + |B| + |C| \neq 0$.

If $A = 0$, then the above equation represents a line.

If $A \neq 0$, we may suppose that $A > 0$. The circle is a really circle, if

$$B^2 + C^2 - 4AD \geq 0. \tag{2.16}$$

We will obtain the complex equation of the circle, if we change the variables (parameters)

$$x^2 + y^2 = z\overline{z}, \quad x = \frac{1}{2}(z + \overline{z}), \quad y = \frac{1}{2i}(z - \overline{z}) = -\frac{i}{2}z + \frac{i}{2}\overline{z},$$

i. e.,

$$Az\overline{z} + B\frac{1}{2}(z + \overline{z}) + C\frac{i}{2}(\overline{z} - z) + D = 0,$$

or

$$Az\overline{z} + \alpha z + \overline{\alpha z} + D = 0 \tag{2.17}$$

where $\alpha = \frac{1}{2}(B - iC), \overline{\alpha} = \frac{1}{2}(B + iC)$.

The condition (2.16), for the equation (2.17) becomes

$$\alpha\overline{\alpha} - AD \geq 0. \tag{2.18}$$

Let us consider the $z = \frac{1}{w}$ transform. Then the relation (2.17) becomes

$$A + \overline{\alpha}w + \alpha\overline{w} + Dw\overline{w} = 0, \tag{2.19}$$

that also represents the equation of a circle. □

Theorem 2.6.3. *The cross ratio is invariant by a bilinear transform T, i. e.,*

$$(z_1, z_2, z_3, z_4) = (Tz_1, Tz_2, Tz_3, Tz_4),$$

where

$$(z_1, z_2, z_3, z_4) = \frac{z_1 - z_2}{z_1 - z_4} : \frac{z_3 - z_2}{z_3 - z_4}.$$

Proof. It is evidently that

$$Tz - Tz' = \frac{az + b}{cz + d} - \frac{az' + b}{cz' + d} = \frac{(ad - bc)(z - z')}{(cz + d)(cz' + d)}.$$

Hence, we have

$$\begin{aligned}
&\frac{Tz_1 - Tz_2}{Tz_1 - Tz_4} : \frac{Tz_3 - Tz_2}{Tz_3 - Tz_4} \\
&= \frac{(ad - bc)(z_1 - z_2)}{(cz_1 + d)(cz_2 + d)} \cdot \frac{(cz_1 + d)(cz_4 + d)}{(ad - bc)(z_1 - z_4)} : \frac{(ad - bc)(z_3 - z_2)}{(cz_3 + d)(cz_2 + d)} \cdot \frac{(cz_3 + d)(cz_4 + d)}{(ad - bc)(z_3 - z_4)} \\
&= \frac{z_1 - z_2}{z_1 - z_4} : \frac{z_3 - z_2}{z_3 - z_4}.
\end{aligned}$$

□

Theorem 2.6.4. *Every arbitrary T bilinear transform is conformal in every $z \in \mathbb{C} \setminus \{-\frac{d}{c}\}$ point.*

Proof. We have

$$\frac{Tz - Tz'}{z - z'} = \frac{ad - bc}{(cz + d)(cz' + d)},$$

and considering $z' \to z$ in this relation, we obtain that $T'(z) = \frac{ad-bc}{(cz+d)^2} \neq 0$. □

Theorem 2.6.5. *If $a, b, c, d \in \mathbb{R}$ and if $ad - bc > 0$, then T maps the upper half-plane into upper half-plane, and the lower half-plane into the lower half-plane.*

If $a, b, c, d \in \mathbb{R}$ and if $ad - bc < 0$, then T maps the upper half-plane into lower half-plane, and the lower half-plane into the upper half-plane.

Proof. Since

$$\overline{Tz} = \overline{\left(\frac{az + b}{cz + d}\right)} = \frac{a\bar{z} + b}{c\bar{z} + d} = T\bar{z},$$

then

$$Tz - \overline{Tz} = Tz - T\bar{z} = \frac{ad - bc}{|cz + d|^2}(z - \bar{z}). \tag{2.20}$$

Hence, if $z \in \mathbb{R} \Rightarrow z - \bar{z} = 0 \Rightarrow Tz - \overline{Tz} = 0 \Rightarrow Tz \in \mathbb{R}$.

If $\operatorname{Im} z > 0$,

$$z - \bar{z} = 2i \operatorname{Im} z \overset{(2.20)}{\Rightarrow} 2i \operatorname{Im} Tz = Tz - \overline{Tz} = \frac{ad - bc}{|cz + d|^2} 2i \operatorname{Im} z.$$

From here, it follows that $\operatorname{Im} Tz > 0$ whenever $ad - bc > 0$, and $\operatorname{Im} Tz < 0$ whenever $ad - bc < 0$. □

Remark 2.6.1. Each bilinear transform depends on **three** complex parameters. Hence, if we know the values of an unknown bilinear transform in three different points, then we may determine this transform.

2.7 The Möbius-type groups

For two given bilinear transforms,

$$T_1 z = \frac{a_1 z + b_1}{c_1 z + d_1} \quad \text{and} \quad T_2 z = \frac{a_2 z + b_2}{c_2 z + d_2}$$

$T_1 \circ T_2$ and $T_2 \circ T_1$ are also bilinear transforms. To an arbitrary T_j bilinear transform, corresponds a matrix of the form

$$\begin{pmatrix} a_j & b_j \\ c_j & d_j \end{pmatrix},$$

obtained with the aid of its coefficients.

Denoting $T_3 = T_1 \circ T_2$, the bilinear transform T_3 has the corresponding matrix given by

$$\begin{pmatrix} a_1 & b_1 \\ c_1 & d_1 \end{pmatrix} \cdot \begin{pmatrix} a_2 & b_2 \\ c_2 & d_2 \end{pmatrix} = \begin{pmatrix} a_1a_2 + b_1c_2 & a_1b_2 + b_1d_2 \\ a_2c_1 + c_2d_1 & b_2c_1 + d_1d_2 \end{pmatrix}.$$

It follows that the matrix of T_3 may be obtained from as the product of the corresponding matrices of T_1 and T_2.

Theorem 2.7.1. *The bilinear transforms set is a group with respect to the functions composition. This group is called* **Möbius-type group.**

Proof. T_j is a bilinear transform $\Leftrightarrow \det\begin{pmatrix} a_j & b_j \\ c_j & d_j \end{pmatrix} \neq 0$. The determinant of the $T_1 \circ T_2$ corresponding matrix is $\det T_1 \cdot \det T_2$, hence the matrix of T_j belongs to the group $GL_2(\mathbb{C})$. From this remark, we deduce that the set of all bilinear transforms represent a group with respect to the composition, which is a subgroup $GL_2(\mathbb{C})$. We call that the bilinear transformation group as it is the holomorphic image of $GL_2(\mathbb{C})$. □

Remark 2.7.1. An arbitrary bilinear transform remains unchanged if we multiply all its coefficients by the same complex number k, $k \neq 0$. That is,

$$\frac{az + b}{cz + d} = \frac{\frac{a}{k}z + \frac{b}{k}}{\frac{c}{k}z + \frac{d}{k}}, \quad \forall z \in \mathbb{C} \setminus \left\{-\frac{d}{c}\right\}.$$

The constant k may be chosen such that $ad - bc = 1$.

Theorem 2.7.2. *Those bilinear transforms with $a, b, c, d \in \mathbb{R}$ and $ad - bc = 1$, represent the subgroup of the Möbius-type group that maps the upper half-plane into the upper half-plane.*

Proof. According to Theorem 2.6.5, if $a, b, , c, d \in \mathbb{R}$ and $ad - bc > 0$, the transform $Tz = \frac{az+b}{cz+d}$ maps the upper half-plane into the upper half-plane. Dividing all its coefficients by $\sqrt{ad - bc}$, and changing the notation, we obtain $ad - bc = 1$.

Conversely, suppose that T maps the upper half-plane into the upper half-plane. Hence, we will deduce that $Tz \in \mathbb{R}$ if $z \in \mathbb{R}$, so T can be written in such a form that $a, b, c, d \in \mathbb{R}$.

Case I: $cd = 0$

1. If $d = 0$, then $Tz = \frac{a}{c} + \frac{b}{cz} \in \mathbb{R}$.
   Further,

$$z \in \mathbb{R}, \quad z \to \infty \Rightarrow \frac{a}{c} \in \mathbb{R} \quad \text{and} \quad z = 1 \Rightarrow \frac{b}{c} \in \mathbb{R}.$$

2. If $c = 0$, then $Tz = \frac{a}{d}z + \frac{b}{d}$.
   Further,

$$z = 0 \Rightarrow \frac{b}{d} \in \mathbb{R} \quad \text{and} \quad z = 1 \Rightarrow \frac{a}{d} \in \mathbb{R}.$$

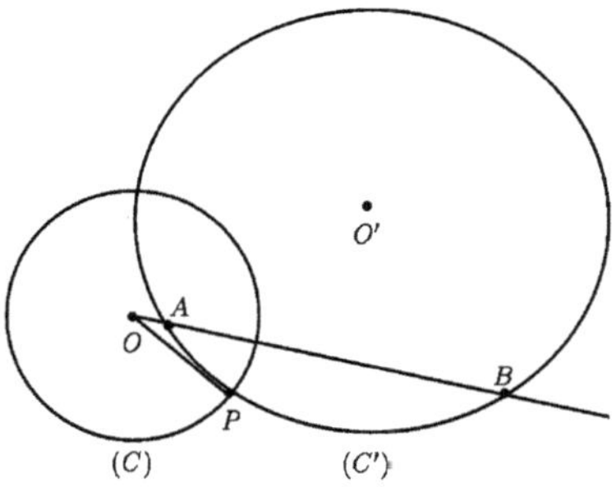

Figure 2.3: Proof of Theorem 2.7.3.

Case II: $cd \neq 0$.

In this case, we obtain

$$z \in \mathbb{R}, \quad z \to \infty \Rightarrow \frac{a}{c} \in \mathbb{R} \quad \text{and} \quad z = 0 \Rightarrow \frac{b}{d} \in \mathbb{R},$$

hence

$$\frac{a}{c} - \frac{b}{d} = \frac{ad - bc}{cd} \in \mathbb{R}.$$

From here, we have

$$Tz = \frac{a}{c} + \frac{\frac{bc-ad}{cd}}{\frac{c}{d}z + 1} \in \mathbb{R}, \quad \text{if } z \in \mathbb{R} \Rightarrow \frac{c}{d}z + 1 \in \mathbb{R}, \quad \text{if } z \in \mathbb{R} \Rightarrow \frac{c}{d} \in \mathbb{R}.$$

Hence, if T maps $\mathbb{R}$ into $\mathbb{R}$, the coefficients a, b, c and d may be chosen to be real numbers. From here, according to Theorem 2.6.5 we obtain that $ad - bc > 0$. □

Definition 2.7.1. Let (C) be the circle in $\mathbb{R}^2$ with the center in O, and with the radius $r > 0$. The points $A, B \in \mathbb{R}^2$ are called **inverse** with respect to the (C) circle, if O, A and B belong to the same half-line with the origin in O, and $OA \cdot OB = r^2$.

Theorem 2.7.3. *The points A and B are inverse points with respect to the circle (C), if and only if every circle (C') which contains both these points is perpendicular (or orthogonal) to (C). (Two circles are called orthogonal if their tangents in the intersection points are orthogonal.)*

Proof. The circles (C) and (C') are orthogonal, if their tangents in the intersection points are orthogonal, which is equivalent to the fact that the radius corresponding to one of the intersection points are orthogonal. This last affirmation is true, if and only if the radius (corresponding to one of the intersection points) of one circle is tangent to the other circle.

Suppose that $(C') \perp (C)$, and let $P \in (C) \cap (C')$ (Figure 2.3).

Then OP is tangent to (C') and $r^2 = OP^2 = OA \cdot OB$ (the power of point O with respect to the circle (C')).

Conversely, let OP the tangent from O to the circle (C'), $P \in (C')$. Then $r^2 = OA \cdot OB = OP^2 \Rightarrow OP = r \Rightarrow P \in (C) \cap (C')$, and since OP is tangent to (C') then OP is orthogonal to the radius $O'P$ of the circle (C'), hence we obtain that $(C) \perp (C')$. □

Theorem 2.7.4. *Let $(C) \subset \mathbb{C}$ be a circle, and let T be a bilinear transform. Then $T((C)) = (C')$ is also a circle.*

The T bilinear transform maps every z_1 and z_2 two inverse points with respect to the (C) circle, into the points $z_1' = Tz_1$ and $z_2' = Tz_2$ that are inverse points with respect to the (C') circle.

Proof. Let (C') be an arbitrary circle that contains the points z_1 and z_2. From Theorem 2.7.3, this arbitrary circle is orthogonal to (C). Since T is a conformal mapping, it maps orthogonal circles onto orthogonal circles.

Let (C_1) and (C_2) be two arbitrary circles that contain the points z_1 and z_2. Then the images $(C_j') = T((C_j))$, $j = 1, 2$ are distinct circles that are orthogonal to (C'), and their intersections are $z_j' = Tz_j$, $j = 1, 2$. But $(C_j') \perp (C')$, $j = 1, 2$, and using again Theorem 2.7.3 it follows that z_1' and z_2' are inverse points with respect to the circle (C'). □

Theorem 2.7.5. *The functions of the group of the bilinear transforms that maps the disc $U(0; r)$ onto itself, has the form*

$$Tz = r^2 e^{i\theta} \frac{z - a}{\overline{a}z - r^2}, \qquad \theta \in \mathbb{R},\ |a| < r.$$

Proof. Let $a \in U(0; r)$ be such a point, so that $Ta = 0$, and let z_1 the inverse of a with respect to the circle $(C) = \{z \in \mathbb{C} : |z| = r\}$. Then

$$|a||z_1| = r^2 \quad \text{and} \quad \arg a = \arg z_1 \Rightarrow z_1 = \frac{r^2}{|a|} e^{i \arg z_1} = \frac{r^2}{|a|} e^{i \arg a} = \frac{r^2}{|a|} \frac{a}{|a|} = \frac{r^2}{\overline{a}}.$$

Since a and $\frac{r^2}{\overline{a}}$ are inverse points, then $Ta = 0$ and $T(\frac{r^2}{\overline{a}})$ are also inverse points with respect to the (C) circle (T maps the circle (C) onto itself). But 0 and $T(\frac{r^2}{\overline{a}})$ could be inverse points if and only if $T(\frac{r^2}{\overline{a}}) = \infty$.

Suppose that $Tz = \frac{a_1 z + b_1}{c_1 z + d_1}$. From $aa_1 + b_1 = 0$ it follows that $a = -\frac{b_1}{a_1}$ and $c_1 \frac{r^2}{\overline{a}} + d_1 = 0$. Hence,

$$Tz = \frac{a_1}{c_1} \frac{z + \frac{b_1}{a_1}}{z + \frac{d_1}{c_1}} = \frac{a_1}{c_1} \frac{z - a}{z - \frac{r^2}{\overline{a}}} = k \frac{z - a}{\overline{a}z - r^2}, \quad \text{where } k = \frac{a_1 \overline{a}}{c_1}.$$

If $|z| = r$, then $|Tz| = r$. We have

$$\begin{aligned} z = r \Rightarrow r^2 = Tr \cdot \overline{Tr} = k\overline{k} \frac{r - a}{r\overline{a} - r^2} \frac{r - \overline{a}}{ra - r^2} = k\overline{k} \frac{r^2 - r(a + \overline{a}) + a\overline{a}}{r^4 - r^3(a + \overline{a}) + r^2 a\overline{a}} = |k|^2 \frac{1}{r^2} \\ \Rightarrow |k|^2 = r^4 \Rightarrow k = r^2 e^{i\theta}, \quad \theta \in \mathbb{R}. \end{aligned}$$

Hence, $Tz = r^2 e^{i\theta} \frac{z - a}{\overline{a}z - r^2}$. □

Remarks 2.7.2.

1. If we multiply the fraction by $-1 = e^{i\pi}$, and we replace $\theta + \pi$ by θ', then

$$Tz = r^2 e^{i\theta'} \frac{z-a}{r^2 - \overline{a}z}, \quad \theta' \in \mathbb{R}, \ |a| < r.$$

2. A special case is that when $r = 1$. Those bilinear transforms that maps the $\mathrm{U} = U(0;1)$ **unit disc** onto itself has the general form

$$Tz = e^{i\theta} \frac{z-a}{1 - \overline{a}z}, \quad \theta \in \mathbb{R}, \ |a| < 1.$$

Example 2.7.1. Determine those bilinear transforms that map the upper half-plane onto the unit disc $\mathrm{U} = U(0;1)$.

Solution. Suppose that the bilinear transform T maps the upper half-plane onto the unit disc U.

Let $a \in \{z \in \mathbb{C} : \operatorname{Im} z > 0\}$ that point, such that $Ta = 0 \in \mathrm{U}$. The function T maps the line $\operatorname{Im} z = 0$ onto the circle $|z| = 1$. The points a and $\overline{a}$ are inverse points with respect to the "circle" $\operatorname{Im} z = 0$, since every arbitrary circle that contain these points are orthogonal to the line $\operatorname{Im} z = 0$. Hence, Ta and $T\overline{a}$ are inverse with respect to the circle $|z| = 1$, and since $Ta = 0$ it is necessary that $T\overline{a} = \infty$. Hence,

$$Tz = k\frac{z-a}{z-\overline{a}}, \quad \text{where } \operatorname{Im} a > 0.$$

But $0 \in \{z \in \mathbb{C} : \operatorname{Im} z = 0\} \Rightarrow T0$ belongs to the boundary of U, hence

$$|T0| = \left| k\frac{0-a}{0-\overline{a}} \right| = |k| = 1,$$

so it follows that $k = e^{i\theta}$, $\theta \in \mathbb{R}$.

We obtained in such a way that the required bilinear transforms have the general form:

$$Tz = e^{i\theta}\frac{z-a}{z-\overline{a}}, \quad \theta \in \mathbb{R}, \ \operatorname{Im} a > 0. \tag{2.21}$$

□

2.8 Multivalued functions

2.8.1 The logarithmic function

The real-valued logarithmic function is the inverse of the exponential function. There exists this inverse because the real-valued exponential function is a bijection between $\mathbb{R}$ and $\mathbb{R}_+ \setminus \{0\}$.

We saw previously that the complex-valued exponential function e^z is a periodical function, with the period equal to $2\pi i$. Thus, the complex-valued exponential function could be injective at most in an open set included into a strip, parallel with the Ox axis, with the width equal to 2π.

For any arbitrary $z \in \mathbb{C}^*$, let us consider the equation $e^w = z$, with the unknown w. Letting $w = u + iv$, then

$$z = e^{u+iv} = e^u e^{iv} = re^{i\theta}, \quad \text{where } z = re^{i\theta}.$$

Hence $e^u = r$, i. e.,

$$u = \ln r = \ln |z| \quad \text{and} \quad v = \theta + 2k\pi,\ k \in \mathbb{Z}.$$

We conclude that any complex number of the form $w_k = \ln r + i(\theta + 2k\pi)$, $k \in \mathbb{Z}$, satisfy the given equation. We will denote by Log z, the set of all the solutions of the equation $e^w = z$, i. e.,

$$\operatorname{Log} z = \{w_k : k \in \mathbb{Z}\} = \{\ln r + i(\theta + 2k\pi) : k \in \mathbb{Z}\},$$

or

$$\operatorname{Log} z = \ln |z| + i \operatorname{Arg} z.$$

Let

$$D = \mathbb{C} \setminus (\{z \in \mathbb{C} : \operatorname{Re} z < 0, \operatorname{Im} z = 0\} \cup \{0\}) = \{z \in \mathbb{C} : |\arg z| < \pi\},$$

where $k \in \mathbb{Z}$ is a given integer. Then we denote

$$\log_k z := \ln |z| + i(\arg z + 2k\pi), \quad |\arg z| < \pi,$$

and the function $\log_k : D \to \mathbb{C}$ is called **a branch of the multivalued** Log **function.**

The $\log_0$ function is called **the main branch of the multivalued** Log **function,** and we will denote it by log, i. e., $\log z := \log_0 z$; we have $\log 1 = 0$ (Figure 2.4).

If we let $z = re^{i\theta}$, $\theta \in (-\pi, \pi)$, then $\log z = \ln r + i\theta$. Denoting $\log z = u(z) + iv(z)$, where $z = x+iy$, then $u(x,y) = \ln \sqrt{x^2 + y^2}$ and $v(x,y) = \arctan \frac{y}{x} + k\pi$, $k \in \mathbb{Z}$. Since $u, v \in C^1(D)$ and these functions satisfy the Cauchy–Riemann conditions on D, it follows that the function log is differentiable on D and

$$(\log z)' = \frac{1}{z}.$$

Thus, computing the derivative of the $e^{\log z} = z$ identity, we deduce that

$$(e^{\log z})' = e^{\log z}(\log z)' = 1 \Rightarrow (\log z)' = \frac{1}{e^{\log z}} = \frac{1}{z}.$$

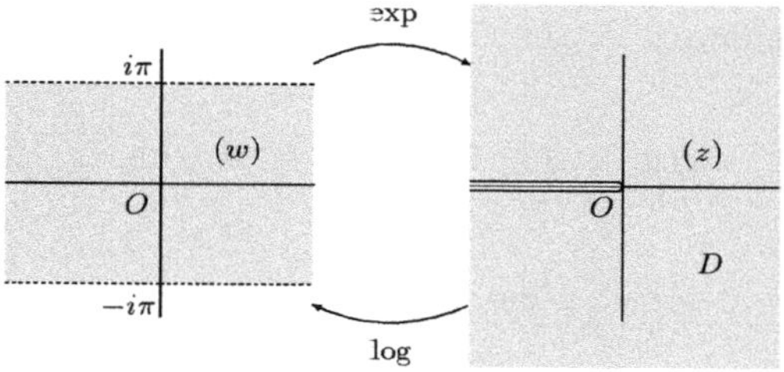

Figure 2.4: The log function.

The next differential equation system (Cauchy-type problem)

$$\begin{cases} f'(z) = \frac{1}{z} \\ f(1) = 0, \quad z \in D \end{cases}$$

has the solution $f = \log$, while the generalized differential equation system

$$\begin{cases} f'(z) = \frac{1}{z} \\ f(1) = 2k\pi i, \quad z \in D,\ k \in \mathbb{Z} \end{cases}$$

has the solution $f = \log_k$.

Since

$$\log_k : D \to \mathbb{C},$$
$$\log_k z = \ln r + i(\theta + 2k\pi), \quad \text{where } z = re^{i\theta},\ \theta \in (-\pi, \pi),$$

it follows that

$$\log_k = \log + 2k\pi i.$$

2.8.2 Inverse trigonometric functions

Let $z \in \mathbb{C}$ be a fixed number, and let us consider the next equation, with the unknown w:

1. $\cos w = z$, where $\cos w = \frac{e^{iw}+e^{-iw}}{2}$.

Hence, from $\frac{e^{iw}+e^{-iw}}{2} = z$ we get $e^{2iw} - 2ze^{iw} + 1 = 0$, i. e., $e^{iw} = z \pm \sqrt{z^2-1}$. It follows that

$$iw = \operatorname{Log}\left(z \pm \sqrt{z^2-1}\right)$$

and

$$\operatorname{Arccos} z := w = -i\operatorname{Log}\left(z \pm \sqrt{z^2-1}\right).$$

Similarly, we will consider the equation:
2. $\sin w = z$, where $\sin w = \frac{e^{iw}-e^{-iw}}{2i}$.
Hence, from $\frac{e^{iw}-e^{-iw}}{2i} = z$, we get $e^{2iw} - 2ize^{iw} - 1 = 0$, i. e., $e^{iw} = iz \pm \sqrt{1-z^2}$. It follows that

$$iw = \operatorname{Log}\,(iz \pm \sqrt{1-z^2})$$

and

$$\operatorname{Arcsin} z := w = -i\operatorname{Log}\,(iz \pm \sqrt{1-z^2}).$$

Like in the previous two cases, let us consider the equation:
3. $\tan w = z$, where $\tan w = \frac{\sin w}{\cos w} = \frac{e^{iw}-e^{-iw}}{i(e^{iw}+e^{-iw})}$.
Hence, from $\frac{e^{iw}-e^{-iw}}{e^{iw}+e^{-iw}} = iz$ we get $e^{2iw} - 1 = iz(e^{2iw}+1) = 0 \Leftrightarrow (1-iz)e^{2iw} = 1+iz$, i. e., $e^{2iw} = \frac{1+iz}{1-iz}$. It follows that

$$2iw = \operatorname{Log}\frac{1+iz}{1-iz}$$

and

$$\operatorname{Arctan} z := w = -\frac{i}{2}\operatorname{Log}\frac{1+iz}{1-iz}.$$

2.8.3 The power function

Letting $\alpha \in \mathbb{C}$, then we will define the function z^α by

$$z^\alpha = e^{\alpha \operatorname{Log} z}.$$

It follows that the function

$$F : \mathbb{C}^* \to \mathcal{P}(\mathbb{C}), \quad F(z) = z^\alpha$$

is a multivalued function.

Considering the main branch log of the multivalued function Log , let us define the function f by

$$f(z) = e^{\alpha \log z}, \quad f(1) = 1.$$

This function is called **the main branch of the multivalued function z^α**, and moreover,

$$f'(z) = \frac{\alpha}{z} e^{\alpha \log z} = \frac{\alpha z^\alpha}{z} = \alpha z^{\alpha-1}.$$

For the special case $\alpha = n \in \mathbb{N}$, we deduce that

$$z^n = e^{n\operatorname{Log} z} = e^{n(\ln r + i(\theta + 2k\pi))} = e^{n\ln r}e^{ni\theta}e^{2kn\pi i} = r^n e^{ni\theta} = (re^{i\theta})^n = z^n.$$

Thus, the new definition of the power functions reduces for the special case $\alpha = n \in \mathbb{N}$ to the well-known definition of the power functions with natural exponents.

Similarly, for $\alpha = \frac{1}{n}$, $n \in \mathbb{N}$, we have

$$\begin{aligned} z^{\frac{1}{n}} &= e^{\frac{1}{n}\operatorname{Log} z} = e^{\frac{1}{n}(\ln r + i(\theta + 2k\pi))} = e^{\frac{1}{n}\ln r}e^{i\frac{\theta+2k\pi}{n}} = \sqrt[n]{r}e^{i\frac{\theta+2k\pi}{n}} \\ &= \sqrt[n]{r}\left(\cos\frac{\theta + 2k\pi}{n} + i\sin\frac{\theta + 2k\pi}{n}\right), \quad k \in \{0, 1, \ldots, n-1\}, \end{aligned}$$

and we obtain the well-known definitions of the nth order root functions.

2.9 Exercises

2.9.1 Real variable complex functions

Exercise 2.9.1. Represent in the complex plane the graphics of the following functions:

1. $f : \mathbb{R} \to \mathbb{C}, f(t) = -3 + i + (1 + i)t;$
2. $f : [1, 3] \to \mathbb{C}, f(t) = 2 - 3i + (1 - 2i)t;$
3. $f : \mathbb{R} \to \mathbb{C}, f(t) = \frac{16}{3-i+(1+2i)t};$
4. $f : \mathbb{R} \to \mathbb{C}, f(t) = \frac{-12+2i-(1-3i)t}{1+3i-(2-i)t}.$

2.9.2 The derivative of a complex function

Exercise 2.9.2. Prove that the function $f : \mathbb{C} \to \mathbb{C}, f(z) = \overline{z}$, is not differentiable in any point of $\mathbb{C}$.

Exercise 2.9.3. Let us consider the function $f : \mathbb{C} \to \mathbb{C}$, $f(z) = z^2 + z\overline{z} - \overline{z}^2 + 2z - \overline{z}$. Determine those points from $\mathbb{C}$ in which f is differentiable, and compute its derivative f' in these points.

Exercise 2.9.4. Determine those points of $\mathbb{C}$ in which the function $f(z) = z\operatorname{Re} z$ is differentiable, and compute its derivative f' in these points.

Exercise 2.9.5. Determine those points of $\mathbb{C}$ in which the function $f(z) = 2\overline{z} + z^2$ is differentiable, and compute its derivative f' in these points.

Exercise 2.9.6. Let us consider the function defined by $f(z) = \ln\sqrt{x^2 + y^2} + i\arctan\frac{y}{x}$, where $z = x + iy$. Prove that $f \in H(\mathbb{C}^*)$, and compute its derivative f'.

Exercise 2.9.7. Prove that the function defined by $f(z) = (e^x - e^{-x})\cos y + i(e^x + e^{-x})\sin y$, where $z = x + iy$, is an entire function, and compute its derivative f'.

Exercise 2.9.8. Determine the values of $a, b, c, d \in \mathbb{R}$, such that the next functions will be differentiable in $\mathbb{C}$:
1. $f : \mathbb{C} \to \mathbb{C}, f(z) = x + ay + i(bx + cy)$;
2. $f : \mathbb{C} \to \mathbb{C}, f(z) = x^2 + axy + by^2 + i(cx^2 + dxy + y^2)$;
3. $f : \mathbb{C} \to \mathbb{C}, f(z) = \cos x(\cosh y + a\sinh y) + i\sin x(\cosh y + b\sinh y)$, where $z = x + iy$.

Exercise 2.9.9. Write the Cauchy–Riemann equation system, and compute the derivative f', in the next three cases:
1. If $z = re^{i\theta}$ and $f(z) = u(z) + iv(z)$, where the variables are r and θ;
2. If $f(z) = R(z)e^{i\Phi(z)}$, $z = x + iy$, where the variables are x and y;
3. If $z = re^{i\theta}$ and $f(z) = R(z)e^{i\Phi(z)}$, where the variables are r and θ,

supposing that the function f is holomorphic, and the functions R and Φ are real differentiable functions.

Exercise 2.9.10. Prove that the function $f(z) = z^n$ is an entire function $\forall n \in \mathbb{N}$, and calculate its derivative.

Exercise 2.9.11. Let $f : D \to \mathbb{C}$ a differentiable non constant function on the domain $D \subset \mathbb{C}$. Prove that the image $f(D)$ cannot be a segment, i. e., there does not exist points $a, b \in \mathbb{C}$, such that

$$f(D) = \{z \in \mathbb{C} : z = (1 - t)a + tb, t \in [0, 1]\}.$$

Exercise 2.9.12. Let G be an open set, and let the function $f : G \to \mathbb{C}, f = u + iv$. Prove that:
1. If for all $z \in G$ there exists $\lim_{\zeta\to z} \operatorname{Re} \frac{f(\zeta)-f(z)}{\zeta-z}$, then there exist the partial derivatives $\frac{\partial u}{\partial x}, \frac{\partial v}{\partial y}$ in G, and $\frac{\partial u(z)}{\partial x} = \frac{\partial v(z)}{\partial y}$, $\forall z \in G$;
2. If for all $z \in G$ there exists $\lim_{\zeta\to z} \operatorname{Im} \frac{f(\zeta)-f(z)}{\zeta-z}$, then there exists the partial derivatives $\frac{\partial u}{\partial y}, \frac{\partial v}{\partial x}$ in G, and $\frac{\partial u(z)}{\partial y} = -\frac{\partial v(z)}{\partial x}$, $\forall z \in G$.

Exercise 2.9.13. Let us define the function e^z as follows:

$$e^z = \lim_{n\to+\infty}\left(1 + \frac{z}{n}\right)^n, \quad \text{where } z = x + iy.$$

Prove that

$$e^z = e^{x+iy} = e^x(\cos y + i\sin y), \quad x, y \in \mathbb{R}.$$

2.9.3 Entire functions

Exercise 2.9.14. Solve the following equations:

1.	$\sin z - \cos z = i$;	2.	$\cosh z - i \sinh z = 1$;	3.	$\sin z = \frac{4i}{3}$;
4.	$\cos z = \frac{3+i}{4}$;	5.	$\sinh z = \frac{i}{2}$;	6.	$\tan z = \frac{5i}{3}$;
7.	$e^{i3z} = -1$;	8.	$e^{\frac{1}{z^2}} = 1$;	9.	$\tanh z = 2$.

Exercise 2.9.15. Find the algebraic form of the complex numbers:

1.	e^i;	2.	$\sinh 2i$;	3.	$\cosh(2+3i)$;	4.	$\cos(1-i)$;
5.	$\tan(1-2i)$;	6.	$\log(-2i)$;	7.	$\log(-3+4i)$;	8.	$\log \frac{1-i}{\sqrt{3}+i}$.

Exercise 2.9.16. Determine the values of the next powers:

1. i^{1-i}; 2. $(1+i\sqrt{3})^i$; 3. 1^{-i}; 4. $e^{\sqrt{i}}$.

Exercise 2.9.17. Prove the following identities:

1. $\sin z = -i \sinh(iz)$, $\sinh(iz) = i \sin z$;
2. $\cosh z = \cos(iz)$, $\cosh(iz) = \cos z$;
3. $\sinh(z_1 \pm z_2) = \sinh z_1 \cosh z_2 \pm \sinh z_2 \cosh z_1$;
4. $\cos(z_1 \pm z_2) = \cos z_1 \cos z_2 \mp \sin z_1 \sin z_2$;
5. $\sin 2z = 2 \sin z \cos z$;
6. $\tan 2z = \frac{2 \tan z}{1-\tan^2 z}$;
7. $\sinh^2 z = \frac{\cosh 2z - 1}{2}$, $\cosh^2 z = \frac{\cosh 2z + 1}{2}$;
8. $|\sin z| = \sqrt{\sin^2 x + \sinh^2 y} = \sqrt{\frac{1}{2}(\cosh 2y - \cos 2x)}$, where $z = x + iy$;
9. $|\cos z| = \sqrt{\cos^2 x + \sinh^2 y} = \sqrt{\frac{1}{2}(\cosh 2y + \cos 2x)}$, where $z = x + iy$;
10. $\tan z = -i \tanh iz$, $\tan iz = i \tanh z$;
11. $\cosh 2z = \cosh^2 z + \sinh^2 z = \frac{1+\tanh^2 z}{1-\tanh^2 z}$.

2.9.4 Bilinear transforms

Exercise 2.9.18. Determine the bilinear transform that maps the points $z_1 = 1 + i$, $z_2 = \infty$ and $z_3 = 1$ into the points $w_1 = i$, $w_2 = -1$ and $w_3 = \infty$, respectively. Then prove that this bilinear transform conformally maps the disc $\{z = x + iy \in \mathbb{C} : (x-1)^2 + (y-1)^2 < 1\}$ onto the half-plane $\{w = u + iv \in \mathbb{C} : u + v > 0\}$.

Exercise 2.9.19. Prove that the function $w = \frac{z-1}{z+1}$ conformally maps the half-plane $\{z \in \mathbb{C} : \operatorname{Re} z > 0\}$ onto the disc $\{w \in \mathbb{C} : |w| < 1\}$.

Exercise 2.9.20. Determine the bilinear transform $w = f(z)$ that conformally maps the half-plane $\{z \in \mathbb{C} : \operatorname{Re} z < 1\}$ onto the disc $\{w \in \mathbb{C} : |w| < R\}$, such that $f(-1) = 0$ and $f'(-1) = 1$. Find also the value of the radius R.

Exercise 2.9.21. Suppose that $\beta < 1$, and let $f(z) = \frac{1+(2\beta-1)z}{1+z}$. Prove that the function f conformally maps the unit disc $\{z \in \mathbb{C} : |z| < 1\}$ onto the half-plane $\{w \in \mathbb{C} : \operatorname{Re} w > \beta\}$.

Exercise 2.9.22. Prove that the function $f(z) = z + az^2$, $a \in \mathbb{C}^*$, is injective on the disc $\{z \in \mathbb{C} : |z| < \frac{1}{2|a|}\}$.

Exercise 2.9.23. Let us define the function $w(z) = z^2 + 5z + 6$.

1. Determine the image of the domain $\{z \in \mathbb{C} : 0 < \operatorname{Re} z < 1, \operatorname{Im} z > 0\}$ by the w function;
2. Prove that the restriction of the function w to $\{z \in \mathbb{C} : \operatorname{Re} z > -\frac{5}{2}\}$ has an inverse, and determine this inverse;
3. Prove that the restriction of the function w to $\{z \in \mathbb{C} : \operatorname{Re} z < -\frac{5}{2}\}$ has an inverse, and determine this inverse.

Exercise 2.9.24. Determine a function $w = f(z)$ that conformally maps the strip

$$\{z \in \mathbb{C} : 0 < \operatorname{Re} z < a\}, \quad a > 0,$$

onto the upper half-plane $\{w \in \mathbb{C} : \operatorname{Im} w > 0\}$.

3 The complex integration

3.1 The homotopic theory of the paths

The homotopic theory of the paths deals with the continuous deformations of the path, and it allows to define and to study the properties of the complex integral.

Definition 3.1.1.
1. Let G be an open set of the complex plane $\mathbb{C}$. The continuous function $\gamma : [0,1] \to G$ is called **a path in** G, where $[0,1]$ is the close real interval between 0 and 1, where the topology of $\mathbb{R}$ is given by the euclidian topology.
2. The point $z_1 = \gamma(0)$ is called the **starting point of the path** γ, while the point $z_2 = \gamma(1)$ is called the **end point of the path** γ.
3. We denote by $\mathcal{D}_G(z_1, z_2)$ the set of all the paths in G with the starting point z_1 and the end point z_2.
4. Whenever $z_1 = \gamma(0) = \gamma(1) = z_2$, we say that the path γ is a **closed path.**
5. The set of all the path in G with the same starting and end point z_0, is called **the set of all the closed paths in** G **starting from** z_0, and it is denoted by $\mathcal{D}_G(z_0)$.
6. The set $\{\gamma\} = \gamma([0,1])$ is called **the image of the path** γ.

Definition 3.1.2. Let $\gamma_1, \gamma_2 \in \mathcal{D}_G(z_1, z_2)$. We say that the path γ_1 **is homotopic with** γ_2 **in** G, if there exists a continuous function $\varphi : S \times T \to G$, where $S = T = [0,1]$, that satisfies the next conditions:

$$\varphi(0,t) = \gamma_1(t), \quad \varphi(1,t) = \gamma_2(t), \quad \forall t \in [0,1]$$

and

$$\varphi(s,0) = z_1, \quad \varphi(s,1) = z_2, \quad \forall s \in [0,1].$$

The function φ is called **the homotopy between the paths** γ_1 **and** γ_2 **(in the set** G**)**, or **the function that establishes the homotopy between the paths** γ_1 **and** γ_2 **(in the set** G**)**.

Hence, for all $s \in S$ we have $\varphi(s,\cdot) \in \mathcal{D}_G(z_1, z_2)$, which geometrically means that the function φ, with the variable s, is a continuous deformation of the path γ_1 to the path γ_2, whenever the argument s runs on the whole interval $[0,1]$.

Notation. We denote by a $\gamma_1 \underset{G}{\sim} \gamma_2$ the fact that the path γ_1 is homotopic with the path γ_2 in the set G.

Theorem 3.1.1. *The relation "$\underset{G}{\sim}$"*[1] *is an equivalence relation in the set of the paths* $\mathcal{D}_G(z_1, z_2)$.

---

1 A relation $\underset{G}{\sim}$ is said to be an *equivalence* relation $\Leftrightarrow$ it is reflexive, transitive and symmetric.

https://doi.org/10.1515/9783110657869-003

Proof. If γ is an arbitrary path in G, then $\gamma \underset{G}{\sim} \gamma$, because the function $\varphi(s,t) = \gamma(t)$, $\varphi : S \times T \to G$ is continuous and satisfies the conditions of the homotopy. Thus, the relation "$\underset{G}{\sim}$" is reflexive, i. e., the path γ is homotopic with itself.

Supposing that $\gamma_1 \underset{G}{\sim} \gamma_2$, we will prove that $\gamma_2 \underset{G}{\sim} \gamma_1$. Let $\varphi : S \times T \to G$ be the homotopy between γ_1 and γ_2, i. e.,

$$\varphi(0,t) = \gamma_1(t), \quad \varphi(1,t) = \gamma_2(t), \quad \forall t \in [0,1]$$

and

$$\varphi(s,0) = z_1, \quad \varphi(s,1) = z_2, \quad \forall s \in [0,1].$$

If we let $\varphi_1(s,t) = \varphi(1-s,t)$, then $\varphi_1 : S \times T \to G$ is continuous, and

$$\varphi_1(0,t) = \varphi(1,t) = \gamma_2(t), \quad \varphi_1(1,t) = \varphi(0,t) = \gamma_1(t), \quad \forall t \in [0,1]$$

and

$$\varphi_1(s,0) = \varphi(1-s,0) = z_1, \quad \varphi_1(s,1) = \varphi(1-s,1) = z_2, \quad \forall s \in [0,1].$$

Hence $\gamma_2 \underset{G}{\sim} \gamma_1$, i. e., the relation "$\underset{G}{\sim}$" is a symmetric relation.

Suppose that $\gamma_1 \underset{G}{\sim} \gamma_2$ and $\gamma_2 \underset{G}{\sim} \gamma_3$. Then there exist the continuous functions $\varphi_1, \varphi_2 : S \times T \to G$, such that

$$\varphi_1(0,t) = \gamma_1(t), \quad \varphi_1(1,t) = \gamma_2(t), \quad \forall t \in [0,1]$$
$$\varphi_1(s,0) = z_1, \quad \varphi_1(s,1) = z_2, \quad \forall s \in [0,1]$$

and

$$\varphi_2(0,t) = \gamma_2(t), \quad \varphi_2(1,t) = \gamma_3(t), \quad \forall t \in [0,1]$$
$$\varphi_2(s,0) = z_1, \quad \varphi_2(s,1) = z_2, \quad \forall s \in [0,1].$$

Let us define the function $\varphi : S \times T \to G$ by

$$\varphi(s,t) = \begin{cases} \varphi_1(2s,t), & s \in [0,\frac{1}{2}] \\ \varphi_2(2s-1,t), & s \in [\frac{1}{2},1]. \end{cases}$$

The function φ is well-defined, and since $\varphi_1(1,t) = \gamma_2(t) = \varphi_2(t)$, $\forall t \in [0,1]$, the function φ is continuous. From the facts,

$$\varphi(0,t) = \gamma_1(t), \quad \varphi(1,t) = \gamma_3(t), \quad \forall t \in [0,1]$$

and

$$\varphi(s,0) = z_1, \quad \varphi(s,1) = z_2, \quad \forall s \in [0,1],$$

it follows that φ is the function that establishes the homotopy $\gamma_1 \underset{G}{\sim} \gamma_3$, thus the relation "$\underset{G}{\sim}$" is a transitive relation. □

Definition 3.1.3.
1. Let $\gamma_1 \in \mathcal{D}_G(z_1, z_2)$, $\gamma_2 \in \mathcal{D}_G(z_2, z_3)$. We denote by $\gamma_1 \cup \gamma_2$ the path defined as follows:

$$(\gamma_1 \cup \gamma_2)(t) = \begin{cases} \gamma_1(2t), & \text{if } t \in [0, \frac{1}{2}] \\ \gamma_2(2t-1), & \text{if } t \in [\frac{1}{2}, 1]. \end{cases}$$

   Then $\gamma_1 \cup \gamma_2 \in \mathcal{D}_G(z_1, z_3)$, and the path $\gamma_1 \cup \gamma_2$ is called **the union of the paths** γ_1 **and** γ_2.
2. We may similarly define the union of many paths. Thus, if $\gamma_1 \in \mathcal{D}_G(z_1, z_2)$, $\gamma_2 \in \mathcal{D}_G(z_2, z_3), \dots, \gamma_n \in \mathcal{D}_G(z_n, z_{n+1})$, then the union of these paths is denoted by $\gamma_1 \cup \gamma_2 \cup \dots \cup \gamma_n$, and is given by

$$\gamma(t) = (\gamma_1 \cup \gamma_2 \cup \dots \cup \gamma_n)(t) = \gamma_{k+1}(nt - k), \quad t \in \left[\frac{k}{n}, \frac{k+1}{n}\right],$$
$$k \in \{0, 1, \dots, n-1\}.$$

Definition 3.1.4. Let γ be a path, and let be $\Delta = (t_0, t_1, \dots, t_n)$ be a division of the interval $[0, 1]$, i. e., $0 = t_0 < t_1 < \dots < t_n = 1$. The path system $(\gamma_1, \gamma_2, \dots, \gamma_n)$ is called **the decomposition of the path** γ **by the division** Δ, if

$$\gamma_k = \gamma \circ h_k,$$

where h_k are the affine functions $h_k : [0, 1] \to [t_{k-1}, t_k]$, defined by $h_k(t) = t_{k-1} + t(t_k - t_{k-1})$, $k \in \{1, \dots, n\}$.

We may see easily that $\{\gamma_k\} = (\gamma \circ h_k)([0, 1]) = \gamma([t_{k-1}, t_k])$.

Theorem 3.1.2. *If γ is a path in G, and $f : [0, 1] \to [0, 1]$ is continuous and surjective function, with $f(0) = 0$, $f(1) = 1$, then the path $\gamma_1 = \gamma \circ f$ is homotopic with the path γ.*

Proof. Letting $\varphi : S \times T \to G$, where $\varphi(s, t) = \gamma(t + s(f(t) - t))$, then φ is a continuous function and

$$\varphi(0, t) = \gamma(t), \quad \varphi(1, t) = (\gamma \circ f)(t), \quad \forall t \in [0, 1],$$
$$\varphi(s, 0) = \gamma(0) = z_1, \quad \varphi(s, 1) = \gamma(1) = z_2, \quad \forall s \in [0, 1],$$

which prove the conclusion.

It is obvious that $\{\gamma\} = \{\gamma \circ f\}$, that is the images of the both of these paths coincide. □

Theorem 3.1.3. *If $(\gamma_1, \gamma_2, \dots, \gamma_n)$ is a decomposition of the path γ in G, then the path γ is homotopic with the path $\gamma_1 \cup \gamma_2 \cup \dots \gamma_n$ in G.*

Proof. Let $(\gamma_1, \gamma_2, \dots, \gamma_n)$ the decomposition of the path γ by the division $\Delta = (t_0, t_1, \dots, t_n)$, i. e., $0 = t_0 < t_1 < \dots < t_n = 1$ and $h_k(t) = t_{k-1} + t(t_k - t_{k-1})$, $t \in [0, 1]$.

Then $\gamma_1 \cup \gamma_2 \cup \cdots \cup \gamma_n = \gamma \circ (h_1 \cup h_2 \cup \cdots \cup h_n)$. Denote by h the function $h : [0,1] \to [0,1]$, with $h = h_1 \cup h_2 \cup \cdots \cup h_n$. Thus, the function h is continuous with $h(0) = h_1(0) = 0$, $h(1) = h_n(1) = 1$, and according to Theorem 3.1.2 we conclude $\gamma \underset{G}{\sim} \gamma \circ h = \gamma_1 \cup \gamma_2 \cup \ldots \gamma_n$. □

Theorem 3.1.4. *Let $\gamma_1 \in \mathcal{D}_G(z_1, z_2)$, $\gamma_2 \in \mathcal{D}_G(z_2, z_3)$, and $\gamma_3 \in \mathcal{D}_G(z_3, z_4)$. Then*

$$(\gamma_1 \cup \gamma_2) \cup \gamma_3 \underset{G}{\sim} \gamma_1 \cup (\gamma_2 \cup \gamma_3).$$

Proof. Define the function $h : [0,1] \to [0,1]$ by

$$h(t) = \begin{cases} 2t, & t \in [0, \frac{1}{4}], \\ t + \frac{1}{4}, & t \in [\frac{1}{4}, \frac{1}{2}], \\ \frac{1+t}{2}, & t \in [\frac{1}{2}, 1]. \end{cases}$$

Then $(\gamma_1 \cup \gamma_2) \cup \gamma_3 = [\gamma_1 \cup (\gamma_2 \cup \gamma_3)] \circ h$, and from Theorem 3.1.2 we deduce that

$$(\gamma_1 \cup \gamma_2) \cup \gamma_3 = [\gamma_1 \cup (\gamma_2 \cup \gamma_3)] \circ h \underset{G}{\sim} \gamma_1 \cup (\gamma_2 \cup \gamma_3).$$ □

Definition 3.1.5. If $\gamma \in \mathcal{D}_G(z_1, z_2)$, then the path $\gamma^- \in \mathcal{D}_G(z_2, z_1)$ defined by $\gamma^-(t) = \gamma(1-t)$, $t \in [0,1]$, is called **the inverse of the path** γ.

Theorem 3.1.5. *If $\gamma_1, \gamma_2 \in \mathcal{D}_G(z_1, z_2)$ and $\gamma_1 \underset{G}{\sim} \gamma_2$, then $\gamma_1^- \underset{G}{\sim} \gamma_2^-$.*

Proof. If $\varphi : S \times T \to G$ is the function that establishes the homotopy $\gamma_1 \underset{G}{\sim} \gamma_2$, then the function $\varphi^-(s,t) = \varphi(s, 1-t)$ will be the homotopy between γ_1^- and γ_2^-. □

Theorem 3.1.6. *Let $\gamma_1, \gamma_3 \in \mathcal{D}_G(z_1, z_2)$ and $\gamma_2, \gamma_4 \in \mathcal{D}_G(z_2, z_3)$. If $\gamma_1 \underset{G}{\sim} \gamma_3$ and $\gamma_2 \underset{G}{\sim} \gamma_4$, then $\gamma_1 \cup \gamma_2 \underset{G}{\sim} \gamma_3 \cup \gamma_4$.*

Proof. Let φ_1 and φ_2 the functions that establish the homotopies $\gamma_1 \underset{G}{\sim} \gamma_3$, and $\gamma_2 \underset{G}{\sim} \gamma_4$ respectively. Then the function

$$\varphi(s,t) = \begin{cases} \varphi_1(s, 2t), & \text{if } t \in [0, \frac{1}{2}], \\ \varphi_2(s, 2t-1), & \text{if } t \in [\frac{1}{2}, 1] \end{cases}$$

is the homotopy between $\gamma_1 \cup \gamma_2$ and $\gamma_3 \cup \gamma_4$. □

Definition 3.1.6. If $z \in G$, then we will denote by e_z **the constant path** defined by $e_z(t) = z$, $\forall t \in [0,1]$. We have that $e_z \in \mathcal{D}_G(z)$.

The path $\gamma \in \mathcal{D}_G(z)$ is said to be **homotopic to a point in** G, if $\gamma \underset{G}{\sim} e_z$.

Theorem 3.1.7. *For all $\gamma \in \mathcal{D}_G(z_1, z_2)$, we have $e_{z_1} \cup \gamma \underset{G}{\sim} \gamma$ and $\gamma \underset{G}{\sim} \gamma \cup e_{z_2}$.*

Proof. Let us define the function $f : [0,1] \to [0,1]$ by

$$f(t) = \begin{cases} 0, & \text{if } t \in [0, \frac{1}{2}], \\ 2t-1, & \text{if } t \in [\frac{1}{2}, 1]. \end{cases}$$

Thus, $e_{z_1} \cup \gamma = \gamma \circ f$, and from Theorem 3.1.2 we have $\gamma \circ f \underset{G}{\sim} \gamma$, hence $e_{z_1} \cup \gamma \underset{G}{\sim} \gamma$. The second conclusion may be proved in a similar way. □

Theorem 3.1.8. *Let $\gamma \in \mathcal{D}_G(z_1, z_2)$. Then $\gamma \cup \gamma^- \underset{G}{\sim} e_{z_1}$, i. e., $\gamma \cup \gamma^-$ is homotopic to a point in G.*

Proof. Let

$$\varphi(s,t) = \begin{cases} \gamma(2st), & \text{if } t \in [0, \frac{1}{2}], \\ \gamma(2s(1-t)), & \text{if } t \in [\frac{1}{2}, 1]. \end{cases}$$

Thus, the function φ establishes the homotopy $e_{z_1} \underset{G}{\sim} \gamma \cup \gamma^-$. □

Theorem 3.1.9. *Let $\gamma_1, \gamma_2 \in \mathcal{D}_G(z_1, z_2)$ and $\gamma_1 \underset{G}{\sim} \gamma_2$. Then $\gamma_1 \cup \gamma_2^- \underset{G}{\sim} e_{z_1}$, that is $\gamma_1 \cup \gamma_2^-$ is homotopic to a point in G.*

Proof. Combining Theorem 3.1.5 with Theorem 3.1.6, we have $\gamma_1 \cup \gamma_2^- \underset{G}{\sim} \gamma_1 \cup \gamma_1^-$. According to the Theorem 3.1.8, we get $e_{z_1} \underset{G}{\sim} \gamma_1 \cup \gamma_1^-$, and the conclusion follows by using the transitivity of the homotopy relation. □

3.1.1 Simply connected domains

Definition 3.1.7.

1. The subset $A \subset \mathbb{C}$ is said to be **arc-connected**, if any arbitrary two points of A can be connected by a path in A (i. e., there exists a path in A, whose starting and end points are any two points in A).
2. If $z_1, z_2 \in \mathbb{C}$, the **linear path that connects the points z_1 and z_2** is the path λ, defined by

$$\lambda(t) = z_1 + t(z_2 - z_1), \quad t \in [0, 1].$$

3. The path γ is called **a broken line**, if there exists such a decomposition of it which elements are linear path.
4. The subset $A \subset \mathbb{C}$ is said to be **connected by broken lines**, if any arbitrary two points of A can be connected by a broken line in A.
   We see immediately that every broken line connected set is a connected by arcs set. The reverse is not true!

Theorem 3.1.10. *The set $D \subset \mathbb{C}$ is a domain, if and only if D is open and a broken line connected set.*

Proof. "$\Leftarrow$"

1. First, we will prove that any open set D, which is connected by arcs will be a domain.

If this is not true, there exists a nonempty subset $A \subset D$, such that A is open and closed in the D metric space. Then the set $D \setminus A = B$ has the same property, i. e., it is open and closed in the D metric space. Let $z_1 \in A$, $z_2 \in B$ and let γ be a path that connects these two points. Then $A \cap \{\gamma\}$ and $B \cap \{\gamma\}$ are nonempty, disjoint and closed subsets of $\{\gamma\}$, and hence the sets $M = \gamma^{-1}(A \cap \{\gamma\})$ and $N = \gamma^{-1}(B \cap \{\gamma\})$ are disjoint, nonempty and closed subsets of $[0, 1]$, with $M \cup N = [0, 1]$. Suppose that $1 \notin M$, and let t_m the maximum (the biggest number) of the set M (this exists, because M is compact). Then $t_m \notin N$ and $(t_m, 1] \subset N$, hence there exists a sequence of points of N that tends to t_m. This contradicts the fact that N is a closed set, and the above property is proved.

2. If D is an open set and it is broken line connected, then D is connected by arcs, and according to the first point, we conclude that it is a domain.

"$\Rightarrow$"

Let z_1, z_2 be two arbitrary points of the domain D. Let define by A, $A \subset D$, the set of all the points of D that can be connected by a broken line with the point z_1.

Let $z \in A$ be an arbitrary point. Since D is open, there exists a number $r > 0$, such that $U(z; r) \subset D$. Any arbitrary point $z' \in U(z; r)$ can be connected by broken lines with z_1 in D (the points z_1 and z can be connected by a broken line, while the points z and z' can be connected by a linear path). It follows that $U(z; r) \subset A$, hence A is open in D.

Let $z \in D$ be the limit of an arbitrary sequence of the set A. Since D is open, there exists a number $r > 0$, such that $U(z; r) \subset D$. Then the above mentioned sequence has an element $z' \in U(z; r)$. The points z_1 and z' can be connected by a broken line, while z' and z can be connected by a linear path in D, hence the points z_1 and z can be connected by a broken line in D. Thus, the set A is closed in the metric space D.

Since D is a domain, it follows that $A = D$, hence $z_2 \in A$, and thus z_1 and z_2 can be connected by a broken line in D. □

Definition 3.1.8.

1. Let γ be a path, and let $\Delta = (t_0, t_1, \ldots, t_n)$ be a division of the interval $[0, 1]$. Then the sum

$$V(\gamma, \Delta) = \sum_{k=1}^{n} \left|\gamma(t_k) - \gamma(t_{k-1})\right|$$

is called **the variation of the path γ corresponding to the division Δ.**

2. If

$$l(\gamma) = \sup\{V(\gamma, \Delta) : \Delta \text{ division}\} < +\infty,$$

then we say that the path γ has **finite variation.** The number $l(\gamma)$ is called **the total variation** of the path γ, it represents **the length** of the path (curve) γ. In such a case, the path γ is said to be **a rectifiable path.**

Corollary 3.1.1. *Since every broken line has finite variation, from Theorem 3.1.10 it follows immediately that any two points of a domain can be connected by a rectifiable path.*

Definition 3.1.9. The domain $D \subset \mathbb{C}$ is called **a simply connected domain,** if any closed path in D is homotopic to a point in D.

Theorem 3.1.11. *The domain $D \subset \mathbb{C}$ is simply connected, if and only if any arbitrary two paths in D, with the same start and the same end points, are homotopic in D.*

Proof. Suppose that the domain D is simply connected, and let $\gamma_1, \gamma_2 \in \mathcal{D}_D(z_1, z_2)$. Then the path $\gamma_1 \cup \gamma_2^- \in \mathcal{D}_D(z_1)$ is closed, and from the above definition we have $\gamma_1 \cup \gamma_2^- \underset{D}{\sim} e_{z_1}$. Using some of the prcperties prove in the previous section, we obtain that

$$\gamma_1 \underset{D}{\sim} \gamma_1 \cup e_{z_2} \underset{D}{\sim} \gamma_1 \cup (\gamma_2^- \cup \gamma_2) \underset{D}{\sim} (\gamma_1 \cup \gamma_2^-) \cup \gamma_2 \underset{D}{\sim} e_{z_1} \cup \gamma_2 \underset{D}{\sim} \gamma_2,$$

thus $\gamma_1 \underset{D}{\sim} \gamma_2$.

Conversely, if any arbitrary two paths in D, with the same start and the same end points, are homotopic in D, then every path $\gamma \in \mathcal{D}_D(z)$ satisfies $\gamma \underset{D}{\sim} e_z$, because γ and e_z have the same start and end points. Hence, the domain D is simply connected. □

Definition 3.1.10.

1. The domain D is said to be **starlike with respect to the point** $z_0 \in D$, if $\forall z \in D$ we have $\{\lambda_z\} \subset D$, where λ_z is the linear path that connects the points z_0 and z.
2. The domain D is said to be **convex,** if it is starlike with respect to any arbitrary point of D.

Remarks 3.1.1.

1. The above last condition is equivalent to the fact, that the linear path that connects any arbitrary point of D is included in D.
2. Any starlike set is a connected by arcs set because every point of the set can be connected by a broken line that consists in the union of two linear paths (each linear paths of the broken line connects the corresponding point with z_0).

Theorem 3.1.12. *Any starlike domain is a simply connected domain.*

Proof. Suppose that the domain D is starlike with respect to the point $z_0 \in D$, and let $\gamma \in \mathcal{D}_D(z_0)$. Then the function $\varphi(s,t) = z_0 + s(\gamma(t) - z_0)$ establishes the $e_{z_0} \underset{D}{\sim} \gamma$ homotopy.

Consider any $z \in D$, and let $\gamma \in \mathcal{D}_D(z)$ be an arbitrary path. If λ is the linear path that connects z_0 and z (there exists, because D is starlike), then $\lambda \cup \gamma \cup \lambda^- \in \mathcal{D}_D(z_0)$ and from some of the previous results $\lambda \cup \gamma \cup \lambda^- \underset{D}{\sim} e_{z_0}$. Using the properties proved in the previous section, we get

$$\gamma \underset{D}{\sim} \lambda^- \cup (\lambda \cup \gamma \cup \lambda^-) \cup \lambda \underset{D}{\sim} \lambda^- \cup e_{z_0} \cup \lambda \underset{D}{\sim} e_z.$$

Thus, we proved that any closed path in D is homotopic to a point, hence D is simply connected. □

3.1.2 Functions of bounded variation and paths

Definition 3.1.11.

1. Let $[a,b] \subset \mathbb{R}, f : [a,b] \to \mathbb{C}$, and let $\Delta = (t_0, t_1, \ldots, t_n)$ be a division of the interval $[a,b]$. Let denote by
$$\|\Delta\| = \max\{t_k - t_{k-1} : k \in \{1,2,\ldots,n\}\}$$
the norm of the division Δ.
The variation of the complex valued function f corresponding to the division Δ is the sum
$$V(f,\Delta) = \sum_{k=1}^{n} |f(t_k) - f(t_{k-1})|.$$
2. The number
$$V(f;[a,b]) = \sup_{\Delta} V(f,\Delta)$$
is called **the total variation of the function f on the interval** $[a,b]$.
3. The function f is said to be **a function with bounded variation,** if $V(f;[a,b]) < +\infty$. The set of all the functions with bounded variation defined on the interval $[a,b]$, is denoted by $BV[a,b]$.

Remarks 3.1.2.

1. If $\gamma : [0,1] \to \mathbb{C}$ and Δ is a division of the interval $[0,1]$, then the variation of f corresponding to Δ, i. e., $V(\gamma,\Delta)$, represents the length of the broken line that connects the points $\gamma(t_k)$, $k \in \{0,1,\ldots,n\}$, that belong to the image $\{\gamma\}$.
2. If $f = u + iv$, from the inequalities
$$|u(t_k) - u(t_{k-1})| \le |f(t_k) - f(t_{k-1})| \le |u(t_k) - u(t_{k-1})| + |v(t_k) - v(t_{k-1})|$$
and
$$|v(t_k) - v(t_{k-1})| \le |f(t_k) - f(t_{k-1})| \le |u(t_k) - u(t_{k-1})| + |v(t_k) - v(t_{k-1})|$$
it follows that the function f is a function with bounded variation, if and only if the real and the imaginary parts of f are functions with bounded variation.

3. If $f, g : [a, b] \to \mathbb{C}$ are two functions with bounded variation, then $\alpha f + \beta g$ with $\alpha, \beta \in \mathbb{C}$ is a function with bounded variation.
   Let $c \in (a, b)$. The function f is a function with bounded variation on the interval $[a, b]$, if f is a function with bounded variation on the intervals $[a, c]$ and $[c, b]$.
   If f is a function with bounded variation, then $|f|$ is also a function with bounded variation.
4. Every path $\gamma : [0, 1] \to \mathbb{C}$ can be uniformly approximated by a broken line consisting of the points $\gamma(t_k)$, $k \in \{0, 1, \dots, n\}$ (that belong to the image $\{\gamma\}$), i. e.,
   $\forall \varepsilon > 0, \exists \gamma_\varepsilon$ a broken line consisting of the points $\gamma(t_k)$, $k \in \{0, 1, \dots, n\}$, such that

$$|\gamma(t) - \gamma_\varepsilon(t)| < \varepsilon, \quad \forall t \in [0, 1].$$

Remark 3.1.3. If $(\gamma_n)_{n\in\mathbb{N}^*}$ is a sequence of paths of G that uniformly converges to the path γ of G, i. e.,

$$\forall \varepsilon > 0,\ \exists n_\varepsilon \in \mathbb{N}^* \quad \text{such that } |\gamma(t) - \gamma_n(t)| < \varepsilon,\ \forall n > n_\varepsilon,\ \forall t \in [0, 1],$$

and $f : G \to \mathbb{C}$ is a continuous function on the open set G, then the function sequence $(f \circ \gamma_n)_{n\in\mathbb{N}^*}$ uniformly converges to the path $f \circ \gamma$.

Proof. Let $\varepsilon > 0$ be an arbitrary number, and let $\eta > 0$ be a sufficiently enough small number, such that

$$U^-(\gamma; \eta) = \Big(\bigcup_{t\in[0,1]} U(\gamma(t); \eta)\Big)^- \subset G.$$

Since $U^-(\gamma; \eta)$ is bounded and closed, then $U^-(\gamma; \eta)$ is compact. According to Cantor theorem (every continuous function on a compact is uniformly continuous on this compact), we have

$$\exists \eta_1 > 0, \eta_1 \le \eta \quad \text{such that } |f(z') - f(z'')| < \varepsilon,\ \forall z', z'' \in U^-(\gamma; \eta),\ |z' - z''| < \eta_1.$$

Let $n_0 \in \mathbb{N}^*$ be an enough big number, such that

$$\forall n > n_0, \quad |\gamma(t) - \gamma_n(t)| < \eta_1,\ \forall t \in [0, 1].$$

Then

$$|f(\gamma(t)) - f(\gamma_n(t))| < \varepsilon, \quad \forall t \in [0, 1] \text{ and } \forall n > n_0.$$

thus the function sequence $(f \circ \gamma_n)$ uniformly converges to the function $f \circ \gamma$. □

3.2 The complex integral

3.2.1 The Riemann–Stieltjes integral for complex valued functions

Definition 3.2.1. Consider the closed interval $[a,b] \subset \mathbb{R}$, let $u, U : [a,b] \to \mathbb{R}$ be two real-valued functions, and define the sum

$$\sigma(u,U,\Delta) = \sum_{k=1}^{n} u(\tau_k)(U(t_k) - U(t_{k-1})),$$

where $\Delta = (t_0, t_1, \dots, t_n)$ is a division of the interval $[a,b]$, and $\tau_k \in [t_{k-1}, t_k]$, $k = \overline{1,n}$ are arbitrary points.

If there exists a real number $I \in \mathbb{R}$, such that

$$\forall \varepsilon > 0,\ \exists \eta > 0 \quad \text{such that } \|\Delta\| < \eta \Rightarrow |I - \sigma(u,U,\Delta)| < \varepsilon,$$

we say that **the function** u **is integrable with respect to the function** U (in the Riemann–Stieltjes sense), and we denote

$$I = \int_a^b u \mathrm{d}U.$$

The number I is called **the Riemann–Stieltjes integral of** u **on the interval** $[a,b]$ **with respect** U.

Definition 3.2.2. Let $f = u + iv$ and $F = U + iV$, $f, F : [a,b] \to \mathbb{C}$ be two given functions. We say that f **is integrable on the interval** $[a,b]$ **with respect to** F (in the Riemann–Stieltjes sense), if u and v are integrable functions on $[a,b]$ with respect to U, and respectively to V.

According to the above definition, the integral of the function f with respect to the function F on the interval $[a,b]$ will be

$$\int_a^b f \mathrm{d}F = \int_a^b u \mathrm{d}U - \int_a^b v \mathrm{d}V + i\left(\int_a^b u \mathrm{d}V + \int_a^b v \mathrm{d}U\right).$$

Theorem 3.2.1. *If f is integrable with respect to F on $[a,b]$, and if for any division $\Delta = (t_0, t_1, \dots, t_n)$ of $[a,b]$ we define the sum*

$$\sigma(f,F,\Delta) = \sum_{k=1}^{n} f(\tau_k)(F(t_k) - F(t_{k-1})),$$

where $\tau_k \in [t_{k-1}, t_k]$, then for all $\varepsilon > 0$ there exists a number $\eta > 0$, such that $\|\Delta\| < \eta$ implies

$$\left|\int_a^b f \mathrm{d}F - \sigma(f,F,\Delta)\right| < \varepsilon.$$

Proof. We have

$$\sigma(f,F,\Delta) = \sigma(u,U,\Delta) - \sigma(v,V,\Delta) + i(\sigma(u,V,\Delta) + \sigma(u,U,\Delta)).$$

Using the relations $\int_a^b u\mathrm{d}U$, etc. together with the above identity, the proof follows immediately. □

Theorem 3.2.2. *Let $f = u + iv$, $F = U + iV$, and $f_n, F_n : [a,b] \to \mathbb{C}$ real variable complex valued functions, and let $\alpha_1, \alpha_2 \in \mathbb{C}$.*

1. *If f is integrable on $[a,b]$ with respect to F, then F is integrable on $[a,b]$ with respect to f, and*

$$\int_a^b f\mathrm{d}F + \int_a^b F\mathrm{d}f = f(b)F(b) - f(a)F(a).$$

2. *If f_1 and f_2 are integrable with respect to F, then $\alpha_1 f_1 + \alpha_2 f_2$ is integrable with respect to F, and*

$$\int_a^b (\alpha_1 f_1 + \alpha_2 f_2)\mathrm{d}F = \alpha_1 \int_a^b f_1\mathrm{d}F + \alpha_2 \int_a^b f_2\mathrm{d}F.$$

3. *If f is continuous and F is a function with bounded variation on $[a,b]$, then f is integrable with respect to F on $[a,b]$.*
4. *Let $(f_n)_{n\in\mathbb{N}^*}$ a sequence of functions that uniformly converges to f, and let $(F_n)_{n\in\mathbb{N}^*}$ a sequence of functions consisting on functions with bounded variation, such that $(F_n)_{n\in\mathbb{N}^*}$ converges pointwise to F and the set of the total variations $V(F_n,[a,b])$ is bounded. Then*

$$\lim_{\substack{n\to\infty \\ k\to\infty}} \int_a^b f_n\mathrm{d}F_k = \int_a^b f\mathrm{d}F.$$

5. *If f is continuous, and F is continuously differentiable on $[a,b]$, then f is integrable on $[a,b]$ with respect to F, and*

$$\int_a^b f\mathrm{d}F = \int_a^b f(t)F'(t)\mathrm{d}t,$$

where the right-hand side integral is given by

$$\int_a^b (u(t)U'(t) - v(t)V'(t))\mathrm{d}t + i\int_a^b (u(t)V'(t) + v(t)U'(t))\mathrm{d}t.$$

6. *If $c \in (a, b)$ and f is integrable with respect to F on the intervals $[a, c]$ and $[c, b]$, then f is integrable with respect to F on $[a, b]$, and*

$$\int_a^b f\mathrm{d}F = \int_a^c f\mathrm{d}F + \int_c^b f\mathrm{d}F.$$

7. *If f is integrable with respect to F on $[a, b]$, and if $h : [a', b'] \to [a, b]$ is a homeomorphism with $h(a') = a$ and $h(b') = b$, then $f \circ h$ is integrable with respect to $F \circ h$ on $[a', b']$, and*

$$\int_a^b f\mathrm{d}F = \int_{a'}^{b'} (f \circ h)\mathrm{d}(F \circ h).$$

Proof. The points 1–6 follow from the corresponding properties of the Riemann–Stieltjes integral of the real-valued functions. We will prove the point 7 of this theorem.

If $f = u + iv$, $F = U + iV$, then $f \circ h = u \circ h + iv \circ h$ and $F \circ h = U \circ h + iV \circ h$. It is sufficient to prove, using the definition of the complex Riemann–Stieltjes integral, that the next formula holds:

$$\int_{a'}^{b'} (u \circ h)\mathrm{d}(U \circ h) = \int_a^b u\mathrm{d}U.$$

Let $I = \int_a^b u\mathrm{d}U$, and let $\varepsilon > 0$ be arbitrary. Then

$$\exists \eta_1 > 0 \quad \text{such that } |I - \sigma(u, U, \Delta)| < \varepsilon, \text{ whenever } \|\Delta\| < \eta_1.$$

Since h is a homeomorphism with $h(a') = a$ and $h(b') = b$, then h is a strictly increasing continuous function, hence it is uniformly continuous. From the definition of the uniform continuity, it follows that

$$\exists \eta > 0, \quad \text{such that } |h(t') - h(t'')| < \eta_1, \ \forall t', t'' \in [a', b'], \ |t' - t''| < \eta.$$

Let $\Delta' = (t'_0, t'_1, \ldots, t'_n)$ a division of the interval $[a', b']$, with $\|\Delta'\| < \eta$, and let $\tau'_k \in [t'_{k-1}, t'_k]$ be arbitrary points. Let denote $t_k = h(t'_k)$ and $\tau_k = h(\tau'_k)$, $k = \overline{1, n}$. Then $\Delta = (t_0, t_1, \ldots, t_n)$ will be a division of the interval $[a', b']$, such that $\|\Delta\| < \eta_1$, and

$$\begin{aligned}\sigma(u \circ h, U \circ h, \Delta') &= \sum_{k=1}^{n} (u \circ h)(\tau'_k)((U \circ h)(t'_k) - (U \circ h)(t'_{k-1})) \\ &= \sum_{k=1}^{n} u(\tau_k)(U(t_k) - U(t_{k-1})) = \sigma(u, U, \Delta).\end{aligned}$$

Thus

$$|I - \sigma(u \circ h, U \circ h, \Delta')| = |I - \sigma(u, U, \Delta)| < \varepsilon,$$

and consequently

$$\int_{a'}^{b'} (u \circ h)\mathrm{d}(U \circ h) = \int_{a}^{b} u\mathrm{d}U.$$ □

Definition 3.2.3. Let $\gamma : [0,1] \to \mathbb{C}$ be a path with finite variation (i. e., a rectifiable path), and let $f : \{\gamma\} \to \mathbb{C}$ be a continuous function. Then the function $f \circ \gamma : [0,1] \to \mathbb{C}$ is continuous, hence it is integrable on the interval $[0,1]$ with respect to the function γ.

Denoting the corresponding integral by

$$\int_0^1 (f \circ \gamma)\mathrm{d}\gamma = \int_\gamma f(z)\mathrm{d}z = \int_\gamma f,$$

this integral will be called the **complex integral** (or the Cauchy integral) **of the function** f **on the path** γ.

Theorem 3.2.3 (The properties of the complex integral). *Let $\gamma \in \mathcal{D}(z_1, z_2)$ and $\gamma_1 \in \mathcal{D}(z_2, z_3)$ be two rectifiable paths, let $f, g : \{\gamma\} \to \mathbb{C}$ be two continuous functions, and let $\alpha, \beta \in \mathbb{C}$. Then:*

1. $\int_\gamma \alpha f + \beta g = \alpha \int_\gamma f + \beta \int_\gamma g$.
2. $\int_{\gamma^-} f = -\int_\gamma f$.
3. *If f is continuous also on $\{\gamma_1\}$, then* $\int_{\gamma \cup \gamma_1} f = \int_\gamma f + \int_{\gamma_1} f$.
4. *If $(\gamma_1, \gamma_2, \ldots, \gamma_n)$ is a division of the path γ, then* $\int_\gamma f = \sum_{k=1}^n \int_{\gamma_k} f$.
5. *If $|f(\gamma(t))| \le M$, $\forall t \in [0,1]$, then $|\int_\gamma f| \le M\, V(\gamma)$, where $V(\gamma)$ is the length of the path γ.*
6. *For the linear path $\gamma(t) = z_1 + t(z_2 - z_1)$, $t \in [0,1]$, we have*

$$\int_\gamma f = (z_2 - z_1) \int_0^1 f(z_1 + t(z_2 - z_1))\mathrm{d}t.$$

7. *If $G \subset \mathbb{C}$ is an open set, the function $f : G \to \mathbb{C}$ is a continuous, $(\gamma_n)_{n \in \mathbb{N}^*}$ is a sequence of rectifiable paths in G that uniformly converges on $[0,1]$ to the path γ, where $\{\gamma\} \subset G$, and the real numbers sequence $(V(\gamma_n))_{n \in \mathbb{N}^*}$ is bounded, then the sequence $(\int_{\gamma_n} f)_{n \in \mathbb{N}^*}$ converges, and*

$$\lim_{n \to \infty} \int_{\gamma_n} f = \int_\gamma f.$$

8. *Let γ be a rectifiable path, and let $f_n : \{\gamma\} \to \mathbb{C}$, $n \in \mathbb{N}^*$ be a sequence of continuous functions that uniformly converges on the set $\{\gamma\}$ to the function f. Then there exists the integral $\int_\gamma f$, and*

$$\lim_{n\to\infty} \int_\gamma f_n = \int_\gamma f.$$

9. *If γ_1 and γ_2 are equivalent paths in G, if the function f is continuous on $\{\gamma_1\}$, and γ_1 is rectifiable, then γ_2 is also rectifiable and*

$$\int_{\gamma_1} f = \int_{\gamma_2} f.$$

Proof. The points 1, 6, 7, 8 and 9 follow immediately from the points 2, 4, 5 and 7 of the Theorem 3.2.2.

2. The integral sums on the path γ^- are

$$\sigma(f \circ \gamma^-, \gamma^-, \Delta) = \sum_{k=1}^{n} f(\gamma(1-\tau_k))(\gamma(1-t_k) - \gamma(1-t_{k-1})).$$

Since $t'_k = 1 - t_{n-k}$, $\tau'_k = 1 - \tau_{n-k+1}$, where $\Delta' = (t'_0, t'_1, \dots, t'_n)$, then $\|\Delta'\| = \|\Delta\|$ and $\tau'_k \in [t'_k, t'_{k-1}]$. If we let $n - k + 1 = m$, then

$$\begin{aligned}\sigma(f \circ \gamma^-, \gamma^-, \Delta) &= \sum_{k=1}^{n} f(\tau'_{n-k+1})(\gamma(t'_{n-k}) - \gamma(t'_{n-k+1})) \\ &= -\sum_{m=1}^{n} f(\gamma(\tau'_m))(\gamma(t'_m) - \gamma(t'_{m-1})) = -\sigma(f \circ \gamma, \gamma, \Delta').\end{aligned}$$

Thus, if $\|\Delta\| = \|\Delta'\| \to 0$ then we get $\int_\gamma f = -\int_{\gamma^-} f$.

3. A simple computation shows that

$$\begin{aligned}\int_{\gamma\cup\gamma_1} f &= \int_0^1 (f \circ (\gamma \cup \gamma_1))(t)\mathrm{d}(\gamma \cup \gamma_1)(t) \\ &= \int_0^{\frac{1}{2}} f(\gamma(2t))\mathrm{d}\gamma(2t) + \int_{\frac{1}{2}}^1 f(\gamma_1(2t-1))\mathrm{d}\gamma_1(2t-1).\end{aligned}$$

According to the point 7 of the Theorem 3.2.2, if $h : [a', b'] \to [a, b]$ is an increasing homeomorphism of the interval $[a', b']$ onto the interval $[a, b]$, then

$$\int_{a'}^{b'} (f \circ h)\mathrm{d}(g \circ h) = \int_a^b f\mathrm{d}g.$$

Letting $h : [0, \frac{1}{2}] \to [0, 1]$, $h(t) = 2t$, then

$$\int_0^{\frac{1}{2}} f(\gamma(2t))\mathrm{d}\gamma(2t) = \int_0^1 f(\gamma(t))\mathrm{d}\gamma(t) = \int_\gamma f,$$

and similarly

$$\int_{\frac{1}{2}}^1 f(\gamma_1(2t-1))\mathrm{d}\gamma_1(2t-1) = \int_{\gamma_1} f.$$

Thus we get

$$\int_{\gamma\cup\gamma_1} f = \int_\gamma f + \int_{\gamma_1} f,$$

and the proof of the point 4 is similar to the above. □

Remarks 3.2.1.

1. The complex integral can be defined by using the second type curve integral. In fact, if we let $f = u + iv$ and $\gamma = \alpha + i\beta$, then

$$\int_\gamma f = \int_0^1 (f \circ \gamma)\mathrm{d}\gamma = \int_0^1 u(\alpha(t), \beta(t))\mathrm{d}\alpha(t) - \int_0^1 v(\alpha(t), \beta(t))\mathrm{d}\beta(t)$$
$$+ i\left(\int_0^1 u(\alpha(t), \beta(t))\mathrm{d}\beta(t) + \int_0^1 v(\alpha(t), \beta(t))\mathrm{d}\alpha(t)\right),$$

   where the integrals that appeared in last sums are second type real line integrals, written with the aid of the real line integrals.
2. The connection between the complex integral and the first type curve integral is given by the next relation:

$$\int_\gamma f(\zeta)|\mathrm{d}\zeta| = \int_0^1 f(\gamma(t))|\gamma'(t)|\mathrm{d}t$$
$$= \int_0^1 f(\gamma(t))\sqrt{\alpha'^2(t) + \beta'^2(t)}\mathrm{d}t = \int_\gamma f\mathrm{d}s,$$

   where

$$\int_\gamma f(\zeta)|\mathrm{d}\zeta| = \lim_{\|\Delta\|\to 0} \sum_{k=1}^n (f \circ \gamma)(\tau_k)|\gamma(t_k) - \gamma(t_{k-1})|.$$

3. Using the previous definition, we deduce that

$$|\sigma(f\circ\gamma,\gamma,\Delta)| = \left|\sum_{k=1}^{n}(f\circ\gamma)(\tau_k)[\gamma(t_k)-\gamma(t_{k-1})]\right|$$
$$\le \sum_{k=1}^{n}|(f\circ\gamma)(\tau_k)||\gamma(t_k)-\gamma(t_{k-1})|,$$

and thus

$$\left|\int_\gamma f(\zeta)d\zeta\right| \le \int_\gamma |f(\zeta)||d\zeta|.$$

4. The complex integral along a closed path does not depend on the starting point. In fact, let $\gamma : [0,1] \to \mathbb{C}$, with $\gamma(0) = \gamma(1) = z_0$, and let $z_1 \in \{\gamma\}$ be an arbitrary point. Then there exists $t_1 \in [0,1]$ such that $\gamma(t_1) = z_1$. Let (γ_1,γ_2) be the decomposition of the path γ by the division $\Delta = (0, t_1, 1)$. Then

$$\int_\gamma f = \int_{\gamma_1\cup\gamma_2} f = \int_{\gamma_1} f + \int_{\gamma_2} f = \int_{\gamma_2} f + \int_{\gamma_1} f = \int_{\gamma_2\cup\gamma_1} f.$$

Remark that the union $\gamma_2\cup\gamma_1$ is well-defined, because the end point of γ_2 is z_0 that it is also the starting point of γ_1, and $\gamma_2\cup\gamma_1 \in \mathcal{D}(z_1)$.

3.3 The Cauchy theorem

3.3.1 The connection between the integral and the primitive function

Definition 3.3.1. Let $G \subset \mathbb{C}$ be an open set, and let f be a complex function, with $f : G \to \mathbb{C}$. The function $g \in H(G)$ is said to be a **primitive function of the function** f, if $g'(z) = f(z)$ for all $z \in G$.

For the real valued functions, the connection between the (definite) integral and the primitive of a function is given by the Newton–Leibniz formula. If f is a real valued real variable continuous function $f : [a,b] \to \mathbb{R}$, then the function

$$F : [a,b] \to \mathbb{R}, \quad F(x) := \int_a^x f(t)dt$$

is a primitive function of f.

For the complex valued complex variable functions, we will try to find a similar property. Let $f \in C(D)$, where $D \subset \mathbb{C}$ domain, and let $\gamma \in \mathcal{D}_D(z_0,z)$ be an arbitrary rectifiable path. A natural question is the following:

Which are the sufficient conditions, such that the function g defined by

$$g(z) = \int_{z_0}^{z} f = \int_{\gamma} f$$

is a primitive function of f, where the rectifiable path γ connects the fixed-point z_0 with the variable point z?

The question has sense, only if the integral $\int_\gamma f$ is independent of the path γ, and depends on only to the end point of γ, which is z.

Remark 3.3.1. The integral of the function f is independent of the integration path, if and only if $\int_\gamma f = 0$ for all arbitrary, closed and rectifiable path γ.

In fact, if $\gamma = \gamma_1 \cup \gamma_2^-$ is an arbitrary decomposition of the closed path γ, then

$$0 = \int_{\gamma} f = \int_{\gamma_1 \cup \gamma_2^-} f = \int_{\gamma_1} f + \int_{\gamma_2^-} f = \int_{\gamma_1} f - \int_{\gamma_2} f \Leftrightarrow \int_{\gamma_1} f = \int_{\gamma_2} f.$$

Conversely, if the integral is independent of the path, then from the above reasons the integral will be equal to zero for all arbitrary, closed and rectifiable path γ.

Theorem 3.3.1 (The connection between the primitive and the integral). *Let $D \subset \mathbb{C}$ be a domain (an open and connected set), and let $f : D \to \mathbb{C}$ be a continuous function. Then the following relations hold:*

1. *If*

$$\int_{\gamma} f = 0 \tag{3.1}$$

for any arbitrary closed and rectifiable path γ, with $\{\gamma\} \subset D$, then the function f has a primitive g in the domain D.

Any other primitive of the function f in the domain D is of the form $g+c$, where $c \in \mathbb{C}$.

2. *If the function f has a primitive function g in the domain D, then for all rectifiable curve γ, with $\{\gamma\} \subset D$, the following Newton–Leibniz formula holds:*

$$\int_{\gamma} f = g(\gamma(1)) - g(\gamma(0)).$$

Thus, if γ a closed path, then $\int_\gamma f = 0$.

Proof. 1. Let $z_0 \in D$ be a given point, and let $z \in D$ be arbitrary. There exists a number $r > 0$, such that $U(z; r) \subset D$. If we consider the point $h \in U(0; r)$, then $z + h \in U(z; r)$.

Let us denote by λ the linear path that connects z with $z+h$, and let γ be an arbitrary rectifiable path that connects z_0 with z, and let γ_1 be an arbitrary rectifiable path that connects z_0 with $z + h$. Then $\gamma \cup \lambda \cup \gamma_1^-$ is a closed path in the domain D, hence from our assumption we get

$$\int_{\gamma \cup \lambda \cup \gamma_1^-} f = 0.$$

If we let $g(z) = \int_\gamma f$, according to the previous remark, it follows that the function g is well-defined, and $g(z + h) = \int_{\gamma_1} f$.

Supposing that $h \neq 0$, from the relation (3.1), we obtain that

$$0 = \int_\gamma f + \int_\lambda f + \int_{\gamma_1^-} f = g(z) + \int_\lambda f - g(z + h),$$

and thus

$$\frac{g(z + h) - g(z)}{h} = \frac{1}{h} \int_\lambda f(\zeta) \mathrm{d}\zeta = \frac{1}{h} \int_\lambda (f(\zeta) - f(z)) \mathrm{d}\zeta + \frac{1}{h} f(z) \int_\lambda \mathrm{d}\zeta. \tag{3.2}$$

Using the fact that $\int_\lambda \mathrm{d}\zeta = h$, because

$$\int_\lambda \mathrm{d}\zeta = \int_0^1 (z + th)_t' \mathrm{d}t = \int_0^1 h \mathrm{d}t = h,$$

the relation (3.2) reduces to

$$\frac{g(z + h) - g(z)}{h} = f(z) + \frac{1}{h} \int_\lambda (f(\zeta) - f(z)) \mathrm{d}\zeta. \tag{3.3}$$

Since the function f is continuous at the point z, then

$$\forall \varepsilon > 0,\ \exists \delta = \delta_\varepsilon > 0 \quad \text{such that } |\zeta - z| \leq |h| < \delta \Rightarrow \left|f(\zeta) - f(z)\right| < \varepsilon.$$

In a such a case, we have

$$\left| \frac{1}{h} \int_\lambda (f(\zeta) - f(z)) \mathrm{d}\zeta \right| \leq \frac{1}{|h|} \varepsilon |h| = \varepsilon.$$

Since $\varepsilon > 0$ is arbitrary, from the above inequality we obtain that

$$\lim_{h \to 0} \frac{1}{h} \int_\lambda (f(\zeta) - f(z)) \mathrm{d}\zeta = 0.$$

Thus, according to the relation (3.3), there exists the limit

$$\lim_{h \to 0} \frac{g(z + h) - g(z)}{h} = \lim_{h \to 0} \frac{1}{h} \int_\gamma f(\zeta) \mathrm{d}\zeta = f(z)$$

which is in fact the derivative $g'(z)$, and it is equal with $f(z)$.

If g_1 is another primitive of the function f on the domain D, then $g_1 - g$ will be a holomorphic function with the derivative identical to zero. Since D is a domain, by using a well-known result concerning the holomorphic functions in a domain with the derivative equal to zero, it follows that $g_1 - g = c$, where c is a constant.

2. Let $\gamma \in \mathcal{D}_D(z_0, z_1)$ be a rectifiable path, let $\Delta = (t_0, t_1, \dots, t_n)$ be a division of the interval $[0,1]$, and let γ_n be the broken line that connects consequently the points $\gamma(t_k)$, $k \in \{0, 1, \dots, n\}$. Then

$$\int_{\gamma_n} f = \int_0^1 (f \circ \gamma_n) \mathrm{d}\gamma_n(t) = \sum_{k=1}^{n} \int_{t_{k-1}}^{t_k} (f \circ \gamma_n) \mathrm{d}\gamma_n = \sum_{k=1}^{n} \int_{t_{k-1}}^{t_k} f(\gamma_n(t))\gamma_n'(t)\mathrm{d}t. \tag{3.4}$$

From the assumption $g'(z) = f(z)$, we have

$$(g(\gamma_n(t)))' = g'(\gamma_n(t))\gamma_n'(t) = f(\gamma_n(t))\gamma_n'(t).$$

Combining this relation with (3.4), we deduce that

$$\begin{aligned}\int_{\gamma_n} f &= \sum_{k=1}^{n} \int_{t_{k-1}}^{t_k} (g(\gamma_n(t)))' \mathrm{d}t = \sum_{k=1}^{n} (g(\gamma_n(t_k)) - g(\gamma_n(t_{k-1})))\\ &= g(\gamma(1)) - g(\gamma(0)) = g(z_1) - g(z_0).\end{aligned}$$

If $\|\Delta\| \to 0$, then γ_n uniformly converges to γ, and thus $\int_{\gamma_n} f \to \int_\gamma f$. Consequently,

$$\int_\gamma f = g(z_1) - g(z_0).$$

For the special case when γ closed, i. e., $z_1 = z_0$, from the above formula we get $\int_\gamma f = 0$. □

Remarks 3.3.2. Consider the function $f(z) = \frac{1}{(z-z_0)^n}$, $n \in \mathbb{Z}$.

1. If $n \in \mathbb{Z} \setminus \{1\}$, then $g(z) = \frac{1}{(1-n)(z-z_0)^{n-1}}$ is a primitive function for f in $D = \mathbb{C} \setminus \{z_0\}$, hence

$$\int_\gamma \frac{1}{(\zeta - z_0)^n} \mathrm{d}\zeta = 0,$$

where $\gamma(t) = z_0 + re^{2\pi i t}$, $t \in [0,1]$, i. e., $\{\gamma\} = \partial U(z_0; r)$.

2. If $n = 1$, then the function $f(z) = \frac{1}{z-z_0}$ has derivative on $D = \mathbb{C} \setminus \{z_0\}$, but it has not a primitive function in D, because

$$\int_\gamma f = \int_\gamma \frac{1}{\zeta - z_0} \mathrm{d}\zeta = 2\pi i,$$

where $\gamma(t) = z_0 + re^{2\pi i t}$, $t \in [0,1]$, i. e., $\{\gamma\} = \partial U(z_0; r)$.

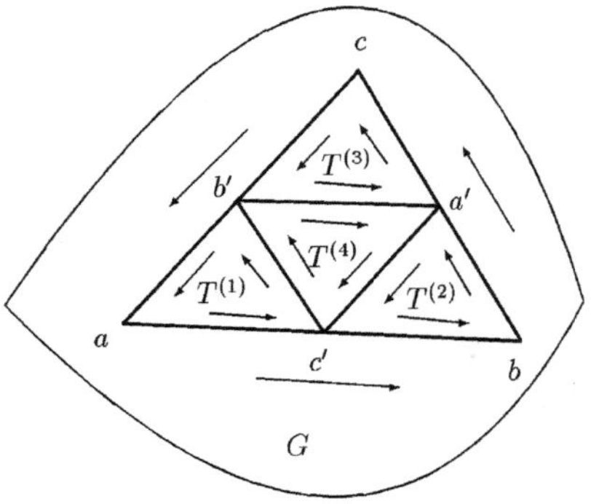

Figure 3.1: Proof of Theorem 3.3.2.

Theorem 3.3.2 (Cauchy theorem for triangles, or Goursat lemma). *Let T be the interior of the triangle with the apex in the points $a, b, c \in \mathbb{C}$ and let $f \in H(T) \cap C(T^-)$. Let $\gamma = \partial T$ denote the path consisting of the boundary of the triangle T. Then $\int_\gamma f = 0$.*

Proof. Suppose that $f \in H(G)$, where $G \subset \mathbb{C}$ is an open set, such that and $T^- \subset G$, and let $I = \int_\gamma f$.

Let denote by a', b' and c' the middle points of the segments bc, ac and, respectively, ab, and let connect these points by segments. We will denote by $T^{(1)}$, $T^{(2)}$, $T^{(3)}$ and $T^{(4)}$ the triangles that we obtained using the above method (Figure 3.1), and let $\gamma^{(k)} = \partial T^{(k)}$, $k = \overline{1,4}$.

With these notations, supposing that all the path $\gamma^{(k)}$ have direct orientation, it follows that

$$I = \int_\gamma f = \sum_{k=1}^{4} \int_{\gamma^{(k)}} f.$$

From this identity, we deduce that

$$|I| \le \sum_{k=1}^{4} \left| \int_{\gamma^{(k)}} f \right|.$$

Thus, there exists at least a triangle $T^{(k)}$ (denoted by T_1, and the boundary by $\gamma_1 = \partial T_1$), such that

$$T_1^- \subset T^-, \quad \gamma_1 = \partial T_1 \quad \text{and} \quad \left| \int_{\gamma_1} f \right| \ge \frac{|I|}{4}.$$

Denoting by $l = l(\gamma)$, the length of the path γ, then $l(\gamma_1) = l_1 = \frac{l}{2}$.

Now, we will apply the above method to the triangle T_1. It follows that there exists a triangle T_2, such that

$$T_2^- \subset T_1^-, \quad \gamma_2 = \partial T_2, \quad l(\gamma_2) = l_2 = \frac{l_1}{2} = \frac{l}{2^2}, \quad \left|\int_{\gamma_2} f\right| \geq \frac{|I|}{4^2}.$$

If we continue this method, we deduce that there exists a sequence of triangles T_k, such that

$$G \supset T^- \supset T_1^- \supset \cdots \supset T_n^- \supset T_{n+1}^- \supset \cdots, \quad \gamma_n = \partial T_n,$$
$$l_n = l(\gamma_n) = \frac{l}{2^n}, \quad \left|\int_{\gamma_n} f\right| \geq \frac{|I|}{4^n}.$$

Using the fact that $\operatorname{diam}(T_n^-) \to 0$, whenever $n \to \infty$, according to the Cantor theorem there exists a point $z_0 \in T_n^-$ for all $n \in \mathbb{N}^*$, and $z_0 \in T^- \subset G$. Since $f \in H(G)$, we have that

$$f(z) = f(z_0) + f'(z_0)(z - z_0) + g(z)(z - z_0),$$

where $\lim_{z\to z_0} g(z) = 0$. Now, using this previous relation we have

$$\int_{\gamma_n} f = [f(z_0) - f'(z_0)z_0] \int_{\gamma_n} \mathrm{d}\zeta + f'(z_0) \int_{\gamma_n} \zeta \mathrm{d}\zeta + \int_{\gamma_n} g(\zeta)(\zeta - z_0)\mathrm{d}\zeta.$$

Since the functions 1 and z have primitives, according to Theorem 3.3.1

$$\int_{\gamma_n} \mathrm{d}\zeta = \int_{\gamma_n} \zeta \mathrm{d}\zeta = 0,$$

hence

$$\int_{\gamma_n} f = \int_{\gamma_n} g(\zeta)(\zeta - z_0)\mathrm{d}\zeta.$$

Let $\varepsilon > 0$ be arbitrary. Since $\lim_{z\to z_0} g(z) = 0$, there exist a number $\delta = \delta_\varepsilon > 0$ and an index $n_0 \in \mathbb{N}^*$, such that

$$z \in G, \quad |z - z_0| < \delta \Rightarrow |g(z)| < \varepsilon \quad \text{and} \quad T_n^- \subset U(z_0;\delta), \quad \forall n > n_0.$$

Consequently,

$$\frac{|I|}{4^n} \leq \left|\int_{\gamma_n} f\right| = \left|\int_{\gamma_n} g(\zeta)(\zeta - z_0)\mathrm{d}\zeta\right| \leq \varepsilon l_n \left|\int_{\gamma_n} \mathrm{d}\zeta\right| \leq \varepsilon l_n^2 = \frac{\varepsilon l^2}{4^n}.$$

Thus, $|I| \leq \varepsilon l^2$, and since $\varepsilon > 0$ was an arbitrary number it follows that $I = 0$.

Let return now to the general case $f \in H(T) \cap C(T^-)$. Let us consider the sequences $(a_n)_{n\in\mathbb{N}^*} \subset T$, $(b_n)_{n\in\mathbb{N}^*} \subset T$, $(c_n)_{n\in\mathbb{N}^*} \subset T$, such that $a_n \to a$, $b_n \to b$, and $c_n \to c$, whenever $n \to \infty$. For the boundaries $\gamma_n = \delta(a_n, b_n, c_n)$ of the triangle (a_n, b_n, c_n), we have $\gamma_n \rightrightarrows \gamma = \delta(a, b, c)$, and thus

$$0 = \int_{\gamma_n} f \to \int_{\gamma} f, \quad \text{hence} \int_{\gamma} f = 0. \qquad \square$$

Theorem 3.3.3 (The connection between the holomorphic functions and the primitives). *Let D be a starlike domain with respect to the point $z_0 \in D$. Let us consider the lines $d_1, d_2, \ldots, d_n$ that contain the point z_0, and let $d = \bigcup_{k=1}^{n} d_k$. If $f : D \to \mathbb{C}$ is a continuous function on the domain D, and holomorphic on $D \setminus d$, i. e., $f \in C(D) \cap H(D \setminus d)$, then f has a primitive function on D.*

Proof. The set $D \setminus d$ is open, hence for every point of $D \setminus d$ there exists a neighborhood that does not intersect the set d.

1. Suppose that $z \in D \setminus d$. If we denote by λ_z the linear path that connects z_0 with z, then the function $g(z) = \int_{\lambda_z} f$ is well-defined.

Letting $U(z; r) \subset D \setminus d$ and $h \in U(0; r)$, then $z + h \in U(z; r)$. Let us consider the path $\gamma = \lambda_z \cup \tilde{\lambda} \cup \lambda_{z+h}^-$, where $\tilde{\lambda}$ is the linear path that connects z with $z + h$. The path γ represents the boundary of the triangle with the apexes z_0, z, and $z + h$. On the interior of this triangle the function f is holomorphic, and it is continuous on the boundary, hence by Theorem 3.3.2 we have that $\int_\gamma f = 0$, i. e.,

$$\int_{\gamma} f = \int_{\lambda_z} f + \int_{\tilde{\lambda}} f + \int_{\lambda_{z+h}^-} f = g(z) + \int_{\tilde{\lambda}} f - g(z+h) = 0.$$

Thus, if $h \neq 0$ we obtain

$$\frac{g(z+h) - g(z)}{h} = \frac{1}{h} \int_{\tilde{\lambda}} f \to f(z), \quad h \to 0.$$

This result is obtained using the same technique like to the proof of the point 1 of the Theorem 3.3.1. Thus, there exists $g'(z)$, and $g'(z) = f(z)$, $\forall z \in D \setminus d$.

2. Suppose that $z \in (d \cap D) \setminus \{z_0\}$, $z \in d_j$. Then there exists a number $r > 0$, such that the disc $U(z; r)$ intersects only the line d_j. Letting $h \in U(0; r)$, then the function f is holomorphic in the interior of the triangle with the apexes in z_0, z and $z + h$, and it is continuous on the boundary of this triangle. In this case, we may use the same method as to the proof of the point 1 of Theorem 3.3.2, hence it follows that the function $g(z) = \int_{\lambda_z} f$ has derivative, and $g'(z) = f(z)$.

3. Finally, we will prove that there exists $g'(z_0)$ and $g'(z_0) = f(z_0)$. From the definition of g, we have $g(z_0) = 0$, and

$$g(z) = \int_{\lambda_z} f = (z - z_0) \int_0^1 f(z_0 + t(z - z_0))dt,$$

hence, for $z \neq z_0$ we have

$$\frac{g(z) - g(z_0)}{z - z_0} = \int_0^1 f(z_0 + t(z - z_0))\mathrm{d}t.$$

We will prove that

$$\lim_{z \to z_0} \int_0^1 f(z_0 + t(z - z_0))\mathrm{d}t = f(z_0).$$

Since the function f is continuous on D and $z_0 \in D$, for all $\varepsilon > 0$ there exists a number $\delta > 0$, such that if $z \in D$ and $|z - z_0| < \delta$, then $|f(z) - f(z_0)| < \varepsilon$.

If $|z - z_0| < \delta$, then $|z_0 + t(z - z_0) - z_0| = |t||z - z_0| \leq |z - z_0| < \delta$, $\forall t \in [0,1]$, hence $|f(z_0 + t(z - z_0)) - f(z_0)| < \varepsilon$ for all $t \in [0,1]$. From here, we deduce that

$$\left|\int_0^1 f(z_0 + t(z - z_0))\mathrm{d}t - f(z_0)\right| = \left|\int_0^1 f(z_0 + t(z - z_0))\mathrm{d}t - \int_0^1 f(z_0)\mathrm{d}t\right|$$

$$= \left|\int_0^1 [f(z_0 + t(z - z_0)) - f(z_0)]\mathrm{d}t\right| \leq \max_{t \in [0,1]} |f(z_0 + t(z - z_0)) - f(z_0)| < \varepsilon,$$

whenever $|z - z_0| < \delta$, hence $\lim_{z \to z_0} \int_0^1 f(z_0 + t(z - z_0))\mathrm{d}t = f(z_0)$.

Thus there exists $g'(z_0)$, and $g'(z_0) = f(z_0)$.

Consequently, the function g is holomorphic on the domain D, and $g' = f$. □

Remarks 3.3.3.

1. According to the result of the point 2 of the Theorem 3.3.1, if the assumptions of the Theorem 3.3.3 hold, then the integral of the function f is independent of the path. As a special case, the integral of f on any closed and rectifiable path is zero.
2. If the function f is continuous on the starlike domain D, and it has derivatives in all the points of D except at most a finite number of points, then the function f has a primitive in D.
3. It seems that the existence of the derivative it is a "too strong" condition for the existence of the primitive. We will see that this condition it is not only sufficient, but it is necessary for the existence of the primitive.

Example 3.3.1. Define the function

$$f(z) = \begin{cases} \frac{\sin(z-i)z}{(z-i)z}, & \text{if } z \in \mathbb{C} \setminus \{0, i\}, \\ 1, & \text{if } z \in \{0, i\}. \end{cases}$$

Since f has derivatives on $\mathbb{C} \setminus \{0, i\}$ and it is continuous on $\mathbb{C}$, using the previous theorem we deduce that there exists a primitive of the function f on the entire complex plane $\mathbb{C}$.

3.3.2 The Cauchy theorem

Using the Goursat theorem (Theorem 3.3.2), we will prove the Cauchy theorem for any arbitrary closed and rectifiable path.

Theorem 3.3.4 (Cauchy theorem). *Let $G \subset \mathbb{C}$ be an open set, and let $f \in H(G)$. Then the next two equivalent affirmations hold:*

(i) *If γ_0 and γ_1 are two rectifiable paths in G, with $\gamma_0 \underset{G}{\sim} \gamma_1$, then*

$$\int_{\gamma_0} f = \int_{\gamma_1} f.$$

(ii) *For any arbitrary closed and rectifiable path in G, that is homotopic to a point in G, that is $\gamma \underset{G}{\sim} 0$, we have*

$$\int_{\gamma} f = 0.$$

Proof. Using a similar proof with those of the Remark 3.3.1, the equivalence between (i) and (ii) follows immediately.

Then it is sufficient to prove the part (i) of the above theorem.

Since γ_0 and γ_1 are homotopic path in G, there exists a continuous function

$$\varphi : S \times T \to G, \quad S = T = [0,1],$$

such that

$$\varphi(0,t) = \gamma_0(t), \quad \varphi(1,t) = \gamma_1(t), \quad \forall t \in T,$$
$$\varphi(s,0) = z_0, \quad \varphi(s,1) = z_1, \quad \forall s \in S,$$

where z_0 and z_1 are the starting and the end points of the paths γ_0 and γ_1, respectively.

Since $S \times T$ is a compact set and φ is a continuous function, the image $K = \varphi(S \times T)$ is a compact subset of G. Thus, $\mathrm{d}(K, \partial G) > 0$. Let $\rho > 0$ be a number, such that $0 < \rho < \mathrm{d}(K, \partial G)$.

The function φ is continuous on the compact set $S \times T$, then φ will be a uniformly continuous function on $S \times T$. Thus, there exists a number $\delta = \delta_\rho > 0$, such that

$$\forall (s',t'), (s'',t'') \in S \times T, \quad \sqrt{(s'-s'')^2 + (t'-t'')^2} < \delta$$
$$\Rightarrow |\varphi(s',t') - \varphi(s'',t'')| < \rho.$$

Let us consider two divisions of T and S, denoted by $\Delta' = (t_0, t_1, \dots, t_n)$, and respectively by $\Delta'' = (s_0, s_1, \dots, s_n)$, such that $t_k = s_k$ for all $k \in \{0, 1, \dots, n\}$, and $\|\Delta'\| = \|\Delta''\| < \frac{\delta}{\sqrt{2}}$.

Let $R_{jk} = [s_j, s_{j+1}] \times [t_k, t_{k+1}]$, $j, k = \overline{0, n-1}$, and

$$z_{jk} = \varphi(s_j, t_k), \quad j, k \in \{0, 1, \dots, n\}.$$

Then

$$z_{0k} = \gamma_0(t_k), \quad z_{nk} = \gamma_1(t_k) \quad \text{(because } s_0 = 0,\ s_n = 1\text{)} \quad \text{and}$$
$$z_{j0} = z_0, \quad z_{jn} = z_1 \quad \text{(because } t_0 = 0,\ t_n = 1\text{)}.$$

Consider $U_{jk} = U(z_{jk}; \rho)$. Then $U_{jk} \subset G$ and $\varphi(R_{jk}) \subset U_{jk}$, because $z_{jk} \in K$ and $d(K, \partial G) > \rho > 0$, and respectively,

$$\forall (s, t) \in R_{jk} \quad \text{we have } |s - s_j| < \frac{\delta}{\sqrt{2}},\ |t - t_k| < \frac{\delta}{\sqrt{2}},$$

hence $\sqrt{(s - s_j)^2 + (t - t_k)^2} < \delta$, and thus $|\varphi(s, t) - \varphi(s_j, t_k)| < \rho$, i. e., $|z - z_{jk}| < \rho$, $\forall z \in \varphi(R_{jk})$.

Let $\Pi_j = (z_0 = z_{j0}, z_{j1}, \dots, z_{jk}, \dots, z_{jn} = z_1)$ be the broken line that connects the previously defined points.

We will prove that $\int_{\Pi_0} f = \int_{\gamma_0} f$. Let $\Delta = (t_0, t_1, \dots, t_n)$ be a division of $[0, 1]$, and let $\gamma_0 = \sigma_1 \cup \sigma_2 \cup \dots \cup \sigma_n$ be the corresponding decomposition of the path γ_0 given by the division Δ. The points z_{0k} and z_{0k+1} together with their connecting path σ_k belong to the disc U_{0k}, because $\{\sigma_k\} \subset \varphi(R_{0k}) \subset U_{0k}$. Since f is a holomorphic function in the disc U_{0k}, which is a starlike domain, according to the Theorem 3.3.3 the function f has a primitive in the disc U_{0k}, and from the point 2 of Theorem 3.3.1 the integral of f is independent of the path, hence

$$\int_{\sigma_k} f = \int_{\lambda_k} f,$$

where λ_k is the linear path that connects the points z_{0k} and z_{0k+1}, so it is a component of the broken line Π_0. Thus,

$$\int_{\gamma_0} f = \sum_{k=1}^{n} \int_{\sigma_k} f = \sum_{k=1}^{n} \int_{\lambda_k} f = \int_{\Pi_0} f. \tag{3.5}$$

From similar reasons, we have

$$\int_{\gamma_1} f = \int_{\Pi_n} f. \tag{3.6}$$

Now we will prove that

$$\int_{\Pi_j} f = \int_{\Pi_{j+1}} f, \quad j \in \{0, 1, \dots, n-1\}. \tag{3.7}$$

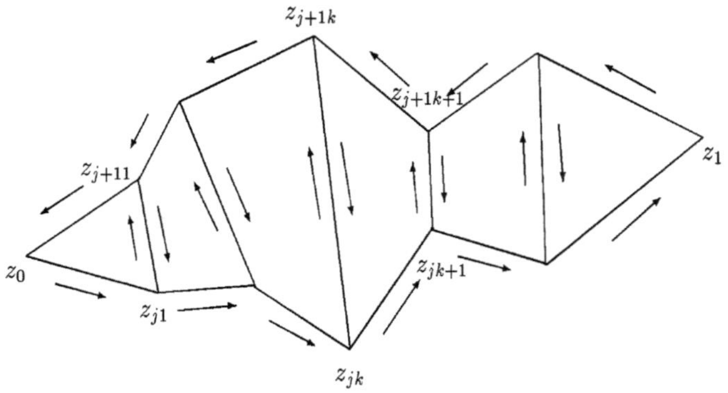

Figure 3.2: Proof of Theorem 3.3.4.

Let us consider the notation used in Figure 3.2, and let us denote by P_{jk} the closed broken line $P_{jk} = (z_{jk}, z_{jk+1}, z_{j+1k+1}, z_{j+1k}, z_{jk})$. Then, from the above assumptions, we have $\{P_{jk}\} \subset U_{jk}$.

Since the function f is holomorphic in the disc U_{jk}, using Theorem 3.3.3 and then the point 2 of Theorem 3.3.1, we deduce that

$$\int_{P_{jk}} f = 0.$$

Taking the orientations of the paths given by our figure, from the above conclusion we obtain that

$$0 = \sum_{k=0}^{n-1} \int_{P_{jk}} f = \int_{\Pi_j} f - \int_{\Pi_{j+1}} f,$$

and thus the relation (3.7) is proved.

Combining the equalities (3.5), (3.6) and (3.7), we conclude that

$$\int_{\gamma_0} f = \int_{\gamma_1} f.$$ □

Corollary 3.3.1. *Let $D \subset \mathbb{C}$ be a simply connected domain, that is any arbitrary closed path γ in D is homotopic to a point in D (i. e., $\gamma \underset{D}{\sim} 0$). Let $f \in H(D)$. Then:*

1. $\int_\gamma f = 0$, *for any arbitrary closed and rectifiable path γ, with $\{\gamma\} \subset D$;*
2. *the function f has a primitive in the domain D.*

Proof. The first point follows from the Cauchy theorem, while the second one follows from the point 1 of Theorem 3.3.1. □

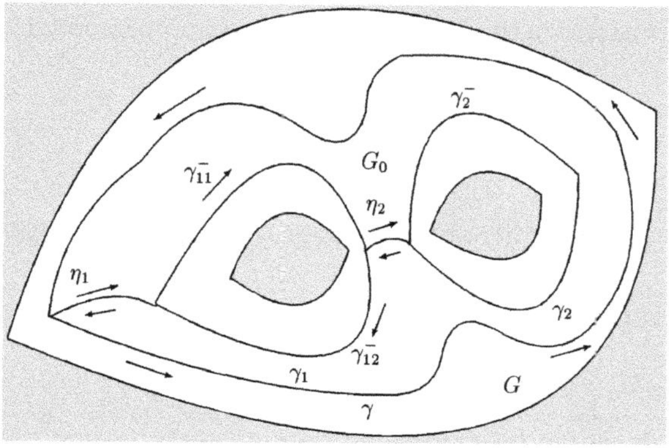

Figure 3.3: Figure for the proof of Corollary 3.3.2.

Corollary 3.3.2. *Let $G \subset \mathbb{C}$ be an open set, and let $f \in H(G)$. Let γ be a closed and rectifiable path in G which is the boundary of a bounded domain that contains the closed and rectifiable paths $\gamma_1, \gamma_2, \ldots, \gamma_n$ of G. Suppose that the closures of all the domains bounded by these paths are disjoint sets, and suppose that the set G_0 which lies between γ and $\gamma_1, \gamma_2, \ldots, \gamma_n$ is a subset of G. If the orientations of these paths are the same, then*

$$\int_{\gamma} f = \int_{\gamma_1} f + \int_{\gamma_2} f + \cdots + \int_{\gamma_n} f.$$

Proof. It is sufficient to prove the above result for $n = 2$, since the method is the same for the general case.

Let us consider Figure 3.3.

Let η_1 be a rectifiable path that connects a point of γ with a point of γ_1, such that $\{\eta_1\} \cup \{\gamma_2\} = \emptyset$. Further, let η_2 be a rectifiable path that connects a point of γ_1 with a point of γ_2, such that $\{\eta_2\}$ does not intersect any other path. From the assumptions, it follows that

$$\vartheta = \gamma \cup \eta_1 \cup \gamma_{\bar{11}} \cup \eta_2 \cup \gamma_{\bar{2}} \cup \eta_{\bar{2}} \cup \gamma_{\bar{12}} \cup \eta_{\bar{1}},$$

where γ_{11} and γ_{12} is a decomposition of the path γ_1, such that the above relation has sense. (We supposed that all the paths have the same orientations, such that the bounded domain in which the boundary is the considered path and is situated on the left-hand side. With this remark, we may easily see that the path ϑ is well-defined.) Hence, the rectifiable path ϑ is homotopic to a point in the set G, and using the point (ii) of Theorem 3.3.4 we deduce

$$\int_{\vartheta} f = \int_{\gamma} f - \int_{\gamma_1} f - \int_{\gamma_2} f = 0,$$

which proves the above result. □

Corollary 3.3.3 (The fundamental theorem of the algebra). *Any complex polynomial*

$$P(z) = a_0 + a_1 z + \cdots + a_n z^n, \quad n \in \mathbb{N}^*, \ a_n \neq 0, \text{ and } a_k \in \mathbb{C}, \ k \in \{0, 1, \ldots, n\},$$

has at least one root in $\mathbb{C}$.

Proof. 1. It is sufficient to prove the above result only for those polynomials that satisfy $P(z) \in \mathbb{R}$, whenever $z \in \mathbb{R}$.

In fact, if P is an arbitrary polynomial, and $\overline{P}$ represents the polynomial

$$\overline{P}(z) = \overline{a}_0 + \overline{a}_1 z + \cdots + \overline{a}_n z^n,$$

then $Q = P \cdot \overline{P}$ will be a polynomial that satisfies the previous condition, i. e., $Q(z) \in \mathbb{R}$, whenever $z \in \mathbb{R}$.

If $Q(z_0) = 0 \Rightarrow P(z_0)\overline{P}(z_0) = 0 \Rightarrow z_0$ is a root for P, or for $\overline{P}$. If $\overline{P}(z_0) = 0$, then $P(\overline{z}_0) = \overline{\overline{P}(z_0)} = 0$, and it follows that P has at least one root.

2. Suppose that $P(z) \in \mathbb{R}$, whenever $z \in \mathbb{R}$, and that $P(z) \neq 0$, $\forall z \in \mathbb{C}$.

It follows that $P(x) \neq 0$, $P(x) \in \mathbb{R}$, $\forall x \in \mathbb{R}$, and since P is continuous, then the sign of the function P will be the same on the whole real axis Ox, and this sign coincides with the sign of $\frac{1}{P(x)}$, $\forall x \in \mathbb{R}$. Hence, there exists the integral

$$I = \int_{-\pi}^{\pi} \frac{d\theta}{P(2\cos\theta)} \neq 0. \tag{3.8}$$

Letting $z = e^{i\theta}$, $\theta \in [-\pi, \pi]$, then the point z runs once all over the unit circle. Let γ denote the unit circle, i. e., $\gamma(\theta) = e^{i\theta}$, $\theta \in [-\pi, \pi]$, and for simplicity we will denote the unit circle with direct (positive) orientation by $|z| = 1$.

Since $z\overline{z} = 1$, $z = \cos\theta + i\sin\theta = e^{i\theta}$, and $z + \overline{z} = 2\cos\theta$, we have

$$2\cos\theta = z + \frac{1}{z}, \quad dz = (e^{i\theta})' d\theta = ie^{i\theta} d\theta \text{ and } dz = izd\theta, \text{ hence } d\theta = -i\frac{dz}{z}.$$

When θ runs all over the segment $[-\pi, \pi]$, then the point z runs all over the unit circle $|z| = 1$, with positive orientation. Hence the real integral I can be written like the complex integral

$$I = -i \int_{|z|=1} \frac{dz}{zP(z + \frac{1}{z})}.$$

But it is easy to see that

$$P\left(z + \frac{1}{z}\right) = a_0 + a_1\left(z + \frac{1}{z}\right) + \cdots + a_n\left(z + \frac{1}{z}\right)^n = \frac{Q(z)}{z^n},$$

where Q is a polynomial, with $Q(0) = a_n \neq 0$.

If $Q(z_0) = 0$ for a $z_0 \neq 0$, then $P(z_0 + \frac{1}{z_0}) = 0$, and this last equality contradicts the assumption that P has not any complex root. So, we conclude that $Q(z) \neq 0$, $\forall z \in \mathbb{C}$, which implies that the function $\frac{1}{Q}$ is holomorphic in $\mathbb{C}$ (because Q is holomorphic in $\mathbb{C}$ and $Q(z) \neq 0$, $\forall z \in \mathbb{C}$). Since the function $\frac{z^{n-1}}{Q(z)}$ is holomorphic in $\mathbb{C}$, according to the Cauchy theorem we obtain that

$$I = -i \int_{|z|=1} \frac{z^{n-1}\mathrm{d}z}{Q(z)} = 0,$$

which contradicts the relation (3.8). □

3.4 The Cauchy formula for the disc

Let γ be a rectifiable path in $\mathbb{C}$, let denote $K = \{\gamma\}$ and let $G \subset \mathbb{C}$, $G \neq \emptyset$, be an open set. Supposing that the function $g : G \times K \to \mathbb{C}$ is continuous, then for any arbitrary point $z \in G$ there exists the integral $\int_\gamma g(z, \zeta)\mathrm{d}\zeta$.

Theorem 3.4.1.

1. *If $g : G \times K \to \mathbb{C}$ is a continuous function, then the function h defined by*

$$h(z) = \int_\gamma g(z, \zeta)\mathrm{d}\zeta$$

is continuous on G.

2. *If there exists the derivative $g'_z(z, \zeta)$, and it is continuous on $G \times K$, then the function h defined to the point* 1. *is holomorphic on G, i. e., $h \in H(G)$, and*

$$h'(z) = \int_\gamma g'_z(z, \zeta)\mathrm{d}\zeta.$$

Proof. 1. Let fix the point $z_0 \in G$. Since G is an open set, there exists a number $r > 0$ such that $U^-(z_0; r) \subset G$, and consider an arbitrary number $\varepsilon > 0$. Since g is a uniformly continuous function on the compact $U^-(z_0; r) \times K$, there exists

$$\eta > 0 \quad \text{such that } |z - z_0| < \eta, \text{ implies } |g(z, \zeta) - g(z_0, \zeta)| < \frac{\varepsilon}{V(\gamma)},$$

where $V(\gamma)$ the length of the path γ.

Thus

$$|h(z) - h(z_0)| \leq \int_\gamma |g(z, \zeta) - g(z_0, \zeta)||\mathrm{d}\zeta| < \frac{\varepsilon}{V(\gamma)} \int_\gamma |\mathrm{d}\zeta| = \varepsilon,$$

hence the function h is continuous at the point z_0.

2. If $g'_z(z,\zeta)$ continuous, and $z_0 \in G$ is a given point, let define the function g_1 by

$$g_1(z,\zeta) = \begin{cases} \frac{g(z,\zeta)-g(z_0,\zeta)}{z-z_0}, & (z,\zeta) \in (G \setminus \{z_0\}) \times K, \\ g'_z(z_0,\zeta), & (z,\zeta) \in \{z_0\} \times K. \end{cases}$$

Then $g_1 : G \times K \to \mathbb{C}$ is a continuous function, and according to the point 1. the function

$$h_1(z) = \int_\gamma g_1(z,\zeta)\mathrm{d}\zeta$$

will be continuous on G, hence $h_1(z_0) = \lim_{z\to z_0} h_1(z)$. On the other hand,

$$h_1(z_0) = \int_\gamma g'_z(z_0,\zeta)\mathrm{d}\zeta \quad \text{and} \quad h_1(z) = \int_\gamma \frac{g(z,\zeta) - g(z_0,\zeta)}{z - z_0}\mathrm{d}\zeta = \frac{h(z) - h(z_0)}{z - z_0}.$$

It follows that there exists the limit

$$\lim_{z\to z_0} \frac{h(z) - h(z_0)}{z - z_0} = h_1(z_0) = \int_\gamma g'_z(z_0,\zeta)\mathrm{d}\zeta,$$

which proves the existence of $h'(z_0)$. □

Corollary 3.4.1. *Let $\varphi : K \to \mathbb{C}$ be a continuous function, let γ be a rectifiable path in $\mathbb{C}$ and let denote $K = \{\gamma\}$ and $G = \mathbb{C} \setminus K$. If we define the function $g : G \times K \to \mathbb{C}$ by $g(z,\zeta) = \frac{\varphi(\zeta)}{\zeta - z}$, then the function h defined by*

$$h(z) = \int_\gamma \frac{\varphi(\zeta)}{\zeta - z}\mathrm{d}\zeta$$

is holomorphic on G, there exists the derivatives $h^{(n)}$, $\forall n \in \mathbb{N}^*$, and moreover,

$$h^{(n)}(z) = n! \int_\gamma \frac{\varphi(\zeta)}{(\zeta - z)^{n+1}}\mathrm{d}\zeta.$$

Theorem 3.4.2 (Cauchy formulas for the disc). *Let $f : U^-(z_0;r) \to \mathbb{C}$ be a continuous function on the closed disc $U^-(z_0;r)$, such that f is holomorphic on the open disc $U(z_0;r)$. Then, for any arbitrary $k \in \mathbb{N}$ there exists the derivatives $f^{(k)}(z)$, $\forall z \in U(z_0;r)$ and*

$$f^{(k)}(z) = \frac{k!}{2\pi i} \int_{\partial U(z_0;r)} \frac{f(\zeta)}{(\zeta - z)^{k+1}}\mathrm{d}\zeta.$$

Proof. 1. First, we will prove the above results for the special case $k = 0$.

Let $z \in U(z_0;r)$ be a given point, and let the number $r_n > 0$ such that $|z - z_0| < r_n < r$. Let denote by γ_n the boundary of the disc $U(z_0;r_n)$, considering a direct (positive) orientation.

It is obvious that

$$I = \frac{1}{2\pi i}\int_{\gamma_n}\frac{f(\zeta)}{\zeta - z}\mathrm{d}\zeta = \frac{1}{2\pi i}\int_{\gamma_n}\frac{f(\zeta)-f(z)}{\zeta - z}\mathrm{d}\zeta + \frac{f(z)}{2\pi i}\int_{\gamma_n}\frac{\mathrm{d}\zeta}{\zeta - z}. \tag{3.9}$$

Now we will compute the integral

$$\int_{\gamma_n}\frac{\mathrm{d}\zeta}{\zeta - z}.$$

Let $\rho > 0$ be an enough small number, such that $U^-(z;\rho) \subset U(z_0;r_n)$, and let $G = U(z_0;r_n) \setminus \{z\}$. Then the function $\frac{1}{\zeta - z}$ is holomorphic with respect to the variable $\zeta \in G$. Since the path $\gamma = \partial U(z;\rho)$ has the equation $\gamma(t) = \rho e^{i2\pi t} + z$, $t \in [0,1]$, it follows that

$$\int_{\gamma}\frac{\mathrm{d}\zeta}{\zeta - z} = \int_0^1 \frac{2\pi i\rho e^{i2\pi t}}{\rho e^{i2\pi t}}\mathrm{d}t = 2\pi i\int_0^1 \mathrm{d}t = 2\pi i.$$

Because $\gamma \underset{G}{\sim} \gamma_n$, from the Cauchy theorem we obtain that

$$\int_{\gamma_n}\frac{\mathrm{d}\zeta}{\zeta - z} = 2\pi i,$$

hence the relation (3.9) reduces to

$$I = \frac{1}{2\pi i}\int_{\gamma_n}\frac{f(\zeta)-f(z)}{\zeta - z}\mathrm{d}\zeta + f(z). \tag{3.10}$$

Since f is a holomorphic function on the disc $U(z_0;r)$, it follows that the function g defined by

$$g(\zeta) = \begin{cases} \frac{f(\zeta)-f(z)}{\zeta - z}, & \zeta \in U(z_0;r) \setminus \{z\}, \\ f'(z), & \zeta = z \end{cases}$$

is holomorphic on $U(z_0;r) \setminus \{z\}$, and it is continuous on $U(z_0;r)$. Using Theorem 3.3.3 for the function g, we deduce that there exists a primitive of g on the disc $U(z_0;r)$, and according to the point 2. of the Theorem 3.3.1 we get $\int_{\gamma_n} g(\zeta)\mathrm{d}\zeta = 0$. Thus, the integral from right-hand side of the equality (3.10) is zero, hence

$$f(z) = \frac{1}{2\pi i}\int_{\gamma_n}\frac{f(\zeta)}{\zeta - z}\mathrm{d}\zeta.$$

If we suppose that $r_n \to r$, then γ_n uniformly converges to $\partial U(z_0;r)$, and

$$f(z) = \frac{1}{2\pi i}\int_{\gamma_n}\frac{f(\zeta)}{\zeta - z}\mathrm{d}\zeta \to \frac{1}{2\pi i}\int_{\partial U(z_0;r)}\frac{f(\zeta)}{\zeta - z}\mathrm{d}\zeta,$$

which proves the formula of the theorem for $k = 0$.

2. Using the Corollary 3.4.1 for the integral

$$f(z) = \frac{1}{2\pi i} \int_{\partial U(z_0;r)} \frac{f(\zeta)}{\zeta - z} \mathrm{d}\zeta,$$

we deduce that $\exists f^{(k)}(z)$, $\forall k \in \mathbb{N}$, and

$$f^{(k)}(z) = \frac{k!}{2\pi i} \int_{\partial U(z_0;r)} \frac{f(\zeta)}{(\zeta - z)^{k+1}} \mathrm{d}\zeta.$$ □

Theorem 3.4.3 (Morera theorem). *If a complex function has a primitive on the open set $G \subset \mathbb{C}$, then the function is holomorphic on this set.*

Proof. Let g be a primitive for the function $f : G \to \mathbb{C}$. Then $g \in H(G)$. Let $z_0 \in G$ be a given point, and let $r > 0$ be a number such that $U^{-}(z_0; r) \subset G$. Applying the Cauchy formula for the function g, we obtain that there exists the derivative $g''(z_0)$. Since $g'(z) = f(z)$, $z \in G$, it follows that there exists $f'(z_0) = g''(z_0)$. Thus, the function f is holomorphic on G. □

Corollary 3.4.2. *The derivative of an holomorphic function is also holomorphic.*

Consequently, any holomorphic function has derivatives of any arbitrary order, and these derivatives are also holomorphic.

Definition 3.4.1. Let $G \subset \mathbb{C}$. A family of lines is said to be **locally finite in the set** G, for every point of G there exists a neighborhood of it that intersects only a finite number of these lines.

For example, the family of the lines $x = \frac{1}{n}$, $n \in \mathbb{N}^*$, is locally finite in the disc $U(1; 1)$, but it is not locally finite in the disc $U(0; 1)$.

Theorem 3.4.4. *Let $G \subset \mathbb{C}$ be an open set, and let d be a family of lines that are locally finite in the set G. If the function f is holomorphic on $G \setminus d$, and it is continuous on G, then $f \in H(G)$.*

Proof. If we let $z_0 \in G$, then there exists a number $r > 0$ such that $U(z_0; r) \subset G$, and such that the disc $U(z_0; r)$ intersects only a finite number of these lines that contain the point z_0. According to the Theorem 3.3.3, the function f has a primitive in the disc $U(z_0; r)$, and then from Theorem 3.4.3 we deduce that $f \in H(U(z_0; r))$. Since z_0 is an arbitrary point of G, we conclude that $f \in H(G)$. □

3.5 The analytical branches of multivalued functions

Theorem 3.5.1 (Existence theorem of the analytical branches of multivalued functions). *Let $D \subset \mathbb{C}$ a simply connected domain in $\mathbb{C}$, and let $g \in H(D)$ a such a function, such that $g(z) \neq 0$, $\forall z \in D$.*

Let $z_1 \in D$ and let $w_1, w_2, \alpha \in \mathbb{C}$ be complex numbers, such that $w_1 \in \operatorname{Log} g(z_1)$, $w_2 \in (g(z_1))^\alpha$.

Then there exists only the functions $f_1, f_2 \in H(D)$, such that:

1. *f_1 is the analytical branch of the multivalued function $\operatorname{Log} g$, with $f_1(z_1) = w_1$,*
2. *f_2 is the analytical branch of the multivalued function g^α, with $f_2(z_1) = w_2$.*

Proof. 1. Let us define $h(z) = \frac{g'(z)}{g(z)}$. Then $h \in H(D)$, since $g(z) \neq 0$, $\forall z \in D$. Because D is a simply connected domain, every arbitrary closed path of D is homotopic with 0 in D, hence on every closed path of D the integral of h is zero. It follows that the function h has a primitive in D, and let denote it by l. Then

$$\left(\frac{e^{l(z)}}{g(z)}\right)' = \frac{(l'(z)g(z) - g'(z))e^{l(z)}}{g^2(z)} = \frac{(g'(z) - g'(z))e^{l(z)}}{g^2(z)} = 0,$$

hence the expression from the first bracket is constant, i. e.,

$$e^{l(z)} = cg(z), \quad c \in \mathbb{C}.$$

Since $e^{l(z)}$ cannot be equal to 0, it follows that $c \neq 0$.

Let choose the constant $k \in \mathbb{C}$, such that the function $f_1(z) = l(z) + k$ satisfies the condition $f_1(z_1) = w_1$, i. e., $k = w_1 - l(z_1)$. Then the function $f_1 \in H(D)$ is the analytical branch of $\operatorname{Log} g$ which we want to find, since

$$e^{f_1(z)} = e^{l(z)}e^k = e^{l(z)}e^{w_1}e^{-l(z_1)} = \frac{cg(z)}{cg(z_1)}e^{w_1} = g(z),$$

because $w_1 \in \operatorname{Log} g(z_1)$, hence $e^{w_1} = g(z_1)$.

Obviously, we have

$$f_1(z) \in \operatorname{Log} g(z), \forall z \in D \quad \text{and} \quad f_1'(z) = l'(z) = \frac{g'(z)}{g(z)},$$

since according to its definition, the function l is a primitive function for $h = \frac{g'}{g}$.

If $\widetilde{f}_1 \in H(D)$ is another branch of $\operatorname{Log} g(z)$ that satisfies the condition $\widetilde{f}_1(z_1) = w_1$, then

$$g(z) = e^{\widetilde{f}_1(z)} = e^{f_1(z)}, \quad \text{hence } e^{\widetilde{f}_1(z) - f_1(z)} = 1.$$

If we differentiate the both sides of this relation, we get

$$(\widetilde{f}_1'(z) - f_1'(z))e^{\widetilde{f}_1(z) - f_1(z)} \equiv 0,$$

so $\widetilde{f}_1'(z) \equiv f_1'(z)$. Since $\widetilde{f}_1(z_1) = f_1(z_1)$, we deduce that $\widetilde{f}_1 = f_1$.

2. Let be the number $w_2 \in \mathbb{C}$, such that $w_2 \in (g(z_1))^\alpha = e^{\alpha \operatorname{Log} g(z_1)}$ and let be $w_1 \in \operatorname{Log} g(z_1)$ a such a number that $w_2 = e^{\alpha w_1}$. Using the first point of the proof, there exists the function $f_1 \in H(D)$, such that $w_1 = f_1(z_1)$ and $f_1(z) \in \operatorname{Log} g(z)$, $\forall z \in D$.

Now define the function $f_2(z) = e^{\alpha f_1(z)}$, $z \in D$. Then

$$f_2(z) \in e^{\alpha \operatorname{Log} g(z)} = (g(z))^\alpha, \quad \forall z \in D, \quad \text{and} \quad f_2(z_1) = e^{\alpha f_1(z_1)} = e^{\alpha w_1} = w_2.$$

Hence there exists a such a function $f_2 \in H(D)$, with $f_2(z) \in (g(z))^\alpha$, $\forall z \in D$, and $f_2(z_1) = w_2$. From here, we have

$$f_2'(z) = \alpha f_1'(z) e^{\alpha f_1(z)} = \alpha \frac{g'(z)}{g(z)} e^{\alpha f_1(z)} \in \alpha g'(z) \frac{(g(z))^\alpha}{g(z)}$$
$$= \alpha g'(z) (g(z))^{\alpha - 1}, \quad \forall z \in D.$$

Let $\widetilde{f}_2 \in H(D)$ another analytic branch of $(g(z))^\alpha$, with $\widetilde{f}_2(z_1) = w_2$. Now define the function $h(z) = \frac{\widetilde{f}_2(z)}{f_2(z)}$. Then $h \in H(D)$ and

$$|h(z)| = \left| e^{\alpha 2\pi i k(z)} \right| = e^{-2\pi k(z) \operatorname{Im} \alpha}, \quad z \in D,$$

where $k(z) \in \mathbb{Z}$, $\forall z \in D$. It follows that $k(z) = -\frac{\ln |h(z)|}{2\pi \operatorname{Im} \alpha}$, $z \in D$. Since $h \in H(D)$ and the function $k : D \to \mathbb{Z}$ is continuous in D, from here we deduce that k is a constant function on D. Hence, according to the above formula, the function $|h|$ will be constant on the domain D, so it follows that h is also constant on D. Since $h(z_1) = 1$, $h(z) = 1$, $\forall z \in D$, we finally get that $\widetilde{f}_2 = f_2$. □

3.6 The index of a path (curve) with respect to a point

Let us consider the following two paths, $\gamma_1(t) = e^{2\pi i t}$ and $\gamma_2(t) = e^{4\pi i t}$. When t runs to the whole interval $[0, 1]$, then the image $\gamma_1(t)$ runs once, while the image $\gamma_2(t)$ runs twice over the unit circle in positive direction. We see that

$$\frac{1}{2\pi i} \int_{\gamma_1} \frac{d\zeta}{\zeta} = 1, \quad \frac{1}{2\pi i} \int_{\gamma_2} \frac{d\zeta}{\zeta} = 2.$$

It follows that the above integrals count however many times the images $\gamma_1(t)$ and $\gamma_2(t)$ run over the image curve; in this case, the unit circle, when the argument runs over the interval $[0, 1]$.

Definition 3.6.1. Let γ be a rectifiable path (curve) in $\mathbb{C}$ (not necessary closed), and let be the point $z_0 \in \mathbb{C} \setminus \{\gamma\}$. **The index of the curve γ with respect to the point z_0** is given by the number

$$\mathrm{n}(\gamma, z_0) = \frac{1}{2\pi i} \int_{\gamma} \frac{d\zeta}{\zeta - z_0}.$$

Definition 3.6.2. The function $f : G \to \mathbb{C}$ is said to be **locally constant** if f is constant in every connected component of G.

If the subset $G \subset \mathbb{C}$ is open, and $f \in H(G)$, then the function f is locally constant on G if and only if $f'(z) = 0$, $\forall z \in G$.

Theorem 3.6.1 (The index theorem). *Let γ be a rectifiable curve in $\mathbb{C}$.*
1. *If $\gamma_1 \underset{\mathbb{C}\setminus\{z_0\}}{\sim} \gamma_2$, then $\mathrm{n}(\gamma_1, z_0) = \mathrm{n}(\gamma_2, z_0)$.*
2. *If $\gamma = \gamma_1 \cup \gamma_2$, then $\mathrm{n}(\gamma, z_0) = \mathrm{n}(\gamma_1, z_0) + \mathrm{n}(\gamma_2, z_0)$.*
3. $\mathrm{n}(\gamma^-, z_0) = -\mathrm{n}(\gamma, z_0)$.

   Let γ be a rectifiable closed curve in $\mathbb{C}$. Then:
4. *If $z_0 \in \mathbb{C} \setminus \{\gamma\}$, then $\mathrm{n}(\gamma, z_0) \in \mathbb{Z}$ (integer number).*
5. *The function $j(z) = \mathrm{n}(\gamma, z)$ is locally constant on $\mathbb{C} \setminus \{\gamma\}$.*
6. *If $\gamma = \partial U(z_0; r)$ and $z \in U(z_0; r)$, then $\mathrm{n}(\gamma, z) = 1$.*
7. *The set $\mathbb{C} \setminus \{\gamma\}$ has exactly one not bounded connected component. If z_0 is an arbitrary point of this not bounded connected component, then $\mathrm{n}(\gamma, z_0) = 0$.*
8. *The set $\{z \in \mathbb{C} : \mathrm{n}(\gamma, z) \neq 0\}$ is bounded.*

Proof. Point 1 follows immediately from the Cauchy's theorem.

Points 2 and 3 are immediate consequences of the properties of the complex integrals.

4. Let $z \in \mathbb{C} \setminus \{\gamma\}$. The function $f(\zeta) = \frac{1}{\zeta - z}$ is holomorphic on the $\mathbb{C} \setminus \{z\}$ set, but it has no primitive, since it is not simply connected set. But it has primitive on every simply connected subset of $\mathbb{C} \setminus \{z\}$. Next, we will divide such an open neighborhood of $\{\gamma\}$ in the corresponding simply connected domains, such that every one of these components of the function f has a primitive.

Let $\rho = d(z; \{\gamma\})$. Since $\{\gamma\}$ is a closed set, and $z \notin \{\gamma\}$, $\rho > 0$, the function γ is uniformly continuous on the closed interval $[0, 1]$, hence

$$\exists \eta > 0 \quad \text{such that, if } |t' - t''| < \eta, \text{ then } |\gamma(t') - \gamma(t'')| < \rho.$$

Let us consider the division $\Delta = (t_0, t_1, \ldots, t_n)$, $0 = t_0 < t_1 < \cdots < t_n = 1$ of the interval $[0, 1]$, such that $\|\Delta\| < \eta$, and let $z_k = \gamma(t_k)$, $k \in \{0, 1, \ldots, n\}$. Then $\gamma([t_k, t_{k+1}]) \subset U(z_k; \rho)$, $k = \overline{0, n-1}$. Let $\gamma_1, \gamma_2, \ldots, \gamma_n$ be the decomposition of the curve γ corresponding to the division Δ. Then $\{\gamma_k\} = \gamma([t_{k-1}, t_k]) \subset U(z_{k-1}; \rho)$, and

$$\mathrm{n}(\gamma, z) = \frac{1}{2\pi i} \int_\gamma \frac{d\zeta}{\zeta - z} = \sum_{k=1}^{n} \frac{1}{2\pi i} \int_{\gamma_k} \frac{d\zeta}{\zeta - z}.$$

Now we will compute the integrals which appears in this sum. We know that $z_0 = \gamma(0)$, and let us consider $w_0 \in \mathrm{Log}\,(z_0 - z)$. Now we will use the existence theorem of the analytical branch for the multivalued function Log , for Log $g(\zeta)$, where $g(\zeta) = \zeta - z$, in the $U(z_0; \rho)$ disc (which is a simply connected domain). It follows that there exists

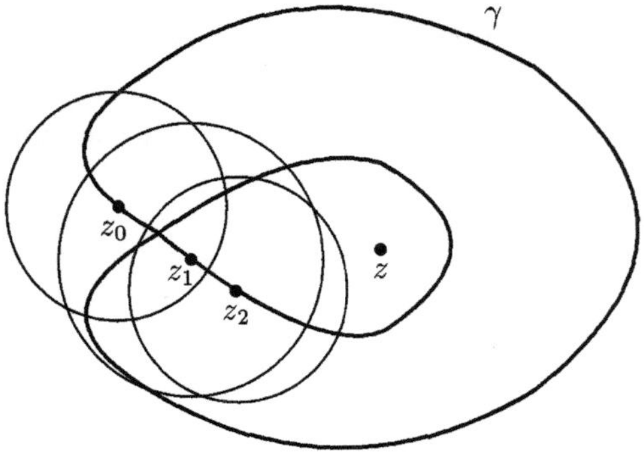

Figure 3.4: Proof of Theorem 3.6.1.

the function $f_0 \in H(U(z_0;\rho))$, such that

$$f_0'(\zeta) = \frac{1}{\zeta - z}, \quad f_0(\zeta) \in \text{Log}\,(\zeta - z), \quad \forall \zeta \in U(z_0;\rho) \text{ and } f_0(z_0) = w_0.$$

Hence $f_0(z_1) \in \text{Log}\,(z_1 - z)$, because $z_1 \in U(z_0;\rho)$ (Figure 3.4). In the disc $U(z_1;\rho)$, we will choose that analytic branch of Log $(\zeta - z)$, denoted by f_1, such that $f_1(z_1) = f_0(z_1)$, and we will repeat this algorithm for all the points $z_2, \dots, z_n$.

Because of the fact that in the domain $U(z_k;\rho)$, the function $\frac{1}{\zeta - z}$ has the primitive $f_k(\zeta) \in \text{Log}\,(\zeta - z)$, it follows that

$$\int_{\gamma_k} \frac{d\zeta}{\zeta - z} = f_k(z_k) - f_k(z_{k-1}).$$

So, we deduce that

$$\begin{aligned} n(\gamma, z) &= \frac{1}{2\pi i} \sum_{k=1}^{n} \int_{\gamma_k} \frac{d\zeta}{\zeta - z} = \frac{1}{2\pi i} \sum_{k=1}^{n} (f_k(z_k) - f_k(z_{k-1})) \\ &= \frac{1}{2\pi i}(f_n(z_n) - f_0(z_0)) = \frac{1}{2\pi i}(f_n(z_0) - f_0(z_0)), \end{aligned}$$

because $f_k(z_k) = f_{k-1}(z_k)$ and $z_n = z_0$ (γ is closed path). Since $f_n(z_0) \in \text{Log}\,(z_0 - z)$ and $f_0(z_0) \in \text{Log}\,(z_0 - z)$,

$$f_n(z_0) - f_0(z_0) = 2k\pi i, \quad \text{for a convenient } k \in \mathbb{Z}.$$

(The number k will be the same, for any arbitrary choose of the starting value $w_0 \in$ Log $(z_0 - z)$, because in its definition the integral is well-defined.)

It follows that

$$n(\gamma, z) = \frac{1}{2\pi i} \cdot 2k\pi i = k \in \mathbb{Z}.$$

5. Let the function $j(z) = \mathrm{n}(\gamma, z) = \frac{1}{2\pi i}\int_\gamma \frac{d\zeta}{\zeta - z}$ defined on $\mathbb{C}\setminus\{\gamma\}$. It is well known that j is differentiable on the whole $\mathbb{C}\setminus\{\gamma\}$ set. On the other hand, from the above point we know that $j(z)$ is an integer number, so j needs to be constant in an enough small neighborhood of z, because it is continuous on z. So, $j'(z) = 0$, and then the function $j(z) = \mathrm{n}(\gamma, z)$ is locally constant on the open set $\mathbb{C}\setminus\{\gamma\}$.

Another proof of this point is the next ____. Let be the function $j(z) = \mathrm{n}(\gamma, z) = \frac{1}{2\pi i}\int_\gamma \frac{d\zeta}{\zeta - z}$ defined on $\mathbb{C}\setminus\{\gamma\}$. It is well known that j is differentiable on the set $\mathbb{C}\setminus\{\gamma\}$, and $j'(z) = \frac{1}{2\pi i}\int_\gamma \frac{d\zeta}{(\zeta - z)^2}$. Since the function $g(\zeta) = \frac{1}{(\zeta - z)^2}$ has a primitive on $\mathbb{C}\setminus\{\gamma\}$, and γ is a closed rectifiable path, it follows that $j'(z) = 0$, $\forall z \in \mathbb{C}\setminus\{\gamma\}$, hence j is locally constant on the open set $\mathbb{C}\setminus\{\gamma\}$.

6. If $\gamma = \partial U(z_0; r)$, then we may easily see that

$$\int_\gamma \frac{d\zeta}{\zeta - z} = 2\pi i, \quad \forall z \in U(z_0; r),$$

hence $\mathrm{n}(\gamma, z) = 1$.

7. Since $\{\gamma\} = \gamma([0, 1])$ is compact, close and bounded, there exists $r > 0$ such that $\{\gamma\} \subset U(0; r)$, and thus $\mathbb{C}\setminus\overline{U}(0; r) \subset \mathbb{C}\setminus\{\gamma\}$. Because $\mathbb{C}\setminus\overline{U}(0; r)$ is a connected domain, it is contained in a connected component of $\mathbb{C}\setminus\{\gamma\}$.

Evidently, this is the unique component of $\mathbb{C}\setminus\{\gamma\}$ that contains the ∞. Using the previous point of the theorem, the function $\mathrm{n}(\gamma, z)$ is constant on the set $\mathbb{C}\setminus\overline{U}(0; r)$.

Let $z_n \in \mathbb{C}\setminus\overline{U}(0; r)$, $z_n \to \infty$. Then

$$0 \le |\mathrm{n}(\gamma, z_n)| = \frac{1}{2\pi}\left|\int_\gamma \frac{d\zeta}{\zeta - z_n}\right| \le \frac{1}{2\pi}\cdot\frac{V(\gamma)}{|z_n| - r} \to 0.$$

Hence $\mathrm{n}(\gamma, z) = 0$, $\forall z \in \mathbb{C}\setminus\overline{U}(0; r)$, and the same is valid for all the points z that belongs to the component of $\mathbb{C}\setminus\{\gamma\}$ that contains the ∞.

8. This point is a direct consequence of the point 7. □

3.7 Cauchy formula for closed curves

Theorem 3.7.1 (Cauchy formulas for closed curves). *If $G \subset \mathbb{C}$, where G is an open set, and if $f \in H(G)$, then the function f has derivatives of arbitrary order in the whole set G.*

For every rectifiable closed γ curve with $\gamma \underset{G}{\sim} 0$, and $\forall z \in G\setminus\{\gamma\}$ the next formula holds:

$$\mathrm{n}(\gamma, z)\cdot f^{(k)}(z) = \frac{k!}{2\pi i}\int_\gamma \frac{f(\zeta)}{(\zeta - z)^{k+1}}d\zeta, \quad \forall k \in \mathbb{N}. \tag{3.11}$$

Proof. Since the differentiability is a local property, using the Cauchy theorem for the disc, the first part of the theorem is true, hence the function f has derivatives of arbitrary order in the whole G set.

Next, we will prove the formula (3.11) for the case $k = 0$. Let $z \in G \setminus \{\gamma\}$ be fixed. Then

$$\frac{1}{2\pi i}\int_{\gamma}\frac{f(\zeta)}{\zeta - z}d\zeta = \frac{1}{2\pi i}\int_{\gamma}\frac{f(\zeta) - f(z)}{\zeta - z}d\zeta + \frac{f(z)}{2\pi i}\int_{\gamma}\frac{d\zeta}{\zeta - z}.$$

The value of the last integral is $2\pi i\, n(\gamma, z)$. The value of the first integral of the right-hand side is zero, since the function g given by

$$g(\zeta) = \begin{cases} \frac{f(\zeta)-f(z)}{\zeta - z}, & \text{if } \zeta \in G \setminus \{z\}, \\ f'(z), & \text{if } \zeta = z \end{cases}$$

is analytic on $G \setminus \{z\}$, and continuous on G.

Hence g has a primitive in a neighborhood of z, and according to the theorem of Morera, it is differentiable at z.

So, we have that $g \in H(G)$, $\gamma \underset{G}{\sim} 0$, and according to Cauchy's integral theorem we obtain $\int_{\gamma} g = 0$. We obtained that the formula (3.11) is true for the case $k = 0$.

Since it is well known that we can differentiate under the integral sign, it follows that the relations (3.11) hold for all $k \in \mathbb{N}$. □

3.8 Some consequences of Cauchy formula

Definition 3.8.1. The real valued function u, defined on the open complex subset G is said to be **harmonic**, if it has second-order continuous partial derivatives on G, which satisfy

$$\frac{\partial^2 u(z)}{\partial x^2} + \frac{\partial^2 u(z)}{\partial y^2} = 0, \quad \forall z = x + iy \in G.$$

Theorem 3.8.1. *Let $G \subset \mathbb{C}$ an open set, and let $f = u + iv$ be holomorphic on G. Then the real and the imaginary parts of f have arbitrary order partial derivatives on G, and both are harmonic functions on G, i. e.,*

$$\frac{\partial^2 u}{\partial x^2} + \frac{\partial^2 u}{\partial y^2} = 0, \qquad \frac{\partial^2 v}{\partial x^2} + \frac{\partial^2 v}{\partial y^2} = 0.$$

Proof. If $f = u + iv$, then $f' = \frac{\partial u}{\partial x} + i\frac{\partial v}{\partial x} = \frac{\partial v}{\partial y} - i\frac{\partial u}{\partial y}$ and since f' is differentiable we have that $\frac{\partial u}{\partial x}, \frac{\partial v}{\partial x}, \frac{\partial u}{\partial y}, \frac{\partial v}{\partial y} \in C(G)$. Using the Cauchy–Riemann theorem, the functions u and v have second-order continuous partial derivatives.

Since $\frac{\partial u}{\partial x} = \frac{\partial v}{\partial y}$ and $\frac{\partial u}{\partial y} = -\frac{\partial v}{\partial x}$, it follows that $\frac{\partial^2 u}{\partial x^2} = \frac{\partial^2 v}{\partial y \partial x}$ and $\frac{\partial^2 u}{\partial y^2} = -\frac{\partial^2 v}{\partial x \partial y}$, and according to the Schwarz theorem we obtain $\frac{\partial^2 u}{\partial x^2} + \frac{\partial^2 u}{\partial y^2} = 0$.

Similarly, it is possible to prove that $\frac{\partial^2 v}{\partial x^2} + \frac{\partial^2 v}{\partial y^2} = 0$. □

Remark 3.8.1. Since the function u and v are connected by the Cauchy–Riemann relations, they are called **harmonically conjugate** functions.

Theorem 3.8.2.
1. *If* $u : D \to \mathbb{R}$ *is a harmonic function on the simply connected domain* $D \subset \mathbb{C}$, *then there exists at least one function* $v : D \to \mathbb{R}$, *such that* $f = u + iv \in H(D)$.
2. *If* $v : D \to \mathbb{R}$ *is a harmonic function on the simply connected domain* $D \subset \mathbb{C}$, *then there exists at least one function* $u : D \to \mathbb{R}$, *such that* $f = u + iv \in H(D)$.

Proof. 1. Let $\widetilde{f} : D \to \mathbb{C}$, $\widetilde{f} = \widetilde{u} + i\widetilde{v}$, where $\widetilde{u} = \frac{\partial u}{\partial x}$, $\widetilde{v} = -\frac{\partial u}{\partial y}$. Then the functions $\widetilde{u}$ and $\widetilde{v}$ have first-order partial derivatives on D, and

$$\frac{\partial \widetilde{u}}{\partial x} = \frac{\partial^2 u}{\partial x^2} = -\frac{\partial^2 u}{\partial y^2} = \frac{\partial \widetilde{v}}{\partial y}$$

$$\frac{\partial \widetilde{u}}{\partial y} = \frac{\partial^2 u}{\partial x \partial y} = \frac{\partial^2 u}{\partial y \partial x} = -\frac{\partial \widetilde{v}}{\partial x}.$$

According to the Cauchy–Riemann theorem, we have $\widetilde{f} \in H(D)$, hence the function $\widetilde{f}$ has primitives on the simply connected domain $D \subset \mathbb{C}$.

If $\overline{f} = \overline{u} + i\overline{v}$ is a primitive for $\widetilde{f}$, hence $\frac{\partial \overline{u}}{\partial x} = \frac{\partial \overline{v}}{\partial y}$ and $\frac{\partial \overline{u}}{\partial y} = -\frac{\partial \overline{v}}{\partial x}$. From the fact

$$\overline{f}\,' = \frac{\partial \overline{u}}{\partial x} + i\frac{\partial \overline{v}}{\partial x} = \widetilde{f} = \frac{\partial u}{\partial x} - i\frac{\partial u}{\partial y},$$

it follows that $\frac{\partial \overline{u}}{\partial x} = \frac{\partial u}{\partial x}$ and $\frac{\partial \overline{u}}{\partial y} = \frac{\partial u}{\partial y}$, so $\overline{u} = u + c$, $c \in \mathbb{R}$, and we obtain $f = \overline{f} - c = u + i\overline{v} \in H(D)$.

2. The proof is similar to the previous point. □

Remark 3.8.2.
1. According to the previous theorem, if $u : D \to \mathbb{R}$ is a harmonic function on the simply connected domain $D \subset \mathbb{C}$, then there exists at least one function $v : D \to \mathbb{R}$, such that $f = u + iv \in H(D)$.
   The function v can be found as follows. Since
   $$\mathrm{d}v = \frac{\partial v}{\partial x}\mathrm{d}x + \frac{\partial v}{\partial y}\mathrm{d}y = -\frac{\partial u}{\partial y}\mathrm{d}x + \frac{\partial u}{\partial x}\mathrm{d}y,$$
   which is a total differential on D, if we integrate it along the curve $[MN] \cup [NP] \subset D$, where $M(x_0, y_0)$, $N(x, y_0)$ and $P(x, y)$, then
   $$v(x,y) - v(x_0,y_0) = -\int_{x_0}^{x} \frac{\partial u(x,y_0)}{\partial y}\mathrm{d}x + \int_{y_0}^{y} \frac{\partial u(x,y)}{\partial x}\mathrm{d}y.$$
2. Similarly, we know that if $v : D \to \mathbb{R}$ is a harmonic function on the simply connected domain $D \subset \mathbb{C}$, then there exists at least one function $u : D \to \mathbb{R}$, such that $f = u + iv \in H(D)$.

The function u can be obtained similarly:

$$\mathrm{d}u = \frac{\partial u}{\partial x}\mathrm{d}x + \frac{\partial u}{\partial y}\mathrm{d}y = \frac{\partial v}{\partial y}\mathrm{d}x - \frac{\partial v}{\partial x}\mathrm{d}y,$$

is a total differential on D, if we integrate it along the curve $[MN] \cup [NP] \subset D$, where $M(x_0, y_0)$, $N(x, y_0)$ and $P(x, y)$, then

$$u(x,y) - u(x_0, y_0) = \int_{x_0}^{x} \frac{\partial v(x, y_0)}{\partial y}\mathrm{d}x - \int_{y_0}^{y} \frac{\partial v(x,y)}{\partial x}\mathrm{d}y.$$

3. If $f \in H(G)$, where $G \subset \mathbb{C}$ is an open set, and $z_0 = x_0 + iy_0 \in G$ such that $f'(z_0) \neq 0$, then

$$\begin{aligned}\frac{\partial(u,v)(x_0,y_0)}{\partial(x,y)} &= \begin{vmatrix} \frac{\partial u(x_0,y_0)}{\partial x} & \frac{\partial v(x_0,y_0)}{\partial x} \\ \frac{\partial u(x_0,y_0)}{\partial y} & \frac{\partial v(x_0,y_0)}{\partial y} \end{vmatrix} \\ &= \frac{\partial u(x_0,y_0)}{\partial x}\frac{\partial v(x_0,y_0)}{\partial y} - \frac{\partial u(x_0,y_0)}{\partial y}\frac{\partial v(x_0,y_0)}{\partial x} \\ &= \left(\frac{\partial u(x_0,y_0)}{\partial x}\right)^2 + \left(\frac{\partial v(x_0,y_0)}{\partial x}\right)^2 = |f'(z_0)|^2 \neq 0.\end{aligned}$$

Since the Jacobian determinant is not zero in $z_0 \in G$, and f has first-order continuous partial derivatives on G, according to the theorem concerning the inverse function, there exist such open sets $U(z_0; r) \subset G$ and $U(f(z_0); \rho) \subset \mathbb{C}$; for those, the function $f : U(z_0; r) \to U(f(z_0; \rho))$ has an inverse $f^{-1} : U(f(z_0; \rho)) \to U(z_0; r)$, which is continuously differentiable and

$$(f^{-1})'(w) = \frac{1}{f'(f^{-1}(w))}, \quad \forall w \in U(f(z_0; \rho)).$$

Theorem 3.8.3 (Cauchy inequalities). *Let M be an upper bound for $|f|$ on the closed disc $U^-(z_0; r) \subset G$, where f is a holomorphic function on open set $G \subset \mathbb{C}$. Then, for every $n \in \mathbb{N}$ we have*

$$|f^{(n)}(z_0)| \leq n!\frac{M}{r^n}.$$

Proof. Using Cauchy formula for the discs, we directly deduce that

$$\begin{aligned}|f^{(n)}(z_0)| &= \frac{n!}{2\pi}\left|\int_{\partial U(z_0;r)} \frac{f(\zeta)}{(\zeta - z_0)^{n+1}}\mathrm{d}\zeta\right| \leq \frac{n!}{2\pi}\int_{\partial U(z_0;r)} \frac{|f(\zeta)|}{|\zeta - z_0|^{n+1}}|\mathrm{d}\zeta| \\ &\leq \frac{n!}{2\pi}\frac{M}{r^{n+1}}\int_{\partial U(z_0;r)} |\mathrm{d}\zeta| = \frac{n!}{2\pi}\frac{M}{r^{n+1}}2\pi r = \frac{n!}{r^n}M.\end{aligned}$$

□

Theorem 3.8.4 (Liouville theorem). *Every bounded entire function is constant.*

Proof. Suppose that $f \in H(\mathbb{C})$ and $|f(z)| < M$, $\forall z \in \mathbb{C}$. Let us consider $z_0 \in \mathbb{C}$, and $r > 0$ be arbitrary. From Theorem 3.8.3, it follows that

$$|f'(z_0)| \leq \frac{M}{r} \to 0, \quad \text{if } r \to \infty.$$

Hence $f'(z_0) = 0$, $\forall z_0 \in \mathbb{C} \Rightarrow f$ is a constant function. □

Corollary 3.8.1 (Fundamental theorem of the algebra). *Every polynomial*

$$P(z) = a_0 + a_1 z + \cdots + a_n z^n, \quad a_k \in \mathbb{C},\ a_n \neq 0,\ n \geq 1$$

has at least a zero (root) in $\mathbb{C}$.

Proof. We are presenting the second proof of the **fundamental theorem of algebra** which is based on the Liouville theorem. Suppose that the conclusion is not true. Then $f(z) = \frac{1}{P(z)}$ is an entire function, and $\lim_{z\to\infty} f(z) = 0$, hence there exists an enough big number $r > 0$, such that $|f(z)| \leq 1$ if $|z| > r$. Let $M = \sup\{|f(z)| : z \in \overline{U}(0;r)\}$. Sine f is holomorphic, then it is continuous on the compact disc $\overline{U}(0;r)$, hence $|f|$ has a finite upper bound $M < +\infty$. Evidently,

$$|f(z)| \leq M + 1, \quad \forall z \in \mathbb{C},$$

hence f is a bounded entire function. According to the Liouville theorem, the function f is constant on $\mathbb{C}$, i. e., $f(z) = c$, $\forall z \in \mathbb{C}$. Hence,

$$P(z) = \frac{1}{c}, \quad \forall z \in \mathbb{C},$$

which contradicts the assumption $n \geq 1$ and $a_n \neq 0$. □

3.9 Schwarz and Poisson formulas

Let f a holomorphic function on the disc $U(z_0;r)$, such that f is continuous on $\overline{U}(z_0;r)$. According to the Cauchy formula, all the values of the function f (and its derivatives) on the disc $U(z_0;r)$ are given by the values of f on the boundary of $U(z_0;r)$ disc.

The corresponding result for the harmonic functions is given by the Poisson formula.

The real part of u of a holomorphic function f on the disc $U(z_0;r)$ enable us to determine the function f, except to a constant, if we know its values on the boundary of $U(z_0;r)$ disc. Schwarz' formula gives this last connection.

To simplify the computations, we assume in both of the cases that $z_0 = 0$.

Theorem 3.9.1 (Poisson formula). *Let u be a harmonic function on the open set $G \subset \mathbb{C}$, such that $\overline{U}(0;R) \subset G$. Then*

$$u(re^{i\theta}) = \frac{1}{2\pi}\int_0^{2\pi} u(Re^{it})\frac{R^2 - r^2}{R^2 - 2Rr\cos(t-\theta) + r^2}\mathrm{d}t, \quad \forall z = re^{i\theta} \in U(0;R).$$

Proof. Since the set $\overline{U}(0;R) \subset G$ is compact, and $\mathbb{C} \setminus G$ is closed, there exists a number $\varepsilon > 0$ such that $D = U(0;R+\varepsilon) \subset G$. Also, there exists a function $f \in H(D)$ such that $\operatorname{Re} f = u$, because D is a simply connected domain and u is a harmonic function on D. Denoting $\gamma = \partial U(0;R)$, using the Cauchy formula for the discs (for the special case $k = 0$), we have

$$f(z) = \frac{1}{2\pi i}\int_\gamma \frac{f(\zeta)}{\zeta - z}\mathrm{d}\zeta = \frac{1}{2\pi i}\int_0^1 \frac{f(\gamma(\tau))}{\gamma(\tau) - z}\gamma'(\tau)\mathrm{d}\tau.$$

Denote $\gamma(\tau) = Re^{2\pi i\tau}$. Using the changing of variables $2\pi\tau = t$, we get

$$f(z) = \frac{1}{2\pi}\int_0^{2\pi} \frac{f(Re^{it})Re^{it}}{Re^{it} - re^{i\theta}}\mathrm{d}t. \tag{3.12}$$

Denote by $z^* = \frac{R^2}{\overline{z}} = \frac{R^2}{r}e^{i\theta}$ the inverse of z with respect to the circle $\partial U(0;R)$, and let us consider the function $g(\zeta) = \frac{f(\zeta)}{\zeta - z^*}$. Since $|z^*| > R$ and $g \in H(U(0;R))$, from the Cauchy integral theorem we have

$$\frac{1}{2\pi i}\int_\gamma \frac{f(\zeta)}{\zeta - z^*}\mathrm{d}\zeta = 0,$$

i. e.,

$$0 = \frac{1}{2\pi}\int_0^{2\pi} \frac{f(Re^{it})re^{it}}{re^{it} - Re^{i\theta}}\mathrm{d}t. \tag{3.13}$$

Subtracting the equality (3.12) from the above relation, we deduce

$$\begin{aligned} f(re^{i\theta}) &= \frac{1}{2\pi}\int_0^{2\pi} f(Re^{it})e^{it}\left[\frac{R}{Re^{it} - re^{i\theta}} - \frac{r}{re^{it} - Re^{i\theta}}\right]\mathrm{d}t \\ &= \frac{1}{2\pi}\int_0^{2\pi} f(Re^{it})\frac{R^2 - r^2}{R^2 - 2Rr\cos(t-\theta) + r^2}\mathrm{d}t, \end{aligned}$$

and from this last formula immediately follows the required relation. □

Remark 3.9.1. For the special case $r = 0$, we obtain

$$u(0) = \frac{1}{2\pi}\int_0^{2\pi} u(Re^{it})\mathrm{d}t = \int_0^1 u(Re^{2\pi i\tau})\mathrm{d}\tau. \tag{3.14}$$

As a consequence, the value to the origin of a disc for any harmonic function, is the mean-value of those from the boundary of the disc.

Theorem 3.9.2 (Schwarz formula). *Let $f = u + iv \in H(G)$, where $G \subset \mathbb{C}$ is an open set, and $\overline{U}(0;R) \subset G$. Let $\gamma = \partial U(0;R)$. Then*

$$f(z) = \frac{1}{2\pi i}\int_\gamma u(\zeta)\frac{\zeta + z}{\zeta - z}\frac{\mathrm{d}\zeta}{\zeta} + iv(0), \quad \forall z \in U(0;R).$$

Proof. Adding the relations (3.12) and (3.13), we obtain

$$\begin{aligned} f(re^{i\theta}) &= \frac{1}{2\pi}\int_0^{2\pi} f(Re^{it})e^{it}\left[\frac{R}{Re^{it} - re^{i\theta}} + \frac{r}{re^{it} - Re^{i\theta}}\right]\mathrm{d}t \\ &= \frac{1}{2\pi}\int_0^{2\pi} f(Re^{it})\left[1 + i\frac{2Rr\sin(\theta - t)}{R^2 - 2Rr\cos(\theta - t) + r^2}\right]\mathrm{d}t, \end{aligned}$$

and hence

$$v(re^{i\theta}) = \frac{1}{2\pi}\int_0^{2\pi} v(Re^{it})\mathrm{d}t + \frac{1}{2\pi}\int_0^{2\pi} u(Re^{it})\frac{2Rr\sin(\theta - t)}{R^2 - 2Rr\cos(\theta - t) + r^2}\mathrm{d}t.$$

Since v is a harmonic function, the first integral of the right-hand side is equal to $v(0)$, and applying Poisson formula we get

$$\begin{aligned} f(re^{i\theta}) &= u(re^{i\theta}) + iv(re^{i\theta}) \\ &= iv(0) + \frac{1}{2\pi}\int_0^{2\pi} u(Re^{it})\frac{R^2 + 2iRr\sin(\theta - t) - r^2}{R^2 - 2Rr\cos(\theta - t) + r^2}\mathrm{d}t \\ &= iv(0) + \frac{1}{2\pi}\int_0^{2\pi} u(Re^{it})\frac{R^2 - r^2 + Rr[e^{i(\theta - t)} - e^{-i(\theta - t)}]}{R^2 - r^2 - Rr[e^{i(\theta - t)} + e^{-i(\theta - t)}]}\mathrm{d}t \\ &= iv(0) + \frac{1}{2\pi}\int_0^{2\pi} u(Re^{it})\frac{Re^{it} + re^{i\theta}}{Re^{it} - re^{i\theta}}\mathrm{d}t. \end{aligned}$$

Since $z = re^{i\theta}$ and $\tau = \frac{t}{2\pi}$, it follows that

$$f(z) = iv(0) + \frac{1}{2\pi i}\int_0^1 u(Re^{2\pi i\tau})\frac{Re^{2\pi i\tau} + z}{Re^{2\pi i\tau} - z}2\pi i\mathrm{d}\tau,$$

and from $\gamma(\tau) = Re^{2\pi i\tau}$,

$$f(z) = iv(0) + \frac{1}{2\pi i}\int_0^1 u(\gamma(\tau))\frac{\gamma(\tau)+z}{\gamma(\tau)-z}\frac{\gamma'(\tau)}{\gamma(\tau)}\mathrm{d}\tau,$$

we conclude

$$f(z) = iv(0) + \frac{1}{2\pi i}\int_\gamma u(\zeta)\frac{\zeta+z}{\zeta-z}\frac{\mathrm{d}\zeta}{\zeta}. \qquad \square$$

3.10 Exercises

3.10.1 The complex integral

Exercise 3.10.1. Calculate the integral

$$\int_\gamma \frac{1}{z}\mathrm{d}z$$

in the following cases:

1. $\{\gamma\} = \widehat{AB}$, where $\widehat{AB}$ is the arc of the circle with the center in O that lies the points $A(-i)$ and $B(i)$, directly oriented;
2. $\{\gamma\} = \widehat{AB}$, where $\widehat{AB}$ is the arc of the circle with the center in O that lies the points $A(-i)$ and $B(i)$, inversely oriented;
3. $\{\gamma\} = \widehat{CD}$, where $\widehat{CD}$ is the arc of the circle with the center in O that lies the points $C(1-i)$ and $D(1+i)$, directly oriented;
4. $\{\gamma\} = \overline{CD}$, where $\overline{CD}$ is linear path that connects the points $C(1-i)$ and $D(1+i)$.

Exercise 3.10.2. Calculate the integral

$$\int_\gamma \frac{z}{|z|}\mathrm{d}z,$$

where $\gamma = \gamma_1 \cup \gamma_2 \cup \gamma_3$, with $\{\gamma_1\} = [-3,-2]$, $\{\gamma_3\} = [2,3]$ are directly oriented segments, and $\{\gamma_2\}$ is the arc of the circle with the center in O that lies the points $A(-2)$ and $B(2)$, inversely oriented.

Exercise 3.10.3. Calculate the integral

$$\int_\gamma \log z\mathrm{d}z,$$

where the path γ has the equation $\gamma(t) = e^{i(2t-1)\frac{\pi}{2}}$, $t \in [0,1]$, and $\log z$ is the main branch of the multivalued function $\mathrm{Log}\, z$, i. e., $\log 1 = 0$.

Exercise 3.10.4. Calculate the integral

$$\int_{\gamma} \frac{1}{\sqrt{z}} dz,$$

where the path γ has the equation $\gamma(t) = e^{i\frac{\pi t}{2}}$, $t \in [0,1]$, and $\sqrt{z}$ is the main branch of the multivalued function $z^{\frac{1}{2}}$, i. e., $\sqrt{1} = 1$.

Exercise 3.10.5. Calculate the integral

$$\int_{\gamma} \frac{1}{z\sqrt{z}} dz$$

in the next cases:

1. $\{\gamma\} = \partial U(0; r^2)$ is a directly oriented path, and $\sqrt{z}$ is the main branch of the multivalued function $z^{\frac{1}{2}}$, i. e., $\sqrt{1} = 1$;
2. $\gamma(t) = r^2 e^{4\pi i t}$, $t \in [0,1]$, and $\sqrt{z}$ is the branch of the multivalued function $z^{\frac{1}{2}}$ with $\sqrt{1} = -1$.

Exercise 3.10.6. Calculate the integral

$$\int_{\gamma} |z| dz$$

in the next cases:

1. $\{\gamma\} = [-1,1]$ is a directly oriented segment;
2. the path γ has the equation $\gamma(t) = e^{\pi i t}$, $t \in [0,1]$;
3. the path γ has the equation $\gamma(t) = e^{\pi i (t-1)}$, $t \in [0,1]$.

Exercise 3.10.7. Calculate the integral

$$\int_{\gamma} \bar{z} dz$$

in the next cases:

1. the path γ has the equation $\gamma(t) = -i + 2it$, $t \in [0,1]$;
2. the path γ has the equation $\gamma(t) = e^{i\frac{\pi(2t-1)}{2}}$, $t \in [0,1]$;
3. $\{\gamma\} = \partial U(0;1)$ is the directly oriented circle.

Exercise 3.10.8. Calculate the integral

$$\int_{\gamma} \operatorname{Re} z dz$$

in the next cases:

1. $\gamma = \gamma_1 \cup \gamma_2$, where γ_1 and γ_2 are given by $\gamma_1(t) = -1 + (i+1)t$, $t \in [0,1]$, and $\gamma_2(t) = i + (1-i)t$, $t \in [0,1]$;
2. $\gamma = \gamma_3$, where $\gamma_3(t) = e^{i\frac{\pi(t-1)}{2}}$, $t \in [0,1]$;
3. $\gamma = \gamma_3 \cup (\gamma_1 \cup \gamma_2)^{-}$.

3.10.2 The Cauchy theorem

Exercise 3.10.9. Calculate the integral

$$\int_\gamma \frac{1}{z-\alpha}\mathrm{d}z$$

in the following cases:
1. $\{\gamma\} = \partial U(\alpha; r)$ is a directly oriented path;
2. $\{\gamma\} = \partial T$, where T is the triangle determined by the points 1, i and -1, directly oriented, such that $\alpha \notin \partial T$.

Exercise 3.10.10. Calculate the integral

$$\int_\gamma (z-a)^n \mathrm{d}z$$

in the following cases:
1. γ is given by $\gamma(t) = a + re^{\pi i t}$, $t \in [0,1]$ and $n \in \mathbb{Z}$;
2. $\{\gamma\} = \partial U(a; r)$ is a directly oriented path, and $n \in \mathbb{Z}$;
3. $\{\gamma\}$ is an arbitrary smooth path, such that $\gamma \in \mathcal{D}(z_0, z_1)$, and $n \in \mathbb{Z} \setminus \{-1\}$;
4. γ is given by $\gamma(t) = a + re^{2\pi i t}$, $t \in [0,1]$ and $n = \frac{1}{2}$.

Exercise 3.10.11. Calculate the integral

$$\int_\gamma \frac{e^z}{\sin z + \cos z}\mathrm{d}z,$$

where $\{\gamma\} = \partial U(0; \frac{1}{4})$ is a directly oriented path.

Exercise 3.10.12. Calculate the integral

$$\int_\gamma (\overline{z} + e^z \sin z)\mathrm{d}z,$$

where $\{\gamma\} = \partial U(0; 2)$ is a directly oriented path.

Exercise 3.10.13. Calculate the integral

$$\int_\gamma \frac{e^z}{\cos z \cos iz} dz,$$

where $\{\gamma\} \subset U(0;1)$ is an arbitrary closed, rectifiable and directly oriented path.

Exercise 3.10.14. Calculate the integral

$$\int_\gamma \frac{1}{(z^2+4)(z^2+16)} dz,$$

where $\{\gamma\} = \partial T$, and T is the triangle determined by the points 1, i and -1, directly oriented.

Exercise 3.10.15. Calculate the integral

$$\int_\gamma \frac{e^{\frac{1}{z-3}} \cos z}{(z-2)^n} dz,$$

where $\{\gamma\} = \partial U(0;1)$ is a directly oriented path, and $n \in \mathbb{Z}$.

Exercise 3.10.16. Calculate the integral

$$\int_\gamma \frac{z}{z^2-1} dz,$$

where $\{\gamma\} = \partial U(0;a)$ is a directly oriented path, and $a \neq 1$.

3.10.3 The Cauchy formula for the disc

Exercise 3.10.17. Calculate the following integral:

$$\int_\gamma \frac{\sin z}{(z-\frac{\pi}{2})(z^2+5)} dz,$$

where $\{\gamma\} = \partial U(0;2)$ is a directly oriented path.

Exercise 3.10.18. Calculate the following integral:

$$\int_\gamma \frac{z^5}{e^z(z+1)^6} dz,$$

where $\{\gamma\} = \partial U(0;2)$ is a directly oriented path.

Exercise 3.10.19. Calculate the following integral:

$$\int_{\gamma} \frac{ze^z}{(z-2)^n} dz,$$

where $\{\gamma\} = \partial U(0;3)$ is a directly oriented path and $n \in \mathbb{N}$.

Exercise 3.10.20. Calculate the following integral:

$$\int_{\gamma} \frac{e^z \sin z}{z-a} dz,$$

where $\{\gamma\} = \partial U(0;r)$ is a directly oriented path, and $r \neq |a|$.

Exercise 3.10.21. Calculate the integral,

$$\int_{\gamma} \frac{1}{z^2+1} dz,$$

in the next cases:
1. $\{\gamma\} = \partial U(i;1)$ is a directly oriented path;
2. $\{\gamma\} = \partial U(-i;1)$ is a directly oriented path;
3. $\{\gamma\} = \partial U(0;2)$ is a directly oriented path.

Exercise 3.10.22. Calculate the following integral:

$$\int_{\gamma} \frac{e^z \cos z}{z^3+8} dz,$$

where $\{\gamma\} = \partial U(0;1)$ is a directly oriented path.

Exercise 3.10.23. Calculate the following integral:

$$\int_{\gamma} \frac{\sin z}{z^k} dz, \quad k \in \mathbb{Z},$$

where $\{\gamma\} = \partial U(0;1)$ is a directly oriented path.

Exercise 3.10.24. Calculate the integral,

$$\int_{\gamma} \frac{e^z}{z-a} dz,$$

in the next cases:
1. $\{\gamma\} = \partial U(a;|a|)$, $a \in \mathbb{C}^*$, is a directly oriented path;
2. $\{\gamma\} = \partial U(-a;|a|)$, $a \in \mathbb{C}^*$, is a directly oriented path.

Exercise 3.10.25. Calculate the following integral:

$$\int_{\gamma} \frac{\sinh z}{z - \frac{\pi i}{2}} \mathrm{d}z,$$

where $\{\gamma\} = \partial U(\frac{\pi i}{2}; 1)$ is a directly oriented path.

Exercise 3.10.26. Calculate the following integral:

$$\int_{\gamma} (z - \alpha)^n \mathrm{d}z, \quad n \in \mathbb{Z},$$

where $\{\gamma\} = \partial U(0; r)$, $|\alpha| \neq r$, is a directly oriented path.

Exercise 3.10.27. Calculate the following integral:

$$\int_{\gamma} \frac{1}{z(z^2 + z - 2)} \mathrm{d}z,$$

in the next cases:
1. $\{\gamma\} = \partial U(0; \frac{3}{2})$ is a directly oriented path;
2. $\{\gamma\} = \partial U(0; 3)$ is a directly oriented path.

Exercise 3.10.28. Calculate the integral

$$\int_{\gamma} \frac{f(z)}{z(z^2 + 1)^n} \mathrm{d}z,$$

in the next cases:
1. $f(z) = 1$, $n = 1$ and $\{\gamma\} = \{z = x + iy : x^2 + 4y^2 - 1 = 0\}$ is a directly oriented path;
2. $f(z) = 1$, $n = 1$ and $\{\gamma\} = \partial U(0; 2)$ is a directly oriented path;
3. $f(z) = e^{iz}$, $n = 1$ and $\{\gamma\} = \{z = x + iy : x^2 + 4y^2 - 1 = 0\}$ is a directly oriented path;
4. $f(z) = e^{iz}$, $n = 1$ and $\{\gamma\} = \partial U(0; 2)$ is a directly oriented path;
5. $f(z) = e^{iz}$, $n = 2$ and $\{\gamma\} = \{z = x + iy : x^2 + 4y^2 - 1 = 0\}$ is a directly oriented path;
6. $f(z) = e^{iz}$, $n = 2$ and $\{\gamma\} = \partial U(0; 2)$ is a directly oriented path.

Exercise 3.10.29. Calculate the integral

$$\int_{\gamma} \frac{\cos \pi z}{z^2 - 1} \mathrm{d}z,$$

where $\{\gamma\} = \{z = x + iy : x^2 + y^2 - 2ax = 0\}$, $a \notin \{-\frac{1}{2}, 0, \frac{1}{2}\}$, is a directly oriented path.

Exercise 3.10.30. Calculate the following integral:

$$\int_{\gamma} \frac{1}{(z^2 + 1)^n} \mathrm{d}z, \quad n \in \mathbb{Z},$$

where $\{\gamma\} = \{z = x + iy : x^2 + y^2 - ay = 0\}$, $a \in \mathbb{R}^* \setminus \{1\}$, is a directly oriented path.

Exercise 3.10.31. Calculate the integral

$$\int_{\gamma} \frac{ze^{\frac{i\pi}{2}z}}{z-1}\mathrm{d}z,$$

where $\{\gamma\} = \partial U(0; a)$, $a \neq 1$, is a directly oriented path.

Exercise 3.10.32. Calculate the integral

$$\int_{\gamma} \frac{z^n}{z-a}\mathrm{d}z, \quad n \in \mathbb{N},$$

where $\{\gamma\} = \partial U(0; 1)$, $|a| \neq 1$ is a directly oriented path.

Exercise 3.10.33. Calculate the following integral:

$$\int_{\gamma} \frac{e^{\pi z}}{z(z-i)}\mathrm{d}z,$$

where $\{\gamma\} = \partial U(0; a)$, $a \neq 1$ is a directly oriented path.

Exercise 3.10.34. Calculate the integral

$$\int_{\gamma} \frac{e^{iz}}{z^2-\pi^2}\mathrm{d}z,$$

where $\{\gamma\} = \partial U(0; a)$, $a \neq \pi$ is a directly oriented path.

Exercise 3.10.35. Calculate the following integral:

$$\int_{\gamma} \frac{\cos \pi z}{(z^2-4)^2}\mathrm{d}z,$$

where $\{\gamma\} = \{z = x + iy : x^2 + y^2 - ax = 0\}$, $a \notin \{-2, 0, 2\}$ is a directly oriented path.

Exercise 3.10.36. Calculate the following integral:

$$\int_{\gamma} \frac{z^{100}e^{i\pi z}}{z^2+1}\mathrm{d}z,$$

where $\{\gamma\} = \{z = x + iy : 4x^2 + y^2 - 4 = 0\}$ is a directly oriented path.

Exercise 3.10.37. Calculate the following integral:

$$\int_{\gamma} \frac{\cosh \frac{\pi z}{2}}{(z+i)^4}\mathrm{d}z,$$

where $\{\gamma\} = \{z \in \mathbb{C} : |z + 2i| = 2\}$ is a directly oriented path.

Exercise 3.10.38. Let $f : \{z \in \mathbb{C} : \operatorname{Re} z \geq 0\} \to \mathbb{C}$ be a continuous function, such that $\lim_{z\to\infty} zf(z) = 0$. Prove that, for all $a \leq 0$, we have

$$\lim_{r\to\infty} \int_{\gamma_r} e^{az} f(z)\mathrm{d}z = 0,$$

where $\gamma_r(t) = re^{i\frac{\pi(2t-1)}{2}}$, $t \in [0,1]$.

Exercise 3.10.39. Let $f : \{z \in \mathbb{C} : \operatorname{Re} z \leq 0\} \to \mathbb{C}$ be a continuous function, such that $\lim_{z\to\infty} zf(z) = 0$. Prove that, for all $a \geq 0$, we have

$$\lim_{r\to\infty} \int_{\gamma_r} e^{az} f(z)\mathrm{d}z = 0,$$

where $\gamma_r(t) = re^{i\frac{\pi(2t+1)}{2}}$, $t \in [0,1]$.

3.10.4 Some consequences of Cauchy formula

Exercise 3.10.40. Determine the entire functions f of the form $f(z) = u(x,y) + iv(x,y)$, $z = x + iy$, if we know that:
1. $v(x,y) = e^x \sin y$ and $f(0) = 1$;
2. $v(x,y) = x^2 - y^2 + xy$ and $f(0) = 0$;
3. $v(x,y) = 2xy$ and $f(0) = 0$;
4. $u(x,y) = e^x(x\cos y - y\sin y)$ and $f(0) = 0$.

Exercise 3.10.41. Prove that there exists a function of the form $f(z) = u(x,y) + iv(x,y)$, $z = x + iy$, which is holomorphic on the domain $D \subset \mathbb{C}$, with $0 \notin D$ and $1 \in D$, such that $v(x,y) = \frac{y}{x^2+y^2}$ and $f(1) = 0$.

Exercise 3.10.42. Did there exists functions of the form $f(z) = u(x,y) + iv(x,y)$, where $z = x + iy$, that are holomorphic on a domain $D \subset \mathbb{C}$, such that:
1. $u(x,y) = \frac{x^2-y^2}{(x^2+y^2)^2}$;
2. $v(x,y) = \ln(x^2 + y^2) - x^2 + y^2$;
3. $u(x,y) = e^{\frac{y}{x}}$?

Exercise 3.10.43. Determine the function f of the form $f(z) = u(x,y) + iv(x,y)$, $z = x + iy$, which is holomorphic on the domain $D \subset \mathbb{C}$, in the following cases:
1. $D = \mathbb{C} \setminus \{z \in \mathbb{C} : \operatorname{Re} z \geq 0, \operatorname{Im} z = 0\}$, $v(x,y) = \frac{y}{x^2+y^2}$ and $f(-2) = 0$;
2. $D = \mathbb{C} \setminus \{z \in \mathbb{C} : \operatorname{Im} z \leq 0, \operatorname{Re} z = 0\}$, $v(x,y) = 3 + x^2 - y^2 - \frac{y}{2(x^2+y^2)}$ and $f(2) = 0$;
3. $D = \mathbb{C} \setminus \{z \in \mathbb{C} : \operatorname{Re} z \leq 0, \operatorname{Im} z = 0\}$, $u(x,y) = \frac{1}{2}\ln(x^2 + y^2)$;
4. $D = \mathbb{C} \setminus \{z \in \mathbb{C} : \operatorname{Re} z \leq 0, \operatorname{Im} z = 0\}$, $v(x,y) = \ln(x^2 + y^2) + x - 2y$;

5. $D = \mathbb{C}, v(x,y) = 1 + xy;$
6. $D = \mathbb{C} \setminus \{z \in \mathbb{C} : \operatorname{Im} z \le 0\}, v(x,y) = y - \frac{y}{x^2+y^2};$
7. $D = \mathbb{C} \setminus (\{z \in \mathbb{C} : \operatorname{Re} z \le 0, \operatorname{Im} z = 1\} \cup \{z \in \mathbb{C} : \operatorname{Re} z \ge 0, \operatorname{Im} z = -1\}),$
   $u(x,y) = \frac{x}{x^2+(y+1)^2} + \frac{x}{x^2+(y-1)^2}$ and $f(1) = 1;$
8. $D = \mathbb{C} \setminus (\{z \in \mathbb{C} : \operatorname{Im} z \le 0, \operatorname{Re} z = -1\} \cup \{z \in \mathbb{C} : \operatorname{Im} z \ge 0, \operatorname{Re} z = 1\}),$
   $v(x,y) = -\frac{y}{(x+1)^2+y^2} - \frac{y}{(x-1)^2+y^2}$ and $f(0) = 0;$
9. $D = \mathbb{C} \setminus (\{z \in \mathbb{C} : \operatorname{Im} z \ge 0, \operatorname{Re} z = -1\} \cup \{z \in \mathbb{C} : \operatorname{Im} z \le 0, \operatorname{Re} z = 1\}),$
   $v(x,y) = -\frac{2xy}{(x^2-y^2-1)^2+4x^2y^2}$ and $f(i) = -\frac{1}{2}.$

3.10.5 Multivalued functions analytical branches

Exercise 3.10.44. Determine the maximal domain $D \subset \mathbb{C}$, such that the following functions are holomorphic in D:

1. $w(z) = \sqrt[3]{\frac{z}{2i-z}};$ 2. $w(z) = \sqrt[3]{z+1};$ 3. $w(z) = \operatorname{Log}(z^2+1);$
4. $w(z) = \operatorname{Log}\frac{z-1}{z+1};$ 5. $w(z) = \frac{\operatorname{Log} z}{z};$ 6. $w(z) = \sqrt{z-i} + \sqrt{z+i}.$

Exercise 3.10.45. Did there exist holomorphic branches of the following functions in the corresponding domain $D \subset \mathbb{C}$?
1. $w(z) = \sqrt[3]{\frac{z+1}{z-1}}$, where $D = \{z \in \mathbb{C} : 1 < |z| < \infty\};$
2. $w(z) = 2\operatorname{Log}\frac{z+1}{z-1}$, where $D = \{z \in \mathbb{C} : 1 < |z| < \infty\};$
3. $w(z) = \sqrt{(z^2-1)(z^2-4)}$, where $D = \{z \in \mathbb{C} : \operatorname{Re} z > 0, |z-3| > \frac{5}{2}\}.$

Exercise 3.10.46. Define the function $f(z) = \sqrt[3]{z+1}$.
1. If $D = \mathbb{C} \setminus T$, where $T = \{z \in \mathbb{C} : \operatorname{Im} z = 0, \operatorname{Re} z \ge -1\}$, then determine the branch f_0 of the function f, such that $f_0 \in H(D)$ and $f_0(-2) = -1;$
2. Compute the values $f_0(i), f_0(-i), f_0'(i)$ and $f_0'(-i);$
3. Denote by $\tilde{f}$ the extension of f_0 to $D \cup T$. Determine the values of the function $\tilde{f}$ on the borders T_a and T_f of the half-line T.

Exercise 3.10.47. Let us define the function $f(z) = \sqrt{z-i} + \sqrt{z+i}$.
1. If $D = \mathbb{C} \setminus (T_1 \cup T_2)$, where $T_1 = \{z \in \mathbb{C} : \operatorname{Re} z = 0, \operatorname{Im} z \le -1\}$, and $T_2 = \{z \in \mathbb{C} : \operatorname{Re} z = 0, \operatorname{Im} z \ge 1\}$, then determine the branch f_0 of the function f, such that $f_0 \in H(D)$ and $f_0(0) = -i\sqrt{2};$
2. Compute the values $f_0(\frac{1}{\sqrt{3}})$ and $f_0(2);$
3. Denote by $\tilde{f}$ the extension of f_0 to $D \cup (T_1 \cup T_2)$. Determine the values of the function $\tilde{f}$ on both of the borders of T_1 and T_2.

Exercise 3.10.48. Let us define the function $f(z) = \operatorname{Log}(z^2+1)$.

1. If $D = \mathbb{C} \setminus (T_1 \cup T_2)$, where $T_1 = \{z \in \mathbb{C} : \operatorname{Re} z = 0, \operatorname{Im} z \le -1\}$, and $T_2 = \{z \in \mathbb{C} : \operatorname{Re} z = 0, \operatorname{Im} z \ge 1\}$, then determine the branch f_0 of the function f, such that $f_0(0) = 0$;
2. Compute the values $f_0(2+i), f_0(-2-i)$ and $f_0(\frac{i}{2})$;
3. Denote by $\tilde{f}$ the extension of f_0 to $D \cup (T_1 \cup T_2)$. Determine the values of the function $\tilde{f}$ on the borders of $T_1 \cup T_2$.

Exercise 3.10.49. Let us define the function $f(z) = \sqrt{z^2+1}$. Determine the holomorphic branches of the function f in the corresponding domains $D \subset \mathbb{C}$:
1. $D = \mathbb{C} \setminus (T_1 \cup T_2)$, where $T_1 = \{z \in \mathbb{C} : \operatorname{Re} z = 0, \operatorname{Im} z \le -1\}$, and $T_2 = \{z \in \mathbb{C} : \operatorname{Re} z = 0, \operatorname{Im} z \ge 1\}$;
2. $D = \mathbb{C} \setminus T_3$, where $T_3 = \{z \in \mathbb{C} : \operatorname{Re} z = 0, \operatorname{Im} z \ge -1\}$;
3. $D = \mathbb{C} \setminus T_4$, where $T_4 = \{z \in \mathbb{C} : \operatorname{Re} z = 0, -1 \le \operatorname{Im} z \le 1\}$.

Exercise 3.10.50. Let us define the function $f(z) = \frac{\operatorname{Log} z}{z}$.
1. If $D = \mathbb{C} \setminus T$, where $T = \{z \in \mathbb{C} : \operatorname{Re} z \le 0, \operatorname{Im} z = 0\}$, then determine the branch f_0 of the function f, such that $f(z) \in \mathbb{R}$ whenever $z > 0$;
2. Compute the values $f_0(i)$ and $f_0(-i)$.

Exercise 3.10.51. Calculate the index $\mathrm{n}(\gamma, z)$ for $z \in \mathbb{C} \setminus \{\gamma\}$, if $\gamma = \gamma_1 \cup \gamma_2$, where γ_1 and γ_2 are given by $\gamma_1(t) = (2t-1)r$, $t \in [0,1]$ and $\gamma_2(t) = re^{\pi i t}$, $t \in [0,1]$.

Exercise 3.10.52. Calculate the index $\mathrm{n}(\gamma, z)$ for $z \in \{z \in \mathbb{C} : r < |z| < \rho, \operatorname{Im} z > 0\}$, if $\gamma = \gamma_1 \cup \gamma_2 \cup \gamma_3 \cup \gamma_4$, where $\gamma_1, \gamma_2, \gamma_3$ and γ_4 are given by $\gamma_1(t) = -\rho(1-t) - rt$, $t \in [0,1]$, $\gamma_2(t) = re^{\pi(1-t)i}$, $t \in [0,1]$, $\gamma_3(t) = r(1-t) + \rho t$, $t \in [0,1]$ and $\gamma_4(t) = \rho e^{\pi i t}$, $t \in [0,1]$.

Exercise 3.10.53. Let $G \subset \mathbb{C}$ be an open set, and let $f : G \to \mathbb{C}$ such that $f \in C(G)$, and $f \in H(G \setminus [a,b])$. Prove that $f \in H(G)$.

Exercise 3.10.54. Let $f \in C(\mathbb{C})$ a bounded function, such that $f \in H(\mathbb{C} \setminus [a,b])$. Prove that the function f is constant on $\mathbb{C}$.

Exercise 3.10.55. Did there exist nonconstant entire functions $f \in H(\mathbb{C})$, such that

$$f(\mathbb{C}) \subset \{w \in \mathbb{C} : \operatorname{Im} w > 0\}?$$

Exercise 3.10.56. Did there exist nonconstant entire functions $f \in H(\mathbb{C})$, such that

$$f(\mathbb{C}) \subset \{w \in \mathbb{C} : \operatorname{Re} w > 0\}?$$

Exercise 3.10.57. Did there exist nonconstant entire functions $f \in H(\mathbb{C})$, such that

$$f(\mathbb{C}) \subset \mathbb{C} \setminus U(0;1)?$$

Exercise 3.10.58. Let the function $f \in H(\mathbb{C})$, such that there exists $\lim_{z\to\infty} f(z) = a \in \mathbb{C}$. Prove that f is a constant function on $\mathbb{C}$.

Exercise 3.10.59. Let $f \in H(\mathbb{C})$ such that $\operatorname{Re} f$ is a bounded function. Prove that f is a constant function on $\mathbb{C}$.

Exercise 3.10.60. Let $f \in C(\mathbb{C})$ such that, for all $r > 0$ we have

$$f(z) = \int_{\partial U(0;r)} \frac{f(\zeta)}{\zeta - z} d\zeta, \quad \forall z \in \mathbb{C} : |z| > r.$$

Prove that $f(z) = 0$, $\forall z \in \mathbb{C}$.

4 Sequences and series of holomorphic functions

4.1 Sequences of holomorphic functions

Definition 4.1.1. Let $(f_n)_{n\in\mathbb{N}^*}$ be a sequence of holomorphic functions defined on the open set $G \subset \mathbb{C}$. We say that the sequence $(f_n)_{n\in\mathbb{N}^*}$ **uniformly converges on compact sets to the function** $f : G \to \mathbb{C}$ (briefly, **compact converges to** f), if for any arbitrary compact set K, with $K \subset G$, the function sequence $(f_n|_K)_{n\in\mathbb{N}^*}$ uniformly converges to $f|_K$. (Denoted by $f_n \rightrightarrows_K f$.)

Remarks 4.1.1.

1. The compact convergence is stronger than the pointwise convergence on the whole set G, but it is weaker than the uniformly convergence on G.
2. The compact convergence can be written as follows:
   The sequence of the functions $(f_n)_{n\in\mathbb{N}^*}$ uniformly converges on compact sets to f, if and only if for any arbitrary number $\varepsilon > 0$, and any arbitrary compact set $K \subset G$, there exists an index $n_0 = n_0(\varepsilon, K) \in \mathbb{N}$, such that for all $n > n_0$ we have
   $$|f_n(z) - f(z)| < \varepsilon, \quad \forall z \in K.$$
3. The next Cauchy uniformly convergence criteria holds:
   The sequence of functions $(f_n)_{n\in\mathbb{N}^*}$ uniformly converges on compact sets, if and only if for any arbitrary number $\varepsilon > 0$, and any arbitrary compact set $K \subset G$, there exists an index $n_0 = n_0(\varepsilon, K) \in \mathbb{N}$, such that for all $n, m > n_0$ we have
   $$|f_n(z) - f_m(z)| < \varepsilon, \quad \forall z \in K.$$
4. Since every compact set $K \subset \mathbb{C}$ can be covered by a finite number of closed discs of G, it follows that the sequence of functions $(f_n)_{n\in\mathbb{N}^*}$ uniformly converges on compact sets, if and only if uniformly converges on any arbitrary closed disc of G.

Theorem 4.1.1. *If the functions $f_n : G \to \mathbb{C}$, $n \in \mathbb{N}^*$, are continuous on the open set G, and $(f_n)_{n\in\mathbb{N}^*}$ compact converges to $f : G \to \mathbb{C}$, then f is a continuous function on G.*

Proof. Let $z_0 \in G$ be an arbitrary point, and let $r > 0$ such that $K = \overline{U}(z_0; r) \subset G$. Since K is compact and $f_n \rightrightarrows_K f$, it follows that f is continuous on K, thus f is continuous at the point z_0. □

Theorem 4.1.2 (Weierstrass lemma). *If the functions $(f_n)_{n\in\mathbb{N}^*}$ are holomorphic on the disc $D = U(z_0; R)$, are continuous on D^-, and the sequence of functions $(f_n)_{n\in\mathbb{N}^*}$ uniformly converges on the boundary ∂D, then:*

1. *$(f_n)_{n\in\mathbb{N}^*}$ is compact convergent on D;*
2. *the limit f of the sequence of functions $(f_n)_{n\in\mathbb{N}^*}$ is holomorphic in D, i. e., $f \in H(D)$;*

https://doi.org/10.1515/9783110657869-004

3. *for any $k \in \mathbb{N}$, the sequence of the derivative functions $(f_n^{(k)})_{n\in\mathbb{N}^*}$ compact converges to $f^{(k)}$ on D.*

Proof. Since $(f_n)_{n\in\mathbb{N}^*} \subset C(\overline{D})$ and $(f_n)_{n\in\mathbb{N}^*}$ uniformly converges on the set ∂D, the limit g of the sequence $(f_n|_{\partial D})_{n\in\mathbb{N}^*}$ is continuous on ∂D. Let $\gamma = \partial D = \partial U(z_0; R)$. Then, according to the Theorem 3.4.1, the function f defined by

$$f(z) = \frac{1}{2\pi i}\int_\gamma \frac{g(\zeta)}{\zeta - z}\mathrm{d}\zeta, \quad z \in D$$

is holomorphic on D, and from Corollary 3.4.1 the function f has derivatives of any order in D. Moreover,

$$f^{(k)}(z) = \frac{k!}{2\pi i}\int_\gamma \frac{g(\zeta)}{(\zeta - z)^{k+1}}\mathrm{d}\zeta, \quad z \in D.$$

We will prove that $(f_n^{(k)})_{n\in\mathbb{N}^*}$ compact converges to $f^{(k)}$ in D, for all natural numbers k.

Using the Cauchy formula for the functions f_n, we have

$$f_n^{(k)}(z) = \frac{k!}{2\pi i}\int_\gamma \frac{f_n(\zeta)}{(\zeta - z)^{k+1}}\mathrm{d}\zeta, \quad z \in D.$$

Hence

$$|f_n^{(k)}(z) - f^{(k)}(z)| = \frac{k!}{2\pi}\left|\int_\gamma \frac{f_n(\zeta) - g(\zeta)}{(\zeta - z)^{k+1}}\mathrm{d}\zeta\right|, \quad z \in D.$$

It is sufficient to check that $(f_n^{(k)})_{n\in\mathbb{N}^*}$ uniformly converges to $f^{(k)}$ on the closed disc $\overline{U}(z_0; r)$, for any arbitrary $r < R$. Let $d = R - r$, and let $\varepsilon > 0$ be an arbitrary number. Since the sequence of functions $(f_n|_{\partial D})_{n\in\mathbb{N}^*}$ uniformly converges to g, there exists an index $n_0 = n_0(\varepsilon) \in \mathbb{N}$, such that $\forall n > n_0$ we have

$$|f_n(\zeta) - g(\zeta)| < \frac{\varepsilon d^{k+1}}{Rk!}, \quad \forall \zeta \in \partial D.$$

It follows that $\forall n > n_0$, and $\forall z \in \overline{U}(z_0; r)$ the next inequality holds:

$$|f_n^{(k)}(z) - f^{(k)}(z)| \le \frac{k!}{2\pi}\int_\gamma \frac{|f_n(\zeta) - g(\zeta)|}{|\zeta - z|^{k+1}}|\mathrm{d}\zeta| \le \frac{k!}{2\pi}\frac{\varepsilon d^{k+1}}{Rk!}\frac{1}{d^{k+1}}2\pi R = \varepsilon,$$

hence $f_n^{(k)} \rightrightarrows_{\overline{U}(z_0;r)} f^{(k)}$, $\forall k \in \mathbb{N}$. □

Theorem 4.1.3 (Weierstrass theorem). *Let $G \subset \mathbb{C}$ be an open set, and let $f_n \in H(G)$, $n \in \mathbb{N}^*$. If the sequence of functions $(f_n)_{n\in\mathbb{N}^*}$ compact converges to $f : G \to \mathbb{C}$ in G, then:*

1. $f \in H(G)$;
2. *the sequence of the derivative functions* $(f_n^{(k)})_{n\in\mathbb{N}^*}$ *compact converges to* $f^{(k)}$ *in* G, $\forall k \in \mathbb{N}$.

Proof. According to the point 4 of the Remark 4.1.1, it is sufficient to prove the compact convergence on an arbitrary closed disc of G. But, using the Weierstrass lemma (Theorem 4.1.2), this last condition holds. □

4.2 Series of functions

Definition 4.2.1.

1. Let $G \subset \mathbb{C}$ be an open set, and let $f_n : G \to \mathbb{C}$, $\forall n \in \mathbb{N}$. The sum

$$\sum f_n = \sum_{n=0}^{\infty} f_n = f_0 + f_1 + \cdots + f_n + \cdots$$

   is called **series of functions.** The notation

$$\sum_{n=0}^{\infty} f_n(z) = f_0(z) + f_1(z) + \cdots + f_n(z) + \cdots$$

   is also used.
2. The finite sum

$$S_n = f_0 + f_1 + \cdots + f_n$$

   is called the **partial sum** of the above series of function.
3. The series of functions $\sum f_n$ **converges, uniformly converges** or **compact converges in** G if the partial sum $(S_n)_{n\in\mathbb{N}}$ converges, uniformly converges, or compact converges in G, respectively.
4. The limit $S = \lim_{n\to\infty} S_n$ is called **the sum of the series of functions,** and it is denoted by the symbol $S = \sum f_n$. The notation

$$S(z) = \sum_{n=0}^{\infty} f_n(z), \quad z \in G$$

   is also used.

Remark 4.2.1. The Cauchy convergence criteria for the series of functions can be written as follows:

The series of functions $\sum f_n$ uniformly converges on the set E, if and only if $\forall \varepsilon > 0$ there exists an index $n_0 = n_0(\varepsilon) \in \mathbb{N}$, such that $\forall n > n_0$ and $\forall p \in \mathbb{N}^*$ we have

$$|f_{n+1}(z) + \cdots + f_{n+p}(z)| < \varepsilon, \quad \forall z \in E.$$

An immediate consequence of this criteria is the next theorem.

Theorem 4.2.1 (Weierstrass uniformly convergence criteria). *If there exists a convergent series of positive numbers $\sum u_n$, and there exists an index $n_0 \in \mathbb{N}$, such that for all $n > n_0$ we have*

$$|f_n(z)| \leq u_n, \quad \forall z \in E,$$

then the series of functions $\sum f_n$ uniformly converges in E.

In fact, the series of functions $\sum f_n$ absolutely converges, in the sense that the series $\sum |f_n|$ (uniformly) converges.

Theorem 4.2.2 (Weierstrass theorem for holomorphic functions series). *Let $G \subset \mathbb{C}$ be an open set, and let $f_n \in H(G)$, $n \in \mathbb{N}$. If the series of functions $\sum f_n$ compact converges in G, then:*
1. *the sum $S = \sum f_n \in H(G)$*
   and
2. *the series of the derivative functions $\sum f_n^{(k)}$ compact converges to $S^{(k)}$ in G, $\forall k \in \mathbb{N}$.*

4.3 Power series

Definition 4.3.1. A **power series** is a series of functions defined as follows:

$$\sum_{n=0}^{\infty} a_n(z - z_0)^n = a_0 + a_1(z - z_0) + \cdots + a_n(z - z_0)^n + \cdots,$$

where z_0 is a fixed number, the variable is z and $a_n \in \mathbb{C}$. The above series of functions is called **the power series with the coefficients a_n, about the point z_0.**

Theorem 4.3.1 (Abel theorem). *If the power series $\sum_{n=0}^{\infty} a_n(z-z_0)^n$ converges in $z_1 \neq z_0$, then it absolutely converges in all the points $z \in \mathbb{C}$, with $|z-z_0| < |z_1-z_0|$, and the series of functions compact converges on the disc $U(z_0; |z_1 - z_0|)$.*

Proof. Let $r = |z-z_0|$, and let denote $r_1 = |z_1-z_0|$. Since the series of complex numbers $\sum a_n(z_1 - z_0)^n$ converges, the general term tends to zero. Hence, there exists an index $n_0 \in \mathbb{N}$ such that

$$|a_n| r_1^n = |a_n(z_1 - z_0)^n| < 1, \quad \forall n > n_0.$$

Considering the series of positive numbers $\sum |a_n| r^n$, for all $n > n_0$ we have $|a_n| r^n < (\frac{r}{r_1})^n$, because $|a_n| < \frac{1}{r_1^n}$.

If $r < r_1$, then the geometric series $\sum (\frac{r}{r_1})^n$ is convergent, and using the Weierstrass criteria (Theorem 4.2.1) we deduce that the series of functions $\sum a_n(z-z_0)^n$ absolutely and uniformly converges on the closed disc $\overline{U}(z_0; r)$. Since we can choose the number r to be arbitrary close to r_1, the above result is proved. □

Theorem 4.3.2 (Cauchy–Hadamard theorem or the radius of convergence theorem). *Let* $\sum_{n=0}^{\infty} a_n(z-z_0)^n$ *be a power series, and define the number* R *using the following* **Cauchy–Hadamard relations***:*

$$\frac{1}{R} = \limsup_{n\to+\infty} \sqrt[n]{|a_n|}, \quad \text{if } \limsup_{n\to+\infty} \sqrt[n]{|a_n|} \in \mathbb{R}^*;$$

$$R = 0, \quad \text{if } \limsup_{n\to+\infty} \sqrt[n]{|a_n|} = +\infty;$$

$$R = +\infty, \quad \text{if } \limsup_{n\to+\infty} \sqrt[n]{|a_n|} = 0.$$

The number R*, called* **the radius of convergence of the power series,** *satisfies the following properties:*

1. *The power series is absolute and compact convergent on the disc* $U(z_0; R)$*;*
2. *the power series is divergent in all the points of* $\mathbb{C} \setminus \overline{U}(z_0; R)$*;*
3. *the sum* S *of the power series is holomorphic on* $U(z_0; R)$*;*
4. *the power series can be derived term by term, and the obtained power series* $\sum_{n=1}^{\infty} na_n(z-z_0)^{n-1}$ *has the radius of convergence equal to* R*, and the sum equal to* S' *(the derivative of the sum* S*), i. e.,*

$$S'(z) = \sum_{n=1}^{\infty} na_n(z-z_0)^{n-1}, \quad \forall z \in U(z_0; R).$$

5. *the power series can be integrated term by term, and the obtained power series* $\sum_{n=1}^{\infty} a_n \int_C (z-z_0)^n dz$ *has the radius of convergence equal to* R*, and the sum equal to* $\int_C S$ *where* C *is smooth piecewise curve which lie inside* $U(z_0, R)$*.*

Proof. Let $L = \limsup_{n\to+\infty} \sqrt[n]{|a_n|}$, and let denote $r = |z - z_0|$. Since $\sqrt[n]{|a_n|r^n} = r\sqrt[n]{|a_n|}$, using the Cauchy root convergence criteria for the series of positive numbers $\sum |a_n|r^n$, we deduce that this series converges if $rL < 1$, and diverges if $rL > 1$. Suppose that $L \neq 0$, and $L \neq +\infty$. Then, for $|z - z_0| = r < R = \frac{1}{L}$ (i. e., if $rL < 1$), the power series $\sum a_n(z-z_0)^n$ absolutely converges.

We will prove that if $|z-z_0| = r > R$, then the power series diverges. In fact, if there exists a number z_1, with $|z_1 - z_0| > R$, such that the series $\sum a_n(z_1 - z_0)^n$ converges, then for the number r with $|z_1 - z_0| > r > R$ we may use a similar proof like of the Abel theorem (Theorem 4.3.1), and we deduce that the series $\sum |a_n|r^n$ is convergent. On the other hand, $r > \frac{1}{L}$, i. e., $Lr > 1$, which contradicts the Cauchy root convergence criteria.

Hence, in this case, the power series absolutely converges in the all the points of the disc $U(z_0; R)$, and by the Abel theorem, it is compact convergent on this open disc. The power series diverges in all the points of the set $\mathbb{C} \setminus \overline{U}(z_0; \text{with } R)$.

If $L = 0$, then $rL < 1$ for all $r > 0$, thus the above results hold if $R = +\infty$. If $L = +\infty$, then $rL > 1$ for all $r > 0$, thus the power series diverges in all the points $z \in \mathbb{C} \setminus \{z_0\}$, so $R = 0$ and the convergence domain will be $\{z_0\}$.

Determine now the convergence radius R' for the series of the derivative functions $\sum_{n=1}^{\infty} na_n(z-z_0)^{n-1}$. Since the nature of the power series remains unchanged if we multiply it by $(z-z_0)$, from the above results it follows that the radius of convergence for the series of the derivative functions is the same with those of

$$\sum_{n=1}^{\infty} na_n(z-z_0)^n.$$

Hence,

$$\frac{1}{R'} = \limsup_{n\to+\infty} \sqrt[n]{n|a_n|} = \lim_{n\to+\infty} \sqrt[n]{n} \cdot \limsup_{n\to+\infty} \sqrt[n]{|a_n|} = \limsup_{n\to+\infty} \sqrt[n]{|a_n|} = \frac{1}{R},$$

and thus $R' = R$. This result could be also obtained by using the Weierstrass theorem for the series of functions (see the Theorem 4.2.2). From this previously mentioned theorem, we conclude that

$$S'(z) = \sum_{n=1}^{\infty} na_n(z-z_0)^{n-1}. \qquad \square$$

Remarks 4.3.1.

1. If there exists the finite or the infinite limit $\lim_{n\to+\infty} \frac{|a_{n+1}|}{|a_n|} = l \in \overline{\mathbb{R}}$, then $l = L$ and
$$R = \lim_{n\to+\infty} \frac{|a_n|}{|a_{n+1}|}.$$
2. If we derive k times the given power series, and then we replace $z = z_0$, it follows that
$$a_k = \frac{S^{(k)}(z_0)}{k!}.$$
3. The disc $U(z_0;R)$ does not represent the domain of convergence of the power series. Theorem 4.3.2 (the Cauchy–Hadamard theorem) does not give any information about the points of the boundary $\partial U(z_0;R)$ (where the convergence of the power series needs to be studied in every given point).
4. For two given power series,
$$S(z) = \sum_{n=0}^{\infty} a_n(z-z_0)^n \quad \text{and} \quad T(z) = \sum_{n=0}^{\infty} b_n(z-z_0)^n,$$
the sum of the series S and T will be
$$\sum_{n=0}^{\infty} (a_n + b_n)(z-z_0)^n,$$

while the product of the series S and T will be

$$\sum_{n=0}^{\infty} c_n(z-z_0)^n, \quad \text{where } c_n = a_0b_n + a_1b_{n-1} + \cdots + a_nb_0.$$

The convergence radius of both of these power series will be greater or equal to the minimum of the convergence radius of S and T.

5. The power series $\sum_{n=0}^{\infty} z^n$ has the convergence radius $R = 1$, and the sum $\frac{1}{1-z}$. The above power series converges if $|z| < 1$, and diverges if $|z| \geq 1$.

Theorem 4.3.3 (The theorem of the identity of the power series coefficients). *Let*

$$S(z) = \sum_{n=0}^{\infty} a_n(z-z_0)^n \quad \textit{and} \quad T(z) = \sum_{n=0}^{\infty} b_n(z-z_0)^n$$

be two power series with the convergence radius at least $r > 0$ (i. e., their sums exist in the disc $U(z_0;r)$). If there exists a subset $E \subset U(z_0;r)$, such that $z_0 \in E'$ and $S(z) = T(z)$, $\forall z \in E$, then the two given power series are identical, i. e., $a_n = b_n$, $\forall n \in \mathbb{N}$.

Proof. The difference of the power series $\sum_{n=0}^{\infty}(a_n - b_n)(z-z_0)^n$ will be $S - T$. If we suppose that $k \in \mathbb{N}$ is the smallest index, such that $a_k \neq b_k$, then

$$\begin{aligned} S(z) - T(z) &= (z-z_0)^k \left(a_k - b_k + \sum_{n=1}^{\infty} (a_{k+n} - b_{k+n})(z-z_0)^n \right) \\ &= (z-z_0)^k Q(z), \quad z \in U(z_0;r). \end{aligned}$$

Let $(z_m)_{m\in\mathbb{N}^*} \subset E \setminus \{z_0\}$ be a sequence, with $\lim_{m\to+\infty} z_m = z_0$.

Since $S(z_m) - T(z_m) = 0$, we have that $Q(z_m) = 0$, $m \in \mathbb{N}^*$. Also, the function Q is continuous at z_0, because it is the sum of a power series about the point z_0. Hence $Q(z_0) = a_k - b_k = \lim_{m\to+\infty} Q(z_m) = 0$, so we get that $a_k = b_k$, which represents a contradiction. □

4.4 The analyticity of holomorphic functions

Theorem 4.4.1 (Taylor series expansion theorem). *Let $D = U(z_0;r)$, with $r > 0$, and let $f \in H(D)$. Then there exists a unique power series $\sum_{n=0}^{\infty} a_n(z-z_0)^n$, with the radius of convergence $R \geq r$, such that*

$$f(z) = \sum_{n=0}^{\infty} a_n(z-z_0)^n, \quad \forall z \in D,$$

and the power series coefficients are given by

$$a_n = \frac{f^{(n)}(z_0)}{n!} = \frac{1}{2\pi i} \int_{\gamma} \frac{f(\zeta)}{(\zeta - z_0)^{n+1}} d\zeta,$$

where $\gamma = \partial U(z_0;\rho)$, with $0 < \rho < r$.

Proof. We will use the Cauchy formula for an arbitrary disc with the radius $\rho \in (0, r)$, and we obtain

$$\frac{f^{(n)}(z_0)}{n!} = \frac{1}{2\pi i} \int_{\gamma} \frac{f(\zeta)}{(\zeta - z_0)^{n+1}} \mathrm{d}\zeta, \quad \gamma = \partial U(z_0; \rho).$$

Thus, the right-hand side integrals are independent on the radius of the circle γ.

Let $z \in D$, and let choose a radius ρ, such that $r_0 = |z - z_0| < \rho < r$. From the Cauchy formula, we get

$$f(z) = \frac{1}{2\pi i} \int_{\gamma} \frac{f(\zeta)}{\zeta - z} \mathrm{d}\zeta.$$

The fraction $\frac{1}{\zeta - z}$ can be written as follows:

$$\frac{1}{\zeta - z} = \frac{1}{\zeta - z_0 + z_0 - z} = \frac{1}{\zeta - z_0} \frac{1}{1 - \frac{z - z_0}{\zeta - z_0}}.$$

If $\zeta \in \{\gamma\}$, from the above choosing method of ρ we have $|\frac{z-z_0}{\zeta - z_0}| = \frac{r_0}{\rho} < 1$. Consequently, from the Weierstrass criteria (Theorem 4.2.1) we deduce that the sum

$$\sum_{n=0}^{\infty} \left(\frac{z - z_0}{\zeta - z_0} \right)^n$$

(where the variable is $\zeta \in \{\gamma\}$) uniformly converges on the set $\{\gamma\}$, because

$$\left| \left(\frac{z - z_0}{\zeta - z_0} \right)^n \right| = \left(\frac{r_0}{\rho} \right)^n \quad \text{and the geometric series} \quad \sum_{n=0}^{\infty} \left(\frac{r_0}{\rho} \right)^n$$

converges. But

$$\sum_{n=0}^{\infty} \left(\frac{z - z_0}{\zeta - z_0} \right)^n = \frac{1}{1 - \frac{z - z_0}{\zeta - z_0}},$$

and if we replace under the integral the fraction $\frac{1}{\zeta - z}$, by using the above relations, then the series

$$\frac{f(\zeta)}{\zeta - z_0} \sum_{n=0}^{\infty} \left(\frac{z - z_0}{\zeta - z_0} \right)^n$$

also uniformly converges on the set $\{\gamma\}$, and its sum will be $\frac{f(\zeta)}{\zeta - z}$. Hence

$$f(z) = \frac{1}{2\pi i} \int_{\gamma} \frac{f(\zeta)}{\zeta - z} \mathrm{d}\zeta = \frac{1}{2\pi i} \int_{\gamma} \frac{f(\zeta)}{\zeta - z_0} \sum_{n=0}^{\infty} \left(\frac{z - z_0}{\zeta - z_0} \right)^n \mathrm{d}\zeta.$$

The uniformly convergence allows us to integrate term by term, and thus

$$f(z)=\sum_{n=0}^{\infty}\left(\frac{1}{2\pi i}\int_{\gamma}\frac{f(\zeta)}{(\zeta-z_0)^{n+1}}\mathrm{d}\zeta\right)(z-z_0)^n.$$

From here, by using the notation from the theorem we obtain

$$f(z)=\sum_{n=0}^{\infty}a_n(z-z_0)^n.$$

From the above proof, we conclude that the power series of the right-hand side converges. Since z is an arbitrary point of the disc $U(z_0;r)$, it follows that the given power series convergence with radius R is greater or equal than r, i. e., $R \geq r$. □

From Theorem 3.3.2, and respectively from Theorem 3.3.3, using the Theorem 4.4.1 we deduce the following results.

Theorem 4.4.2. *Let $D = U(z_0;r)$, and let denote by A the set of all the complex numbers sequences $(a_n)_{n\in\mathbb{N}} \subset \mathbb{C}$, such that the power series $\sum_{n=0}^{\infty} a_n(z-z_0)^n$ converges, $\forall z \in D$. Then there exists a unique bijective and linear function $L : H(D) \to A$, such that if $f \in H(D)$ and if $L(f) = (a_n)$, then $f(z) = \sum_{n=0}^{\infty} a_n(z-z_0)^n$, $\forall z \in D$. The coefficients a_n, $n \in \mathbb{N}$, are given by the formulas*

$$a_n=\frac{f^{(n)}(z_0)}{n!}.$$

Definition 4.4.1.
1. Let $G \subset \mathbb{C}$ open set, and let $f : G \to \mathbb{C}$. We say that the function f **may be expanded in Taylor series about the point** $z_0 \in G$, if there exists a number $r > 0$ with $U(z_0;r) \subset G$, and there exists a power series $\sum_{n=0}^{\infty} a_n(z-z_0)^n$ which converges on the disc $U(z_0;r)$, such that

$$f(z)=\sum_{n=0}^{\infty}a_n(z-z_0)^n,\quad \forall z\in U(z_0;r).$$

2. If the above property holds for any arbitrary point $z_0 \in G$, then we say that **the function f is analytic on** G.

Theorem 4.4.3 (The analyticity of holomorphic functions). *The complex valued function f defined on the open set $G \subset \mathbb{C}$ is holomorphic on G, if and only if it is analytic on G.*

Proof. Let $f \in H(G)$ and let $z_0 \in G$. Then there exists a number $r > 0$, such that $U(z_0;r) \subset G$, and the function f is holomorphic on this disc. Then, from the Theorem 4.4.1 the function f may be expanded in Taylor series on the disc $U(z_0;r)$, and thus f is analytic in z_0. Since $z_0 \in G$ is arbitrary, then function f will be analytic on G.

If the function f is analytic on G, then $\forall z_0 \in G$, $\exists r > 0$, such that $U(z_0; r) \subset G$ and the function f may be expanded in Taylor series on the disc $U(z_0; r)$. Hence, the function f is the sum of the power series, and from the Cauchy–Hadamard theorem it will be holomorphic on this disc. So, there exists the derivative $f'(z_0) \in \mathbb{C}$, and since $z_0 \in G$ is arbitrary we conclude that $f \in H(G)$. □

Examples 4.4.1.

1. Let us consider the geometric power series

$$\frac{1}{1-z} = \sum_{n=0}^{\infty} z^n, \quad z \in U(0;1).$$

Replacing in this formula the variable z by $-z$, we obtain that

$$\frac{1}{1+z} = \sum_{n=0}^{\infty} (-1)^n z^n, \quad z \in U(0;1).$$

From the Weierstrass lemma, we can derive term by term the above relation, and we deduce that

$$\frac{1}{(1-z)^2} = \sum_{n=1}^{\infty} nz^{n-1}, \quad \frac{1}{(1+z)^2} = \sum_{n=1}^{\infty} (-1)^{n-1} nz^{n-1}, \quad z \in U(0;1).$$

2. Since e^z, $\cos z$, and $\sin z$ are integer functions, from the Theorem 4.4.1 we get that

$$e^z = \sum_{n=0}^{\infty} \frac{z^n}{n!}, \quad \cos z = \sum_{n=0}^{\infty} (-1)^n \frac{z^{2n}}{(2n)!}, \quad \sin z = \sum_{n=0}^{\infty} (-1)^n \frac{z^{2n+1}}{(2n+1)!}, \quad z \in \mathbb{C}.$$

3. Let $f(z) = \log(1+z)$ be those branch of the multivalued function Log $(1+z)$, such that $f(0) = 0$. Since $f \in H(U(0;1))$, from the Theorem 4.4.1 we obtain the following formula:

$$\log(1+z) = \sum_{n=1}^{\infty} (-1)^{n+1} \frac{z^n}{n}, \quad z \in U(0;1).$$

4. Let $\alpha \in \mathbb{C}$, and let $f(z) = (1+z)^\alpha$ be those branch of the multivalued power function $(1+z)^\alpha$, such that $f(0) = 1$. We deduce similarly that

$$(1+z)^\alpha = 1 + \sum_{n=1}^{\infty} \frac{\alpha(\alpha-1)\cdot \ldots \cdot(\alpha-n+1)}{n!} z^n, \quad z \in U(0;1),$$

and this last series is called the **binomial series.**

4.5 The zeros of holomorphic functions

Definition 4.5.1. Let $G \subset \mathbb{C}$ an open set, and let $f \in H(G)$. The number $a \in G$ is said to be an ***n*th order zero** for the f function, if

$$f(a) = f'(a) = \cdots = f^{(n-1)}(a) = 0, \quad f^{(n)}(a) \neq 0.$$

Theorem 4.5.1 (Characterization of the zeros of holomorphic functions). *If G is an open subset of $\mathbb{C}$, then $a \in G$ is an nth order zero for the holomorphic function f, if and only if there exists a function $g \in H(G)$, such that*

$$f(z) = (z-a)^n g(z), \quad \forall z \in G \text{ and } g(a) \neq 0.$$

Proof. Since f is analytic on G, there exists the number $r > 0$ such that $U(a;r) \subset G$, and

$$\begin{aligned} f(z) &= f(a) + \frac{f'(a)}{1!}(z-a) + \cdots + \frac{f^{(n-1)}(a)}{(n-1)!}(z-a)^{n-1} + \frac{f^{(n)}(a)}{n!}(z-a)^n + \cdots \\ &= (z-a)^n\left(\frac{f^{(n)}(a)}{n!} + \frac{f^{(n+1)}(a)}{(n+1)!}(z-a) + \cdots\right), \quad \forall z \in U(a;r). \end{aligned} \tag{4.1}$$

Let us define the function

$$g(z) = \begin{cases} \frac{f(z)}{(z-a)^n}, & \text{if } z \in G \setminus \{a\}, \\ \frac{f^{(n)}(a)}{n!}, & \text{if } z = a. \end{cases}$$

From the relation (4.1), it follows that

$$\lim_{z \to a} g(z) = \frac{f^{(n)}(a)}{n!} = g(a) \neq 0,$$

hence g is continuous at a. Since g is holomorphic on $G \setminus \{a\}$, we obtain evidently that g is differentiable also in a, so $g \in H(G)$.

Conversely, if f has the form of the conclusion of the theorem, then f may be written as in the above expansion, and then the number a is an nth order zero for f. □

Theorem 4.5.2 (The set of the zeros of holomorphic functions). *Let f be an holomorphic function on the domain $D \subset \mathbb{C}$, and let introduce the following notation:*

$$A = A(f) = \{a \in D : f(a) = 0\}$$

and

$$B = B(f) = \{a \in D : f^{(k)}(a) = 0, \forall k \in \mathbb{N}\}.$$

Then the next statements are equivalent:

(i) $f \equiv 0$;

(ii) $B \neq \emptyset$;

(iii) $A' \cap D \neq \emptyset$

(where A' represents the set of all accumulation points of A).

Proof. It is obvious that from (i) it follows (ii) and (iii).

We will prove that (ii) ⇒ (i). For this purpose, we will show that B is a closed, and in the same time an open in D.

We see that, if $a \in B$, from the analyticity of f it follows that there exists a number $r > 0$, such that

$$f(z) = \sum_{n=0}^{\infty} a_n (z-a)^n, \quad \forall z \in U(a;r),$$

where $a_n = \frac{f^{(n)}(a)}{n!} = 0$, $\forall n \in \mathbb{N}$, because $a \in B$. Then $f(z) = 0$, $\forall z \in U(a;r)$, and consequently, $f^{(n)}(z) = 0$, $\forall n \in \mathbb{N}$ and $\forall z \in U(a;r)$. Thus $z \in B$, $\forall z \in U(a;r)$, that is the set B is open in D.

Let $(z_n)_{n\in\mathbb{N}} \subset B$ and $a = \lim_{n\to+\infty} z_n \in D$. Since $f^{(k)} \in H(D)$, the function $f^{(k)}$ is continuous at a, $\forall k \in \mathbb{N}$. Hence, $\lim_{n\to+\infty} f^{(k)}(z_n) = f^{(k)}(a)$. But $f^{(k)}(z_n) = 0$, because $z_n \in B$, and thus $f^{(k)}(a) = 0$, $\forall k \in \mathbb{N}$, so $a \in B$. Hence, the set B is closed in D. From the fact that D is connected, we conclude that $B = D$, and then $f \equiv 0$.

Now we will prove that (iii) ⇒ (ii). Let $a \in A' \cap D$. From the continuity of f, we have $f(a) = 0$. We will prove that $a \in B$. If this would not be true, then there exists the smallest $m \in \mathbb{N}$ natural number, such that $f^{(m)}(a) \neq 0$. Then the number $a \in D$ will be an mth order zero for f, and by Theorem 4.5.1 there exists a function $g \in H(D)$, such that $f(z) = (z-a)^m g(z)$ and $g(a) \neq 0$. Since $a \in A'$, there exists a sequence $(a_n)_{\in\mathbb{N}} \subset A \setminus \{a\}$ such that $\lim_{n\to+\infty} a_n = a$. Since $f(a_n) = 0$ and $a_n - a \neq 0$, $\forall n \in \mathbb{N}$, $g(a_n) = 0$, $\forall n \in \mathbb{N}$, from the continuity of g it follows that $g(a) = \lim_{n\to+\infty} g(a_n) = 0$, which represents a contradiction. □

Corollary 4.5.1 (Properties of the zeros holomorphic functions). *If $D \subset \mathbb{C}$ is a domain, $f \in H(D)$, $f \not\equiv 0$ and $a \in D$ is a zero for the function f, then:*
1. *$\exists r > 0$ such that $f(z) \neq 0$, $\forall z \in \dot{U}(a;r) = U(a;r) \setminus \{a\}$;*
2. *the number a is a finite order zero for the function f;*
3. *$\exists n \in \mathbb{N}^*$ and $\exists g \in H(D)$, such that $f(z) = (z-a)^n g(z)$, $\forall z \in D$ and $g(a) \neq 0$.*

Theorem 4.5.3 (Uniqueness theorem of holomorphic functions). *Let $D \subset \mathbb{C}$ a domain, and let $f, g \in H(D)$. Suppose that one of the next assumptions holds:*
1. *$\exists E \subset D$, such that $E' \cap D \neq \emptyset$ and $f|_E = g|_E$*
   (that is $\{z \in D : f(z) = g(z)\}' \cap D \neq \emptyset$);
2. *$\exists a \in D$, such that $f^{(n)}(a) = g^{(n)}(a)$, $\forall n \in \mathbb{N}$.*
   Then $f \equiv g$.

Proof. We will apply Theorem 4.5.2 for the $f - g$ function. □

Remarks 4.5.1.
1. From the Cauchy formula, it follows that the function $f \in H(D)$ is well determined if we know its values on a closed curve (path).

2. According to Theorem 4.5.3, the function $f \in H(D)$ is well determined if we know its values on an arbitrary sequence that converges in D.

Corollary 4.5.2. *Let $D \subset \mathbb{C}$ be a domain. Then the set $H(D)$ is a domain of integrity with respect to the addition and the multiplication of functions.*

Proof. Suppose that $f, g \in H(D)$ are such functions, that $f \cdot g \equiv 0$. If $f \not\equiv 0$, then the set $A = A(f)$ (the set of f function zeros) consists from isolated points. In order that $f(z)g(z) = 0$ holds $\forall z \in D$, it is necessary that $g(z) = 0$, $\forall z \in D \setminus A$. Since $(D\setminus A)' \cap D \neq \emptyset$, from Theorem 4.5.3, we deduce that $g \equiv 0$. Hence, the assumption $f \cdot g \equiv 0$ implies that $f \equiv 0$ or $g \equiv 0$. □

Theorem 4.5.4 (The principle of the continuation of holomorphic functions). *Let D and $\widetilde{D}$ two domains in $\mathbb{C}$, and $D \subset \widetilde{D}$, $D \neq \emptyset$. If $f \in H(D)$ and $\widetilde{f} \in H(\widetilde{D})$ are two functions, such that $\widetilde{f}|_D = f$, then the function $\widetilde{f}$ is the unique continuation (extension) of f to the domain $\widetilde{D}$.*

The function $\widetilde{f}$ is called the **holomorphic (analytic) continuation (extension)** *of f.*

Proof. In fact, if $\widetilde{f}_1 \in H(\widetilde{D})$ and $\widetilde{f}_1|_D = f$, then $\widetilde{f}_1|_D = \widetilde{f}|_D$. Since $D' \cap \widetilde{D} \neq \emptyset$, from Theorem 4.5.3 we have $\widetilde{f}_1 = \widetilde{f}$. □

4.6 The maximum principle of the holomorphic functions

Theorem 4.6.1 (The maximum principle of the holomorphic functions). *If $D \subset \mathbb{C}$ is a domain, and $f \in H(D)$ is a nonconstant function, then $|f|$ cannot take its maximum value in the domain D.*

Proof. Suppose that there exists $z_0 \in D$, such that $|f(z_0)| \geq |f(z)|$, $\forall z \in D$. Let $r > 0$ such that $\overline{U}(z_0; r) \subset D$, and let denote $\gamma = \partial U(z_0; r)$. Thus $\{\gamma\} \subset D$, and using Cauchy formula we get

$$f(z_0) = \frac{1}{2\pi i} \int_\gamma \frac{f(\zeta)}{\zeta - z_0} d\zeta.$$

We know that $\gamma(t) = z_0 + re^{2\pi i t}$, $t \in [0, 1]$, hence

$$\begin{aligned} f(z_0) &= \frac{1}{2\pi i} \int_0^1 \frac{f(z_0 + re^{2\pi i t})}{re^{2\pi i t}} d(re^{2\pi i t}) \\ &= \frac{1}{2\pi i} \int_0^1 \frac{f(\gamma(t))}{re^{2\pi i t}} \cdot r2\pi i e^{2\pi i t} dt = \int_0^1 f(\gamma(t)) dt. \end{aligned}$$

It follows

$$|f(z_0)| = \left|\int_0^1 f(\gamma(t))\mathrm{d}t\right| \le \int_0^1 |f(\gamma(t))|\mathrm{d}t,$$

or

$$0 \le \int_0^1 (|f(\gamma(t))| - |f(z_0)|)\mathrm{d}t \le \int_0^1 (|f(z_0)| - |f(z_0)|)\mathrm{d}t = 0.$$

This yields that $|f(\gamma(t))| = |f(z_0)|$, $\forall t \in [0, 1]$.

Hence, on the set $\{\gamma\}$ we have $|f(z)| = |f(z_0)|$. But the radius of γ can be changed, to be arbitrary and less or equal to r. Thus we deduce that $|f(z)|$ is constant for all $z \in U(z_0; r)$. From a well-known previous result, we get that f is constant on the $U(z_0; r)$ disc. Then, according to the principle of the continuation of holomorphic functions, we conclude that f is constant on the whole domain D. □

Remark 4.6.1. The previous theorem has the next equivalent form:

If $D \subset \mathbb{C}$ is a domain, $f \in H(D)$ and there exists $z_0 \in D$ such that $|f(z)| \le |f(z_0)|$, $\forall z \in D$, then f is constant on D.

Corollary 4.6.1. *If $f \in H(D)$ is a nonconstant function, and if $f(z) \ne 0$, $\forall z \in D$, where D is a domain, then $|f|$ cannot take its minimum value in the domain D.*

Proof. From the assumption, we have $g = \frac{1}{f} \in H(D)$, and since $|g|$ cannot take its maximum value in D, it follows that $|f|$ cannot take its minimum value in D. □

Corollary 4.6.2. *Let $D \subset \mathbb{C}$ be a bounded domain, and let $f \in C(\overline{D}) \cap H(D)$. Then*

$$\max_{z \in \overline{D}} |f(z)| = \max_{z \in \partial D} |f(z)|.$$

Proof. Since $f \in C(\overline{D})$, there exists $z_0 \in \overline{D}$ such that $|f(z)| \le |f(z_0)|$, $\forall z \in \overline{D}$. If the function f is nonconstant, from Theorem 4.6.1 we get that $z_0 \notin D$, hence $z_0 \in \partial D$. □

The next result is one of the extensions of Liouville theorem.

Theorem 4.6.2. *If $f \in H(\mathbb{C})$, that is f is an entire function, and if*

$$|f(z)| \le \frac{1}{|\operatorname{Im} z|}, \quad \forall z \in \mathbb{C}, \tag{4.2}$$

then $f(z) = 0$, $\forall z \in \mathbb{C}$.

Proof. Evidently, the function f is bounded on the set $\mathbb{C} \setminus \Delta_\varepsilon$ for all arbitrary $\varepsilon > 0$, where $\Delta_\varepsilon = \{z \in \mathbb{C} : |\operatorname{Im} z| < \varepsilon\}$, but it is not sure that it is bounded on Δ_ε.

Let $r > 0$, and let define the function $g(z) = (z^2 - r^2)f(z)$, $z \in \mathbb{C}$.

If $z \in \mathbb{C}$ with $|z| = r$ and $\operatorname{Re} z \geq 0$, then there exists $\theta(z) \in [0, \frac{\pi}{4}]$ such that $\cos\theta(z) = \frac{|\operatorname{Im} z|}{|z-r|}$. Then, according to relation (4.2), we get

$$|(z-r)f(z)| \leq \frac{|z-r|}{|\operatorname{Im} z|} = \frac{1}{\cos\theta(z)} \leq \sqrt{2}, \quad \forall z \in \mathbb{C} : |z| = r,\ \operatorname{Re} z \geq 0. \tag{4.3}$$

If $z \in \mathbb{C}$ with $|z| = r$ and $\operatorname{Re} z < 0$, similarly we obtain

$$|(z+r)f(z)| \leq \frac{|-z-r|}{|\operatorname{Im}(-z)|} = \frac{1}{\cos\theta(-z)} \leq \sqrt{2}, \quad \forall z \in \mathbb{C} : |z| = r,\ \operatorname{Re} z < 0. \tag{4.4}$$

Now, from (4.3) and (4.4), we respectively deduce that

$$|g(z)| = |(z-r)f(z)||z+r| \leq 2\sqrt{2}r, \quad \text{if } |z| = r,\ \operatorname{Re} z \geq 0$$
$$|g(z)| = |(z+r)f(z)||z-r| \leq 2\sqrt{2}r, \quad \text{if } |z| = r,\ \operatorname{Re} z < 0,$$

hence $|g(z)| \leq 2\sqrt{2}r$, if $z \in \mathbb{C}$ and $|z| = r$. From here, using Theorem 4.6.1, we conclude that

$$|g(z)| \leq 2\sqrt{2}r, \quad \forall z \in \mathbb{C} : |z| \leq r,$$

hence

$$|f(z)| \leq \frac{2\sqrt{2}r}{|z^2 - r^2|}, \quad \forall z \in \mathbb{C} : |z| < r,$$

and if in this last inequality we take $r \to +\infty$, it follows that $f(z) = 0$, $\forall z \in \mathbb{C}$. □

Theorem 4.6.3. *Let* $\Delta = \{z \in \mathbb{C} : 0 < \operatorname{Re} z < 1\}$, *let* $f \in H(\Delta)$ *such that* $f \in C(\overline{\Delta})$, *and suppose that* f *is bounded on the compact set* $\overline{\Delta}$. *For* $\theta \in [0,1]$, *let define the function*

$$M_\theta(f) = \sup\{|f(\theta + iy)| : y \in \mathbb{R}\}.$$

Then

$$M_\theta(f) \leq [M_0(f)]^{1-\theta} \cdot [M_1(f)]^{\theta}.$$

Proof. Without loss of the generality, suppose that $M_0(f) > 0$ and $M_1(f) > 0$. Moreover, if we replace the function f by $f(z) \cdot [M_0(f)]^{\theta-1} \cdot [M_1(f)]^{-\theta}$, we may suppose that $M_0(f) = M_1(f) = 1$, and in this last case we need to prove that $|f(z)| \leq 1$, $\forall z \in \overline{\Delta}$.

Let $M = \sup\{|f(z)| : z \in \overline{\Delta}\} < +\infty$, and define the function

$$f_n(z) = \frac{f(z)}{1 + \frac{M}{n}z}, \quad z \in \overline{\Delta},\ n \in \mathbb{N}^*.$$

If we denote by ∂R_n the boundary of the rectangle with corners in $-in$, $1-in$, $1+in$ and in, i. e.,

$$R_n = \{z \in \mathbb{C} : 0 < \operatorname{Re} z < 1, |\operatorname{Im} z| < n\},\ n \in \mathbb{N}^*,$$

then we may easily prove that

$$|f_n(z)| \le 1, \quad \forall z \in \partial\Delta : |\operatorname{Im} z| \le n, \text{ or } \forall z \in \Delta : |\operatorname{Im} z| = n.$$

It follows that $|f_n(z)| \le 1$, $\forall z \in \partial R_n$, $\forall n \in \mathbb{N}^*$. From here, according to Theorem 4.6.1, we deduce that

$$|f_n(z)| \le 1, \quad \forall z \in \overline{R}_n, \ \forall n \in \mathbb{N}^*,$$

and since $\lim_{n\to+\infty} f_n(z) = f(z)$, $\forall z \in \overline{\Delta}$, we conclude that $|f(z)| \le 1$, $\forall z \in \overline{\Delta}$. □

Theorem 4.6.4 (Schwarz lemma). *If $f \in H(U(0;1))$ satisfies the conditions $f(0) = 0$ and $|f(z)| < 1$, $\forall z \in U(0;1)$, then:*
1. $|f(z)| \le |z|$, $\forall z \in U(0;1)$;
2. $|f'(0)| \le 1$.

If there exists $z_0 \in U(0;1)$ such that $|f(z_0)| = |z_0|$, or such that $|f'(0)| = 1$, then f has the form $f(z) = cz$, where $c \in \mathbb{C}$, $|c| = 1$.

Proof. Define the function g by

$$g(z) = \begin{cases} \frac{f(z)}{z}, & \text{if } z \in \dot{U} = U(0;1) \setminus \{0\}, \\ f'(0), & \text{if } z = 0. \end{cases}$$

Then $g \in C(U) \cap H(\dot{U})$, and according to a known result we have $g \in H(U)$. Let r be an arbitrary number, such that $0 < r < 1$. From Corollary 4.6.2, it follows

$$\max_{z\in\overline{U}(0;r)} |g(z)| = \max_{z\in\partial U(0;r)} |g(z)|.$$

If $z \in \partial U(0;r)$, from the assumption of the theorem we obtain

$$|g(z)| = \frac{|f(z)|}{|z|} \le \frac{1}{r},$$

and according to the previous remarks

$$|g(z)| \le \frac{1}{r}, \quad \forall z \in \overline{U}(0;r).$$

Letting $r \to 1$, we get

$$|g(z)| \le 1, \quad \forall z \in U(0;1),$$

and from the definition of g we have

$$|f(z)| \le |z|, \quad \forall z \in U(0;1).$$

If $z = 0$, then

$$|g(0)| \le 1, \quad \text{that is } |f'(0)| \le 1.$$

Suppose that there exists $z_0 \in U$ such that $|f(z_0)| = |z_0|$. Then $|g(z_0)| = 1$ and since $|g(z)| \le 1$, $\forall z \in U$, according to the Theorem of the module maximum of the holomorphic functions it follows that $|g(z)| = 1$, $\forall z \in U$. So, the holomorphic function g has a constant module on the domain U, hence g is constant on U, i. e.,

$$g(z) = c, \quad \forall z \in U,$$

where $|c| = 1$. From the definition of g, we now deduce that

$$f(z) = cz, \quad \forall z \in U.$$

We may will use the same proof if $|f'(0)| = 1$, because $g(0) = f'(0)$. □

Remark 4.6.2. The geometric interpretation of Schwarz lemma is the next:

Let $f \in H(U(0;1))$ a such a function, with $f(0) = 0$ and $f(U(0;1)) \subset U(0;1)$. Then the inclusion $f(U(0;r)) \subset U(0;r)$ holds for every $r \in (0,1)$. There exists a number $r_* \in (0,1)$ such that $f(U(0;r_*)) = U(0;r_*)$ if and only if the function f has the form $f(z) = cz$, where $|c| = 1$.

The Schwarz lemma can be generalized as follows.

Theorem 4.6.5 (Generalized Schwarz lemma). *If the function $f \in H(U(z_0;r))$ satisfies the condition $|f(z) - f(z_0)| < R$, $\forall z \in U(z_0;r)$, then:*
1. $|f(z) - f(z_0)| \le \frac{R}{r}|z - z_0|$, $\forall z \in U(z_0;r)$;
2. $|f'(z_0)| \le \frac{R}{r}$.

If there exists $z_* \in U(z_0;r)$ such that $|f(z_*) - f(z_0)| = \frac{R}{r}|z_* - z_0|$, or such that $|f'(z_0)| = \frac{R}{r}$, then the function f has the form $f(z) = f(z_0) + c\frac{R}{r}(z - z_0)$, where $c \in \mathbb{C}$, $|c| = 1$.

Proof. Let us define the function $g(\zeta) = \frac{1}{R}[f(z) - f(z_0)]$, where $z = z_0 + r\zeta$ and $\zeta \in U$. It follows that $g \in H(U)$, $g(0) = 0$ and $|g(\zeta)| < 1$, $\forall \zeta \in U$. Using the Schwarz lemma, we get $|g(\zeta)| \le |\zeta|$, $\forall \zeta \in U$ and $|g'(0)| \le 1$, hence

$$|f(z) - f(z_0)| \le \frac{R}{r}|z - z_0|, \quad \forall z \in U(z_0;r) \text{ and } |f'(z_0)| \le \frac{R}{r}.$$

If there exists $z_* \in U(z_0;r)$ such that $|f(z_*) - f(z_0)| = \frac{R}{r}|z_* - z_0|$, or such that $|f'(z_0)| = \frac{R}{r}$, then $|g(\zeta_*)| = |\zeta_*|$ or $|g'(0)| = 1$, where $z_* = z_0 + r\zeta_*$. Applying the Schwarz lemma, we deduce that g has the form $g(\zeta) = c\zeta$, where $c \in \mathbb{C}$ and $|c| = 1$, i. e., $f(z) = f(z_0) + c\frac{R}{r}(z - z_0)$, with $c \in \mathbb{C}$, $|c| = 1$. □

Remark 4.6.3. The generalized Schwarz lemma has the following geometric interpretation:

Let $f \in H(U(z_0;r))$ a function that satisfies $f(U(z_0;r)) \subset U(f(z_0);R)$. Then $f(U(z_0;r_1)) \subset U(f(z_0);R_1)$ for all $r_1 \in (0,r)$, with $\frac{R_1}{R} = \frac{r_1}{r}$. There exists a number $r_1 \in (0,r)$ such that $\frac{R_1}{R} = \frac{r_1}{r}$ and $f(U(z_0;r_1)) = U(f(z_0);R_1)$, if and only if the function f has the form $f(z) = f(z_0) + c\frac{R}{r}(z-z_0)$, $|c| = 1$.

4.7 Laurent series

Definition 4.7.1.

1. If $z_0 \in \mathbb{C}$ is a given point, then **a Laurent series about** z_0 will be the sum of the form

$$\sum_{n=-\infty}^{\infty} a_n(z-z_0)^n = \cdots + \frac{a_{-n}}{(z-z_0)^n} + \cdots + \frac{a_{-1}}{z-z_0} + a_0 + \cdots + a_n(z-z_0)^n + \cdots,$$

   where $a_n \in \mathbb{C}$, $n \in \mathbb{Z}$, are called **the Laurent series coefficients.**
2. If $a_n = 0$, $\forall n < 0$, then the Laurent series will be a power series.
3. The sum

$$P(z) = \sum_{n=1}^{\infty} a_{-n}(z-z_0)^{-n}$$

   is called **the main part of the Laurent series.**
4. The sum

$$T(z) = \sum_{n=0}^{\infty} a_n(z-z_0)^n$$

   is called **the Taylor part of the Laurent series.**
5. The Laurent series is said to be **pointwise convergent (or uniformly convergent)** on the set $E \subset \mathbb{C} \setminus \{z_0\}$, if the main and the Taylor parts are pointwise convergent (or respectively, uniformly convergent) on the set E.

Theorem 4.7.1 (Theorem of the annulus of convergence). *Let us consider the Laurent series*

$$S(z) = \sum_{n=-\infty}^{\infty} a_n(z-z_0)^n = P(z) + T(z),$$

and let

$$r = \limsup_{n\to+\infty} \sqrt[n]{|a_{-n}|}, \quad \frac{1}{R} = \limsup_{n\to+\infty} \sqrt[n]{|a_n|},$$

where for the values of R we used the conventions from the Cauchy–Hadamard theorem (Theorem 4.3.2).

If $r < R$, then:
1. *on the circular ring $U(z_0; r, R)$ the Laurent series S is absolute and compact convergent;*
2. *the Laurent series S diverges in all the points of the set $\mathbb{C} \setminus \overline{U}(z_0; r, R)$;*
3. *the sum $S(z) = \sum_{n=-\infty}^{\infty} a_n (z - z_0)^n$ is holomorphic on the circular ring $U(z_0; r, R)$.*

Proof. The Taylor part of the Laurent series S has a radius of convergence R, given by the Cauchy–Hadamard theorem. Also, the Taylor part $T(z)$ compactly converges on the disc $U(z_0; R)$, and diverges in all the points of $\mathbb{C} \setminus \overline{U}(z_0; R)$. Moreover, the sum $T(z)$ is a holomorphic function in the disc $U(z_0; R)$.

In the main part $P(z)$ of the Laurent series $S(z)$, let use the substitution $w = \frac{1}{z-z_0}$. The power series obtained in this way, i. e.,

$$\sum_{n=1}^{\infty} a_{-n} w^n,$$

has the disc of convergence $U(0; \frac{1}{r})$, where it is absolute and compact convergent, and the sum will be a holomorphic with respect to the variable w in $U(0; \frac{1}{r})$. The inequality $|w| < \frac{1}{r}$ is equivalent to the assumption $|z - z_0| > r$. It follows that the main part is absolute and compact convergent on the open set $\mathbb{C} \setminus \overline{U}(z_0; r)$, and thus, the sum of the main part is holomorphic on $\mathbb{C} \setminus \overline{U}(z_0; r)$. Also, the main part diverges in all the points that satisfy the inequality $|z - z_0| < r$.

We conclude that the Laurent series S is absolute and compact convergent on the circular ring $U(z_0; r, toR)$, and its sum is holomorphic in this open set.

In all the points of the set $\mathbb{C} \setminus \overline{U}(z_0; r, R)$, at least one of the parts $P(z)$ or $T(z)$ are divergent, hence the Laurent series S diverges in these points. □

Theorem 4.7.2 (The identity of the Laurent series coefficients). *If the next two Laurent series $\sum_{n=-\infty}^{\infty} a_n (z - z_0)^n$ and $\sum_{n=-\infty}^{\infty} b_n (z - z_0)^n$ have the same sums on the circular ring $U(z_0; r, R)$, with $0 \le r < R$, then $a_n = b_n$, $\forall n \in \mathbb{Z}$.*

Proof. Denoting $c_n = a_n - b_n$, then the sum of the Laurent series $\sum_{n=-\infty}^{\infty} c_n (z - z_0)^n$ will be identically zero on the circular ring $U(z_0; r, R)$. Multiplying this by $(z - z_0)^{-n-1}$, we deduce that the sum of this new Laurent series will be also identically zero, i. e.,

$$\sum_{k=-\infty}^{\infty} c_k (z - z_0)^{k-n-1} = 0, \quad \forall z \in U(z_0; r, R).$$

Letting $\gamma = \partial U(z_0; \frac{r+R}{2})$, since the above series uniformly converges on the compact $\{\gamma\}$, it can be integrated term by term, and thus,

$$\sum_{k=-\infty}^{\infty} c_k \int_{\gamma} (z - z_0)^{k-n-1} dz = 2\pi i c_n = 0,$$

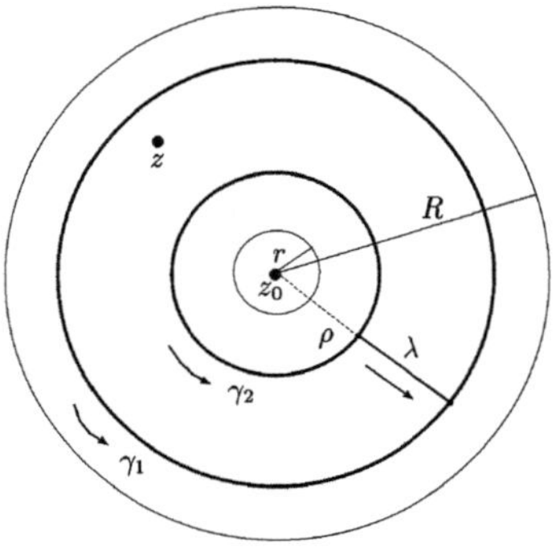

Figure 4.1: Proof of Theorem 4.7.3.

i. e., $c_n = a_n - b_n = 0$. Since $n \in \mathbb{Z}$ was chosen arbitrary, we conclude that

$$a_n = b_n, \quad \forall n \in \mathbb{Z}.$$

□

Theorem 4.7.3 (Laurent series expansion theorem). *Let f be a holomorphic function on the circular ring $D = U(z_0; r, R)$ $(0 \le r < R)$. Then the Laurent series*

$$\sum_{n=-\infty}^{\infty} a_n(z - z_0)^n,$$

where

$$a_n = \frac{1}{2\pi i} \int_{\gamma} \frac{f(\zeta)}{(\zeta - z_0)^{n+1}} d\zeta$$

(with $\gamma = \partial U(z_0; \rho)$, $r < \rho < R$) has a ring of convergence that contains the domain D, and the sum of the Laurent series in the domain D is equal to f, i. e.,

$$f(z) = \sum_{n=-\infty}^{\infty} a_n(z - z_0)^n, \quad \forall z \in D.$$

Proof. We remark that according to the Cauchy theorem, the integrals that appeared in the definitions of the coefficients a_n are independent of ρ, hence the only assumption for ρ is $r < \rho < R$.

Let $z \in D$ be a given arbitrary point, let choose the real numbers $r_1, r_2 \in \mathbb{R}$ such that

$$r < r_2 < |z - z_0| < r_1 < R,$$

and let denote $\gamma_1 = \partial U(z_0; r_1)$, $\gamma_2 = \partial U(z_0; r_2)$.

Let us consider the linear path λ that connects the paths γ_2 with γ_1, such that $z_0 \in \{\gamma\}$. Then the path $\tilde{\gamma} = \lambda \cup \gamma_1 \cup \lambda^-$ satisfies the condition $\tilde{\gamma} \underset{D}{\sim} \gamma_2$ (Figure 4.1).

Considering the function

$$g(\zeta) = \begin{cases} \frac{f(\zeta)-f(z)}{\zeta-z}, & \text{if } \zeta \in D \setminus \{z\}, \\ f'(z), & \text{if } \zeta = z, \end{cases}$$

then $g \in H(D)$, and thus,

$$\int_{\tilde{\gamma}} g = \int_{\gamma_2} g \Rightarrow \int_{\lambda} g + \int_{\gamma_1} g + \int_{\lambda^-} g = \int_{\gamma_2} g,$$

i. e.,

$$\int_{\gamma_1} g = \int_{\gamma_2} g. \tag{4.5}$$

Further,

$$\int_{\gamma_1} g = \int_{\gamma_1} \frac{f(\zeta) - f(z)}{\zeta - z} \mathrm{d}\zeta = \int_{\gamma_1} \frac{f(\zeta)}{\zeta - z} \mathrm{d}\zeta - f(z) \underbrace{\int_{\gamma_1} \frac{\mathrm{d}\zeta}{\zeta - z}}_{2\pi i} \tag{4.6}$$

and

$$\int_{\gamma_2} g = \int_{\gamma_2} \frac{f(\zeta) - f(z)}{\zeta - z} \mathrm{d}\zeta = \int_{\gamma_2} \frac{f(\zeta)}{\zeta - z} \mathrm{d}\zeta - f(z) \underbrace{\int_{\gamma_2} \frac{\mathrm{d}\zeta}{\zeta - z}}_{0}. \tag{4.7}$$

Computing the value of $f(z)$ from (4.6), and using the relations (4.5) and (4.7), we get

$$f(z) = \frac{1}{2\pi i} \int_{\gamma_1} \frac{f(\zeta)}{\zeta - z} \mathrm{d}\zeta - \frac{1}{2\pi i} \int_{\gamma_1} g = \frac{1}{2\pi i} \int_{\gamma_1} \frac{f(\zeta)}{\zeta - z} \mathrm{d}\zeta - \frac{1}{2\pi i} \int_{\gamma_2} \frac{f(\zeta)}{\zeta - z} \mathrm{d}\zeta. \tag{4.8}$$

(Note that the formula (4.8) is called **the Cauchy formula for circular rings.**)

In the first integral of the right-hand side of (4.8), we will use the same computation as to the proof of the Taylor series expansion theorem:

$$\frac{1}{\zeta - z} = \frac{1}{\zeta - z_0} \frac{1}{1 - \frac{z-z_0}{\zeta-z_0}}, \quad \text{where } \zeta \in \gamma_1, \text{ and thus } \left|\frac{z - z_0}{\zeta - z_0}\right| < 1,$$

hence

$$\frac{f(\zeta)}{\zeta - z} = \frac{f(\zeta)}{\zeta - z_0} \sum_{n=0}^{\infty} \left(\frac{z - z_0}{\zeta - z_0}\right)^n.$$

The series of functions from the right-hand side is uniformly convergent on the compact $\{\gamma_1\}$, consequently

$$\frac{1}{2\pi i}\int_{\gamma_1}\frac{f(\zeta)}{\zeta-z}\mathrm{d}\zeta=\frac{1}{2\pi i}\sum_{n=0}^{\infty}\int_{\gamma_1}f(\zeta)\frac{(z-z_0)^n}{(\zeta-z_0)^{n+1}}\mathrm{d}\zeta=\sum_{n=0}^{\infty}a_n(z-z_0)^n,$$

where

$$a_n=\frac{1}{2\pi i}\int_{\gamma_1}\frac{f(\zeta)}{(\zeta-z_0)^{n+1}}\mathrm{d}\zeta.$$

Now we will calculate the last integral that appeared in the formula (4.8). Since $\zeta\in\{\gamma_2\}$, it follows that $|\frac{\zeta-z_0}{z-z_0}|<1$, and then we deduce that

$$-\frac{1}{\zeta-z}=\frac{1}{z-\zeta}=\frac{1}{z-z_0}\frac{1}{1-\frac{\zeta-z_0}{z-z_0}}=\frac{1}{z-z_0}\sum_{n=0}^{\infty}\left(\frac{\zeta-z_0}{z-z_0}\right)^n,$$

where the power series from the last term is uniformly convergent (with the variable ζ) on the compact $\{\gamma_2\}$. Hence,

$$\begin{aligned}-\frac{1}{2\pi i}\int_{\gamma_2}\frac{f(\zeta)}{\zeta-z}&=\sum_{n=0}^{\infty}\frac{1}{(z-z_0)^{n+1}}\frac{1}{2\pi i}\int_{\gamma_2}f(\zeta)(\zeta-z_0)^n\mathrm{d}\zeta\\&=\sum_{m=1}^{\infty}a_{-m}(z-z_0)^{-m},\end{aligned}$$

where

$$a_{-m}=\frac{1}{2\pi i}\int_{\gamma_2}\frac{f(\zeta)}{(\zeta-z_0)^{-m+1}}\mathrm{d}\zeta.$$

Since the integrals that appear in the definitions of a_n and a_{-m} are independent on the values of r_1 and r_2, we have that

$$a_n=\frac{1}{2\pi i}\int_{\gamma}\frac{f(\zeta)}{(\zeta-z_0)^{n+1}}\mathrm{d}\zeta,\quad\forall n\in\mathbb{Z},\tag{4.9}$$

where $\gamma=\partial U(z_0;\rho)$, with $r<\rho<R$, and ρ is the above choose arbitrary number.

Finally, from the relation (4.8) we conclude that

$$f(z)=\sum_{n=-\infty}^{\infty}a_n(z-z_0)^n,$$

where the coefficients a_n are given by the formulas (4.9). □

4.8 Isolated singular points

Definition 4.8.1.

1. Let $f \in H(G)$, where $G \subset \mathbb{C}$ is an open set. The number $z_0 \in \mathbb{C}$ is said to be **an isolated singular point** for f, if $z_0 \notin G$, but there exists $r > 0$ such that $\dot{U}(z_0; r) \subset G$, where $\dot{U}(z_0; r) = U(z_0; r) \setminus \{z_0\}$.
2. An isolated singular point z_0 for $f \in H(G)$ is said to be **a removable** isolated singular point, if there exists a function $\widetilde{f} \in H(\widetilde{G})$, where $\widetilde{G} = G \cup \{z_0\}$ such that $\widetilde{f}|_G = f$, (it means that f has a holomorphic extension (continuation) in z_0).
3. The points of G together with all the removable isolated singular points of f are called **regular** points for f.

Theorem 4.8.1 (The first removability criterion). *An isolated singular point $z_0 \in \mathbb{C}$ for the function $f \in H(G)$, where $G \subset \mathbb{C}$ is an open set, is removable if and only if there exists $\lim_{z\to z_0} f(z) = l \in \mathbb{C}$.*

Proof. If z_0 is removable, then there exists a function $\widetilde{f} \in H(\widetilde{G})$, where $\widetilde{G} = G \cup \{z_0\}$ and $\widetilde{f}|_G = f$. It follows that $\widetilde{f}$ is continuous on z_0, hence $\lim_{z\to z_0} \widetilde{f}(z) = \widetilde{f}(z_0) = l \in \mathbb{C}$, i. e., $\lim_{z\to z_0} f(z) = l \in \mathbb{C}$.

Suppose that there exists the limit from the assumption of the theorem. Then, if we define

$$\widetilde{f}(z) = \begin{cases} f(z), & \text{if } z \in G, \\ l = \lim_{z\to z_0} f(z), & \text{if } z = z_0, \end{cases}$$

it follows that $\widetilde{f} \in H(\widetilde{G} \setminus \{z_0\})$ and $\widetilde{f} \in C(\widetilde{G})$, where $\widetilde{G} = G \cup \{z_0\}$. According to a well-known result, we obtain that $\widetilde{f} \in H(\widetilde{G})$ and $\widetilde{f}|_G = f$, hence z_0 is removable. □

Examples 4.8.1.

1. Let $f(z) = \frac{\sin z}{z}$, $z \in \mathbb{C}^* = \mathbb{C} \setminus \{0\}$. Since

$$\sin z = z - \frac{1}{3!}z^3 + \frac{1}{5!}z^5 - \cdots \Rightarrow \frac{\sin z}{z} = 1 - \frac{1}{3!}z^2 + \frac{1}{5!}z^4 - \cdots \quad \text{and}$$
$$\lim_{z\to 0} \frac{\sin z}{z} = 1,$$

it follows that $z_0 = 0$ is a removable isolated singular point.

2. Let $f(z) = \frac{1}{z}$, $z \in \mathbb{C}^*$. Then $\lim_{z\to 0} f(z) = \infty \notin \mathbb{C}$, hence $z_0 = 0$ is not a removable isolated singular point.
3. Let $f(z) = \sin\frac{1}{z}$, $z \in \mathbb{C}^*$. Like we know, $\nexists \lim_{z\to 0} f(z)$, because $\nexists \lim_{\substack{x\to 0\\ x\in\mathbb{R}}} \sin\frac{1}{x}$, hence $z_0 = 0$ is not a removable isolated singular point.

Theorem 4.8.2 (The second removability criterion). *Let $f \in H(G)$, where $G \subset \mathbb{C}$ is an open set. An isolated singular point $z_0 \in \mathbb{C}$ for the function f is removable, if and only if* *the main part of the Laurent-series expansion*

$$f(z) = \sum_{n=-\infty}^{\infty} a_n (z - z_0)^n, \quad z \in \dot{U}(z_0; R), \tag{4.10}$$

vanishes, i. e., $a_n = 0$ for all $n < 0$.

Proof. If z_0 is removable and $\widetilde{f}$ is the holomorphic extension (continuation) of f to the set $\widetilde{G} = G \cup \{z_0\}$, then $\widetilde{f}$ can be expanded in Taylor-series in a $U(z_0; R) \subset \widetilde{G}$ disc. This Taylor expansion need to be identical to the Laurent-series expansion given by (4.10) (according to the uniqueness of Laurent-series coefficients), since $\widetilde{f} = f$ in the circular ring $\dot{U}(z_0; R)$.

Conversely, if the Laurent expansion (4.10) has only Taylor part, then its sum is an holomorphic function in a disc $U(z_0; R)$, and it is equal to f on the whole set $\dot{U}(z_0; R)$. So, the function f has an holomorphic extension on z_0, thus z_0 is a removable isolated singular point. □

Theorem 4.8.3 (The third, or Cauchy–Riemann, removability criterion). *Let $f \in H(G)$, where $G \subset \mathbb{C}$ is an open set. An isolated singular point $z_0 \in \mathbb{C}$ for the function f is removable, if and only if there exists $M > 0$ and $R > 0$ such that $|f(z)| < M$, $\forall z \in \dot{U}(z_0; R) \subset G$.*

Proof. Let $0 < r < R$. Then the function f may be expanded in the circular ring $\dot{U}(z_0; R)$ in a Laurent expansion of the form (4.10), and according to a well-known theorem, these coefficients are given by

$$a_n = \frac{1}{2\pi i} \int_{\gamma} \frac{f(\zeta)}{(\zeta - z_0)^{n+1}} d\zeta, \quad \forall n \in \mathbb{Z}.$$

Using the same proof as for Cauchy inequalities, for all integers $n \in \mathbb{Z}$ we have

$$|a_n| \le \frac{M}{r^n}.$$

If $n < 0$, then $r^{-n} \to 0$, if $r \to 0$, hence

$$|a_n| \le \frac{M}{r^n} \to 0, \quad \text{if } r \to 0.$$

Since we can choose the arbitrary number r as small is possible, it follows that

$$|a_n| = 0, \quad \forall n < 0,$$

hence the main part of (4.10) is identically zero.

Conversely, if z_0 is a removable isolated singular point for f, there exists $\widetilde{f} \in H(\widetilde{G})$ such that $\widetilde{f}|_G = f$ and $\widetilde{G} = G \cup \{z_0\}$. Since $\exists R > 0$ such that $\overline{U}(z_0; R) \subset \widetilde{G}$, the function $\widetilde{f}$ is bounded on the compact set $\overline{U}(z_0; R)$, hence f is bounded on $\dot{U}(z_0; R)$. □

Corollary 4.8.1. *If $z_0 \in \mathbb{C}$ is not a removable isolated singular point for the function $f \in H(G)$, where $G \subset \mathbb{C}$ is an open set, then there exists a sequence $(z_n)_{n\in\mathbb{N}^*} \subset G$, $z_n \to z_0$, such that $\lim_{n\to+\infty} f(z_n) = \infty$.*

Definition 4.8.2. Let $z_0 \in \mathbb{C}$ be a nonremovable isolated singular point for $f \in H(G)$, where $G \subset \mathbb{C}$ is an open set.
1. If $\lim_{z\to z_0} f(z) = \infty$, then z_0 is said to be **a pole** for f;
2. if $\nexists \lim_{z\to z_0} f(z)$, then z_0 is said to be **an essential isolated singular point** for f.

Theorem 4.8.4 (Characterization theorem for the poles). *Let $f \in H(G)$ where $G \subset \mathbb{C}$ is an open set, and suppose that $z_0 \in \mathbb{C}$ is an isolated singular point for f. The next statements are equivalent:*

(i) *z_0 is a pole for f;*

(ii) *z_0 is a zero for $\frac{1}{f}$, in the sense that $\lim_{z\to z_0} \frac{1}{f(z)} = 0$;*

(iii) *there exists a unique $n \in \mathbb{N}^*$, and a number $r > 0$, such that $\dot{U}(z_0;r) \subset G$ and*

$$f(z) = \frac{a_{-n}}{(z-z_0)^n} + \cdots + \frac{a_{-1}}{z-z_0} + a_0 + a_1(z-z_0) + \cdots + a_n(z-z_0)^n + \cdots, \quad \forall z \in \dot{U}(z_0;r),$$

where $a_{-n} \neq 0$;

(iv) *there exists a unique $n \in \mathbb{N}^*$, and a function $g \in H(\widetilde{G})$, where $\widetilde{G} = G \cup \{z_0\}$, such that $g(z_0) \neq 0$ and*

$$f(z) = \frac{g(z)}{(z-z_0)^n}, \quad \forall z \in G.$$

Proof. (i) ⇒ (ii) Since z_0 is a pole, $\lim_{z\to z_0} f(z) = \infty$, then there exists a number $R > 0$ such that $\dot{U}(z_0;R) \subset G$, and $|f(z)| > 1$, if $z \in \dot{U}(z_0;R)$. Hence, $h = \frac{1}{f} \in H(\dot{U}(z_0;R))$ and $|h(z)| < 1$, if $z \in \dot{U}(z_0;R)$. Using Theorem 4.8.3, the number z_0 is a removable isolated singular point for the function $h = \frac{1}{f}$, and $\lim_{z\to z_0} h(z) = 0 = h(z_0)$.

(ii) ⇒ (iii) From the fact that $\lim_{z\to z_0} f(z) = \infty$, there exists a number $R > 0$ such that $|f(z)| > 1$, $\forall z \in \dot{U}(z_0;R)$. Hence, $h = \frac{1}{f} \in H(\dot{U}(z_0;R))$. Since $\lim_{z\to z_0} \frac{1}{f(z)} = 0$, the number z_0 is a removable isolated singular point for the function $\frac{1}{f}$. Also, for the holomorphic extension $h = \frac{1}{f}$ the number z_0 is a zero. From the theorem of characterization of holomorphic function zeros, there exists $n \in \mathbb{N}^*$ and $h_1 \in H(U(z_0;R))$, such that

$$h(z) = \frac{1}{f(z)} = h_1(z)(z-z_0)^n, \quad \forall z \in U(z_0;R), \text{ where } h_1(z_0) \neq 0.$$

Since $h_1(z_0) \neq 0$, the function h_1 does not vanish in a small enough neighborhood of z_0. In this neighborhood, the function $\frac{1}{h_1}$ is holomorphic, hence it can be expanded in a Taylor series:

$$\frac{1}{h_1(z)} = a_{-n} + a_{-n+1}(z-z_0) + \cdots + a_0(z-z_0)^n + \cdots, \quad \text{and} \quad 0 \neq \frac{1}{h_1(z_0)} = a_{-n}.$$

Hence, we have

$$f(z) = \frac{1}{h_1(z)(z-z_0)^n} = \frac{a_{-n}}{(z-z_0)^n} + \frac{a_{-n+1}}{(z-z_0)^{n-1}} + \cdots$$
$$+ a_0 + a_1(z-z_0) + a_2(z-z_0)^2 + \cdots.$$

From the theorem of characterization of the zeros of holomorphic functions, the number $n \in \mathbb{N}^*$ is unique.

(iii) $\Rightarrow$ (iv) Lets define

$$g(z) = \begin{cases} (z-z_0)^n f(z), & \text{if } z \in G, \\ a_{-n}, & \text{if } z = z_0. \end{cases}$$

From the expansion of f about the point z_0, given in the previous point of the theorem, it follows that $g \in H(G) \cap C(\widetilde{G})$, where $\widetilde{G} = G \cup \{z_0\}$. Hence, from a well-known result, we obtain that $g \in H(\widetilde{G})$ and $g(z_0) = a_{-n} \neq 0$.

(iv) $\Rightarrow$ (i) From $\lim_{z\to z_0} g(z) = g(z_0) \neq 0$, we have that $\lim_{z\to z_0} f(z) = \infty$, hence z_0 is a pole for f. □

Definition 4.8.3. The natural number $n \in \mathbb{N}^*$ that appears in the previous theorem is called **the order of pole** z_0.

Corollary 4.8.2. *Let $f \in H(G)$, where $G \subset \mathbb{C}$ is an open set. The number $z_0 \in \mathbb{C}$ is an nth order pole for the function f, if and only if z_0 is an nth order zero for the holomorphic extension (continuation) at z_0 of the function $\frac{1}{f}$.*

Theorem 4.8.5. *Let $z_0 \in \mathbb{C}$ be an essential isolated singular point for the function $f \in H(G)$, where $G \subset \mathbb{C}$ is an open set. Then z_0 is an essential isolated singular point for the function $\frac{1}{f}$, or it is an accumulation point for the poles of $\frac{1}{f}$.*

Proof. 1. Suppose that z_0 is not an accumulation point for the zeros of f. Then $\exists r > 0$ such that $\dot{U}(z_0; r) \subset G$ and $f(z) \neq 0$, $\forall z \in \dot{U}(z_0; r)$. Hence, $g = \frac{1}{f}$ is holomorphic in $\dot{U}(z_0; r)$.

Now we will prove that the number z_0 is an essential isolated singular point for $g = \frac{1}{f}$. Supposing that there exists $\lim_{z\to z_0} g(z) = l \in \mathbb{C}_\infty$, then we have the next two cases:

(i) If $l \in \mathbb{C}^*_\infty$, then $\lim_{z\to z_0} f(z) = \frac{1}{l} \in \mathbb{C}$, hence z_0 is a removable isolated singular point for f, which contradicts the assumption;

(ii) If $l = 0$, then $\lim_{z\to z_0} f(z) = \infty$, hence z_0 is a pole for f, which also contradicts the assumption.

In conclusion, it follows that in such a case the number z_0 is an essential isolated singular point for $g = \frac{1}{f}$.

2. Suppose that z_0 is an accumulation point for the zeros of f. These zeros of f are poles for $g = \frac{1}{f}$, hence z_0 is an accumulation point for the poles of $\frac{1}{f}$. □

Example 4.8.1. If $f(z) = \frac{1}{\sin\frac{1}{z}}$, then $z_0 = 0$ is an accumulation point for the poles of f, because $f \in H(\mathbb{C}^* \setminus \{\frac{1}{k\pi} : k \in \mathbb{Z}^*\})$, where $z_k = \frac{1}{k\pi}$, $k \in \mathbb{Z}^*$, are poles for f and $z_0 = 0$ is an accumulation point for these z_k, $k \in \mathbb{Z}^*$ points.

Theorem 4.8.6 (Casorati–Weierstrass theorem). *If $z_0 \in \mathbb{C}$ is an essential isolated singular point for the function $f \in H(G)$, where $G \subset \mathbb{C}$ is an open set, then $\forall w_0 \in \mathbb{C}_\infty$, $\exists (z_n)_{n\in\mathbb{N}^*} \subset G$ sequence, such that $\lim_{n\to+\infty} z_n = z_0$ and $\lim_{n\to+\infty} f(z_n) = w_0$.*

Proof. 1. If $w_0 = \infty$, the conclusion follows from Corollary 4.8.1.

2. If $w_0 \in \mathbb{C}$, according to the Theorem 4.8.5, the number z_0 is an essential isolated singular point for the function g, or it is an accumulation point for the poles of g, where

$$g(z) = \frac{1}{f(z) - w_0}$$

(because z_0 is an essential isolated singular point also for the function $f - w_0$).

(i) If z_0 is an accumulation point for the poles of g, then z_0 is an accumulation point for the zeros of $f - w_0$. Hence $\exists (z_n)_{n\in\mathbb{N}^*} \subset G$, $z_n \to z_0$, such that $f(z_n) - w_0 = 0$, $\forall n \in \mathbb{N}^*$, i. e., $\lim_{n\to+\infty} f(z_n) = w_0$.

(ii) If z_0 is an essential isolated singular point for the function g, from Corollary 4.8.1, it follows that $\exists (z_n)_{n\in\mathbb{N}^*} \subset G$, $z_n \to z_0$, such that $\lim_{n\to+\infty} g(z_n) = \infty$. Hence,

$$\lim_{n\to+\infty} (f(z_n) - w_0) = 0 \Rightarrow \lim_{n\to+\infty} f(z_n) = w_0.$$

□

Definition 4.8.4.
1. Let $z_0 = \infty$ and $f \in H(G)$, where $G \subset \mathbb{C}$ is an open set. We say that ∞ **is an isolated singular point for** f, if $\exists R > 0$ such that $\{z \in \mathbb{C} : |z| > R\} \subset G$.
2. Making the substitution $\zeta = \frac{1}{z}$, we see that for $z_0 = \infty$ corresponds the point $\zeta_0 = 0$. Define the function

$$\varphi(\zeta) = f\left(\frac{1}{\zeta}\right).$$

We say that the point $z_0 = \infty$ is a **removable isolated singular point, a pole** or an **essential isolated singular point** for f, if $\zeta_0 = 0$ is a **removable isolated singular point, a pole** or an **essential isolated singular point** for φ, respectively.

4.9 Meromorphic functions

Definition 4.9.1. Let $\widetilde{G} \subset \mathbb{C}_\infty$ be an open set. We say that the function $f \in H(\widetilde{G} \setminus E)$ **is meromorphic on** $\widetilde{G}$, if the set $E \subset \widetilde{G}$ consists only in the removable isolated singular points and the poles of the function f.

Remarks 4.9.1.
1. If we denote by G the set of the points where f is regular, and by B the set of the f function poles, then $\widetilde{G} = G \cup B$.
2. Evidently, the set G is open if $\widetilde{G}$ is open, and G is connected if $\widetilde{G}$ is connected.

Remarks 4.9.2.
1. The set E is at most countable.
2. The set B is at most countable, and cannot have any accumulation points in $\widetilde{G}$.
3. The function $f : \widetilde{G} \to \mathbb{C}_\infty$ is continuous as a function with the images in the whole set $\mathbb{C}_\infty$, where $f(b) = \infty$, if $b \in B$.
4. The function $f : \widetilde{G} \to \mathbb{C}_\infty$ is meromorphic, if and only if about $\forall z_0 \in \widetilde{G}$ the function f can be expanded in such a Laurent-series, whose main part contains a finite number of terms.
   Every point $z_0 \in G$ is characterized by the fact that the Laurent expansion of f about z_0 has a vanishing main part.

Notation. We denote by $M(\widetilde{G})$ the set of all meromorphic functions on $\widetilde{G}$.

Remark 4.9.3. If $f \in M(\widetilde{G})$, then this function has a continuous extension $\widetilde{f} : \widetilde{G} \to \mathbb{C}_\infty$. It is usual to write f instead of $\widetilde{f}$, and we will use this notation. Evidently, $H(\widetilde{G}) \subset M(\widetilde{G})$.

Examples 4.9.1.
1. Every rational function (the quotient of two polynomials) is meromorphic on $\mathbb{C}_\infty$.
2. The function $\cot z$ is meromorphic on $\mathbb{C}$ and the points $z_k = k\pi$, $k \in \mathbb{Z}$, are its poles. The ∞ point is an accumulation point for the poles z_k, hence $\cot z$ cannot be a meromorphic function on $\mathbb{C}_\infty$.
3. The function $\tan \frac{1}{z}$ is meromorphic on $\mathbb{C}_\infty \setminus \{0\}$, because ∞ is a regular point, and $z_k = \frac{2}{(2k+1)\pi}$, $k \in \mathbb{Z}$, are its poles, and 0 is the accumulation point for the poles z_k.

Theorem 4.9.1. *The set $M(\widetilde{G})$ is a commutative field with respect to the addition and multiplication.*

Proof. Evidently, if $f, g \in M(\widetilde{G})$, then their product fg has only regular points or poles in $\widetilde{G}$.

If $f \in M(\widetilde{G})$, then the function $\frac{1}{f}$ has only regular points or poles in $\widetilde{G}$, hence $\frac{1}{f} \in M(\widetilde{G})$. □

Theorem 4.9.2. *A function is meromorphic on $\mathbb{C}_\infty$, if and only if it is a rational function (the quotient of two polynomials).*

Proof. 1. Evidently, any rational function f is meromorphic on $\mathbb{C}_\infty$.

2. First, we will see that any meromorphic function on $\mathbb{C}_\infty$ could have only a finite number of poles. Contrary, if there exists a meromorphic function with an infinite

number of poles on $\mathbb{C}_\infty$, then those poles will have an accumulation point in $\mathbb{C}_\infty$, and this accumulation point cannot be regular point or pole.

Let $B = \{b_1, \dots, b_n\}$ denote the set of all poles of the function f.

(i) Suppose that $\infty \notin B$.

The Laurent expansion of f about the point $b_k \in B$ has the form $f(z) = P_k(z)+T_k(z)$, where P_k is the main part. For the function $g = f - \sum_{k=1}^n P_k$, the points b_k are removable isolated singular points, because about every b_k point the function g can be expanded in a Taylor series (its Laurent expansion about b_k consists only from the Taylor part). Since all these b_k are regular points for g, if $\tilde{g}$ is the holomorphic extension of g, then $\tilde{g}$ is holomorphic on $\mathbb{C}_\infty$. But ∞ is a regular point for $\tilde{g}$, and using Definition 4.8.4 and Theorem 4.8.3 it follows that $\tilde{g}$ is a bounded function in a neighborhood of ∞. From here, we deduce that $\tilde{g}$ is a bounded function on $\mathbb{C}$, and according to the Liouville theorem the function $\tilde{g}$ will be constant. Hence, $f - \sum_{k=1}^n P_k$ is a constant function, and if we denote this constant by c, then

$$f = c + \sum_{k=1}^{n} P_k.$$

Since P_k are rational fractions $\forall k = \overline{1,n}$, it follows that f is also a rational function.

(ii) Suppose now, that $\infty \in B$.

That means ∞ is a pole for f, and if $a \in \mathbb{C}$ is a regular point for f, then for the function

$$h(z) = f\left(\frac{1}{z} + a\right)$$

the point $z = \infty$ is a regular point, and h is meromorphic on $\mathbb{C}_\infty$. From the previous part of the proof, the function h will be a rational function, and thus we proved that

$$f(z) = h\left(\frac{1}{z-a}\right)$$

is also a rational function. □

Remarks 4.9.4.

1. Let $R : \mathbb{C}_\infty \to \mathbb{C}_\infty$, $R(z) = \frac{P(z)}{Q(z)}$, where P and Q are such polynomials which don't have any common zeros. Hence, R is a rational function and $R \in H(\mathbb{C} \setminus \{b_1, b_2, \dots, b_m\})$, where $\{b_1, b_2, \dots, b_m\}$ is the set of all zeros of the Q polynomial. If

$$R(z) = \frac{\alpha_0 + \alpha_1 z + \cdots + \alpha_n z^n}{\beta_0 + \beta_1 z + \cdots + \beta_m z^m}, \quad \alpha_0 \neq 0, \beta_0 \neq 0, \tag{4.11}$$

then the function

$$S(z) = R\left(\frac{1}{z}\right)$$

will have the form

$$S(z) = z^{m-n} \frac{\alpha_0 z^n + \alpha_1 z^{n-1} + \cdots + \alpha_n}{\beta_0 z^m + \beta_1 z^{m-1} + \cdots + \beta_m},$$

Hence, from Definition 4.8.4 we have $R(\infty) = S(0)$.

2. Every rational function has the property that the number of its zeros and the number of its poles on $\mathbb{C}_\infty$ are the same, every zero and every pole will be counted together with its order. This number is called **the order of the rational function.**

The next table shows the discussion about the number of the zeros and of the poles on $\mathbb{C}_\infty$, of a rational function of the form (4.11):

Degrees comparison	The number of zeros in $\mathbb{C}$	The order of ∞ zero	The number of zeros in $\mathbb{C}_\infty$	The number of poles in $\mathbb{C}$	The order of ∞ pole	The number of poles in $\mathbb{C}_\infty$
$m > n$	n	$m - n$	m	m	–	m
$m = n$	n	–	n	m	–	m
$m < n$	n	–	n	m	$n - m$	n

4.10 Exercises

4.10.1 Power series

Exercise 4.10.1. Determine the radius of convergence of the following power series:

1. $\sum_{n=1}^{\infty} \frac{z^n}{n}$;
2. $\sum_{n=1}^{\infty} \frac{z^n}{n^2}$;
3. $\sum_{n=0}^{\infty} \frac{z^n}{n!}$;
4. $\sum_{n=0}^{\infty} n! z^n$;
5. $\sum_{n=1}^{\infty} \frac{1 \cdot 2 \cdot \ldots \cdot n}{3 \cdot 5 \cdot \ldots \cdot (2n+1)} z^n$;
6. $\sum_{n=1}^{\infty} \frac{z^n}{n\sqrt{n}}$;
7. $\sum_{n=0}^{\infty} e^{\alpha n} z^n$, $\alpha = a + ib$;
8. $\sum_{n=0}^{\infty} (\cos n\alpha) z^n$, $\alpha = a + ib, b \neq 0$;
9. $\sum_{n=0}^{\infty} [3 + (-1)^n]^n z^n$;
10. $\sum_{n=0}^{\infty} \left[\frac{n}{2} + \frac{1 + (-1)^{n+1}}{4}\right] z^n$;
11. $\sum_{n=1}^{\infty} \left[\frac{1 + (-1)^n}{2} n + \frac{1 - (-1)^n}{2} \ln n\right] z^n$;
12. $\sum_{n=1}^{\infty} a_n z^n$, where $a_n = \sum_{k=1}^{n} \frac{1}{k}$;
13. $\sum_{n=1}^{\infty} (n + \alpha^n) z^n$, $\alpha \in \mathbb{C}$;
14. $\sum_{n=0}^{\infty} P(n) z^n$, where P is a polynomial, $\deg P = p$;

15. $\sum_{n=0}^{\infty} \frac{(n!)^2}{(3n)!} z^n$;

16. $\sum_{n=1}^{\infty} \left(\frac{n+1}{n}\right)^n z^n$;

17. $\sum_{n=1}^{\infty} \frac{3^n + (-2)^n}{n} z^n$;

18. $\sum_{n=0}^{\infty} [3 + 2(-1)^n] z^n$;

19. $\sum_{n=0}^{\infty} a_n z^n$, where

$$a_n = \begin{cases} a^n, & n = 2m,\ m \in \mathbb{N}, \\ b^n, & n = 2m+1,\ m \in \mathbb{N}; \end{cases}$$

20. $\sum_{n=1}^{\infty} a_n z^n$, where

$$a_n = \begin{cases} \frac{1}{n}, & n = 3m,\ m \neq 0, \\ (1 - \frac{1}{n})^n, & n = 3m+1,\ m \in \mathbb{N}, \\ 2^n, & n = 3m+2,\ m \in \mathbb{N}. \end{cases}$$

Exercise 4.10.2. Determine the set of convergence of the following series of functions, and then determine a formula for their sums on the convergence set:

1. $1 + \sum_{n=1}^{\infty} \frac{z^n}{n}$;
2. $\sum_{n=1}^{\infty} n z^n$;
3. $\sum_{n=1}^{\infty} n^2 z^n$;
4. $\sum_{n=2}^{\infty} n(n-1) z^{n-2}$;
5. $\sum_{n=2}^{\infty} (-1)^n \frac{n}{n^2 - 1} z^n$;
6. $\sum_{n=1}^{\infty} n z^{n-1}$;
7. $\sum_{n=1}^{\infty} \frac{z^{2n+1}}{2n+1}$;
8. $\sum_{n=1}^{\infty} (-1)^{n+1} \frac{z^n}{n}$;
9. $\sum_{n=0}^{\infty} (\frac{z}{z+1})^n$.

Exercise 4.10.3. Prove the formulas:
1. $\frac{1}{z^2+a^2} = \sum_{n=0}^{\infty} (-1)^n a^{-2n-2} z^{2n}$, if $|z| < a$, $a \neq 0$;
2. $\frac{z(z+a)}{(a-z)^3} = \sum_{n=1}^{\infty} \frac{n^2 z^n}{a^{n+1}}$, if $|z| < a$, $a \neq 0$;
3. $1 + z = \sum_{n=0}^{\infty} (\frac{z}{z+1})^n$, if $\operatorname{Re} z > -\frac{1}{2}$.

Exercise 4.10.4. Prove the formulas:
1. $\sum_{n=1}^{\infty} \frac{\cos n\varphi}{n} = -\ln(2 \sin \frac{\varphi}{2})$, $0 < \varphi < 2\pi$;
2. $\sum_{n=1}^{\infty} \frac{\sin n\varphi}{n} = \frac{\pi - \varphi}{2}$, $0 < \varphi < 2\pi$;
3. $\sum_{n=0}^{\infty} \frac{\cos(2n+1)\varphi}{2n+1} = \frac{1}{2} \ln(\cot \frac{\varphi}{2})$, $0 < \varphi < \pi$;
4. $\sum_{n=0}^{\infty} \frac{\sin(2n+1)\varphi}{2n+1} = \frac{\pi}{4}$, $0 < \varphi < \pi$.

4.10.2 Taylor and Laurent series

Exercise 4.10.5. Expand in a Taylor series, respectively, in the Laurent series, about the point z_0, the following functions:
1. $f(z) = \frac{z}{z+1}$, $z_0 = 1$;
2. $f(z) = \frac{z^2}{(z+1)^2}$, $z_0 = 1$;

3. $f(z) = \frac{z-1}{z-2}$, $z_0 = 0$ and $z_0 = i$;
4. $f(z) = \frac{1}{z^2+z+1}$, $z_0 = 1$ and $z_0 = \infty$;
5. $f(z) = \frac{z+3}{z^2-8z+15}$, $z_0 = 4$ and $z_0 = \infty$;
6. $f(z) = \sin \frac{z}{1-z}$, $z_0 = 0$;
7. $f(z) = \cosh^2 z$, $z_0 = 0$;
8. $f(z) = \cos^3 z$, $z_0 = 0$;
9. $f(z) = \sin^3 z$, $z_0 = 0$;
10. $f(z) = e^z \sin z$, $z_0 = 0$;
11. $f(z) = \sqrt[3]{z}$, where $f(1) = \frac{-1+i\sqrt{3}}{2}$, $z_0 = 1$;
12. $f(z) = \sqrt[5]{z-2i}$, where $f(0) = \sqrt[5]{2}e^{i\frac{7\pi}{10}}$, $z_0 = 0$ and $z_0 = 2$;
13. $f(z) = \log z$, where $f(1+i) = \frac{1}{2}\ln 2 - i\frac{7\pi}{4}$, $z_0 = -i$;
14. $f(z) = \frac{\log(1+z)}{1+z}$, where $f(0) = -4\pi i$, $z_0 = 0$.

Exercise 4.10.6. Expand in the Laurent series about the point $z_0 = 0$, in the corresponding domains, for the functions:
1. $f(z) = \frac{1}{z^2-5z+6}$, (a) $|z| < 2$, (b) $2 < |z| < 3$, (c) $|z| > 3$;
2. $f(z) = \frac{1}{z(z+1)(z+2)}$, (a) $0 < |z| < 1$, (b) $1 < |z| < 2$, (c) $|z| > 2$;
3. $f(z) = \frac{2z^2-3z+3}{z^3-2z^2+z-2}$, (a) $|z| < 1$, (b) $1 < |z| < 2$, (c) $|z| > 2$;
4. $f(z) = \log \frac{1-z}{1+z}$, where $f(0) = 0$, (a) $|z| < 1$, (b) $|z| > 1$.

Exercise 4.10.7. In the corresponding domains, expand in the Laurent series about $z_0 = \infty$ the following functions:
1. $f(z) = \frac{1}{z^5-z^2}$, $|z| > 1$;
2. $f(z) = \frac{z^{15}}{(z^2+1)^5(z^4-16)}$, $|z| > 2$;
3. $f(z) = a_0 + a_1 z + \cdots + a_n z^n$, $z \in \mathbb{C}$;
4. $f(z) = z^3 \sin \frac{1}{z}$, $|z| > 0$;
5. $f(z) = \log \frac{z-a}{z-b}$, where $\log 1 = 2k\pi i$, $|z| > \max\{|a|, |b|\}$;
6. $f(z) = e^{\frac{2z}{z+1}}$, $|z| > 1$.

Exercise 4.10.8. Expand the following functions in the Laurent series about the point z_0, and then determine the type of the singularity z_0:
1. $f(z) = \frac{1}{(z^2-1)^2}$, $z_0 = 1$ and $z_0 = \infty$;
2. $f(z) = \sin \frac{1}{1-z}$, $z_0 = 1$ and $z_0 = \infty$;
3. $f(z) = \frac{1}{z \sin z}$, $z_0 = 0$ and $z_0 = \pi$;

4. $f(z) = e^{\frac{2z}{z+1}}$, $z_0 = \infty$;
5. $f(z) = e^{i\pi\frac{z+i}{z-i}}$, $z_0 = i$;
6. $f(z) = \cot z$, $z_0 = 0$ and $z_0 = k\pi$, $k \in \mathbb{Z}$;
7. $f(z) = \frac{\sin^2 z}{z^5}$, $z_0 = 0$.

Exercise 4.10.9. Expand in the Laurent series about the point $z_0 = 0$, in the domain $1 < |z| < 2$, the next functions:
1. $f(z) = \frac{1}{(z-1)^2(z+2)}$;
2. $f(z) = \frac{z}{(z^2+1)^2(z+2)}$.

Exercise 4.10.10. In the given circular ring D, expand in the Laurent series about the corresponding point z_0 in the following functions:
1. $f(z) = \frac{1}{z(z-3)^2}$, $z_0 = 1$, $D = U(1;1,2)$;
2. $f(z) = \frac{1}{(z^2-1)(z^2+4)^2}$, $z_0 = 0$, $D = U(0;2,\infty)$;
3. $f(z) = z^3 e^{\frac{1}{z}}$, $z_0 = 0$, $D = U(0;0,\infty)$;
4. $f(z) = z^3 \cos\frac{1}{z-2}$, $z_0 = 2$, $D = U(2;0,\infty)$;
5. $f(z) = \frac{1}{z^2(1-z)\sin z}$, $z_0 = 0$, $D = U(0;0,1)$;
6. $f(z) = \cos^2\frac{1}{z}$, $z_0 = 0$, $D = U(0;0,\infty)$.

Exercise 4.10.11. Write the first three terms of the Laurent series expansion about the point $z_0 = 0$ for the next functions:
1. $f(z) = \frac{1}{\sin z}$;
2. $f(z) = \frac{1}{e^z-1}$.

Exercise 4.10.12. Define on the set $\mathbb{C} \setminus [-i,i]$ the holomorphic branch of the multivalued function $f(z) = \sqrt{1+z^2}$, for which $f(\frac{3}{4}) = \frac{5}{4}$. Expand this branch of the function f in a Laurent series around the point $z_0 = 0$, on the circular ring $|z| > 1$.

Exercise 4.10.13. Prove that

$$\frac{1}{\sqrt{z^2-1}} = \frac{1}{z} + \sum_{n=-\infty}^{-1} \frac{\frac{1}{2}\cdot\frac{3}{2}\cdot\ldots\cdot(\frac{1}{2}-n-1)}{(-n)!} z^{2n-1}, \quad |z| > 1,$$

where we considered the branch of the multivalued function $f(z) = \sqrt{z^2-1}$ that satisfies $f(\sqrt{5}) = 2$.

Exercise 4.10.14. Prove that the expansion in the Laurent series of the function $f(z) = \frac{1}{1-z-z^2}$ about the point $z_0 = 0$ has the form $f(z) = \sum_{n=0}^{\infty} a_n z^n$, where

$$a_n = \frac{1}{\sqrt{5}}\left[\left(\frac{1+\sqrt{5}}{2}\right)^{n+1} - \left(\frac{1-\sqrt{5}}{2}\right)^{n+1}\right].$$

Exercise 4.10.15. Determine the main part of the Laurent series expansion about the point z_0 for the functions:
1. $f(z) = \frac{z}{(z+2)^2}, z_0 = -2;$
2. $f(z) = \frac{e^z+1}{e^z-1}, z_0 = 0$ and $z_0 = 2\pi i;$
3. $f(z) = \frac{z-1}{\sin^2 z}, z_0 = 0.$

4.10.3 Isolated singular points

Exercise 4.10.16. Prove that z_0 is a removable isolated singular point for the corresponding functions:
1. $f : \mathbb{C} \setminus \{1\} \to \mathbb{C}, f(z) = \frac{z^2-1}{z-1}, z_0 = 1;$
2. $f : \mathbb{C} \setminus \{0\} \to \mathbb{C}, f(z) = \frac{\sin z}{z}, z_0 = 0;$
3. $f : \mathbb{C} \setminus (\{k\pi : k \in \mathbb{Z}\} \cup \{\frac{\pi}{2} + k\pi : k \in \mathbb{Z}\}) \to \mathbb{C}, f(z) = \frac{z}{\tan z}, z_0 = 0;$
4. $f : \mathbb{C} \setminus \{\frac{\pi}{2} + k\pi : k \in \mathbb{Z}\} \to \mathbb{C}, f(z) = \frac{1}{\cos^2 z} - \frac{1}{(z-\frac{\pi}{2})^2}, z_0 = \frac{\pi}{2};$
5. $f : \mathbb{C} \setminus \{0\} \to \mathbb{C}, f(z) = \frac{1-\cos z}{z^2}, z_0 = 0;$
6. $f : \mathbb{C} \setminus (\{k\pi : k \in \mathbb{Z}\} \cup \{2k\pi i : k \in \mathbb{Z}\}) \to \mathbb{C}, f(z) = \frac{1}{e^z-1} - \frac{1}{\sin z}, z_0 = 0;$
7. $f : \mathbb{C} \setminus \{z \in \mathbb{C} : z^3 + 1 = 0\} \to \mathbb{C}, f(z) = \frac{z^2-1}{z^3+1}, z_0 = -1.$

Exercise 4.10.17. Prove that the point z_0 is a pole for the corresponding functions:
1. $f : \mathbb{C} \setminus \{0\} \to \mathbb{C}, f(z) = \frac{1}{z}, \quad z_0 = 0;$
2. $f : \mathbb{C} \setminus \{-i, i\} \to \mathbb{C}, f(z) = \frac{1}{(z^2+1)^2}, z_0 = i;$
3. $f : \mathbb{C} \setminus \{-1\} \to \mathbb{C}, f(z) = \frac{z^2+1}{z+1}, z_0 = -1;$
4. $f : \mathbb{C} \setminus \{2k\pi : k \in \mathbb{Z}\} \to \mathbb{C}, f(z) = \frac{1}{1-\cos z}, z_0 = 0;$
5. $f : \mathbb{C} \setminus \{-i, i\} \to \mathbb{C}, f(z) = \frac{1}{(z^2+1)^2}, z_0 = -i;$
6. $f : \mathbb{C} \setminus \{2k\pi : k \in \mathbb{Z}\} \to \mathbb{C}, f(z) = \frac{z}{1-\cos z}, z_0 = 2\pi;$
7. $f : \mathbb{C} \setminus \{z \in \mathbb{C} : e^z + 1 = 0\} \to \mathbb{C}, f(z) = \frac{z}{e^z+1}, z_0 = \pi i.$

Exercise 4.10.18. Prove that the point z_0 is an essential isolated singular point for the corresponding functions:
1. $f : \mathbb{C} \setminus \{0\} \to \mathbb{C}, f(z) = \sin \frac{\pi}{z^2}, z_0 = 0;$
2. $f : \mathbb{C} \setminus \{1\} \to \mathbb{C}, f(z) = (z-1)^2 \cos \frac{\pi}{z-1}, z_0 = 1;$
3. $f : \mathbb{C} \setminus \{-1\} \to \mathbb{C}, f(z) = e^{\frac{1}{z+1}}, z_0 = -1;$
4. $f : \mathbb{C} \setminus \{-1\} \to \mathbb{C}, f(z) = \cos \frac{z}{z+1}, z_0 = -1;$
5. $f : \mathbb{C} \setminus \{-i, i\} \to \mathbb{C}, f(z) = \sin \frac{\pi}{z^2+1}, z_0 = i;$
6. $f : \mathbb{C} \setminus \{0\} \to \mathbb{C}, f(z) = \sin e^{\frac{1}{z}}, z_0 = 0.$

Exercise 4.10.19. Let $D \subset \mathbb{C}$ be a domain, and let $f, g \in H(D)$ be such a function that satisfies $f(a) = g(a) = 0$, and $g'(a) \neq 0$. Prove that the point a is a removable isolated singular point of the function

$$h : D \setminus \{z \in D : g(z) = 0\} \to \mathbb{C}, \quad h(z) = \frac{f(z)}{g(z)}.$$

Exercise 4.10.20. Let $f : U(a;r) \setminus \{a\} \to \mathbb{C}^*$ a holomorphic function, such that the point a is a pole for f. Prove that the function

$$h : U(a;r) \to \mathbb{C}, \quad h(z) = \begin{cases} \frac{1}{f(z)}, & \text{if } z \in U(a;r) \setminus \{a\}, \\ 0, & \text{if } z = a \end{cases}$$

is holomorphic in the disc $U(a;r)$.

4.10.4 The module maximum of the holomorphic functions

Exercise 4.10.21. Let $f \in H(D)$ be a nonconstant function, where $D \subset \mathbb{C}$ is a domain. Prove that the functions $g = \operatorname{Re} f$ and $h = \operatorname{Im} f$ have no locally maximum or minimum points in the domain D.

Exercise 4.10.22 (Carathéodory inequality). Let $f \in H(U(0;r))$, such that $f(0) = 0$ and $\operatorname{Re} f(z) \leq A$, $\forall z \in U(0;r)$. Prove that

$$|f(z)| \leq \frac{2A|z|}{r - |z|}, \quad \forall z \in U(0;r).$$

Exercise 4.10.23. Let $f \in H(U(0;1))$, such that $|f(z)| < 1$, $\forall z \in U(0;1)$. If there exist the numbers $a, b \in U(0;1)$, $a \neq b$, such that $f(a) = a$ and $f(b) = b$, then prove that $f(z) = z$, $\forall z \in U(0;1)$.

Exercise 4.10.24. Suppose that $f \in H(U(0;1))$ and there exists $\alpha \in U(0;1)$, such that $f(\alpha) = 0$ and $|f(z)| \leq 1$, $\forall z \in U(0;1)$. Prove that

$$|f(z)| \leq \left| \frac{z - \alpha}{1 - \bar{\alpha} z} \right|, \quad \forall z \in U(0;1).$$

Exercise 4.10.25. Did there exist functions $f \in H(\mathbb{C} \setminus \{-1\})$, such that

$$f\left(\frac{1}{n}\right) = \frac{1}{n+1}, \quad \forall n \in \mathbb{N}^*?$$

If there exist, how many functions with this property did there exist?

Exercise 4.10.26. Did there exist functions $f \in H(U(\frac{1}{2}; \frac{5}{6}))$, such that its values on the points $z_n = \frac{1}{n}$, $n \in \mathbb{N}^*$, are given by the next sequences:

1. $0, 1, 0, \frac{1}{2}, 0, \frac{1}{3}, 0, \frac{1}{4}, \ldots, 0, \frac{1}{k}, \ldots$
2. $\frac{1}{2}, \frac{1}{2}, \frac{1}{3}, \frac{1}{3}, \frac{1}{4}, \frac{1}{4}, \ldots \frac{1}{k}, \frac{1}{k}, \ldots$?

Exercise 4.10.27. Did there exist functions $f \in H(U(0;1))$, such that

1. $f(\frac{1}{n}) = f(-\frac{1}{n}) = \frac{1}{n^2}, \forall n \in \mathbb{N}^*$,
2. $f(\frac{1}{n}) = f(-\frac{1}{n}) = \frac{1}{n^3}, \forall n \in \mathbb{N}^*$?

5 Residue theory

5.1 Residue theorem

One of the most important consequences of the Laurent series expansion theorem is the method of the calculation of the integrals on a closed rectifiable curve that is given by the *residue theorem*. Using this result, we can elegantly compute some famous integrals that appeared in physics and engineering. On the other side, the residue theorem has some profound theoretical consequences. In this part of our book, we will present both important implications.

Let $G \subset \mathbb{C}$ be an open set, let $f \in H(G)$ and let S the set of all nonremovable isolated singular points of the function f. Then $\widetilde{G} = G \cup S$ is also an open set. The function f may be expanded in Laurent series about every point z_0 of $\widetilde{G}$, i. e., $\forall z_0 \in \widetilde{G}$, $\exists r > 0$ such that $\dot{U}(z_0; r) \subset G$ and

$$f(z) = \sum_{n=-\infty}^{\infty} a_n (z - z_0)^n, \quad \forall z \in \dot{U}(z_0; r). \tag{5.1}$$

Definition 5.1.1. The coefficient a_{-1} that appears in the Laurent series expansion (5.1) is called **the residue of f at the point** z_0, and it is denoted by $\operatorname{Res}(f; z_0) = a_{-1}$.

Remark 5.1.1. If $z_0 \in \widetilde{G}$ is a regular point for f, then $\operatorname{Res}(f; z_0) = 0$.

Theorem 5.1.1 (The residue theorem). *Let $f \in H(G)$, where $G \subset \mathbb{C}$ is an open set, and let B the set of all isolated singular points of the function f. Let $\widetilde{G} = G \cup B$ and let γ be a closed rectifiable curve (path) in G, i. e., homotopic with zero in $\widetilde{G}$, that is $\gamma \underset{\widetilde{G}}{\sim} 0$.*

Then the sum $\sum_{z \in \widetilde{G}} \mathrm{n}(\gamma, z) \operatorname{Res}(f; z)$ is finite, and

$$\int_{\gamma} f = 2\pi i \sum_{z \in \widetilde{G}} \mathrm{n}(\gamma, z) \operatorname{Res}(f; z).$$

Proof. Let $S = T = [0, 1]$ and let $\varphi : S \times T \to \widetilde{G}$ be the continuous deformation of γ to the constant path γ_0. The set $K = \varphi(S \times T)$ is compact, and thus $\varepsilon = \frac{1}{2}\mathrm{d}(K, \mathbb{C} \setminus G) > 0$. It follows that the set $D = \bigcup\{U(z; \varepsilon) : z \in K\}$ is bounded, hence D^- is compact, and from its definition we have $K \subset D \subset D^- \subset \widetilde{G}$ (Figure 5.1).

We see that the set $D^- \cap B$ is finite. Contrary, if $D^- \cap B$ is not finite, then $D^- \cap B$ needs to have an accumulation point in $\widetilde{G}$, and this point cannot be an isolated singular point for f.

Let $D \cap B = \{b_1, \dots, b_n\}$, and denote by $P_k(z)$ the main part of the Laurent expansion of f about b_k. (P_k exists and it is holomorphic on $\mathbb{C} \setminus \{b_k\}$, because it exists and it is holomorphic in a small enough neighborhood of b_k, or for enough small values of $|z - b_k|$, such that for greater values of $|z - b_k|$ it also exists and is holomorphic, because these terms appear to be the fractions' denominators.)

https://doi.org/10.1515/9783110657869-005

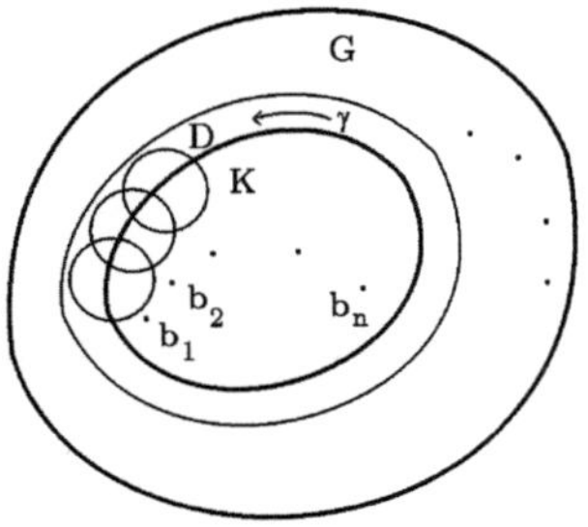

Figure 5.1: Proof of Theorem 5.1.1.

The function $g = f - \sum_{k=1}^{n} P_k$ has a holomorphic extension to b_k point, hence g can be considered as a holomorphic function on D, and $\gamma \underset{D}{\sim} 0$. According to the Cauchy integral theorem, we have $\int_\gamma g = 0$, hence

$$\int_\gamma f = \sum_{k=1}^{n} \int_\gamma P_k.$$

Let

$$P_k(z) = \sum_{m=1}^{\infty} \frac{a_{-m}^{(k)}}{(z - b_k)^m}.$$

Since P_k converges uniformly on each anact of $\mathbb{C} \setminus \{b_k\}$ (because P_k is holomorphic on the set $\mathbb{C} \setminus \{b_k\}$), it can be integrated term by term on the curve γ. But

$$\int_\gamma \frac{dz}{(z - b_k)^m} = 0, \quad \text{if } m > 1$$

(since in this case the function has a primitive) and

$$\int_\gamma \frac{dz}{z - b_k} = 2\pi i \mathrm{n}(\gamma, b_k).$$

Hence,

$$\int_\gamma P_k = 2\pi i \mathrm{n}(\gamma, b_k) \operatorname{Res}(f; b_k).$$

To complete the proof of the theorem, we need to show that $\mathrm{n}(\gamma, z_0) \operatorname{Res}(f; z_0) = 0$, $\forall z_0 \in \widetilde{G} \setminus (D \cap B)$. For the case $z_0 \notin B$, we have that z_0 is a regular point and then $\operatorname{Res}(f; z_0) = 0$. For the case $z_0 \notin D$, then $\mathbb{C} \setminus \{\gamma\}$ has a such a connected component that contains the ∞ point, hence $\mathrm{n}(\gamma, z_0) = 0$. It follows that

$$\int_\gamma f = \sum_{k=1}^{n} \int_\gamma P_k = 2\pi i \sum_{k=1}^{n} \mathrm{n}(\gamma, b_k) \operatorname{Res}(f; b_k) = 2\pi i \sum_{z \in \widetilde{G}} \mathrm{n}(\gamma, z) \operatorname{Res}(f; z). \qquad \square$$

Theorem 5.1.2 (Residue value in a pole). *If $z_0 \in \widetilde{G}$ is a pole of order k for the function f, then*

$$\operatorname{Res}(f; z_0) = \frac{1}{(k-1)!} \lim_{z \to z_0} \left[(z - z_0)^k f(z)\right]^{(k-1)}.$$

Proof. It is well known that, in a neighborhood of z_0 we have

$$f(z) = \frac{a_{-k}}{(z - z_0)^k} + \cdots + \frac{a_{-1}}{z - z_0} + a_0 + a_1(z - z_0) + \cdots, \quad \forall z \in \dot{U}(z_0; r),$$

hence

$$(z - z_0)^k f(z) = a_{-k} + a_{-k+1}(z - z_0) + \cdots + a_{-1}(z - z_0)^{k-1} + \cdots, \quad \forall z \in \dot{U}(z_0; r),$$

and

$$\left[(z - z_0)^k f(z)\right]^{(k-1)} = (k-1)! a_{-1} + (z - z_0)(k! a_0 + \cdots), \quad \forall z \in \dot{U}(z_0; r).$$

From here, it follows that

$$a_{-1} = \frac{1}{(k-1)!} \lim_{z \to z_0} \left[(z - z_0)^k f(z)\right]^{(k-1)}.$$ □

Corollary 5.1.1. *If $z_0 \in G$ is a simple pole for f, and f has the form $f(z) = \frac{g(z)}{h(z)}$, where g and h are holomorphic functions on G, with $g(z_0) \neq 0$, $h(z_0) = 0$, and $h'(z_0) \neq 0$, then*

$$\operatorname{Res}(f; z_0) = \frac{g(z_0)}{h'(z_0)} = a_{-1}.$$

Proof. In this special case,

$$\frac{g(z)}{h(z)} = \frac{a_{-1}}{z - z_0} + a_0 + a_1(z - z_0) + \cdots, \quad \forall z \in \dot{U}(z_0; r),$$

hence

$$a_{-1} = \lim_{z \to z_0} (z - z_0) \frac{g(z)}{h(z)} = \lim_{z \to z_0} \frac{g(z)}{\frac{h(z) - h(z_0)}{z - z_0}} = \frac{g(z_0)}{h'(z_0)}.$$ □

Definition 5.1.2 (The residue in the $z_0 = \infty$). Let $z_0 = \infty$ be an isolated singular point for the holomorphic function f. Then there exists $r > 0$ such that f is holomorphic in the circular ring $U(0; r, +\infty)$, so f may be expanded in Laurent series on this circular ring, i. e.,

$$f(z) = \cdots + \frac{a_{-n}}{z^n} + \cdots + \frac{a_{-1}}{z} + a_0 + a_1 z + \cdots + a_n z^n + \cdots, \quad \forall z \in U(0; r, +\infty).$$

Let $\gamma^- = \partial U(0;\rho)$, where $r < \rho < +\infty$. Then γ runs with direct orientation about the ∞ point (because it runs with inverse orientation around the point 0), hence

$$\frac{1}{2\pi i}\int_\gamma f = \frac{1}{2\pi i}\int_\gamma\left(\sum_{n=-\infty}^{\infty} a_n\zeta^n\right)d\zeta = \frac{1}{2\pi i}\sum_{n=-\infty}^{\infty} a_n\int_\gamma \zeta^n d\zeta = -a_{-1},$$

because of the fact that the Laurent series converges uniformly on the compact $\{\gamma\}$, and since the orientation of γ is inverse. In order that the Theorem 5.1.1 will be true also in this case, we need to denote

$$\operatorname{Res}(f;\infty) = -a_{-1}.$$

Theorem 5.1.3. *Let f be a holomorphic function on the domain $D = \mathbb{C}\setminus\{z_1,\ldots,z_n\}$, and let $z_{n+1} = \infty$. Then*

$$\sum_{k=1}^{n+1}\operatorname{Res}(f;z_k) = 0. \tag{5.2}$$

Proof. Let $\gamma = \partial U(0;r)$, where $r > \max\{|z_1|,\ldots,|z_n|\}$. Since the γ closed curve has the property $\mathrm{n}(\gamma,z_j) = 1$, $j = \overline{1,n}$, and thus, according to Theorem 5.1.1 we have

$$\frac{1}{2\pi i}\int_\gamma f = \sum_{k=1}^{n}\operatorname{Res}(f;z_k).$$

From the definition of the residue in the ∞ point, we have

$$\frac{1}{2\pi i}\int_\gamma f = -\operatorname{Res}(f;\infty),$$

and using the above relations we obtain the conclusion (5.2). □

5.2 Applications of the residue theorem to the calculation of the integrals

Exercise 5.2.1. Let $R(u,v)$ a rational function, where the numerator and the denominator are two real variables polynomials with real coefficients, such that the denominator does not vanish on the unit circle $u^2 + v^2 = 1$. Then the integral

$$I = \int_0^{2\pi} R(\sin x, \cos x)dx$$

can be calculated by using the next formula:

$$I = 2\pi\sum_{|z|<1}\operatorname{Res}(g;z), \quad \text{where } g(z) = \frac{1}{z}R\left(\frac{z-z^{-1}}{2i},\frac{z+z^{-1}}{2}\right).$$

Solution. Let $\gamma(t) = e^{2\pi i t}$, $t \in [0,1]$. Using the residue theorem, we get

$$\int_\gamma g(z)dz = 2\pi i \sum_{|z|<1} \operatorname{Res}(g; z). \tag{5.3}$$

On the other hand,

$$\begin{aligned}
\int_\gamma g(z)dz &= \int_0^1 (g \circ \gamma)(t)\gamma'(t)dt \\
&= \int_0^1 e^{-2\pi i t} R\left(\frac{e^{2\pi i t} - e^{-2\pi i t}}{2i}, \frac{e^{2\pi i t} + e^{-2\pi i t}}{2}\right) 2\pi i e^{2\pi i t} dt \\
&= 2\pi i \int_0^1 R\left(\frac{e^{2\pi i t} - e^{-2\pi i t}}{2i}, \frac{e^{2\pi i t} + e^{-2\pi i t}}{2}\right) dt \\
&= i \int_0^{2\pi} R\left(\frac{e^{ix} - e^{-ix}}{2i}, \frac{e^{ix} + e^{-ix}}{2}\right) dx.
\end{aligned} \tag{5.4}$$

Combining the relations (5.3) and (5.4), we finally obtain

$$i \int_0^{2\pi} R(\sin x, \cos x)dx = 2\pi i \sum_{|z|<1} \operatorname{Res}(g; z). \qquad \square$$

Remark 5.2.1. Using a similar proof that for the above relation, we can easily obtain the next result:

Let $R(u, v)$ a rational function, where the numerator and the denominator are two real variables polynomials with real coefficients, such that the denominator does not vanish on the unit circle $u^2 + v^2 = 1$. Then:

1. $\int_0^{2\pi} R(\sin x, \cos x) \sin mx dx = 2\pi \sum_{|z|<1} \operatorname{Res}(g_1; z)$, where $g_1(z) = \frac{1}{z} R(\frac{z-z^{-1}}{2i}, \frac{z+z^{-1}}{2}) \times \frac{z^m - z^{-m}}{2i}$, $m \in \mathbb{N}^*$,
2. $\int_0^{2\pi} R(\sin x, \cos x) \cos mx dx = 2\pi \sum_{|z|<1} \operatorname{Res}(g_2; z)$, where $g_2(z) = \frac{1}{z} R(\frac{z-z^{-1}}{2i}, \frac{z+z^{-1}}{2}) \times \frac{z^m + z^{-m}}{2}$, $m \in \mathbb{N}^*$.

Lemma 5.2.1. *Let $f \in C(S_0(\theta_1, \theta_2)^-)$, where*

$$S_0(\theta_1, \theta_2) = \{z \in \mathbb{C} : \theta_1 < \arg z < \theta_2\} \quad \textit{and} \quad 0 \le \theta_1 < \theta_2 \le 2\pi.$$

Let $\gamma_r(t) = re^{i[(1-t)\theta_1 + t\theta_2]}$, $t \in [0,1]$ (Figure 5.2).

1. *If $\lim_{z\to\infty} zf(z) = 0$, then $\lim_{r\to+\infty} \int_{\gamma_r} f = 0$.*
2. *If $\lim_{z\to 0} zf(z) = 0$, then $\lim_{r\to 0} \int_{\gamma_r} f = 0$.*

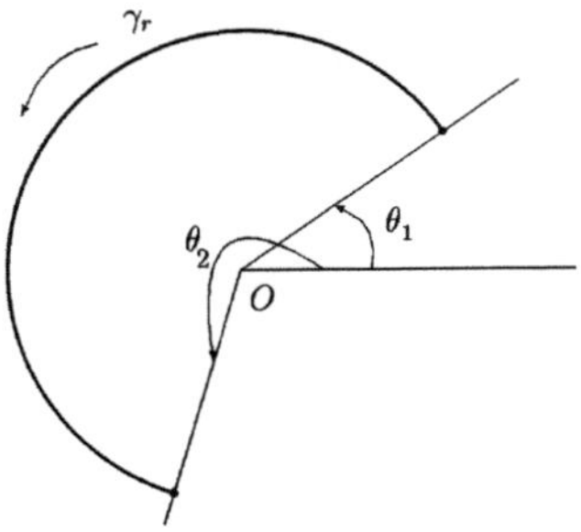

Figure 5.2: The $S_0(\theta_1, \theta_2)$ angular sector.

3. *For* $\theta_1 = 0, \theta_2 = \frac{\pi}{p}$, *where* $p > 0$, *and* $\lim_{z\to\infty} z^{1-p} f(z) = 0$, *then*

$$\lim_{r\to+\infty} \int_{\gamma_r} f(z) e^{imz^p}\,\mathrm{d}z = 0, \quad \textit{for all } m > 0.$$

4. *If* $f \in H(S_0(\theta_1, \theta_2)^- \setminus \{0\})$ *and* $z_0 = 0$ *is a simple pole of the function* f, *then*

$$\lim_{r\to 0} \int_{\gamma_r} f = (\theta_2 - \theta_1) i \operatorname{Res}(f; 0).$$

Proof. We may prove the points 1 and 2 of lemma, by using the inequality

$$\left|\int_{\gamma_r} f\right| \le M(r) r(\theta_2 - \theta_1), \quad \text{where } M(r) = \sup\{|f(\gamma_r(t))| : t \in [0,1]\}.$$

In fact, if $\lim_{\substack{z\to\infty \\ z\in\{\gamma_r\}}} zf(z) = \lim_{z\to\infty} zf(z) = 0$, $\lim_{r\to+\infty} rM(r) = 0$ and if $\lim_{\substack{z\to 0 \\ z\in\{\gamma_r\}}} zf(z) = \lim_{z\to 0} zf(z) = 0$, $\lim_{r\to 0} rM(r) = 0$. From these relations, we respectively obtain the points 1 and 2 of the lemma.

3. If $\theta_1 = 0$ and $\theta_2 = \frac{\pi}{p}$, then $\gamma_r(t) = re^{i\frac{\pi}{p}t}$, $t \in [0,1]$. It follows that

$$\begin{aligned}
\left|\int_{\gamma_r} f(z) e^{imz^p}\,\mathrm{d}z\right| &= \left|\int_0^1 f(re^{i\frac{\pi}{p}t}) e^{imr^p(\cos \pi t + i \sin \pi t)} ri\frac{\pi}{p} e^{i\frac{\pi}{p}t}\,\mathrm{d}t\right| \\
&\le \int_0^1 \left|f(re^{i\frac{\pi}{p}t}) e^{mr^p(-\sin \pi t + i\cos \pi t)} ri\frac{\pi}{p} e^{i\frac{\pi}{p}t}\right| \mathrm{d}t = \int_0^{\frac{\pi}{p}} |f(re^{iu})| e^{-mr^p \sin pu} r\,\mathrm{d}u \\
&\le M(r) r \int_0^{\frac{\pi}{p}} e^{-mr^p \sin pu}\,\mathrm{d}u = M(r) r \int_0^{\pi} e^{-mr^p \sin\varphi} \frac{1}{p}\,\mathrm{d}\varphi \\
&= 2\frac{r}{p} M(r) \int_0^{\frac{\pi}{2}} e^{-mr^p \sin\varphi}\,\mathrm{d}\varphi.
\end{aligned}$$

Since $\frac{2}{\pi} \leq \frac{\sin\varphi}{\varphi} \leq 1$, $\varphi \in [0, \frac{\pi}{2}]$ and $m > 0$, we get

$$\int_0^{\frac{\pi}{2}} e^{-mr^p \sin\varphi} d\varphi \leq \int_0^{\frac{\pi}{2}} e^{-mr^p \frac{2\varphi}{\pi}} d\varphi = \frac{\pi}{2mr^p} e^{-\frac{2mr^p}{\pi}\varphi}\Big|_{\frac{\pi}{2}}^{0} \leq \frac{\pi}{2mr^p},$$

hence

$$\left|\int_{\gamma_r} f(z)e^{imz^p} dz\right| \leq \frac{\pi M(r)}{mpr^{p-1}}.$$

Using the assumption $\lim_{\substack{z\to\infty \\ z\in\{\gamma_r\}}} z^{1-p}f(z) = \lim_{z\to\infty} z^{1-p}f(z) = 0$, we deduce that $\lim_{r\to+\infty} \frac{M(r)}{r^{p-1}} = 0$, and from the above inequality we conclude that

$$\lim_{r\to+\infty} \int_{\gamma_r} f(z)e^{imz^p} dz = 0.$$

4. If $z_0 = 0$ is a simple pole of the function f, then

$$f(z) = \frac{a_{-1}}{z} + a_0 + a_1 z + \cdots, \quad \forall z \in S_0(\theta_1, \theta_2)^- \setminus \{0\},$$

and from this relation we get

$$\lim_{r\to 0} \int_{\gamma_r} f = \operatorname{Res}(f; 0) \lim_{r\to 0} \int_{\gamma_r} \frac{1}{z} dz + \sum_{n=0}^{\infty} a_n \left(\lim_{r\to 0} \int_{\gamma_r} z^n dz\right).$$

From point 1, we have $\lim_{r\to 0} \int_{\gamma_r} z^n dz = 0$, for all $n \in \mathbb{N}$. Using elementary computations, we deduce that $\lim_{r\to 0} \int_{\gamma_r} \frac{1}{z} dz = (\theta_2 - \theta_1)i$, hence

$$\lim_{r\to 0} \int_{\gamma_r} f = (\theta_2 - \theta_1)i \operatorname{Res}(f; 0). \qquad \square$$

Remark 5.2.2. For the special case $p = 1$ and $m = 1$, point 3 of Lemma 5.2.1 is well known as the Jordan lemma.[1]

---

1 Jordan lemma: Consider a complex-valued, continuous function f, defined on a semicircular contour

$$C = \{Re^{i\theta} : \theta \in [0, \pi]\}$$

of positive radius R lying in the upper half-plane, centered at the origin. If function f is of the form

$$f(z) = e^{iaz} g(z), \quad z \in C$$

with a positive parameter a, then Jordan's lemma states the following upper bound for the contour integral

$$\left|\int_{C_R} f(z)dz\right| \leq \frac{\pi}{a} M_R \quad \text{where } M_R := \max_{\theta\in[0,\pi]} \left|g\left(Re^{i\theta}\right)\right|$$

where equal sign is when g vanishes everywhere. An analogous statement for a semicircular contour in the lower half-plane holds when $a < 0$.

Remark 5.2.3. Point 3 of Lemma 5.2.1 may be generalized as follows:

Let $f \in C(S_0(\theta_1,\theta_2)^-)$, where

$$S_0(\theta_1,\theta_2) = \{z \in \mathbb{C} : \theta_1 < \arg z < \theta_2\} \quad \text{and} \quad 0 \le \theta_1 < \theta_2 \le 2\pi.$$

Let $\gamma_r(t) = re^{i[(1-t)\theta_1+t\theta_2]}$, $t \in [0,1]$. If $\theta_1 = 0$, $\theta_2 = \frac{\pi}{np}$, where $n \in \{1\} \cup [2,+\infty)$, $p > 0$ and $\lim_{z\to\infty} z^{1-p}f(z) = 0$, then

$$\lim_{r\to+\infty} \int_{\gamma_r} f(z)e^{imz^p}\,dz = 0, \quad \text{for all } m > 0.$$

Proof. The proof is similar to the proof of point 3 of Lemma 5.2.1.

If $\theta_1 = 0$ and $\theta_2 = \frac{\pi}{np}$, where $n \ge 2$, then $\gamma_r(t) = re^{i\frac{\pi}{np}t}$, $t \in [0,1]$, hence

$$\begin{aligned}
\left|\int_{\gamma_r} f(z)e^{imz^p}\,dz\right| &= \left|\int_0^1 f(re^{i\frac{\pi}{np}t})e^{imr^p(\cos\frac{\pi}{n}t+i\sin\frac{\pi}{n}t)}ri\frac{\pi}{np}e^{i\frac{\pi}{np}t}\,dt\right| \\
&\le \int_0^1 \left|f(re^{i\frac{\pi}{np}t})e^{mr^p(-\sin\frac{\pi}{n}t+i\cos\frac{\pi}{n}t)}ri\frac{\pi}{np}e^{i\frac{\pi}{np}t}\right|dt \\
&= \int_0^{\frac{\pi}{np}} |f(re^{iu})|e^{-mr^p\sin pu}r\,du \le M(r)r\int_0^{\frac{\pi}{np}} e^{-mr^p\sin pu}\,du \\
&= M(r)r\int_0^{\frac{\pi}{n}} e^{-mr^p\sin\varphi}\frac{1}{p}\,d\varphi = \frac{M(r)r}{p}\int_0^{\frac{\pi}{n}} e^{-mr^p\sin\varphi}\,d\varphi.
\end{aligned}$$

Since $\frac{2}{\pi} \le \frac{\sin\varphi}{\varphi} \le 1$, $\varphi \in [0,\frac{\pi}{n}] \subseteq [0,\frac{\pi}{2}]$ and $m > 0$, we get

$$\int_0^{\frac{\pi}{n}} e^{-mr^p\sin\varphi}\,d\varphi \le \int_0^{\frac{\pi}{n}} e^{-mr^p\frac{2\varphi}{\pi}}\,d\varphi = \frac{\pi}{2mr^p}e^{-\frac{2mr^p}{\pi}\varphi}\Big|_{\frac{\pi}{n}}^{0} \le \frac{\pi}{2mr^p},$$

hence

$$\left|\int_{\gamma_r} f(z)e^{imz^p}\,dz\right| \le \frac{\pi M(r)}{2mpr^{p-1}}.$$

Using the assumption $\lim_{\substack{z\to\infty \\ z\in\{\gamma_r\}}} z^{1-p}f(z) = \lim_{z\to\infty} z^{1-p}f(z) = 0$, we deduce that $\lim_{r\to+\infty}\frac{M(r)}{r^{p-1}} = 0$, and from the above inequality we conclude that

$$\lim_{r\to+\infty} \int_{\gamma_r} f(z)e^{imz^p}\,dz = 0. \qquad \square$$

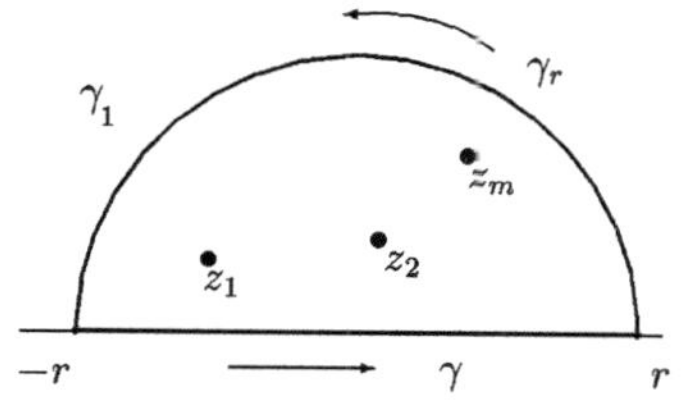

Figure 5.3: Proof of Problem 5.2.2.

Exercise 5.2.2. Compute the integrals of the type $\int_{-\infty}^{+\infty} \frac{P(x)}{Q(x)} dx$, where P and Q are two real variables polynomials with real coefficients, such that the denominator Q has no real zeros (roots), and $\deg Q \geq \deg P + 2$. Then the above integral may be computed using the formula

$$\int_{-\infty}^{+\infty} \frac{P(x)}{Q(x)} dx = 2\pi i \sum_{\operatorname{Im} z_k > 0} \operatorname{Res}\left(\frac{P}{Q}; z_k\right).$$

Solution. Let $z_1, \ldots, z_m$ be the roots of the Q polynomial (Figure 5.3). We will choose the number $r > 0$ such that it satisfies $r > \max\{|z_k| : k = \overline{1,m}\}$. Let $\gamma_r(t) = re^{i\pi t}$, $t \in [0, 1]$, and let us denote by $\gamma(t) = -r(1-t) + tr$, $t \in [0, 1]$, the linear path that connects the point $-r$ with r. Thus,

$$\int_{\gamma_1 = \gamma \cup \gamma_r} f(z) dz = \int_{-r}^{r} \frac{P(x)}{Q(x)} dx + \int_{\gamma_r} f = 2\pi i \sum_{\operatorname{Im} z_k > 0} \operatorname{Res}\left(\frac{P}{Q}; z_k\right),$$

since the index the $\gamma_1 = \gamma \cup \gamma_r$ curve with respect to every z_k point is equal to 1 (this curve is a boundary curve), where the paths are presented in Figure 5.3.

Hence,

$$\lim_{z \to \infty} zf(z) = \lim_{z \to \infty} \frac{zP(z)}{Q(z)} = 0 \overset{\text{Lemma}}{\Rightarrow} \lim_{r \to \infty} \int_{\gamma_r} f = 0$$

$$\Rightarrow \int_{-\infty}^{+\infty} \frac{P(x)}{Q(x)} dx = 2\pi i \sum_{\operatorname{Im} z_k > 0} \operatorname{Res}\left(\frac{P}{Q}; z_k\right). \qquad \square$$

Remark 5.2.4. If P and Q are two real variables polynomials with real coefficients, such that the denominator Q has no real zeros (roots), and $\deg Q \geq \deg P + 2$, then by using a similar proof we obtain the formula:

$$\int_{-\infty}^{+\infty} \frac{P(x)}{Q(x)} dx = -2\pi i \sum_{\operatorname{Im} z_k < 0} \operatorname{Res}\left(\frac{P}{Q}; z_k\right),$$

with $\gamma_2 = \gamma \cup \gamma_r$, where the curve $\gamma(t) = (1-t)r - tr$, $t \in [0, 1]$, is the linear path that connects r with $-r$, and $\gamma_r(t) = re^{(1+t)i\pi}$, $t \in [0, 1]$ (Figure 5.4).

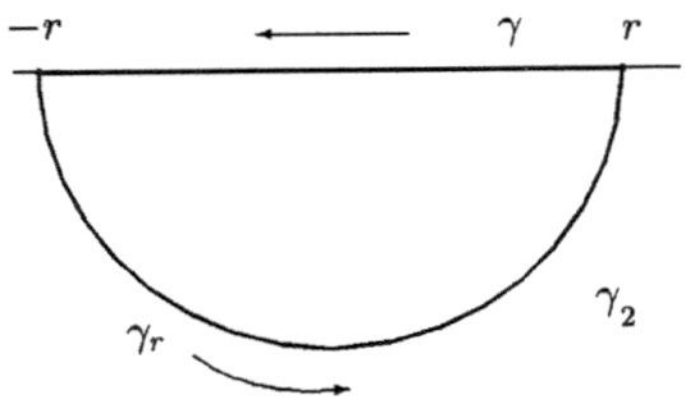

Figure 5.4: Figure for Remark 5.2.4 and Remark 5.2.5.

Exercise 5.2.3. Compute the integrals of the type $\int_{-\infty}^{+\infty} \frac{P(x)}{Q(x)} e^{ix} dx$, where P and Q are two real variables polynomials with real coefficients, such that the denominator Q has no real zeros (roots), and $\deg Q \geq \deg P + 1$. Then the integral $\int_{-\infty}^{+\infty} \frac{P(x)}{Q(x)} e^{ix} dx$ is convergent and may be computed using the formula

$$\int_{-\infty}^{+\infty} \frac{P(x)}{Q(x)} e^{ix} dx = 2\pi i \sum_{\operatorname{Im} z_k > 0} \operatorname{Res}\left(\frac{P(z)}{Q(z)} e^{iz}; z_k\right).$$

Solution. There exist the numbers $\eta > \xi > 0$ such that the function $R = \frac{P}{Q}$ is monotone on the $[\xi, \eta]$ interval. Using now the second mean-value theorem for the integral calculation, there exist the points $\xi_1, \xi_2 \in [\xi, \eta]$ such that

$$\int_{\xi}^{\eta} R(x) e^{ix} dx = \int_{\xi}^{\eta} R(x) \cos x dx + i \int_{\xi}^{\eta} R(x) \sin x dx$$
$$= R(\xi) \int_{\xi}^{\xi_1} \cos x dx + R(\eta) \int_{\xi_1}^{\eta} \cos x dx + i \left[R(\xi) \int_{\xi}^{\xi_2} \sin x dx + R(\eta) \int_{\xi_2}^{\eta} \sin x dx \right].$$

Since the modules of all the integrals are bounded, and since $\lim_{\xi \to +\infty} R(\xi) = \lim_{\eta \to +\infty} R(\eta) = 0$, it follows that the integral $\int_0^{+\infty} R(x) e^{ix} dx$ is convergent. Similarly, we may prove that the integral $\int_{-\infty}^{0} R(x) e^{ix} dx$ also converges, hence $\int_{-\infty}^{+\infty} R(x) e^{ix} dx$ is convergent.

Let $z_1, \ldots, z_m$ the zeros (roots) of the Q polynomial. Let us choose a number $r > 0$, such that $r > \max\{|z_k| : k = \overline{1, m}\}$. Let $\gamma_r(t) = re^{i\pi t}$, $t \in [0, 1]$, and denote by $\gamma(t) = -r(1-t) + tr$, $t \in [0, 1]$ the linear path that connects $-r$ with r (Figure 5.3). Thus,

$$\int_{\gamma_1 = \gamma \cup \gamma_r} R(z) e^{iz} dz = \int_{-r}^{r} \frac{P(x)}{Q(x)} e^{ix} dx + \int_{\gamma_r} R(z) e^{iz} dz$$
$$= 2\pi i \sum_{\operatorname{Im} z_k > 0} \operatorname{Res}\left(\frac{P(z)}{Q(z)} e^{iz}; z_k\right),$$

since the index the $\gamma_1 = \gamma \cup \gamma_r$ curve with respect to every z_k point is equal to 1 (this curve is a boundary curve).

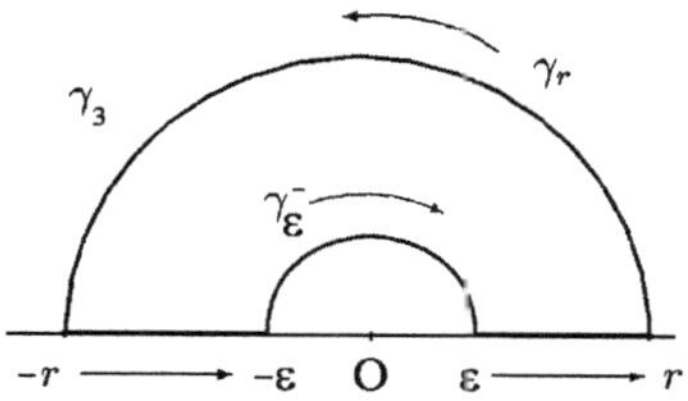

Figure 5.5: Proof of Problem 5.2.3.

Using point 3 of the lemma that extends the Jordan lemma, for the special case $p = 1$ and $m = 1$, we deduce that $\lim_{z\to\infty} R(z) = 0$, so it follows that $\lim_{r\to+\infty} \int_{\gamma_r} R(z) \times e^{iz}\,dz = 0$, hence

$$\int_{-\infty}^{+\infty} \frac{P(x)}{Q(x)} e^{ix}\,dx = 2\pi i \sum_{\operatorname{Im} z_k>0} \operatorname{Res}\left(\frac{P(z)}{Q(z)} e^{iz}; z_k\right). \qquad \square$$

Remark 5.2.5. If P and Q are two real variables polynomials with real coefficients, such that the denominator Q has no real zeros (roots), and $\deg Q \geq \deg P + 1$, then

$$\int_{-\infty}^{+\infty} \frac{P(x)}{Q(x)} e^{ix}\,dx = -2\pi i \sum_{\operatorname{Im} z_k<0} \operatorname{Res}\left(\frac{P(z)}{Q(z)} e^{iz}; z_k\right),$$

where we followed the same proof as the above, but for the curve $\gamma_2 = \gamma \cup \gamma_r$, where $\gamma(t) = (1-t)r - tr$, $t \in [0,1]$, is the linear path that connects r with $-r$, and $\gamma_r(t) = re^{(1+t)i\pi}$, $t \in [0,1]$ (Figure 5.4).

Remark 5.2.6. If P and Q are two real variables polynomials with real coefficients, such that the denominator Q has no real zeros (roots), and $\deg Q \geq \deg P$, then

$$\int_0^{+\infty} \left[\frac{P(x)}{Q(x)} e^{ix} - \frac{P(-x)}{Q(-x)} e^{-ix}\right] \frac{dx}{x} = 2\pi i \left[\frac{1}{2}\frac{P(0)}{Q(0)} + \sum_{\operatorname{Im} z_k>0} \operatorname{Res}\left(\frac{P(z)}{zQ(z)} e^{iz}; z_k\right)\right].$$

Proof. Let $z_1, \ldots, z_m$ be the zeros (roots) of the Q polynomial. Let $r > 0$ such a number, that $r > \max\{|z_k| : k = \overline{1,m}\}$, and let $\varepsilon > 0$ a number with $\varepsilon < \min\{|z_k| : k = \overline{1,m}\}$. Define the curve $\gamma_3 = \gamma_{[-r,-\varepsilon]} \cup \gamma_\varepsilon^- \cup \gamma_{[\varepsilon,r]} \cup \gamma_r$, where $\gamma_{[-r,-\varepsilon]}(t) = -(1-t)r - t\varepsilon$, $\gamma_\varepsilon(t) = \varepsilon e^{t\pi i}$, $\gamma_{[\varepsilon,r]}(t) = (1-t)\varepsilon + tr$ and $\gamma_r(t) = re^{t\pi i}$, $t \in [0,1]$ (Figure 5.5).

If we use the notation $g(z) = \frac{P(z)}{zQ(z)} e^{iz}$, we have

$$\begin{aligned}\int_{\gamma_3} g &= \int_{-r}^{-\varepsilon} \frac{P(x)}{xQ(x)} e^{ix}\,dx - \int_{\gamma_\varepsilon} \frac{P(z)}{zQ(z)} e^{iz}\,dz + \int_{\varepsilon}^{r} \frac{P(x)}{xQ(x)} e^{ix}\,dx + \int_{\gamma_r} \frac{P(z)}{zQ(z)} e^{iz}\,dz \\ &= 2\pi i \sum_{\operatorname{Im} z_k>0} \operatorname{Res}(g; z_k).\end{aligned}$$

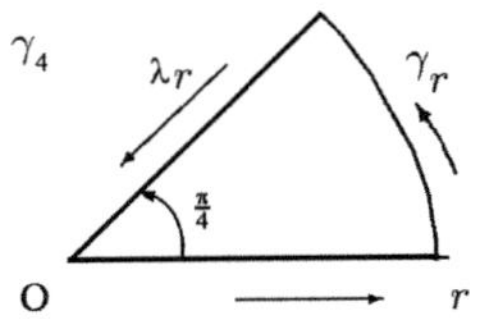

Figure 5.6: Proof of Remark 5.2.6 and Problem 5.2.4.

Using point 3 of the lemma that extends the Jordan lemma, we get

$$\lim_{r\to+\infty}\int_{\gamma_r}\frac{P(z)}{zQ(z)}e^{iz}\,dz = 0,$$

and from point 4 of the same lemma we have

$$\lim_{\varepsilon\to 0}\int_{\gamma_\varepsilon}\frac{P(z)}{zQ(z)}e^{iz}\,dz = \pi i\,\mathrm{Res}\Big(\frac{P(z)}{zQ(z)}e^{iz};0\Big) = \frac{\pi i P(0)}{Q(0)}.$$

If we make in the first integral the variable changing $y = -x$, we conclude that

$$\int_0^{+\infty}\Big[\frac{P(x)}{Q(x)}e^{ix} - \frac{P(-x)}{Q(-x)}e^{-ix}\Big]\frac{dx}{x} = 2\pi i\Big[\frac{1}{2}\frac{P(0)}{Q(0)} + \sum_{\mathrm{Im}\,z_k>0}\mathrm{Res}\Big(\frac{P(z)}{zQ(z)}e^{iz};z_k\Big)\Big]. \qquad \square$$

Remark 5.2.7. For the special case $P(z) = Q(z) = 1$, the result of Remark 5.2.6 becomes

$$\int_0^{+\infty}\frac{\sin x}{x}\,dx = \frac{\pi}{2}.$$

Exercise 5.2.4. Prove the following identity:

$$\int_0^{+\infty}\cos x^2\,dx = \int_0^{+\infty}\sin x^2\,dx = \frac{1}{2}\sqrt{\frac{\pi}{2}}.$$

(The above integrals are called **Fresnel-type integrals.**)

Solution. Let us define the curve $\gamma_4 = \gamma_{[0,r]}\cup\gamma_r\cup\lambda_r$, where $\gamma_{[0,r]}(t) = rt$, $\gamma_r(t) = re^{ti\frac{\pi}{4}}$ and $\lambda_r(t) = (1-t)re^{i\frac{\pi}{4}}$, $t\in[0,1]$ (Figure 5.6).

Denoting $g(z) = e^{iz^2}\in H(\mathbb{C})$, then

$$\int_0^r e^{ix^2}\,dx + \int_{\gamma_r} e^{iz^2}\,dz + \int_{\lambda_r} e^{iz^2}\,dz = 0.$$

Now we will use the generalization of the Jordan lemma for the special case $n = 2$, $p = 2$, $m = 1$ and $f(z) = 1$, and we get

$$\lim_{r\to+\infty} \int_{\gamma_r} e^{iz^2} dz = 0.$$

Since

$$\int_{\lambda_r} e^{iz^2} dz = \int_0^1 e^{i^2(1-t)^2 r^2} (-r) e^{i\frac{\pi}{4}} dt = -e^{i\frac{\pi}{4}} \int_0^r e^{-u^2} du,$$

it follows that

$$\int_0^{+\infty} e^{ix^2} dx = e^{i\frac{\pi}{4}} \int_0^{+\infty} e^{-u^2} du = e^{i\frac{\pi}{4}} \frac{\sqrt{\pi}}{2}.$$

Using the fact that the above inequality holds if and only if both real, and imaginary parts are respectively identical, we obtain the required Fresnel-type formulas. □

Exercise 5.2.5. Compute the integrals of the type $\int_0^{+\infty} \frac{P(x)}{Q(x)} dx$, where P and Q are two real variables polynomials with real coefficients, such that the denominator Q has no nonnegative zeros (roots), and $\deg Q \geq \deg P + 2$. Then the above integral may be computed using the formula

$$\int_0^{+\infty} \frac{P(x)}{Q(x)} dx = -\sum_{z_k \in \mathbb{C}^*} \operatorname{Res}\left(\frac{P(z)}{Q(z)} \log z; z_k\right), \quad \text{where } \log 1 = 0.$$

Solution. Define the function $f(z) = \frac{P(z)}{Q(z)} \log z$, where $\log z = \ln|z| + i \arg z$, $0 < \arg z < 2\pi$. Let $z_1, \ldots, z_m$ be the zeros (roots) of the Q polynomial. Then there exist the numbers $\varepsilon > 0$ and $r > 0$, such that

$$0 < \varepsilon < \min\{|z_k| : k = \overline{1,m}\} \leq \max\{|z_k| : k = \overline{1,m}\} < r.$$

Let $\gamma_5 = \lambda \cup \gamma_r \cup \lambda^- \cup \gamma_\varepsilon^-$, where this curve is presented in Figure 5.7, i. e., $\lambda(t) = (1-t)\varepsilon + tr$, $\gamma_r(t) = re^{2\pi i t}$, $\gamma_\varepsilon(t) = \varepsilon e^{2\pi i t}$, $t \in [0,1]$.

Thus,

$$\begin{aligned}\int_{\gamma_5} f &= \int_\varepsilon^r \frac{P(x)}{Q(x)} \ln x dx + \int_{\gamma_r} f + \int_r^\varepsilon \frac{P(x)}{Q(x)} [\ln x + 2\pi i] dx - \int_{\gamma_\varepsilon} f \\ &= 2\pi i \sum_{k=1}^m \operatorname{Res}(f; z_k) = -2\pi i \int_\varepsilon^r \frac{P(x)}{Q(x)} dx + \int_{\gamma_r} f - \int_{\gamma_\varepsilon} f.\end{aligned}$$

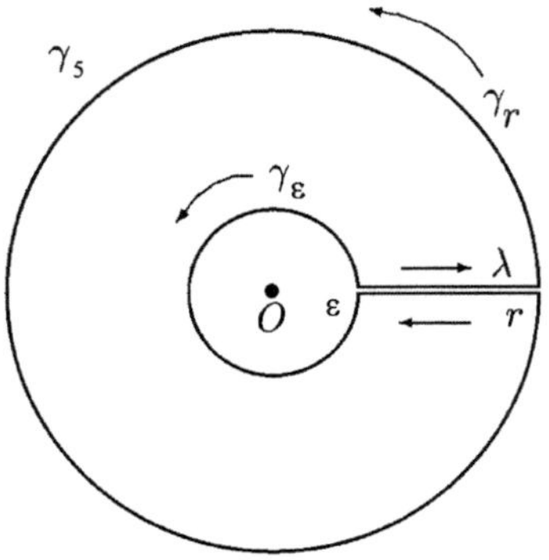

Figure 5.7: Proof of Problem 5.2.5 and Problem 5.2.7.

But

$$|zf(z)| = \left|\frac{zP(z)}{Q(z)}\log z\right| = \left|\frac{z^2P(z)}{Q(z)}\right|\left|\frac{\log z}{z}\right| \to 0, \quad \text{if } z \to \infty,$$

and

$$|zf(z)| = \left|\frac{P(z)}{Q(z)}\right||z\log z| \to 0, \quad \text{if } z \to 0,$$

because the ratio $\frac{P}{Q}$ is continuous at 0 and $\lim_{z\to 0} z\log z = 0$.

Taking $\varepsilon \to 0$ and $r \to \infty$, from points 1 and 2 of the lemma that extends the Jordan lemma, we get

$$\int_0^{+\infty} \frac{P(x)}{Q(x)}\mathrm{d}x = -\sum_{k=1}^{m} \operatorname{Res}(f; z_k). \qquad \square$$

Exercise 5.2.6. Compute the integrals of the type $\int_0^{+\infty} \frac{P(x)}{Q(x)} \ln x\mathrm{d}x$, where P and Q are two real variables polynomials with real coefficients, such that the denominator Q has no nonnegative zeros (roots), and $\deg Q \geq \deg P + 2$. Then the integral $\int_0^{+\infty} \frac{P(x)}{Q(x)} \ln x\mathrm{d}x$ is convergent, and

$$\int_0^{+\infty} \frac{P(x)}{Q(x)} \ln x\mathrm{d}x = -\frac{1}{2}\operatorname{Re}\left[\sum_{z_k\in\mathbb{C}^*} \operatorname{Res}\left(\frac{P(z)}{Q(z)}(\log z)^2; z_k\right)\right], \quad \text{where } \log 1 = 0.$$

Solution. Since $\deg Q \geq \deg P + 2$, it follows that the given integral is convergent. Let us define the function $g(z) = R(z)(\log z)^2$, where $\log z = \ln|z| + i\arg z$, $0 < \arg z < 2\pi$ and $R(z) = \frac{P(z)}{Q(z)}$.

Let $z_1, \ldots, z_m$ be the zeros (roots) of the Q polynomial. Then there exist the numbers $\varepsilon > 0$ and $r > 0$, such that

$$0 < \varepsilon < \min\{|z_k| : k = \overline{1,m}\} \leq \max\{|z_k| : k = \overline{1,m}\} < r.$$

Let $\gamma_5 = \lambda \cup \gamma_r \cup \lambda^- \cup \gamma_\varepsilon^-$, where this curve is presented in Figure 5.7, i. e., $\lambda(t) = (1-t)\varepsilon + tr$, $\gamma_r(t) = re^{2\pi i t}$, $\gamma_\varepsilon(t) = \varepsilon e^{2\pi i t}$, $t \in [0,1]$.

Thus,

$$\int_\varepsilon^r R(x)\ln^2 x dx + \int_{\gamma_r} g + \int_r^\varepsilon R(x)[\ln x + 2\pi i]^2 dx - \int_{\gamma_\varepsilon} g = 2\pi i \sum_{z_k \in \mathbb{C}^*} \operatorname{Res}(g; z_k).$$

But

$$|zg(z)| = \left|\frac{zP(z)}{Q(z)}(\log z)^2\right| = \left|\frac{z^2 P(z)}{Q(z)}\right|\left|\frac{(\log z)^2}{z}\right| \to 0, \quad \text{if } z \to \infty,$$

and

$$|zg(z)| = \left|\frac{P(z)}{Q(z)}\right| |z(\log z)^2| \to 0, \quad \text{if } z \to 0,$$

because the ratio $\frac{P}{Q}$ is continuous at 0 and $\lim_{z\to 0} z(\log z)^2 = 0$.

Letting now $\varepsilon \to 0$ and $r \to \infty$, from points 1 and 2 of the lemma that extends the Jordan lemma, we get

$$\int_0^{+\infty} R(x)\ln^2 x dx - \int_0^{+\infty} R(x)[\ln x + 2\pi i]^2 dx = 2\pi i \sum_{z_k \in \mathbb{C}^*} \operatorname{Res}(g; z_k).$$

Identifying in this last relation the real and the imaginary parts, we finally deduce that

$$\int_0^{+\infty} \frac{P(x)}{Q(x)} \ln x dx = -\frac{1}{2} \operatorname{Re}\left[\sum_{z_k \in \mathbb{C}^*} \operatorname{Res}\left(\frac{P(z)}{Q(z)}(\log z)^2; z_k\right)\right]$$

and

$$\int_0^{+\infty} \frac{P(x)}{Q(x)} dx = -\frac{1}{2\pi} \operatorname{Im}\left[\sum_{z_k \in \mathbb{C}^*} \operatorname{Res}\left(\frac{P(z)}{Q(z)}(\log z)^2; z_k\right)\right]. \qquad \square$$

Remark 5.2.8. The last relation obtained in the above proof shows that if P and Q are two real variables polynomials with real coefficients, such that the denominator Q has no nonnegative zeros (roots), and $\deg Q \geq \deg P + 2$, then

$$\int_0^{+\infty} \frac{P(x)}{Q(x)} dx = -\frac{1}{2\pi} \operatorname{Im}\left[\sum_{z_k \in \mathbb{C}^*} \operatorname{Res}\left(\frac{P(z)}{Q(z)}(\log z)^2; z_k\right)\right], \quad \text{where } \log 1 = 0.$$

Exercise 5.2.7. Compute the integrals of the type $\int_0^{+\infty} \frac{P(x)}{Q(x)} \frac{1}{x^a} dx$, where P and Q are two real variables polynomials with real coefficients, such that the denominator Q has no nonnegative zeros (roots), and $\deg Q \geq \deg P+1$ and $\alpha \in (0,1)$. Then the above integral may be computed using the formula

$$\int_0^{+\infty} \frac{P(x)}{Q(x)}\frac{1}{x^\alpha}\,dx = \frac{\pi e^{\alpha\pi i}}{\sin(\alpha\pi)} \sum_{z_k\in\mathbb{C}^*} \operatorname{Res}\left(\frac{P(z)}{Q(z)}\frac{1}{z^\alpha}; z_k\right), \quad \text{where } \log 1 = 0.$$

Solution. Let $g(z) = \frac{R(z)}{z^\alpha}$, where $\log z = \ln|z| + i\arg z$, $0 < \arg z < 2\pi$ and $R(z) = \frac{P(z)}{Q(z)}$.

Let $z_1, \ldots, z_m$ be the zeros (roots) of the Q polynomial. Then there exist the numbers $\varepsilon > 0$ and $r > 0$, such that

$$0 < \varepsilon < \min\{|z_k| : k = \overline{1,m}\} \leq \max\{|z_k| : k = \overline{1,m}\} < r.$$

Let $\gamma_5 = \lambda \cup \gamma_r \cup \lambda^- \cup \gamma_\varepsilon^-$, where this curve is presented in Figure 5.7, i. e., $\lambda(t) = (1-t)\varepsilon + tr$, $\gamma_r(t) = re^{2\pi it}$, $\gamma_\varepsilon(t) = \varepsilon e^{2\pi it}$, $t \in [0,1]$.

Thus,

$$\int_\varepsilon^r \frac{R(x)}{x^\alpha}\,dx + \int_{\gamma_r} g + \int_r^\varepsilon \frac{R(x)}{x^\alpha} e^{-2\alpha\pi i}\,dx - \int_{\gamma_\varepsilon} g = 2\pi i \sum_{z_k\in\mathbb{C}^*} \operatorname{Res}(g; z_k),$$

hence

$$\int_{\gamma_r} g - \int_{\gamma_\varepsilon} g + (1 - e^{-2\alpha\pi i}) \int_\varepsilon^r \frac{R(x)}{x^\alpha}\,dx = 2\pi i \sum_{z_k\in\mathbb{C}^*} \operatorname{Res}(g; z_k).$$

But

$$|zg(z)| = \left|\frac{zP(z)}{Q(z)}\right|\left|\frac{1}{z^\alpha}\right| \to 0, \quad \text{if } z \to \infty,$$

and

$$|zg(z)| = \left|\frac{P(z)}{Q(z)}\right||z^{1-\alpha}| \to 0, \quad \text{if } z \to 0,$$

because the ratio $\frac{P}{Q}$ is continuous at 0 and $\lim_{z\to 0}|z^{1-\alpha}| = 0$.

Letting now $\varepsilon \to 0$ and $r \to \infty$, from the points 1 and 2 of the lemma that extends the Jordan lemma, we get

$$(1 - e^{-2\alpha\pi i}) \int_0^{+\infty} \frac{R(x)}{x^\alpha}\,dx = 2\pi i \sum_{z_k\in\mathbb{C}^*} \operatorname{Res}(g; z_k),$$

i. e.,

$$\int_0^{+\infty} \frac{P(x)}{Q(x)}\frac{1}{x^\alpha}\,dx = \frac{\pi e^{\alpha\pi i}}{\sin(\alpha\pi)} \sum_{z_k\in\mathbb{C}^*} \operatorname{Res}\left(\frac{P(z)}{Q(z)}\frac{1}{z^\alpha}; z_k\right). \qquad \square$$

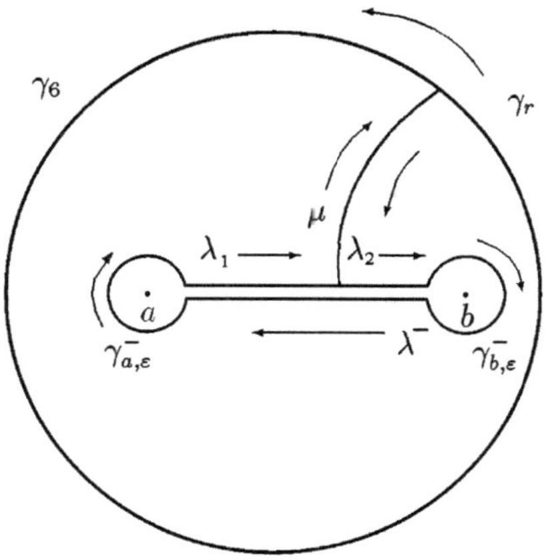

Figure 5.8: Proof of Problem 5.2.8.

Exercise 5.2.8. Compute the integrals of the type $\int_a^b \frac{P(x)}{Q(x)(x-a)^\alpha(b-x)^\beta}\mathrm{d}x$, where P and Q are two real variables polynomials with real coefficients, such that the denominator Q has no zeros (roots) on the interval $[a,b]$, $\deg Q \geq \deg P + 2$, and $\alpha, \beta > 0$ satisfy the condition $\alpha + \beta = 1$. Then the above integral may be computed using the formula

$$\int_a^b \frac{P(x)}{Q(z)(x-a)^\alpha(b-x)^\beta}\mathrm{d}x = \frac{\pi e^{2\alpha\pi i}}{\sin(2\alpha\pi)} \sum_{z_k \in \mathbb{C}\setminus[a,b]} \operatorname{Res}\left(\frac{P(z)}{Q(z)(z-a)^\alpha(b-z)^\beta}; z_k\right),$$

where we consider the main branch for all the powers, i. e., $\log 1 = 0$.

Solution. Let $g(z) = \frac{R(z)}{(z-a)^\alpha(b-z)^\beta}$, where $\log z = \ln|z| + i\arg z$, $0 < \arg z < 2\pi$ and $R(z) = \frac{P(z)}{Q(z)}$.

Let $z_1, \ldots, z_m$ be the zeros (roots) of the Q polynomial. Then there exist the numbers $\varepsilon > 0$ and $r > 0$, such that all the roots of Q lies in those bounded domain, with the boundary given by the γ_6 curve, where this curve is presented in Figure 5.8, i. e., $\lambda(t) = (1-t)(a+\varepsilon)+t(b-\varepsilon)$, $\gamma_r(t) = re^{2\pi i t}$, $\gamma_{a,\varepsilon}(t) = a + \varepsilon e^{2\pi i t}$, $\gamma_{b,\varepsilon}(t) = b+\varepsilon e^{2\pi i t}$, $t \in [0,1]$. Let μ be a such a rectifiable curve that connects two arbitrary points from $\{\lambda\}$ and $\{\gamma_r\}$, and don't contain any root of Q.

Thus,

$$\int_{\gamma_r} g - \int_{\gamma_{a,\varepsilon}} g - \int_{\gamma_{b,\varepsilon}} g + \int_{a+\varepsilon}^{b-\varepsilon} \frac{P(x)}{Q(x)(x-a)^\alpha(b-x)^\beta}\mathrm{d}x$$
$$- \int_{a+\varepsilon}^{b-\varepsilon} \frac{P(x)}{Q(x)(x-a)^\alpha(b-x)^\beta} e^{2\beta\pi i} e^{-2\alpha\pi i}\mathrm{d}x = 2\pi i \sum_{z \in \mathbb{C}\setminus[a\, b]} \operatorname{Res}(g; z_k).$$

Letting $\varepsilon \to 0$ and $r \to \infty$, from points 1 and 2 of the lemma that extends the Jordan lemma, we get

$$\lim_{\varepsilon\to 0}\int_{\gamma_{a,\varepsilon}} g = \lim_{\varepsilon\to 0}\int_{\gamma_{b,\varepsilon}} g = 0 \quad \text{and} \quad \lim_{r\to+\infty}\int_{\gamma_r} g = 0.$$

Hence, it follows that

$$\int_a^b \frac{P(x)}{Q(z)(x-a)^\alpha(b-x)^\beta}\,dx = \frac{\pi e^{2\alpha\pi i}}{\sin(2\alpha\pi)} \sum_{z_k\in\mathbb{C}\setminus[a,b]} \operatorname{Res}\left(\frac{P(z)}{Q(z)(z-a)^\alpha(b-z)^\beta}; z_k\right). \qquad \square$$

5.3 The study of meromorphic functions with the residue theorem

Definition 5.3.1. Let $\widetilde{G} \subset \mathbb{C}$ an open set, and let $f \in M(\widetilde{G})$. The **order of the function f in the $z_0 \in \widetilde{G}$ point** is the biggest integer number $n \in \mathbb{Z}$ that satisfies the next property:

$$\exists r > 0,\ \exists g \in H(U(z_0;r)) \quad \text{such that } g(z_0) \neq 0 \text{ and}$$
$$\forall z \in \dot{U}(z_0;r), f(z) = (z-z_0)^n g(z).$$

This integer number is denoted by the symbol $\Theta(f, z_0)$.

Remarks 5.3.1. From the theorem of characterization of holomorphic function zeros and the theorem of characterization of the poles, it follows that:
1. The point $z_0 \in \widetilde{G}$ is an nth order zero for the function $f \Leftrightarrow \Theta(f, z_0) = n$;
2. The point $z_0 \in \widetilde{G}$ is a pth order pole for the function $f \Leftrightarrow \Theta(f, z_0) = -p$;
3. The point $z_0 \in \widetilde{G}$ with $f(z_0) \neq 0$ is a regular point for the function $f \Leftrightarrow \Theta(f, z_0) = 0$.

Proof. If $z_0 \in \widetilde{G}$ is an nth order zero for the function f, then

$$f(z) = (z-z_0)^n(a_n + a_{n+1}(z-z_0) + \cdots), \quad \forall z \in U(z_0;r) \text{ and } a_n \neq 0,$$

hence $f(z_0) = f'(z_0) = \cdots = f^{(n-1)}(z_0) = 0, f^{(n)}(z_0) \neq 0$. Then it follows that $\Theta(f, z_0) = n$.

If $z_0 \in \widetilde{G}$ is a pth order pole for the function f, then

$$f(z) = \frac{a_{-p}}{(z-z_0)^p} + \cdots + \frac{a_{-1}}{z-z_0} + a_0 + \cdots, \quad \forall z \in \dot{U}(z_0;r) \text{ and } a_{-p} \neq 0,$$

hence $\Theta(f, z_0) = -p$.

If $z_0 \in \widetilde{G}$ is a regular point for the function f such that $f(z_0) \neq 0$, then

$$f(z) = a_0 + a_n(z-z_0) + \cdots, \quad \text{and} \quad a_0 = f(z_0) \neq 0,$$

so we get $\Theta(f, z_0) = 0$. $\square$

Evidently, the next results hold.

Theorem 5.3.1. *For all $f, g \in M(\widetilde{G})$, where $\widetilde{G} \subset \mathbb{C}$ is an open set and for all $z_0 \in \widetilde{G}$, the following relations are true:*

$$\Theta(f \cdot g, z_0) = \Theta(f, z_0) + \Theta(g, z_0),$$
$$\Theta\left(\frac{f}{g}, z_0\right) = \Theta(f, z_0) - \Theta(g, z_0).$$

Definition 5.3.2. Let $f \in M(\widetilde{G})$, where $\widetilde{G} \subset \mathbb{C}$ is an open set, and let $D \subset \widetilde{G}$. If the sum $\sum_{z\in D} \Theta(f, z)$ is finite, then this sum represents **the order of the function f in the set** D and it is denoted by the symbol $\Theta(f, D)$.

Remarks 5.3.2.
1. We have $\Theta(f, D) = N - P$, where N represents the number of all zeros of the function f that lie in D, where every zero will be counted together with its order, and P represents the number of all poles of the function f that lie in D, where every pole will be counted together with its order.
2. If p_n is a nth order polynomial, then $p_n \in H(\mathbb{C}) \subset M(\mathbb{C})$ and $\Theta(p_n, \mathbb{C}) = n$.

Lemma 5.3.1. *Let $f \in M(\widetilde{G})$, where $\widetilde{G} \subset \mathbb{C}$ is an open set, let $g \in H(\widetilde{G})$, $a \in \widetilde{G}$, and denote $h = \frac{f'}{f}g$. Then*

$$\operatorname{Res}(h; a) = \Theta(f, a)g(a).$$

Proof. 1. If $a \in \widetilde{G}$ is an m-th order pole of the function f, then there exists a function f_1 holomorphic in a neighborhood of a, such that $f_1(a) \neq 0$ and $f(z) = \frac{f_1(z)}{(z-a)^m}$. Hence,

$$\frac{f'(z)}{f(z)} = \frac{f_1'(z)(z-a)^m - m(z-a)^{m-1}f_1(z)}{(z-a)^{2m}} \cdot \frac{(z-a)^m}{f_1(z)} = \frac{f_1'(z)}{f_1(z)} - m\frac{1}{z-a},$$

where $\frac{f_1'(z)}{f_1(z)}$ is holomorphic in the point a, then the point a is a first-order pole of $\frac{f'}{f}$ and $\operatorname{Res}(\frac{f'}{f}; a) = -m$. Hence, it follows that

$$\operatorname{Res}\left(\frac{f'}{f}g; a\right) = -mg(a).$$

2. If $a \in \widetilde{G}$ is an nth order zero of the function f, then there exists a function f_1 holomorphic in a neighborhood of a, such that $f_1(a) \neq 0$ and $f(z) = (z-a)^n f_1(z)$. Hence,

$$\frac{f'(z)}{f(z)} = \frac{n(z-a)^{n-1}f_1(z) + (z-a)^n f_1'(z)}{(z-a)^n f_1(z)} = \frac{n}{z-a} - \frac{f_1'(z)}{f_1(z)},$$

where $\frac{f_1'}{f_1}$ is holomorphic in the point a. then the point a is a first-order pole of $\frac{f'}{f}$ and $\operatorname{Res}(\frac{f'}{f}; a) = n$. Thus, we get

$$\operatorname{Res}\left(\frac{f'}{f} g; a\right) = ng(a).$$

We obtained that in the both cases $\operatorname{Res}(h, a) = \Theta(f, a)g(a)$.

3. If a is a regular point, then we may suppose $f(a) \neq 0$, and taking in the second part of our proof $n = 0$ we deduce the required result. □

Theorem 5.3.2 (Cauchy theorem related to the zeros and poles). *Let $\widetilde{G} \subset \mathbb{C}$ an open set, $f \in M(\widetilde{G})$, $g \in H(\widetilde{G})$, where $f \not\equiv 0$ and let γ a such a rectifiable closed curve (path) in $\widetilde{G}$, which does not contain nor any zero nor any pole of the function f, and $\gamma \underset{\widetilde{G}}{\sim} 0$. Then the sum*

$$\sum_{z \in \widetilde{G}} g(z)\Theta(f, z)\mathrm{n}(\gamma, z) \quad \textit{is finite, and}$$

$$\frac{1}{2\pi i} \int_{\gamma} g\frac{f'}{f} = \sum_{z \in \widetilde{G}} g(z)\Theta(f, z)\mathrm{n}(\gamma, z).$$

Proof. Let us denote $h = g\frac{f'}{f}$. Then, h is meromorphic in $\widetilde{G}$, and the set of all singular points of h, denoted by $S = S(h)$. It coincides with $A \cup B$, where A is the set of all the zeros of the function f, and B is the set of all the poles of the function f. The above remarks follow from the proof of the Lemma 5.3.1.

Let $G = \widetilde{G} \setminus S$. Then G is an open set and $h \in H(G)$. Since $\{\gamma\}$ lies in the G set, applying the residue theorem for the function h, we have

$$\int_{\gamma} g\frac{f'}{f} = \int_{\gamma} h = 2\pi i \sum_{z \in \widetilde{G}} \operatorname{Res}(h; z)\mathrm{n}(\gamma, z). \tag{5.5}$$

The index $\mathrm{n}(\gamma, z)$ does not vanish only on the bounded domain with the boundary given by γ, and in rest we have $\operatorname{Res}(h; z) \neq 0 \Rightarrow z \in A \cup B$. Since in the above mentioned bounded domain with the boundary given by γ, the set $A \cup B$ has only a finite number of points; the sum (5.5) is finite. From Lemma 5.3.1, it follows that if $z_0 \in A \cup B$, then $\Theta(f, z_0) \neq 0$, and $\operatorname{Res}(h; z_0) = \Theta(f, z_0)g(z_0)$. Hence,

$$\frac{1}{2\pi i} \int_{\gamma} g\frac{f'}{f} = \sum_{z \in \widetilde{G}} g(z)\Theta(f, z)\mathrm{n}(\gamma, z).$$ □

Theorem 5.3.3. *Let γ be a boundary curve in $\widetilde{G}$, where $\widetilde{G} \subset \mathbb{C}$ is an open set and $(\gamma)^{-} \subset \widetilde{G}$ (where (γ) denotes that bounded domain, which is bounded by the curve $\{\gamma\}$). Let f a such a meromorphic function on $\widetilde{G}$, which have nor any zero nor any pole in the $\{\gamma\}$ set. Let $a_1, \dots, a_N$ denote the all zeros of the function f that lie in (γ), and let $b_1, \dots, b_P$ denote* *the all poles of the function f that lie in (γ) (where every zero and every pole is counted together with its order). If $g \in H(\widetilde{G})$, then*

$$\frac{1}{2\pi i}\int_{\gamma} g\frac{f'}{f} = \sum_{j=1}^{N} g(a_j) - \sum_{k=1}^{P} g(b_k).$$

Proof. Since γ is a boundary curve, then $n(\gamma, a_j) = n(\gamma, b_k) = 1$, $\forall j = \overline{1,N}$, $\forall k = \overline{1,P}$, and the proof follows from Theorem 5.3.2 and from Lemma 5.3.1. □

Corollary 5.3.1. *From Theorem 5.3.3, we have*

$$N - P = \frac{1}{2\pi i}\int_{\gamma} \frac{f'}{f} = \Theta(f, D), \quad \text{where } D = (\gamma)$$

and N represents the number of all the zeros of the function f that lie in D, where every zero is counted together with its order, and P represents the number of all the poles of the function f that lie in D, where every pole is counted together with its order.

Theorem 5.3.4 (Argument principle). *Let $f \in M(\widetilde{G})$ be a non constant function, where $\widetilde{G} \subset \mathbb{C}$ is an open set, let γ a curve (path) in $\widetilde{G}$, such that there exists γ' which is continuous, and $\{\gamma\}$ does not contain any pole of the function f. Let $\Gamma = f \circ \gamma$, and let define the function*

$$\mathrm{I} : \mathbb{C} \setminus \{\Gamma\} \to \mathbb{C}, \quad \mathrm{I}(w) = \frac{1}{2\pi i}\int_{\gamma} \frac{f'}{f - w}.$$

Thus,
1. $\mathrm{I}(w) = \mathrm{n}(\Gamma, w)$;
2. *If Γ is a closed curve (path), then I is locally constant, and $\mathrm{I}(w) \in \mathbb{Z}$, $\forall w \in \mathbb{C} \setminus \{\Gamma\}$;*
3. *If $D = U(z_0; r)$ and $\gamma = \partial D$, $D^- \subset \widetilde{G}$, then $\mathrm{I}(w) = \mathrm{n}(\Gamma, w) = \Theta(f - w, D)$.*

Proof. 1. From the definition, we have

$$\begin{aligned} \mathrm{n}(\Gamma, w) &= \frac{1}{2\pi i}\int_{\Gamma} \frac{d\zeta}{\zeta - w} = \frac{1}{2\pi i}\int_{0}^{1} \frac{\Gamma'(t)}{\Gamma(t) - w}dt \\ &= \frac{1}{2\pi i}\int_{0}^{1} \frac{f'(\gamma(t))\gamma'(t)}{f(\gamma(t)) - w}dt = \frac{1}{2\pi i}\int_{\gamma} \frac{f'}{f - w} = \mathrm{I}(w). \end{aligned}$$

2. Since $\mathrm{I}(w) = \mathrm{n}(\Gamma, w)$, from the Index theorem it follows that I is locally constant, and $\mathrm{I}(w) \in \mathbb{Z}$, $\forall w \in \mathbb{C} \setminus \{\Gamma\}$.

3. Since

$$\mathrm{I}(w) = \frac{1}{2\pi i}\int_{\gamma} \frac{f'}{f - w},$$

from Corollary 5.3.1 we get

$$\frac{1}{2\pi i}\int_{\gamma}\frac{f'}{f-w}=\Theta(f-w,D),$$

hence $\mathrm{I}(w)=\mathrm{n}(\Gamma,w)=\Theta(f-w,D)$. □

Corollary 5.3.2 (Special case of the principle of argument variation). *Let $f\in M(\widetilde{G})$, where $\widetilde{G}\subset\mathbb{C}$ is an open set, and let γ be a boundary curve in $\widetilde{G}$ such that there exists γ' which is continuous, and $\{\gamma\}$ does not contain any zero or any pole of the function f.*

Let N be the number of all the zeros of the function f that lie in (γ) (every zero is counted together with its order), and let P be the number of all the poles of the function f that lie in (γ) (every pole is counted together with its order), and let $\Gamma=f\circ\gamma$. Then the next relation holds

$$\Theta(f,D)=N-P=\mathrm{n}(\Gamma,0),\quad \textit{where } D=(\gamma),$$

which we usually write as

$$\frac{1}{2\pi i}\int_{\gamma}\frac{f'}{f}=\mathrm{n}(\Gamma,0)=\frac{1}{2\pi}\{\operatorname{Arg}f(z)\}_{\gamma},$$

where the right-hand side symbol represents "the variation" of the argument of the function f when z runs over the curve (path) γ.

Remark 5.3.3. The name of the previous corollary follows from the fact that

$$\log f(z)=\ln|f(z)|+i\arg f(z),$$

where $\log f$ is the primitive of the function $\frac{f'}{f}$. Using the fact that $\gamma(0)=\gamma(1)$ together with $\ln|f(\gamma(0))|=\ln|f(\gamma(1))|$, it follows that

$$\frac{1}{2\pi i}\int_{\gamma}\frac{f'}{f}=\frac{1}{2\pi i}\log f(z)|_{\gamma(0)}^{\gamma(1)}=\frac{1}{2\pi i}i[\arg f(\gamma(1))-\arg f(\gamma(0))]=\frac{1}{2\pi}\{\operatorname{Arg}f(z)\}_{\gamma}.$$

Theorem 5.3.5 (Rouché theorem). *Let $f,g\in H(G)$, where $G\subset\mathbb{C}$ is an open set, and let γ be a boundary curve in G such that there exists γ' which is continuous. Let $D=(\gamma)$ (where (γ) denotes that bounded domain, which is bounded by the curve $\{\gamma\}$), and suppose that $D^{-}\subset G$. If*

$$|g(\zeta)|<|f(\zeta)|,\quad \forall\zeta\in\{\gamma\},$$

then the next two equations have the same numbers of zeros in D:

1. $f(z)=0$;
2. $f(z)+g(z)=0$,

i. e.,

$$\Theta(f, D) = \Theta(f + g, D).$$

Proof. We need to prove that $\Theta(f, D) = \Theta(f + g, D)$.

Since for all the points $\zeta \in \{\gamma\} = \partial D$, we have $|g(\zeta)| < |f(\zeta)|$. It follows that $f(\zeta) \neq 0$, $\forall \zeta \in \{\gamma\}$. On the other hand, $h = \frac{g}{f} \in M(D_1)$, where D_1 is a connected component of G that contains the set D. Hence,

$$\Theta(f + g, D) - \Theta(f, D) = \Theta\left(\frac{f + g}{f}, D\right) = \Theta(h + 1, D).$$

If $\Gamma = h \circ \gamma$, from Theorem 5.3.4 we have

$$\Theta(h + 1, D) = \mathrm{n}(\Gamma, -1).$$

If $t \in [0, 1]$, then $|\Gamma(t)| = |h(\gamma(t))| = |\frac{g(\gamma(t))}{f(\gamma(t))}| < 1$, hence $\{\Gamma\} \subset U(0; 1)$, and from here we deduce $F = \mathbb{C} \setminus \overline{U}(0; 1) \subset \mathbb{C} \setminus \{\Gamma\}$. Since F is not bounded connected set, then F it is a subset of the $\mathbb{C} \setminus \{\Gamma\}$ unbounded component, and from the index theorem we get $\mathrm{n}(\Gamma, w) = 0$, for all $w \in F$. Thus, we have $\mathrm{n}(\Gamma, -1) = 0$, because $-1 \in F$, and hence $\Theta(h + 1, D) = 0$. □

Corollary 5.3.3 (Fundamental theorem of the algebra). *Let*

$$p(z) = a_0 + a_1 z + \cdots + a_n z^n, \quad n \in \mathbb{N}^*,\ a_n \neq 0.$$

Then the polynomial p has n roots in $\mathbb{C}$.

Proof. Let us consider the next two equations:

$$a_n z^n = 0; \tag{5.6}$$

$$p(z) = 0. \tag{5.7}$$

Let us denote $f(z) = a_n z^n$, and $g(z) = a_0 + a_1 z + \cdots + a_{n-1} z^{n-1}$. Since $\lim_{z \to \infty} \frac{g(z)}{f(z)} = 0$, it follows that

$$\exists r_0 > 0 \quad \text{such that } \forall r > r_0, \text{ and } \forall \zeta \in \{\gamma\},\ \gamma = \partial U(0; r) \text{ we have } \left|\frac{g(\zeta)}{f(\zeta)}\right| < 1,$$

i. e.,

$$|g(\zeta)| < |f(\zeta)|, \quad \forall \zeta \in \{\gamma\}.$$

According to the Rouché theorem, equations (5.6) and (5.7) have the same number of zeros in $U(0; r)$. But the equation (5.6) has exactly n zeros in 0. □

Theorem 5.3.6 (The principle of the domain conservation). *Let $f \in H(D)$, where $D \subset \mathbb{C}$ is a domain (i. e., an open connected set). If the function f is nonconstant, then $\Delta = f(D)$ is also a domain.*

Proof. Since the function f is continuous and the set D is connected, it follows easily that Δ is also connected.

We need to prove that $\Delta = f(D)$ is an open set. We will use two different methods to prove this.

1. Let $w_0 \in f(D)$. Then $\exists z_0 \in D$ such that $f(z_0) = w_0$, hence z_0 is a zero of the function $f - w_0$. But $f - w_0$ is a nonconstant function, because f is nonconstant, hence the point z_0 is an isolated zero. Thus, $\exists r > 0$ such that $U^-(z_0; r) \subset D$ and $\forall \zeta \in \{\gamma\}$, where $\gamma = \partial U(z_0; r)$, we have $f(\zeta) - w_0 \neq 0$.

Since the set $\{\gamma\}$ is compact and the function f is continuous, we deduce that

$$\rho = \min_{\zeta \in \{\gamma\}} |f(\zeta) - w_0| > 0.$$

Let $w_1 \in U(w_0; \rho)$ be an arbitrary number, and consider the two equations:

$$f(z) - w_0 = 0; \tag{5.8}$$

$$f(z) - w_1 = 0. \tag{5.9}$$

The second one may be written in the equivalent form:

$$(f - w_0) + (w_0 - w_1) = 0. \tag{5.10}$$

From the definition of the number ρ and from the way we have choose w_1, we get

$$|w_0 - w_1| < \rho \leq |f(\zeta) - w_0|, \quad \forall \zeta \in \{\gamma\}.$$

Now, according to Rouché theorem, equations (5.8) and (5.10) have the same number of zeros in the set $(\gamma) = U(z_0; r)$. Since $z_0 \in U(z_0; r)$ is the zero of the equation (5.8), it follows that the equation (5.9) has at least one zero in $U(z_0; r)$, hence $\exists z_1 \in U(z_0; r)$ such that $f(z_1) = w_1$.

Since w_1 is an arbitrary element of the open disc $U(w_0; \rho)$, then

$$U(w_0; \rho) \subset f(U(z_0; r)) \subset f(D).$$

Hence, the point w_0 is an interior point of $f(D)$, thus $\Delta = f(D)$ is an open set.

2. Let $w_0 \in f(D)$. Then $\exists z_0 \in D$ such that $f(z_0) = w_0$ and $\exists r_1 > 0$ such that $U(z_0; r_1) \subset D$. Hence, the number z_0 is a zero of the function $f - w_0$, and the function $f - w_0$ is a nonconstant function, because f is nonconstant. Thus, the point z_0 is an isolated zero, hence

$$\exists r > 0 : U^-(z_0; r) \subset U(z_0; r_1) \quad \text{and} \quad f(\zeta) - w_0 \neq 0, \ \forall \zeta \in \{\gamma\},$$

where $\gamma = \partial U(z_0; r)$. It follows that $w_0 \in \mathbb{C} \setminus \{\Gamma\}$, where $\Gamma = f \circ \gamma$.

According to the principle of the argument variation (Theorem 5.3.6), we have that the function

$$\mathrm{I}(w) = \Theta(f - w, U(z_0; r))$$

is locally constant, an since $\mathrm{I}(w_0) > 0$, because $f(z_0) = w_0$, it follows that

$$\exists r_0 > 0 : \Theta(f - w, U(z_0; r)) = \Theta(f - w_0, U(z_0; r)) = \mathrm{I}(w_0) > 0, \quad \forall w \in U(w_0; r_0).$$

Thus,

$$w \in f(U(z_0; r)) \subset f(D), \quad \forall w \in U(w_0; r_0) \Leftrightarrow U(w_0; r_0) \subset f(U(z_0; r)) \subset f(D).$$

Hence, the point w_0 is an interior point of $f(D)$, so $\Delta = f(D)$ is an open set. □

Similarly, we can prove the following generalization of the above theorem.

Theorem 5.3.7. *Let $f \in H(G)$, where $G \subset \mathbb{C}$ is an open set. Suppose that the function f is not locally constant.*

Then, for all open subset $G_1 \subset G$, its image $f(G_1)$ is also an open set.

5.4 Exercises

5.4.1 Residue theorem

Exercise 5.4.1. Calculate the residue of the following functions in the corresponding points:
1. $f : \mathbb{C} \setminus \{2\} \to \mathbb{C}, f(z) = \frac{\sin z}{z-2}, z_0 = 2$;
2. $f : \mathbb{C} \setminus \{1, 2\} \to \mathbb{C}, f(z) = \frac{z}{(z-1)(z-2)^2}, z_0 = 1$ and $z_1 = 2$;
3. $f : \mathbb{C} \setminus \{z_1, z_2\} \to \mathbb{C}, f(z) = \frac{z}{(z-z_1)^p(z-z_2)}$, where $p \in \mathbb{N}^*$, z_1 and z_2;
4. $f : \mathbb{C} \setminus \{-1, 0, 1\} \to \mathbb{C}, f(z) = \frac{1}{z^3-z^5}, z_0 = -1, z_1 = 0$ and $z_2 = 1$;
5. $f : \mathbb{C} \setminus \{-i, i\} \to \mathbb{C}, f(z) = \frac{z^2}{z^2+1}, z_0 = -i$ and $z_1 = i$;
6. $f : \mathbb{C} \setminus \{-1\} \to \mathbb{C}, f(z) = \frac{z^{2n}}{(z+1)^n}, z_0 = -1$ and $z_1 = \infty$;
7. $f : \mathbb{C} \setminus \{2k\pi i : k \in \mathbb{Z}\} \to \mathbb{C}, f(z) = \frac{1}{1-e^z}, z_k = 2k\pi i, k \in \mathbb{Z}$.

Exercise 5.4.2. Calculate the residue of the following functions in their isolated singular points:

$$1.\ f(z) = \frac{z^{2n}}{1 + z^n}, \quad n \in \mathbb{N}^*; \quad 2.\ f(z) = \frac{e^{az}}{(1 + e^{\frac{z}{2}})^2}, \quad a \in \mathbb{C};$$

$$3.\ f(z) = e^{z - \frac{1}{z}}; \quad 4.\ f(z) = \frac{z^n e^{\frac{1}{z}}}{1 + z}, \quad n \in \mathbb{N}^*.$$

Exercise 5.4.3. Calculate the residue of the following functions in their isolated singular points:

1. $f(z) = \dfrac{e^{iaz}}{\sinh z}$, $a \in \mathbb{C}$; 2. $f(z) = \dfrac{e^{\frac{1}{z}}}{(z-1)^2}$; 3. $f(z) = \dfrac{1}{(z^2+1)^n}$, $n \in \mathbb{N}^*$;
4. $f(z) = \dfrac{1}{z^2 \sin z}$; 5. $f(z) = z^3 e^{\frac{1}{1-z}}$; 6. $f(z) = \dfrac{e^{iaz}}{\cosh^2 z}$, $a \in \mathbb{C}$.

Exercise 5.4.4. Using the residue theorem, calculate the next integrals on the corresponding paths:

1. $\int_\gamma \frac{z+2}{z(z^2+4)^2} dz$, where
   (a) $\{\gamma\} = \partial U(0;1)$,
   (b) $\{\gamma\} = \partial U(0;5)$ are directly oriented;
2. $\int_\gamma \frac{1}{(z-1)^2(z^2+1)} dz$, where $\{\gamma\} = \partial U(1+i;2)$ is directly oriented;
3. $\int_\gamma \frac{1}{z^n+2} dz$, $n \in \mathbb{N}^*$, where $\gamma(t) = (3+\varepsilon)e^{2\pi i t} + 1$, $t \in [0,1]$, and $\varepsilon > 0$;
4. $\int_\gamma \frac{1}{(z^n-1)(z^3-1)} dz$, $n \in \mathbb{N}^*$, where $\{\gamma\} = \partial U(0;r), r < 1$, is directly oriented;
5. $\int_\gamma \frac{1}{(z-1)(z^2+1)} dz$, where $\gamma(t) = \sqrt{2}e^{2\pi i t} + (1+i)$, $t \in [0,1]$;
6. $\int_\gamma \frac{1}{z^3(z^{10}-1)} dz$, where $\{\gamma\} = \partial U(0;2)$ is directly oriented;
7. $\int_\gamma e^{\frac{1}{1-z}} dz$, where $\{\gamma\} = \partial U(0;2)$ is directly oriented;
8. $\int_\gamma \tan \pi z dz$, where $\{\gamma\} = \partial U(0;n)$, $n \in \mathbb{N}^*$, is directly oriented;
9. $\int_\gamma \frac{\tan z}{z^2} dz$, where $\{\gamma\} = \partial U(0;2)$ is directly oriented;
10. $\int_\gamma \frac{1}{z^4+1} dz$, where $\{\gamma\} = \partial U(1;1)$ is directly oriented;
11. $\int_\gamma \frac{1}{z} \sin \frac{1}{(z-1)^2} dz$, where $\{\gamma\} = \partial U(0;2)$ is directly oriented;
12. $\int_\gamma \frac{(z-1)^3}{z} e^{\frac{1}{z-1}} dz$, where $\{\gamma\} = \partial U(0;2)$ is directly oriented;
13. $\int_\gamma \frac{1}{z^{111}+z^{11}+z+1} dz$, where $\{\gamma\} = \partial U(0;r)$ is directly oriented, and $r > 0$ is enough big, such that the disc $U(0;r)$ contains all the zeros of the function $z^{111} + z^{11} + z + 1$;
14. $\int_\gamma \frac{1}{z-1} e^{\frac{1}{z-1}} dz$, where $\{\gamma\} = \partial U(0;r)$, $r \neq 1$, is directly oriented;
15. $\int_\gamma \frac{z^n e^{\frac{1}{z}}}{z^2-1} dz$, $n \in \mathbb{N}^*$, where $\{\gamma\} = \partial U(0;3)$ is directly oriented;
16. $\int_\gamma \sin^n \frac{1}{z} dz$, $n \in \mathbb{N}^*$, where $\{\gamma\} = \partial U(0;r)$, $r > 0$, is directly oriented;
17. $\int_\gamma (1+z+z^2)(e^{\frac{1}{z}} + e^{\frac{1}{z-1}} + e^{\frac{1}{z-2}}) dz$, where $\{\gamma\} = \partial U(0;3)$ is directly oriented;
18. $\int_\gamma \frac{1}{(1+e^z)^2} dz$, where $\{\gamma\} = \partial U(0;r)$, $(2n-1)\pi < r < (2n+1)\pi$, $n \in \mathbb{N}^*$, is directly oriented.

Exercise 5.4.5. Calculate the following integrals, where the corresponding paths are directly oriented:

1. $\int_\gamma \frac{\cosh\frac{\pi z}{2}}{(z+i)^4}\,dz$, where $\{\gamma\} = \{z \in \mathbb{C} : |z+2i| = 2\}$;
2. $\int_\gamma \frac{z^{100}e^{i\pi z}}{z^2+1}\,dz$, where $\{\gamma\} = \{z = x+iy \in \mathbb{C} : 4x^2+y^2-4=0\}$;
3. $\int_\gamma \frac{\sqrt[5]{\frac{z+3i}{3-z}}}{z^3}\,dz$, where $\{\gamma\} = \partial U(0;r)$, $r<3$, where $\sqrt[5]{\frac{z+3i}{3-z}}\big|_{z=3i} = \sqrt[10]{2}e^{i\frac{19\pi}{20}}$;
4. $\int_\gamma \frac{1}{z\cos z^2}\,dz$, where $\{\gamma\} = \{z = x+iy \in \mathbb{C} : x^2+y^2-2y-3=0\}$;
5. $\int_\gamma z^2 e^{\frac{2z}{z+1}}\,dz$, where $\{\gamma\} = \{z = x+iy \in \mathbb{C} : x^2+y^2+2x=0\}$;
6. $\int_\gamma \frac{z^{13}}{(z-2)^4(z^5+3)^2}\,dz$, where $\{\gamma\} = \{z = x+iy \in \mathbb{C} : 4x^2+9y^2-36=0\}$;
7. $\int_\gamma \frac{1}{z\sqrt{4z^2+12z+13}}\,dz$, where $\{\gamma\} = \partial U(0;r)$, $0<r<\frac{\sqrt{13}}{2}$, and $\sqrt{4z^2+12z+13}\big|_{z=0} = \sqrt{13}$;
8. $\int_\gamma \frac{\log(z-a)}{z^2}\,dz$, where $\{\gamma\} = \{z \in \mathbb{C} : z = Re^{i\theta}, 0 \le \theta < 2\pi\}$, $R > a > 0$, and $\log 1 = 0$.

Exercise 5.4.6. Using the residue theorem calculate the following integrals on the corresponding directly oriented paths:

1. $\int_\gamma \frac{1}{z\sin z}\,dz$, where $\{\gamma\} = \{z \in \mathbb{C} : |z| = r\}$, $r \ne k\pi$, $k \in \mathbb{Z}$;
2. $\int_\gamma \frac{1}{3\sin z - \sin 3z}\,dz$, where $\{\gamma\} = \{z \in \mathbb{C} : |z| = 4\}$;
3. $\int_\gamma \frac{e^{\frac{1}{z}}}{(z-1)^2}\,dz$, where
   (a) $\{\gamma\} = \{z \in \mathbb{C} : |z| = \frac{\sqrt{2}}{2}\}$,
   (b) $\{\gamma\} = \{z \in \mathbb{C} : |z| = 2\}$;
4. $\int_\gamma \frac{\log\frac{z-1}{z+1}}{(z^2+1)(z^2-4)}\,dz$, where
   (a) $\{\gamma\} = \{z \in \mathbb{C} : |z| = \frac{1}{2}\}$,
   (b) $\{\gamma\} = \{z \in \mathbb{C} : |z| = \sqrt{2}\}$,
   (c) $\{\gamma\} = \{z = x+iy \in \mathbb{C} : 3x^2+y^2-2=0\}$,
   (d) $\{\gamma\} = \{z \in \mathbb{C} : |z| = 3\}$, where $\log\frac{z-1}{z+1}\big|_{z=0} = \pi i$;
5. $\int_\gamma \frac{\sin z}{z^2(z^4+1)}\,dz$, where $\{\gamma\} = \{z \in \mathbb{C} : |z| = 2\}$;
6. $\int_\gamma \frac{1}{z^2(z-1)\sin z}\,dz$, where $\{\gamma\} = \{z \in \mathbb{C} : |z| = r\}$, $n\pi < r < (n+1)\pi$, $n \in \mathbb{N}$;
7. $\int_\gamma \frac{z}{\sin z(1-\cos z)}\,dz$, where $\{\gamma\} = \{z \in \mathbb{C} : |z| = 4\}$;
8. $\int_\gamma \frac{z^3 e^{\frac{1}{z}}}{z+1}\,dz$, where
   (a) $\{\gamma\} = \{z \in \mathbb{C} : |z| = \frac{\sqrt{2}}{2}\}$,
   (b) $\{\gamma\} = \{z \in \mathbb{C} : |z| = \sqrt{2}\}$;
9. $\int_\gamma \frac{z^n e^{\frac{1}{z}}}{z^2-1}\,dz$, $n \in \mathbb{N}$, where
   (a) $\{\gamma\} = \{z \in \mathbb{C} : |z| = \frac{1}{2}\}$,
   (b) $\{\gamma\} = \{z \in \mathbb{C} : |z| = 2\}$,
   (c) $\{\gamma\} = \{z = x+iy \in \mathbb{C} : x^2+y^2-2x-\frac{5}{4}=0\}$;
10. $\int_\gamma \frac{e^{\frac{\pi}{z-1}}}{z^2+1}\,dz$, where $\{\gamma\} = \{z \in \mathbb{C} : |z| = 2\}$;

11. $\int_\gamma \frac{1}{z(z^2+a^2)^2}\mathrm{d}z$, where $\{\gamma\} = \{z \in \mathbb{C} : |z| = r\}$, $r > a > 0$;
12. $\int_\gamma \frac{1}{z^n+1}\mathrm{d}z$, $n \in \mathbb{N}^*$, where $\{\gamma\} = \{z \in \mathbb{C} : |z| = r\}$, $r \neq 1$;
13. $\int_\gamma \frac{1}{z^2 \sin z}\mathrm{d}z$, where $\{\gamma\} = \{z \in \mathbb{C} : |z| = r\}$, $n\pi < r < (n+1)\pi$, $n \in \mathbb{N}^*$;
14. $\int_\gamma \frac{\sqrt{1+z^2}}{(1-z^2)(z^2-4)}\mathrm{d}z$, where
    (a) $\{\gamma\} = \{z \in \mathbb{C} : |z| = \frac{\sqrt{2}}{2}\}$,
    (b) $\{\gamma\} = \{z = x + iy \in \mathbb{C} : \frac{x^2}{2} + 4y^2 - 1 = 0, x \geq 0\} \cup \{z = x + iy \in \mathbb{C} : x^2 + y^2 - \frac{1}{4} = 0, x < 0\}$;
15. $\int_\gamma \frac{\sqrt{z^2-1}\log\frac{z-1}{z+1}}{(z^2+1)(z^2+4)}\mathrm{d}z$, where
    (a) $\{\gamma\} = \{z \in \mathbb{C} : |z| = \frac{\sqrt{2}}{2}\}$,
    (b) $\{\gamma\} = \{z = x + iy \in \mathbb{C} : 36x^2 + 4y^2 - 9 = 0\}$,
    (c) $\{\gamma\} = \{z = x + iy \in \mathbb{C} : 10x^2 + y^2 - 5 = 0\}$,
    (d) $\{\gamma\} = \{\gamma\} = \{z = x + iy \in \mathbb{C} : x^2 + y^2 - 5y = 0\}$, where $\sqrt{z^2-1}|_{z=0} = i$ and $\log\frac{z-1}{z+1}|_{z=0} = \pi i$.

5.4.2 Applications of the residue theorem to the calculation of the trigonometric integrals

Exercise 5.4.7. Calculate the following real integrals by using the *residue theorem*:
1. $\int_0^{2\pi} \frac{1}{2-\sin^2\theta}\mathrm{d}\theta$;
2. $\int_0^{2\pi} \frac{1+\sin\theta}{2+\cos\theta}\mathrm{d}\theta$;
3. $\int_0^{2\pi} \frac{2+\sin\theta}{2+\cos\theta}\mathrm{d}\theta$;
4. $\int_0^{2\pi} \frac{1}{a+\cos\theta}\mathrm{d}\theta$, $a > 1$;
5. $\int_0^{2\pi} \frac{1}{1+a\sin\theta}\mathrm{d}\theta$, $-1 < a < 1$;
6. $\int_0^{2\pi} \frac{1}{(a+b\cos\theta)^2}\mathrm{d}\theta$, $a > b > 0$;
7. $\int_0^{2\pi} \frac{1+\cos\theta}{(13-5\cos\theta)^2}\mathrm{d}\theta$;
8. $\int_0^{2\pi} \frac{1}{(17+8\cos\theta)^2}\mathrm{d}\theta$;
9. $\int_0^{2\pi} \frac{1}{1-2p\cos\theta+p^2}\mathrm{d}\theta$, $0 < p < 1$;
10. $\int_0^{2\pi} \frac{\cos^2 2\theta}{1-2p\cos\theta+p^2}\mathrm{d}\theta$, $0 < p < 1$;
11. $\int_0^{2\pi} \frac{\cos^2 3\theta}{1-2p\cos 2\theta+p^2}\mathrm{d}\theta$, $0 < p < 1$.

Exercise 5.4.8. Using the *residue theorem*, calculate the following real integrals:
1. $\int_0^{2\pi} \frac{\cos n\theta}{1-2a\cos\theta+a^2}\mathrm{d}\theta$, $\int_0^{2\pi} \frac{\cos n\theta}{1-2a\sin\theta+a^2}\mathrm{d}\theta$, $a > 1$, $n \in \mathbb{N}^*$;
2. $\int_0^{2\pi} \frac{\sin\theta\sin n\theta}{5-4\cos\theta}\mathrm{d}\theta$, $\int_0^{2\pi} \frac{\sin\theta\cos n\theta}{5-4\cos\theta}\mathrm{d}\theta$, $n \in \mathbb{N}^*$.

Exercise 5.4.9. Prove the formulas:
1. $\int_0^{\pi} \frac{\cos n\theta}{b - ia\cos\theta} d\theta = \frac{\pi i^n}{\sqrt{a^2+b^2}} \frac{a^n}{(\sqrt{a^2+b^2}+b^2)^n}$, $a > 0$, $b > 0$, $n \in \mathbb{N}$;
2. $\int_0^{2\pi} \frac{1}{a + b\cos^2\theta} d\theta = \frac{2\pi}{\sqrt{a(a+b)}}$, $a > 0$, $b > 0$.

5.4.3 Applications of the residue theorem to the calculation of the improper integrals

Exercise 5.4.10. Calculate the following real improper integrals:
1. $\int_0^{+\infty} \frac{\sin x}{x} dx$;
2. $\int_{-\infty}^{+\infty} \frac{1}{x^4+1} dx$;
3. $\int_0^{+\infty} \frac{1}{(a+bx^2)^n} dx$, $a, b > 0$, $n \in \mathbb{N}^*$;
4. $\int_{-\infty}^{+\infty} \frac{1}{1+x^{2n}} dx$, $n \in \mathbb{N}^*$;
5. $\int_0^{+\infty} \frac{\ln x}{(x+a)^2+b^2} dx$, $a, b > 0$;
6. $\int_0^{+\infty} \frac{\cos ax}{(x^2+b^2)^2} dx$, $\quad a, b \in \mathbb{R}$, $b > 0$;
7. $\int_0^{+\infty} \frac{\cos ax - \cos bx}{x^2} dx$, $a, b \in \mathbb{R}$;
8. $\int_{-\infty}^{+\infty} e^{-ax^2} \cos bx dx$, $a > 0$ (Poisson-type integral).

5.4.4 The study of meromorphic functions using the residue theorem

Exercise 5.4.11. Determine the number of the roots of the following equations that lies in the disc $U(0; 1)$:
1. $z^8 - 4z^5 + z^2 - 1 = 0$;
2. $z^8 - 11z + 1 = 0$;
3. $23z^5 - 10z^3 + 6z - 5 = 0$;
4. $e^z - 3iz = 0$;
5. $z^n + 8z^2 + 1 = 0$, $n \in \mathbb{N}$, $n \geq 3$;
6. $z^n + 3z^2 + 1 = 0$, $n \in \mathbb{N}$, $n \geq 3$;
7. $e^z - 4z^n + 1 = 0$, $n \in \mathbb{N}^*$;
8. $z^n + pz^2 + qz + r = 0$, $n \in \mathbb{N}$, $n \geq 3$, where $|p| > |q| + |r| + 1$;
9. $z^{10} - 9z^6 + 3z^3 + z^2 - 2 = 0$.

Exercise 5.4.12. Determine the number of the roots of the equation

$$az^n = e^z, \quad n \in \mathbb{N}^*,$$

that belong to the disc $U(0; r)$, if $|a| > \frac{e^r}{r^n}$ and $r > 0$ are given numbers.

Exercise 5.4.13. Determine the number of the roots of the following equations that lies in the corresponding circular ring:

1. $z^4 - 9z + 1 = 0$, $U(0; 1, 3)$;
2. $3z^5 - 7z^3 + z^2 - 2 = 0$, $U(0; 1, 2)$;
3. $z^4 + z^3 - 4z + 1 = 0$, $U(0; 1, 3)$.

Exercise 5.4.14. If $0 < a < \frac{1}{e}$, prove that:

1. the equation $z = ae^z$ has only one root in the disc $U(0; 1)$;
2. the equation $z^2 = ae^z$ has two roots in the disc $U(0; 1)$.

6 Conformal representations

6.1 Special classes of holomorphic functions

Definition 6.1.1.

1. Denote by $C(G)$ the class of all continuous functions in an open set $G \subset \mathbb{C}$. We say that the subclass $\mathcal{F} \subset C(G)$ is **uniformly continuous at the point** $z_0 \in G$, if
$$\begin{aligned}&\forall \varepsilon > 0,\ \exists \delta = \delta(\varepsilon) > 0, \quad \text{such that}\\&\forall z \in G,\ |z - z_0| < \delta \Rightarrow |f(z) - f(z_0)| < \varepsilon,\ \forall f \in \mathcal{F}.\end{aligned}$$
(Here, the number $\delta > 0$ is independent of the function f.)
2. We say that the subset $\mathcal{F} \subset C(G)$ is **uniformly continuous** (on the set G), if $\mathcal{F}$ is uniformly continuous in all the points $z \in G$.

Theorem 6.1.1. *Let $\mathcal{F} \subset C(G)$ a set of uniformly continuous functions on the open subset $G \subset \mathbb{C}$. Suppose that $(f_n)_{n\in\mathbb{N}} \subset \mathcal{F}$ and $f_n \xrightarrow{E} f$, where $E \subset G \subset E'$ (E' is the set of all accumulation points of E).*

Then the function f has an extension to G, such that $f_n \xrightarrow{G} f, f \in C(G)$ and $f_n \rightrightarrows_K f$, $\forall K \subset G$, where K is compact (i. e., the sequence $(f_n)_{n\in\mathbb{N}}$ uniformly converges on the compacts to f).

Proof. 1. First, we will prove that $(f_n)_{n\in\mathbb{N}}$ converges on G.

Let $z_0 \in G$ and $\varepsilon > 0$ be arbitrary. Since $\mathcal{F}$ is uniformly continuous, we get
$$\exists r > 0, \quad \text{such that } U(z_0;r) \subset G \text{ and } |f_n(z) - f_n(z_0)| < \frac{\varepsilon}{3},\ \forall n \in \mathbb{N},\ \forall z \in U(z_0;r).$$
The set E is dense in G, then there exists $z_1 \in E \cap U(z_0;r)$. Since $(f_n(z_1))_{n\in\mathbb{N}}$ converges, then
$$\begin{aligned}&\exists n_0 \in \mathbb{N} : \forall n, m > n_0 \Rightarrow |f_n(z_1) - f_m(z_1)| < \frac{\varepsilon}{3}\\&\Rightarrow |f_n(z_0) - f_m(z_0)| = |f_n(z_0) - f_n(z_1) + f_n(z_1) - f_m(z_1) + f_m(z_1) - f_m(z_0)|\\&\le |f_n(z_0) - f_n(z_1)| + |f_n(z_1) - f_m(z_1)| + |f_m(z_1) - f_m(z_0)|\\&< \frac{\varepsilon}{3} + \frac{\varepsilon}{3} + \frac{\varepsilon}{3} = \varepsilon, \quad \forall n, m > n_0,\end{aligned}$$
hence the sequence $(f_n(z_0))_{n\in\mathbb{N}}$ converges. Let $f(z_0) = \lim_{n\to\infty} f_n(z_0)$.

2. Second, we will prove that the function f is continuous on G.

Let $\varepsilon > 0$ and $z_0 \in G$ be arbitraries. From the fact that $\mathcal{F}$ is uniformly continuous, it follows that
$$\exists r > 0 : \overline{U}(z_0;r) \subset G \quad \text{and} \quad |f_n(z) - f_n(z_0)| < \frac{\varepsilon}{3}, \quad \forall n \in \mathbb{N}, \forall z \in \overline{U}(z_0;r).$$

https://doi.org/10.1515/9783110657869-006

Letting $n \to +\infty$, then

$$|f(z) - f(z_0)| \le \frac{\varepsilon}{3}, \quad \forall z \in \overline{U}(z_0; r),$$

hence $f \in C(G)$.

3. Finally, we will prove that $f_n \rightrightarrows_K f$, $\forall K \subset G$, where K is compact.

Let $K \subset G$ be a compact set and $z_0 \in K$. For a given number $\varepsilon > 0$, as in the previous steps, it corresponds the number $r > 0$ such that $U(z_0; r) \subset G$. Since $f_n \overset{G}{\to} f$, we have

$$\begin{aligned}
&\exists n_0 \in \mathbb{N} : \forall n > n_0, \ |f_n(z_0) - f(z_0)| < \frac{\varepsilon}{3} \\
&\quad \Rightarrow |f_n(z) - f(z)| = |f_n(z) - f_n(z_0) + f_n(z_0) - f(z_0) + f(z_0) - f(z)| \\
&\quad \le |f_n(z) - f_n(z_0)| + |f_n(z_0) - f(z_0)| + |f(z_0) - f(z)| < \frac{\varepsilon}{3} + \frac{\varepsilon}{3} + \frac{\varepsilon}{3} = \varepsilon, \\
&\quad \forall z \in U(z_0; r), \ \forall n > n_0.
\end{aligned}$$

For all the points $z_j \in K$, it corresponds to a number $r_j > 0$ with the above property, and an index $n_j \in \mathbb{N}$. All the discs of the form $U(z_j; r_j)$ represent a covering of the set K, and from the fact that K is compact it follows that it contains a finite covering $U(z_1; r_1), \dots, U(z_m; r_m)$. If $n > \max\{n_1, \dots, n_m\}$, then $|f_n(z) - f(z)| < \varepsilon$, $\forall z \in K$, hence $f_n \rightrightarrows_K f$. □

Definition 6.1.2. We say that the function set $\mathcal{F} \subset C(G)$ is **bounded** (i. e., all the functions of $\mathcal{F}$ are uniformly bounded on the compacts of G), where $G \subset \mathbb{C}$ is an open set, if

$$\forall K \subset G,\ K \text{ compact},\ \exists M = M(K) > 0, \quad \text{such that } \forall z \in K,\ \forall f \in \mathcal{F} \Rightarrow |f(z)| \le M.$$

(The number M is the same for all the functions of $\mathcal{F}$.)

Theorem 6.1.2. *If $\mathcal{F} \subset H(G)$ is bounded, where $G \subset \mathbb{C}$ is an open set, then the set $\mathcal{F}' = \{f' : f \in \mathcal{F}\}$ is also bounded.*

Proof. Let $K \subset G$ be an arbitrary compact, and let $0 < R < \mathrm{d}(K, \partial G)$, $0 < r < R$, where R and r are fixed numbers. Thus, $\overline{U}(z_0; R) \subset G$, $\forall z_0 \in K$, hence $\gamma = \partial U(z_0; R) \subset G$.

Letting an arbitrary $z \in \overline{U}(z_0; r)$, then $\forall \zeta \in \{\gamma\}$ we have $|\zeta - z| \ge R - r > 0$. From the Cauchy formula, we get

$$f'(z) = \frac{1}{2\pi i} \int_\gamma \frac{f(\zeta)}{(\zeta - z)^2} \mathrm{d}\zeta.$$

Since the set $\mathcal{F}$ is bounded, then $|f(\zeta)| \le M$, $\forall \zeta \in \{\gamma\}$, $\forall f \in \mathcal{F}$, and from here it follows that

$$|f'(z)| \le \frac{1}{2\pi} \frac{M}{(R - r)^2} 2\pi R = M', \quad \forall z \in \overline{U}(z_0; r).$$

Since K is compact, it may be covered by a finite number of the discs of the form $\overline{U}(z_j;r)$, with $|f'(z)| \leq M_j'$, $\forall z \in \overline{U}(z_j;r)$. Hence, the set $\{f' : f \in \mathcal{F}\} = \mathcal{F}'$ is uniformly bounded on the compacts of G, and thus the function set $\mathcal{F}'$ is bounded. □

Theorem 6.1.3. *If $\mathcal{F} \subset H(G)$ is a bounded function set, where $G \subset \mathbb{C}$ is an open set, then $\mathcal{F}$ is uniformly continuous in all the points of G.*

Proof. Let $z_0 \in G$ and $\varepsilon > 0$ be arbitraries, and let $r > 0$ such that $\overline{U}(z_0;r) \subset G$. Since $\mathcal{F}$ is bounded, from the above theorem the function set $\mathcal{F}'$ is also bounded. Hence, $\exists M > 0$ such that $|f'(z)| \leq M$, $\forall z \in \overline{U}(z_0;r)$, $\forall f \in \mathcal{F}$.

Let us consider the linear path $\lambda(t) = (1-t)z_0 + tz$, $t \in [0,1]$, connecting the points $z \in U(z_0;r)$ and z_0. Then

$$f(z) - f(z_0) = \int_{z_0}^{z} f'(\zeta)\mathrm{d}\zeta \quad \text{and} \quad |f(z) - f(z_0)| \leq M|z - z_0|,$$

hence, for all points $z \in U(z_0;r)$, with $|z - z_0| < \delta = \frac{\varepsilon}{M}$, we have $|f(z) - f(z_0)| < \varepsilon$, $\forall f \in \mathcal{F}$. □

Corollary 6.1.1. *Let $\mathcal{F} \subset H(G)$, where $\mathcal{F}$ is bounded and $G \subset \mathbb{C}$ is an open set. If $(f_n)_{n\in\mathbb{N}} \subset \mathcal{F}$ converges to f on the set E, where $E \subset G \subset E'$, then the function f has a such an holomorphic extension to G that satisfies the condition*

$$f_n \rightrightarrows_K f, \quad \forall K \subset G,\ K \text{ compact}.$$

Proof. Since the function set $\mathcal{F} \subset H(G)$ is bounded, from the above theorem, we have that $\mathcal{F}$ is uniformly continuous on G. Now, according to Theorem 6.1.1, there exists the continuous function $f : G \to \mathbb{C}$ that satisfies the above uniformly convergence on the compacts sets. But according to the Weierstrass theorem, the function f will be holomorphic on G. □

Definition 6.1.3. Let $\mathcal{F} \subset C(G)$, where $G \subset \mathbb{C}$ is an open set. We say that the function set $\mathcal{F}$ is **relatively compact** (in $C(G)$), if $\forall (f_n)_{n\in\mathbb{N}} \subset \mathcal{F}$ sequence $\exists (f_{n_k})_{k\in\mathbb{N}} \subset (f_n)_{n\in\mathbb{N}}$, such that $(f_{n_k})_{k\in\mathbb{N}}$ uniformly converges on compact sets to $f \in C(G)$.

Remark 6.1.1. From the Bolzano–Weierstrass theorem, we get: in the Euclidian spaces the relatively compact sets coincide to the bounded sets. This property also holds for the set $H(G)$ of holomorphic functions on the open set G.

Theorem 6.1.4 (Montel theorem). *Let $\mathcal{F} \subset H(G)$, where $G \subset \mathbb{C}$ is an open set. Then*

$$\mathcal{F} \text{ is bounded} \Leftrightarrow \mathcal{F} \text{ is relatively compact in } H(G).$$

Proof. "$\Rightarrow$" Let E be the subsets of all the points of G with rational coordinates. Then $E \subset G \subset E'$ and the set E is countable, so it may be written as $E = \{z_j : j \in \mathbb{N}^*\}$. Let $(f_n)_{n\in\mathbb{N}^*} \subset \mathcal{F}$ be an arbitrary function sequence. Since $\mathcal{F}$ is bounded, then $(f_n(z_1))_{n\in\mathbb{N}^*}$

is a bounded sequence in $\mathbb{C}$, hence, according to the Bolzano–Weierstrass theorem, it contains a convergent subsequence

$$f_{11}(z_1),\ f_{21}(z_1),\ \ldots,\ f_{n1}(z_1),\ \ldots.$$

The sequence $(f_{n1}(z_2))_{n\in\mathbb{N}^*}$ is a bounded sequence, hence it contains a convergent subsequence

$$f_{11}(z_2),\ f_{22}(z_2),\ f_{32}(z_2),\ \ldots,\ f_{n2}(z_2),\ \ldots.$$

The sequence $(f_{n2}(z_3))_{n\in\mathbb{N}^*}$ is a bounded sequence, hence it contains a convergent subsequence, and we continue this algorithm. After n steps, we may choose the subsequence

$$f_{11},\ f_{22},\ f_{33},\ \ldots,\ f_{nn},\ f_{n+1n},\ f_{n+2n},\ \ldots$$

that converges on the $\{z_1, z_2, \ldots, z_n\}$ set. Continuing in the same way, we obtain the subsequence $(f_{nn})_{n\in\mathbb{N}^*}$ that converges in all the points of the E set. Hence, $(f_{nn})_{n\in\mathbb{N}^*} \subset H(G)$, $f_{nn} \xrightarrow{E} f$, where $f(z_k) = \lim_{n\to\infty} f_{nn}(z_k)$, $\forall k \in \mathbb{N}^*$. From Corollary 6.1.1 of Theorem 6.1.3, the function f has such a holomorphic extension on G, which satisfies that $(f_{nn})_{n\in\mathbb{N}^*}$ uniformly converges on compacts to f.

"$\Leftarrow$" Suppose that $\mathcal{F}$ is not bounded. Then there exist $K \subset G$, K compact, and there exists $f_n \in \mathcal{F}$, such that $\exists z_n \in K : |f_n(z_n)| > n$. From the sequence $(f_n)_{n\in\mathbb{N}^*}$, we may choose such a subsequence $(f_{n_k})_{k\in\mathbb{N}^*}$ that uniformly converges on compacts to $f \in H(G)$. Let $z_0 \in K$ be an accumulation point for the sequence $(z_n)_{n\in\mathbb{N}^*} \subset K$. Thus, $\exists (z_{n_k})_{k\in\mathbb{N}^*} \subset (z_n)_{n\in\mathbb{N}^*}$ is a subsequence, such that $\lim_{k\to\infty} z_{n_k} = z_0$ and $\lim_{k\to\infty} f(z_{n_k}) = f(z_0)$. Now, since $(f_{n_k})_{k\in\mathbb{N}^*}$ uniformly converges on compacts to f, we get $\lim_{k\to\infty} |f_{n_k}(z_{n_k}) - f(z_{n_k})| = 0$, hence

$$|f_{n_k}(z_{n_k})| \le |f_{n_k}(z_{n_k}) - f(z_{n_k})| + |f(z_{n_k})| \to |f(z_0)|, \quad k \to \infty.$$

But this last inequality cannot be true, because $|f_{n_k}(z_{n_k})| > n_k \to \infty$, $k \to \infty$. □

Examples 6.1.1.

1. Let G and G_1 be two open sets in $\mathbb{C}$, such that G_1 is bounded. Thus, $\mathcal{F} = \{f \in H(G) : f(G) \subset G_1\}$ is bounded, hence it is relatively compact in $H(G)$.
2. Let us consider the set $\mathrm{U} = U(0;1)$, and let

   $$\mathcal{F} = \left\{f \in H(\mathrm{U}) : |f(z)| \le \frac{1}{1-|z|}, z \in \mathrm{U}\right\}.$$

   Thus $\mathcal{F} \ne \emptyset$, because $g(z) = \frac{1}{1+z} \in \mathcal{F}$. The set $\mathcal{F}$ is bounded, because for all $r \in [0,1)$ we have

   $$\forall z \in \overline{U}(0;r), \quad \forall f \in \mathcal{F} \Rightarrow |f(z)| \le \frac{1}{1-r},$$

   and according to the Montel theorem the set $\mathcal{F}$ is relatively compact in $H(\mathrm{U})$.

3. If $f \in H(G)$, where $G \subset \mathbb{C}$ is an open set, and $\mathcal{F} = \{f + c : c \in \mathbb{C}\}$, then $\mathcal{F}$ is not bounded, hence it is not relatively compact in $H(G)$.

Theorem 6.1.5 (Vitali theorem). *Let $D \subset \mathbb{C}$ be a domain, and let $(f_n)_{n\in\mathbb{N}} \subset H(D)$ be a bounded sequence such that $f_n \xrightarrow{E} f$, where $E \subset D$ and $E' \cap D \neq \emptyset$. Then*

$$f_n \rightrightarrows_K f, \quad \forall K \subset D,\ K \text{ compact}$$

(i. e., the sequence $(f_n)_{n\in\mathbb{N}}$ uniformly converges to f on compact sets in D).

Proof. According to the Corollary 6.1.1, it is sufficient to prove that $f_n \xrightarrow{D} f$. If this is not true, then there exists the point $z_0 \in D$ such that the sequence $(f_n(z_0))_{n\in\mathbb{N}}$ diverges. Since $(f_n)_{n\in\mathbb{N}}$ is a bounded sequence, the set $(f_n(z_0))_{n\in\mathbb{N}}$ has two different limit points w_1 and w_2. It follows that the sequence $(f_n(z_0))_{n\in\mathbb{N}}$ contains two subsequences $(f_{n_1}(z_0))_{n_1\in\mathbb{N}}$ and $(f_{n_2}(z_0))_{n_2\in\mathbb{N}}$, such that

$$\lim_{n_j\to\infty} f_{n_j}(z_0) = w_j, \quad j = 1, 2.$$

From the Montel theorem (Theorem 6.1.4), the sequences $(f_{n_1})_{n_1\in\mathbb{N}}$ and $(f_{n_2})_{n_2\in\mathbb{N}}$ have such subsequences that uniformly converges on compacts in D to g, respectively to h, where g and h are holomorphic functions. Since $(f_{n_1})_{n_1\in\mathbb{N}}, (f_{n_2})_{n_2\in\mathbb{N}} \subset (f_n)_{n\in\mathbb{N}}$ and $(f_n)_{n\in\mathbb{N}}$ converges on E, it follows that $g|_E = h|_E$. But $E' \cap D \neq \emptyset$, and according to the theorem of holomorphic functions identity we get $g = h$, that contradicts the fact that (from the definitions of g and h) $g(z_0) = w_1 \neq w_2 = h(w_0)$. □

6.2 Univalent functions

Definition 6.2.1. An holomorphic and injective function on the domain $D \subset \mathbb{C}$ is said to be an **univalent** function. The set of all univalent functions on the domain D is denoted by $H_u(D)$.

Examples 6.2.1.

1. Any circular transform is univalent on the set $\mathbb{C} \setminus \{z_0\}$, where z_0 is the pole of the function.
2. The function $f(z) = \frac{z}{(1+z)^2}$ is univalent in the unit disc $\mathrm{U} = U(0; 1)$, and

$$f(\mathrm{U}) = \mathbb{C} \setminus \left[\frac{1}{4}, +\infty\right).$$

   Since $f(e^{i\theta}) = \frac{1}{(2\cos\frac{\theta}{2})^2}$, $\theta \in \mathbb{R}$, it follows that $f(\partial \mathrm{U}) = [\frac{1}{4}, +\infty) \cup \{\infty\}$.
3. If $f_1, f_2 \in H_u(D)$, then $f_1 \circ f_2 \in H_u(D)$.
4. If $f(z) = \log \frac{1-z}{1+z}$, where $\log 1 = 0$, then $f \in H_u(\mathrm{U})$ and

$$D = f(\mathrm{U}) = \left\{w \in \mathbb{C} : -\frac{\pi}{2} < \operatorname{Im} w < \frac{\pi}{2}\right\}.$$

Letting $g(z) = \frac{1-z}{1+z} \in H_u(\mathrm{U})$, then $g(\mathrm{U}) = \Delta = \{\zeta \in \mathbb{C} : \operatorname{Re}\zeta > 0\}$ and $h(\zeta) = \log\zeta \in H_u(\Delta)$, $h(\Delta) = D$, thus $f = h \circ g \in H_u(\mathrm{U})$ and $f(\mathrm{U}) = D$.

Theorem 6.2.1. *If $f \in H_u(D)$, where $D \subset \mathbb{C}$ is a domain, then $f'(z) \neq 0$, $\forall z \in D$.*

Proof. Suppose that there exists a point $z_0 \in D$, such that $f'(z_0) = 0$. From the univalence of f it follows that f and f' are not constant functions, and from the theorem of holomorphic functions zeros, there exists a number $r > 0$, such that $\overline{U}(z_0;r) \subset D$, $f'(z) \neq 0$, $\forall z \in \dot{U}(z_0;r)$ and $f(z) \neq f(z_0)$, $\forall z \in C = \partial U(z_0;r)$.

Since $f - f(z_0)$ is a continuous function, and since $f(z) - f(z_0) \neq 0$, $\forall z \in C$, where C is a compact set, it follows that

$$\min\{|f(z) - f(z_0)| : z \in C\} = m > 0.$$

From the continuity of the function f in the point z_0, there exists $z_1 \in \dot{U}(z_0;r)$ such that $|f(z_1) - f(z_0)| < m$, hence $|f(z_1) - f(z_0)| < |f(z) - f(z_0)|$, $\forall z \in C$. From here, according to the Rouché theorem, we obtain that the functions $f - f(z_0)$ and

$$f - f(z_1) = f - f(z_0) + [f(z_0) - f(z_1)]$$

have the same number of zeros in the $U(z_0;r)$ disc.

Since $f'(z_0) = 0$, the function $f - f(z_0)$ has at least two zeros in $U(z_0;r)$, hence the function $f - f(z_1)$ also has at least two zeros in $U(z_0;r)$. From the fact that $f'(z_1) \neq 0$, the number z_1 is a simple zero for $f - f(z_1)$. Hence, there exists a point $z_2 \in U(z_0;r)\setminus\{z_1\}$ which is a zero for $f - f(z_1)$, and thus $f(z_2) = f(z_1)$, with $z_2 \neq z_1$. These last relations contradict the injectivity of f. □

Remarks 6.2.1.
1. If $f \in H_u(D)$, where $D \subset \mathbb{C}$ is a domain, then the function f is a conformal mapping of D.
2. Let $D \subset \mathbb{C}$ be a domain. Contrary as in the case of real functions, the condition $f'(z) \neq 0$, $\forall z \in D$ is a necessary, but not a sufficient condition for the injectivity of the function f. For example, the function $f(z) = e^z$, $z \in \mathbb{C}$, is not injective in $\mathbb{C}$, but $f'(z) \neq 0$, $\forall z \in \mathbb{C}$.

Theorem 6.2.2 (Pompeiu theorem). *Let $f \in H(D)$, where $D \subset \mathbb{C}$ is a domain, and let $z_0 \in D$. Then, in any disc $U(z_0;r) \subset D$ there exist two distinct points $z_1, z_2 \in U(z_0;r)$, such that*

$$f(z_1) - f(z_2) = f'(z_0)(z_1 - z_2).$$

Proof. 1. Suppose that $f'(z_0) = 0$. If $f' \equiv 0$, then the conclusion is evident. If $f' \not\equiv 0$, then f and f' are not constant functions, and similarly like in the proof of Theorem 6.2.1 there exist two distinct points $z_1, z_2 \in U(z_0;r)$, such that $f(z_1) - f(z_2) = 0$.

2. Suppose that $f'(z_0) \neq 0$. Let us define the function $g(z) = f(z) - f'(z_0)z$, $z \in D$. Thus, $g'(z_0) = 0$. According to the previous step, there exist two points $z_1, z_2 \in U(z_0; r)$, $z_1 \neq z_2$, such that $g(z_1) = g(z_2)$, so it follows that $f(z_1) - f(z_2) = f'(z_0)(z_1 - z_2)$. □

Theorem 6.2.3. *Let $f \in H(U(z_0; r))$, where $z_0 \in \mathbb{C}$ and $r > 0$. If*

$$|f'(z) - f'(z_0)| < |f'(z_0)|, \quad \forall z \in U(z_0; r) \setminus \{z_0\}, \tag{6.1}$$

then the function f is univalent in $U(z_0; r)$.

Proof. Let $z_1, z_2 \in U(z_0; r)$ such that $z_1 \neq z_2$. Let us denote by γ the linear path that connects the points z_1 and z_2, i. e., $\gamma(t) = (1-t)z_1 + tz_2$, $t \in [0,1]$. Hence, $\{\gamma\} = [z_1, z_2] \subset U(z_0; r)$, and

$$\begin{aligned} |f(z_1) - f(z_2)| &= |f(\gamma(0)) - f(\gamma(1))| = \left|\int_\gamma f'(\zeta)\mathrm{d}\zeta\right| \\ &\geq \left|\int_\gamma f'(z_0)\mathrm{d}\zeta\right| - \left|\int_\gamma (f'(\zeta) - f'(z_0))\mathrm{d}\zeta\right|. \end{aligned} \tag{6.2}$$

Since

$$\left|\int_\gamma f'(z_0)\mathrm{d}\zeta\right| = |f'(z_0)||z_1 - z_2|$$

and

$$\left|\int_\gamma (f'(\zeta) - f'(z_0))\mathrm{d}\zeta\right| \leq |z_1 - z_2| \sup\{|f'(\zeta) - f'(z_0)| : \zeta \in [z_1, z_2]\},$$

from the inequality (6.2) it follows that

$$|f(z_1) - f(z_2)| \geq |z_1 - z_2|(|f'(z_0)| - \sup\{|f'(\zeta) - f'(z_0)| : \zeta \in [z_1, z_2]\}). \tag{6.3}$$

1. If there exists $a \in [z_1, z_2] \setminus \{z_0\}$ such that

$$|f'(a) - f'(z_0)| = \sup\{|f'(\zeta) - f'(z_0)| : \zeta \in [z_1, z_2]\},$$

from the relations (6.3) and (6.1) we get

$$|f(z_1) - f(z_2)| \geq |z_1 - z_2|(|f'(z_0)| - |f'(a) - f'(z_0)|) > 0,$$

hence f is univalent in $U(z_0; r)$.

2. If $z_0 \in [z_1, z_2]$ and $0 = |f'(z_0) - f'(z_0)| = \sup\{|f'(\zeta) - f'(z_0)| : \zeta \in [z_1, z_2]\}$, then $f'(\zeta) = f'(z_0)$, $\forall \zeta \in [z_1, z_2]$. From here, according to a previous theorem, it follows that $f'(\zeta) = f'(z_0)$, $\forall \zeta \in U(z_0; r)$, thus

$$f(\zeta) = f(z_0) + f'(z_0)(\zeta - z_0), \quad \forall \zeta \in U(z_0; r).$$

Since from (6.1) we have $f'(z_0) \neq 0$, using the above relation we deduce that f is univalent in $U(z_0; r)$. □

Theorem 6.2.4. *Let $f \in H(D)$, where $D \subset \mathbb{C}$ is a simply connected domain. Suppose that there exists a function $g \in H_u(D)$, such that*

$$\text{(i)} \quad g(D) \quad \textit{is a convex domain}$$

and

$$\text{(ii)} \quad \operatorname{Re}\frac{f'(z)}{g'(z)} > 0, \quad \forall z \in D.$$

Then the function f is univalent in D, i. e., $f \in H_u(D)$.

Proof. Let us denote $\Delta = g(D)$. Since $g \in H_u(D)$, there exists $g^{-1} : \Delta \to D$ and $g^{-1} \in H_u(\Delta)$. Thus, $h = f \circ g^{-1} \in H(\Delta)$, and

$$\operatorname{Re} h'(w) = \operatorname{Re}\frac{f'(z)}{g'(z)} > 0, \quad \forall w = g(z) \in \Delta.$$

Let us consider $w_1, w_2 \in \Delta$, $w_1 \neq w_2$. The set $\Delta = g(D)$ is convex, hence $[w_1, w_2] \subset \Delta$. Computing the next integral on this segment, we have

$$h(w_2) - h(w_1) = \int_{w_1}^{w_2} h'(w)\mathrm{d}w = (w_2 - w_1)\int_0^1 h'[w_1 + t(w_2 - w_1)]\mathrm{d}t,$$

which implies

$$\operatorname{Re}\frac{h(w_2) - h(w_1)}{w_2 - w_1} = \int_0^1 \operatorname{Re} h'[w_1 + t(w_2 - w_1)]\mathrm{d}t > 0.$$

Hence, the function h is injective in Δ, so it follows that $f = h \circ g$ is also injective in D. □

Corollary 6.2.1. *If $D \subset \mathbb{C}$ is a convex domain and the function $f \in H(D)$ satisfies the condition $\operatorname{Re} f'(z) > 0$, $\forall z \in D$, then f is injective in D, i. e., $f \in H_u(D)$.*

Corollary 6.2.2. *Let $\mathrm{U} = U(0;1)$ and let $f \in H(\mathrm{U})$, such that*

$$\operatorname{Re}[(1 - z^2)f'(z)] > 0, \quad \forall z \in \mathrm{U}.$$

Then the function f is univalent in U, i. e., $f \in H_u(\mathrm{U})$.

Proof. Let us define the function $g(z) = \log\frac{1+z}{1-z}$, $z \in \mathrm{U}$, where we choose the principal branch of the logarithmic function, i. e., $\log 1 = 0$. According to the fourth point of Example 6.2.1, we have $g \in H_u(\mathrm{U})$ and the image

$$g(\mathrm{U}) = \left\{w \in \mathbb{C} : -\frac{\pi}{2} < \operatorname{Im} w < \frac{\pi}{2}\right\}$$

is a convex domain. Since

$$\operatorname{Re}\frac{f'(z)}{g'(z)} = \frac{1}{2}\operatorname{Re}[(1-z^2)f'(z)] > 0, \quad \forall z \in \mathrm{U},$$

from the above corollary we deduce $f \in H_u(\mathrm{U})$. □

Theorem 6.2.5 (Hurwitz theorem). *Let $D \subset \mathbb{C}$ be a domain, and let $(f_n)_{n\in\mathbb{N}} \subset H_u(D)$ be a sequence of univalent functions in D, that uniformly converges on compacts to f in D, and the function f is not constant. Then $f \in H_u(D)$, i. e., the function f is univalent in D.*

Proof. From the Weierstrass theorem, we have $f \in H(D)$. Suppose that the function f is not injective, then there exist the points $z_1, z_2 \in D$, $z_1 \neq z_2$, such that $f(z_1) = f(z_2) = w_0$. Since f is not constant, according to the theorem of the zeros of holomorphic functions, there exists a number $r > 0$ such that

$$\overline{U}(z_1;r) \cup \overline{U}(z_2;r) \subset D, \quad \overline{U}(z_1;r) \cap \overline{U}(z_2;r) = \emptyset,$$

and

$$\forall z \in C_1 \cup C_2, \quad \text{where } C_1 = \partial U(z_1;r), C_2 = \partial U(z_2;r) \Rightarrow f(z) - w_0 \neq 0.$$

Let $m = \min\{|f(z)-w_0 \ : z \in C_1 \cup C_2\} > 0$. Since the function sequence $(f_n)_{n\in\mathbb{N}}$ uniformly converges on compacts to f in D, there exists $n_0 \in \mathbb{N}$ such that

$$\forall n > n_0, \quad \forall z \in C_1 \cup C_2 \Rightarrow |f_n(z) - f(z)| < m.$$

Hence, $\forall n > n_0$ and $\forall z \in C_1 \cup C_2$ we have $|f_n(z) - f(z)| < |f(z) - w_0|$, and according to the Rouché theorem, we get that $\forall n > n_0$, the functions $f - w_0$ and $f_n - w_0 = f - w_0 + (f_n - f)$ have the same number of zeros in $U(z_1;r)$ and $U(z_2;r)$, respectively. It follows that $\forall n > n_0$ the function $f_n - w_0$ has zeros in both of the above discs. From here, if $\forall n > n_0$ then the functions f_n cannot be injective, which contradicts the assumption. □

6.3 The problem of conformal representation

According to the principle of the domain conservation, we have: if D is a domain and $f \in H(D)$, then $\Delta = f(D)$ is also a domain. The function f is conformal in every points $z \in D$ with $f'(z) \neq 0$. From the geometric interpretation of the derivative, It is well known that if f belongs to the class C^1 (i. e., the real and the imaginary parts of f have continuous first-order partial derivatives), and it is first and second type conformally in a point z, then we have $f'(z) \neq 0$.

If the function f is univalent in the domain D, then f is a homeomorphism between D and $\Delta = f(D)$, and f^{-1} is univalent in the domain Δ. From a previous result, for all $f \in H_u(D)$ the relation $f'(z) \neq 0$, $\forall z \in D$ holds, hence $(f^{-1})'(w) \neq 0$, $\forall w \in \Delta$. So, we deduce that f is a conformal mapping in all the points of D, while f^{-1} is a conformal mapping in all the point of Δ.

Definition 6.3.1.
1. The univalent function f in the domain $D \subset \mathbb{C}$ is said to be a **conformal isomorphism** between the domains D and $\Delta = f(D)$.
2. The domains D and Δ are called to be **conformally equivalent**, if there exists a conformal isomorphism between D and Δ.
3. The conformal isomorphisms of the domain D to itself are called the **conformal automorphisms** of D.
4. The set of all the conformal automorphisms of the domain D together with the function composition represents a group, is called the **conformal group** of D, and it is denoted by $A(D)$.

Remarks 6.3.1.
1. The **direct problem** of the conformal representation is: for a given function $f \in H_u(D)$, determine the image $f(D)$ (i. e., the image of the domain D by the function f).
2. The **inverse problem** of the conformal representation, that in fact is the **problem of the conformal representation:**
   Let D and Δ two given domains. Determine, if there exists, a conformal representation between these domains, i. e., determine, if there exists, a function $f \in H_u(D)$ such that $f(D) = \Delta$.
   A such kind of conformal mapping does not exist in every case. For example, if D is a simply connected domain, but the domain Δ is not simply connected, the above problem has no solution, because for all D simply connected domain the image $f(D)$ is also simply connected, whenever $f \in H_u(D)$. Really, if γ is an arbitrary closed path in $f(D)$, then $f^{-1} \circ \gamma$ is a closed path in D, because f^{-1} is a homeomorphism. If φ is the continuous deformation in D of the closed path $f^{-1} \circ \gamma$ to the constant path e_{z_0}, then $f \circ \varphi$ is the continuous deformation in $f(D)$ of the path γ to the constant path $e_{f(z_0)}$. Hence, the domain $f(D)$ is a simply connected domain.

Theorem 6.3.1. *If the function f_0 is a conformal representation of the domain D onto the domain Δ, then the set of all the conformal representations of D onto Δ is given by*

$$\{f : f = \varphi \circ f_0, \varphi \in A(\Delta)\},$$

where $A(\Delta)$ represents the set of the conformal automorphisms of Δ.

Proof. Let the function f be such a representation. Then $\varphi = f \circ f_0^{-1} \in A(\Delta)$, hence $f = \varphi \circ f_0$. Conversely, if $\varphi \in A(\Delta)$, then the function $f = \varphi \circ f_0$ is a conformal representation of D onto Δ. □

Theorem 6.3.2. *Every conformal representation $f : D \to \Delta$ induces an isomorphism function f^* between the groups $A(D)$ and $A(\Delta)$, where*

$$f^* : A(D) \to A(\Delta), \quad f^*(\varphi) = f \circ \varphi \circ f^{-1}, \quad \varphi \in A(D).$$

Proof. If $\varphi \in A(D)$, then $f^*(\varphi) \in A(\Delta)$. Conversely, if $\psi \in A(\Delta)$, then $\varphi = f^{-1} \circ \psi \circ f \in A(D)$ and $f^*(\varphi) = \psi$. If $f^*(\varphi_1) = f^*(\varphi_2)$, then $\varphi_1 = \varphi_2$, hence the function f^* is a bijection. If $\varphi_1, \varphi_2 \in A(D)$, then

$$f^*(\varphi_1) \circ f^*(\varphi_2) = (f \circ \varphi_1 \circ f^{-1}) \circ (f \circ \varphi_2 \circ f^{-1}) = f \circ (\varphi_1 \circ \varphi_2) \circ f^{-1} = f^*(\varphi_1 \circ \varphi_2),$$

hence the function f^* is a homomorphism. □

Theorem 6.3.3. *Denote by* $\mathrm{U} = U(0;1)$ *the unit open disc. Then the functions of the conformal group* $A(\mathrm{U})$ *(i. e., all the conformal isomorphisms of the unit disc* U *onto itself) are of the form:*

$$\varphi(z) = e^{i\theta}\frac{z-a}{1-\overline{a}z}, \quad \theta \in \mathbb{R},\ a \in \mathrm{U}. \tag{6.4}$$

Proof. All the functions φ of the form (6.4) are conformal automorphisms of U like we know from the theory of the circular transforms.

Let $f \in A(\mathrm{U})$, and denote $f(0) = w_0$. Let us define the function $g = \varphi \circ f$, where $\varphi \in A(\mathrm{U})$ is given by

$$\varphi(w) = \frac{w - w_0}{1 - \overline{w}_0 w}, \quad w \in \mathrm{U}.$$

Since $g(0) = 0$ and $|g(z)| < 1$, $\forall z \in \mathrm{U}$, from the Schwarz lemma we get $|g(z)| \le |z|$, for all $z \in \mathrm{U}$. The function $g^{-1} = f^{-1} \circ \varphi^{-1}$ also satisfies the conditions of the assumption of the Schwarz lemma, hence $|g^{-1}(w)| \le |w|$, $\forall w \in \mathrm{U}$. Letting $w = g(z)$, then $|z| \le |g(z)|$, $\forall z \in \mathrm{U}$, and hence $|g(z)| = |z|$, $\forall z \in \mathrm{U}$.

According to the Schwarz lemma, the last equality holds if and only if $g(z) = e^{i\theta}z$, $\theta \in \mathbb{R}$. Since $g(z) = \varphi(f(z))$, then $f(z) = \varphi^{-1}(e^{i\theta}z)$. Letting $a = -e^{-i\theta}w_0$, from the relation $\varphi^{-1}(z) = \frac{z+w_0}{1+\overline{w}_0 z}$ we have that

$$\varphi^{-1}(e^{i\theta}z) = \frac{e^{i\theta}z + w_0}{1 + \overline{w}_0 e^{i\theta}z} = e^{i\theta}\frac{z + w_0 e^{-i\theta}}{1 + \overline{w_0 e^{-i\theta}}z} = e^{i\theta}\frac{z-a}{1-\overline{a}z},$$

and it follows

$$f(z) = \varphi^{-1}(e^{i\theta}z) = e^{i\theta}\frac{z-a}{1-\overline{a}z}.$$ □

6.4 The Riemann mapping theorem

Let us denote by U the open unit disc, i. e., $\mathrm{U} = U(0;1)$, and let $D \subset \mathbb{C}$ be a simply connected domain. Let us put the next natural question: did there exist a conformal representation of D onto U?

Theorem 6.4.1. *The complex plane* $\mathbb{C}$ *and the unit disc* $\mathrm{U} = U(0;1)$ *are not conformally equivalent.*

Proof. If $f \in H(\mathbb{C})$ and $f(\mathbb{C}) = \mathrm{U}$, then $|f(z)| < 1$, $\forall z \in \mathbb{C}$, and from the Liouville theorem we have $f(z) = c$, $\forall z \in \mathbb{C}$, hence the function f is not univalent. □

Theorem 6.4.2. *If* D *is a simply connected domain of* $\mathbb{C}$, *and if the function* $f_0 : D \to \mathrm{U}$ *is a conformal representation of* D *onto* U, *then all the conformal representations of the domain* D *onto the unit disc* U *have the form:*

$$f(z) = e^{i\theta} \frac{f_0(z) - a}{1 - \overline{a} f_0(z)}, \qquad \theta \in \mathbb{R},\ a \in \mathrm{U}.$$

Proof. This result is an immediate consequence of Theorem 6.3.1 and Theorem 6.3.3. □

Theorem 6.4.3 (Riemann theorem). *Every simply connected domain* $D \subset \mathbb{C}$, *where* $D \neq \mathbb{C}$, *is conformally equivalent with the unit disc* U.

Proof. 1. Let $z_0 \in D$ be a fixed point, and let us define the $\mathcal{F}$ set by

$$\mathcal{F} = \{f \in H_u(D) : f(z_0) = 0, f(D) \subset \mathrm{U}\}.$$

We will prove that $\mathcal{F}$ is a nonempty set.

Since $D \neq \mathbb{C}$, let $a \in \mathbb{C} \setminus D$. From the multivalued functions analytical branches existence theorem, there exists a function $g \in H(D)$ such that $(g(z))^2 = z - a$, $\forall z \in D$ (i. e., the function g is a branch of the multivalued function $(z-a)^{\frac{1}{2}}$, where the function $z-a$ does not vanish in D). If $z_1, z_2 \in D$ and $g(z_1) = \pm g(z_2)$, then $(g(z_1))^2 = (g(z_2))^2$, hence $z_1 = z_2$. From here, it follows that $g \in H_u(D)$.

Since $g(D)$ is a domain, there exists a number $r > 0$ such that $U(g(z_0); r) \subset g(D)$.

We will prove that $U(-g(z_0); r) \cap g(D) = \emptyset$. Contrary, if it is not true, i. e., $U(-g(z_0); r) \cap g(D) \neq \emptyset$, then

$$\begin{aligned} &\exists z_1 \in D : g(z_1) \in U(-g(z_0); r) \Rightarrow |g(z_1) + g(z_0)| < r \Leftrightarrow |-g(z_1) - g(z_0)| < r \\ &\quad \Leftrightarrow -g(z_1) \in U(g(z_0); r) \subset g(D). \end{aligned}$$

Then $\exists z_2 \in D$ such that $g(z_2) = -g(z_1)$, and from the above proved property of g we deduce that $z_1 = z_2$. Hence, $g(z_1) = -g(z_1)$, i. e., $g(z_1) = 0$, and from here we get $0 = (g(z_1))^2 = z_1 - a$, so $z_1 = a$. This contradicts the fact that $a \in \mathbb{C} \setminus D$.

It follows that the function $g_1 = \frac{r}{g + g(z_0)}$ is well-defined on the domain D (since $g(z) \neq -g(z_0)$, $\forall z \in D$, because $g(D) \cap U(-g(z_0); r) = \emptyset$). Moreover, the function g_1 is univalent on D, because g is univalent. Further, $g_1(D) \subset \mathrm{U}$, because $|g(z) + g(z_0)| \geq r$, $\forall z \in D$, thus $|g_1(z)| \leq 1$, $\forall z \in D$. Since $g_1(D)$ is an open set, it follows that $g_1(D) \subset \mathrm{U}$.

Let us choose the function $\varphi \in A(\mathrm{U})$ such that

$$\varphi(w) = \frac{w - g_1(z_0)}{1 - \overline{g_1(z_0)}w}.$$

Then $\varphi(g_1(z_0)) = 0$, and the function $g_2 = \varphi \circ g_1$ is univalent on the domain D. Also, $g_2(z_0) = 0$, $g_2(D) \subset \mathrm{U}$, i. e., $g_2 \in \mathcal{F}$, and thus $\mathcal{F} \neq \emptyset$.

2. To all functions $f \in \mathcal{F}$, it corresponds the strictly positive number $|f'(z_0)|$. Since this number represents the coefficient of the linear deformation of f in z_0, we expect that the domain $f(D)$ will be maximal if $|f'(z_0)|$ will be maximal. This reason will be used for the following steps of the proof.

Let $M = \sup\{|f'(z_0)| : f \in \mathcal{F}\}$. From the definition, the set $\mathcal{F}$ is a bounded function set, and by a known theorem the set $\mathcal{F}' = \{f' : f \in \mathcal{F}\}$ will be also bounded. Hence, $M < +\infty$. From the definition of M, we have

$$\forall n \in \mathbb{N}^*, \exists f_n \in \mathcal{F} : M - \frac{1}{n} < |f_n'(z_0)| \leq M,$$

hence $\lim_{n\to+\infty} |f_n'(z_0)| = M$.

Since $\mathcal{F}$ is bounded, from the Montel theorem, we have that $\mathcal{F}$ is relatively compact in $H(D)$. Thus, the function sequence $(f_n)_{n\in\mathbb{N}^*}$ contains the subsequence $(f_{n_k})_{k\in\mathbb{N}^*}$ that uniformly converges on compacts to a function $f_0 \in H(D)$. Thus, let $f_0(z_0) = \lim_{k\to+\infty} f_{n_k}(z_0) = 0$.

According to the Weierstrass theorem, the function sequence $(f_{n_k}')_{k\in\mathbb{N}^*}$ uniformly converges on compacts to the function f_0', hence $M = \lim_{k\to+\infty} |f_{n_k}'(z_0)| = |f_0'(z_0)|$.

The function f_0 cannot be constant on D, since contrary, from $f_0(z_0) = 0$ it follows that $f_0 \equiv 0$. Thus, the relations $f_0 \equiv 0$ and $|f_0'(z_0)| = M > 0$ are in contradiction. From the Hurwitz theorem, we get that $f_0 \in H_u(D)$. On the other hand, $|f_{n_k}(z)| < 1$, $\forall z \in D$, $\forall k \in \mathbb{N}^*$, and by letting $k \to +\infty$ we deduce that $|f_0(z)| \leq 1$, $\forall z \in D$.

Since the function f_0 is not constant, from the theorem of the module maximum of the holomorphic functions, we obtain that the function $|f_0|$ cannot take its maximum value on the domain D, thus it follows that $|f_0(z)| < 1$, $\forall z \in D$, hence $f_0(D) \subset \mathrm{U}$. So, we obtained that $f_0 \in \mathcal{F}$ and

$$|f_0'(z_0)| = M = \sup\{|f'(z_0)| : f \in \mathcal{F}\}.$$

3. In the last step, we will prove $f_0(D) = \mathrm{U}$. Let us suppose that this is not true. Then $\exists \alpha \in \mathrm{U} \setminus f_0(D)$, and let us define the function $\psi \in A(\mathrm{U})$ as follows:

$$\psi(w) = \frac{w - \alpha}{1 - \overline{\alpha}w}, \quad w \in \mathrm{U}.$$

The function $\psi \circ f_0$ does not vanish in D, because $f_0(z) \neq \alpha$, $\forall z \in D$. So, we may use the multivalued functions analytical branches existence theorem, which implies that: there exists a function h holomorphic in the domain D, such that $h^2 = \psi \circ f_0$. Since $(\psi \circ f_0)(D) \subset \mathrm{U}$, it follows that $h(D) \subset \mathrm{U}$. Moreover, $h \in H_u(D)$, because if this is not true, then the function $\psi \circ f_0$ will not be univalent, which is a contradiction with the facts that $\psi \in A(\mathrm{U})$ and $f_0 \in \mathcal{F}$. Further, we have $|h(z_0)|^2 = |-\alpha| = |\alpha|$ and

$$\begin{aligned} 2h(z_0)h'(z_0) &= \psi'(f_0(z_0))f_0'(z_0) = \psi'(0)f_0'(z_0) \\ &= \left(\frac{w-\alpha}{1-\overline{\alpha}w}\right)'\Bigg|_{w=0} f_0'(z_0) = (1-|\alpha|^2)f_0'(z_0), \end{aligned}$$

i. e.,

$$|h'(z_0)| = \left|\frac{1}{2h(z_0)}\psi'(0)f_0'(z_0)\right| = \frac{1-|\alpha|^2}{2\sqrt{|\alpha|}}|f_0'(z_0)| = \frac{1-|\alpha|^2}{2\sqrt{|\alpha|}}M. \tag{6.5}$$

We will define the function $\chi \in A(\mathrm{U})$ as follows:

$$\chi(w) = \frac{w-h(z_0)}{1-\overline{h(z_0)}w}, \quad w \in \mathrm{U}.$$

The function $f = \chi \circ h$ will be univalent in the domain D, and $f(D) \subset \mathrm{U}$, $f(z_0) = 0$. Hence, $f \in \mathcal{F}$. On the other hand,

$$\begin{aligned} f'(z_0) &= \chi'(h(z_0))h'(z_0) \\ &= \frac{1-\overline{h(z_0)}w+\overline{h(z_0)}(w-h(z_0))}{(1-\overline{h(z_0)}w)^2}\Bigg|_{w=h(z_0)} h'(z_0) = \frac{1}{1-|h(z_0)|^2}h'(z_0), \end{aligned}$$

and thus, using the inequality (6.5), we have

$$|f'(z_0)| = \frac{1}{1-|\alpha|}\frac{1-|\alpha|^2}{2\sqrt{|\alpha|}}M = \frac{1+|\alpha|}{2\sqrt{|\alpha|}}M.$$

But $1+|\alpha| > 2\sqrt{|\alpha|}$, and we deduce that $|f'(z_0)| > M$, which contradicts the choice of M. This contradiction proves that $f_0(D) = \mathrm{U}$. □

Corollary 6.4.1. *If D is a simply connected domain in $\mathbb{C}$ with $D \neq \mathbb{C}$, and if $z_0 \in D$, then there exists a unique function $f \in H_u(D)$ such that $f(z_0) = 0$ and $f'(z_0) = 1$, which conformally maps the domain D onto an open disc with the center at the origin 0.*

The radius of this open disc is called the **conformal radius** *of the domain D at the z_0 point.*

Proof. Let the function f_0 be a conformal representation of the domain D onto the unit disc U that satisfies the conditions $f_0(z_0) = 0$ and $f_0'(z_0) > 0$.

We will show that there exists a such a function, and this is the unique function with these properties. Let $g \in H_u(D)$ a function with $g(z_0) = 0$ and $g(D) = \mathrm{U}$. Then all the other functions that satisfy the above properties are of the form $f = e^{i\theta}g$, $\theta \in \mathbb{R}$ (because from Theorem 6.4.2, we have $f = e^{i\theta}\frac{g-a}{1-\bar{a}g}$, and from $f(z_0) = 0$ we get $a = 0$). On the other hand, $f'(z_0) = e^{i\theta}g'(z_0)$ and $g'(z_0) \neq 0$, because the function g is univalent. Supposing that $f'(z_0) > 0$, then $\arg(e^{i\theta}g'(z_0)) = 0$, hence $\theta + \arg g'(z_0) = 0$, or $\theta = -\arg g'(z_0)$. From here, it follows immediately that there exists a such a function f_0, and this is the unique function with these properties.

Let $R = \frac{1}{f_0'(z_0)}$. Then the function $f = Rf_0$ conformally maps the domain D onto the disc $U(0;R)$. The uniqueness of the function f follows from the uniqueness of f_0, because if f satisfies the assumptions of the Corollary 6.4.1, then the function $\frac{1}{R}f$ will satisfy the previous conditions on f_0, hence $\frac{1}{R}f = f_0$. □

Let D be a simply connected domain, such that $0 \in D$ and $D \neq \mathbb{C}$. Let us define the function set

$$\mathcal{F} = \{f \in H(D) : f(0) = 0, f'(0) = 1\},$$

which is nonempty, because in contains the function $f(z) = z$.

Let us introduce the functional $M : \mathcal{F} \to [0, +\infty]$, by

$$M(f) = \sup\{|f(z)| : z \in D\}, \quad f \in \mathcal{F}.$$

Corollary 6.4.2. *The infimum value of the functional M on the domain D is the conformal radius of the domain D in the origin 0, denoted by R. The infimum of the functional M is attained for that function $f_1 \in H_u(D)$ which conformally transforms the domain D onto the disc $U(0;R)$, i. e.,*

$$M(f_1) = R = \min\{M(f) : f \in \mathcal{F}\}.$$

Proof. Let $f_0 \in H_u(D)$ that unique conformal mapping of the domain D onto the unit disc U, which satisfies $f_0(0) = 0$ and $f_0'(0) > 0$ (see the proof of Corollary 6.4.1). Letting $g = f_0^{-1}$, we see that it is sufficient to search the infimum of the functional M on the subset of $\mathcal{F}$ for which $M(f) < +\infty$, i. e., $\mathcal{F}_0 = \{f \in \mathcal{F} : M(f) < +\infty\}$, because $\exists f_0 \in \mathcal{F}$ such that $f_0(D) = U(0;R)$.

Let $f \in \mathcal{F}_0$ be arbitrary, and let us define the function $h = \frac{1}{M(f)}f \circ g$. The function h is holomorphic in the unit disc U, and satisfies the conditions

$$h(0) = 0$$

and

$$|h(w)| = \left|\frac{1}{M(f)}f(g(w))\right| = \frac{1}{M(f)}|f(z)| \leq 1, \quad \forall w \in \mathrm{U},$$

since $g(w) = z \in D$. From the fact that the set $h(D)$ is open, we deduce that $|h(w)| < 1$, $\forall w \in \mathrm{U}$. Thus, $h(0) = 0$ and $|h(w)| < 1$, $\forall w \in \mathrm{U}$, and using the Schwarz lemma we get

$$|h(w)| \leq |w|, \quad \forall w \in \mathrm{U} \text{ and } |h'(0)| \leq 1.$$

From here, using the relation

$$h'(0) = \frac{1}{M(f)}f'(g(0))g'(0) = \frac{1}{M(f)}f'(0)\frac{1}{f_0'(0)},$$

it follows that

$$|h'(0)| = \frac{1}{M(f)}|f'(0)|\frac{1}{f_0'(0)} \le 1 \Rightarrow M(f) \ge \frac{1}{f_0'(0)} = R,$$

where R is the conformal radius of the domain D in the origin 0.

Hence, we obtained that $\forall f \in \mathcal{F}_0$ we have $M(f) \ge R$, and this inequality holds also for all $f \in \mathcal{F}$, since $\mathcal{F}_0 \subset \mathcal{F}$. But, the function f determined in the Corollary 6.4.1 belongs to $\mathcal{F}$, and for this function the relation $M(f) = R$ holds, because $\sup\{|f(z)| : z \in D\} = R$. □

Remarks 6.4.1.

1. From the Riemann theorem, it follows that any two simply connected domains of $\mathbb{C}$, different to $\mathbb{C}$, is conformally equivalent (conformally isomorph).
   The simply connected domains of $\mathbb{C}$ can be divided into two classes: the first one is the entire set $\mathbb{C}$, while the second ones consist of the all simply connected domains of $\mathbb{C}$, different to $\mathbb{C}$.
2. From the Riemann theorem and Theorem 6.3.2, we deduce the next: if $D \neq \mathbb{C}$ is a simply connected domain, then the group of all the conformal automorphisms of D, denoted by $A(D)$, is isomorphic with $A(\mathrm{U})$. Hence, according to the point 1 of these remarks, we deduce that the groups of conformal automorphisms of any two simply connected domains of $\mathbb{C}$, not equal to $\mathbb{C}$, are isomorph groups.

Theorem 6.4.4. *The functions of the group $A(\mathbb{C})$ of all the conformal automorphisms of $\mathbb{C}$, have the form $f : \mathbb{C} \to \mathbb{C}$, where $f(z) = az + b$, $a, b \in \mathbb{C}$, $a \neq 0$.*

Proof. 1. All the functions of f of the above form is a conformal automorphism of $\mathbb{C}$.

2. Conversely, let $f \in A(\mathbb{C})$. Then the function f is an entire function, so it may be expanded in Taylor series around the origin 0, i. e.,

$$f(z) = a_0 + a_1 z + a_2 z^2 + \cdots, \quad \forall z \in \mathbb{C}.$$

Let us consider the neighborhood $V = \mathbb{C}_\infty \setminus \overline{U}(0;r)$ of the ∞ point, where $r > 0$ is an arbitrary number. Since $f^{-1}(\overline{U}(0;r))$ is compact, then it is closed and bounded, hence we get that $W = \mathbb{C}_\infty \setminus f^{-1}(\overline{U}(0;r))$ is a neighborhood of ∞.

But

$$f(W \cap \mathbb{C}) = f(\mathbb{C} \setminus f^{-1}(\overline{U}(0;r))) = f(\mathbb{C}) \setminus \overline{U}(0;r) \subset V \Rightarrow \lim_{z\to\infty} f(z) = \infty,$$

hence $z = \infty$ is a pole for f, and hence we deduce that 0 is a pole for the function $\varphi(z) = f(\frac{1}{z})$. But

$$\varphi(z) = f\left(\frac{1}{z}\right) = a_0 + a_1\frac{1}{z} + a_2\frac{1}{z^2} + \cdots, \quad \forall z \in \mathbb{C}^*,$$

and since $z = 0$ is its pole, the sum needs to have a finite number of nonvanishing terms, which implies that

$$f(z) = a_0 + a_1 z + \cdots + a_n z^n, \quad \forall z \in \mathbb{C},$$

hence f is a polynomial.

If the degree of f is greater than 1, i. e., $\deg f \geq 2$, then $\deg f' \geq 1$, hence the equation $f'(z) = 0$ has at least a zero in $\mathbb{C}$. Then, according to a well-known theorem, we have $f \notin A(\mathbb{C})$. It follows that f is a first degree polynomial, i. e., $f(z) = az + b$, $a \neq 0$. □

6.5 Exercises

Exercise 6.5.1. Consider the half-planes $D = \{z \in \mathbb{C} : \operatorname{Re} z > 0\}$ and $\Omega = \{z \in \mathbb{C} : \operatorname{Im} z > 0\}$.

1. Prove that the function $f : D \to \Omega$, $f(z) = iz$ maps conformally the half-plane D onto the half-plane Ω;
2. Prove that the function $g : \Omega \to D$, $g(z) = -iz$ maps conformally the half-plane Ω onto the half-plane D.

Exercise 6.5.2. Let

$$D = \{z \in \mathbb{C} : |z| < 1, \operatorname{Re} z > 0\} \quad \text{and} \quad \Omega = U(0;1) \setminus [-1, 0].$$

Prove that the function $f : D \to \Omega$, $f(z) = z^2$ maps conformally the domain D onto the domain Ω.

Exercise 6.5.3. Let $D = \{z \in \mathbb{C} : \operatorname{Im} z > 0\}$, $\Omega = U(0;1)$ and $g : \mathbb{C} \setminus \{-i\} \to \mathbb{C}$, $g(z) = \frac{z-i}{z+i}$. Prove that the function $f = g|_D$ conformally maps the half-plane D onto the disc Ω.

Exercise 6.5.4. Prove that the function $f(z) = i\frac{1+z}{1-z}$ conformally maps the disc $D = U(0;1)$ onto the half-plane $\Omega = \{z \in \mathbb{C} : \operatorname{Im} z > 0\}$.

Exercise 6.5.5. Prove that the function $f(z) = \frac{z-i}{z+i}$ conformally maps the domain $D = \{z \in \mathbb{C} : \operatorname{Re} z < 0, \operatorname{Im} z > 0\}$ onto the domain $\Omega = \{z \in \mathbb{C} : |z| < 1, \operatorname{Im} z > 0\}$.

Exercise 6.5.6. Prove that the Joukowski function $f(z) = \frac{1}{2}(z + \frac{1}{z})$ conformally maps the domain $D = \{z \in \mathbb{C} : |z| < 1, \operatorname{Im} z > 0\}$ onto the half-plane $\Omega = \{z \in \mathbb{C} : \operatorname{Im} z < 0\}$.

Exercise 6.5.7. Let $g : \mathbb{C}^* \to \mathbb{C}$, $g(z) = z + \frac{1}{z}$, and let us define the domain

$$D = \{z \in \mathbb{C} : |z| > 1, \operatorname{Im} z > 0\}.$$

Prove that the function $f = g|_D$ conformally maps the domain D onto the half-plane $\Omega = \{z \in \mathbb{C} : \operatorname{Im} z > 0\}$.

Exercise 6.5.8. Determine a function that conformally maps the domain $D = \{z \in \mathbb{C}_\infty : |z| > 1, \operatorname{Im} z > 0\}$ onto the disc $\Omega = U(0;1)$.

Exercise 6.5.9. Let us define the domains $D = \{z \in \mathbb{C} : |z| > 1\}$ and $\Omega = \mathbb{C} \setminus [-1,1]$. Prove that the Joukowski function $f : D \to \Omega$, $f(z) = \frac{1}{2}(z + \frac{1}{z})$, conformally maps the domain D onto the domain Ω.

Exercise 6.5.10. Let

$$D = \{z \in \mathbb{C} : \operatorname{Re} z > 0\} \quad \text{and} \quad \Omega = \mathbb{C} \setminus \{z \in \mathbb{C} : \operatorname{Im} z = 0, \operatorname{Re} z \leq 0\}.$$

Prove that the function $f : D \to \Omega$, $f(z) = z^2$, conformally maps the half-plane D onto the domain Ω.

Exercise 6.5.11. Determine a function that conformally maps the domain $D = \{z \in \mathbb{C} : |z| < 1, \operatorname{Im} z > 0\}$ onto the unit disc $\Omega = U(0;1)$.

Exercise 6.5.12. Determine one function that conformally maps the domain

$$D = \left\{z \in \mathbb{C} : |z| < \sqrt[3]{2}, 0 < \arg z < \frac{\pi}{3}\right\}$$

onto the half-plane $\Omega = \{z \in \mathbb{C} : \operatorname{Im} z > 0\}$.

Exercise 6.5.13. Let $a, b \in \mathbb{R}$ such that $a < b$ and $D = \{z \in \mathbb{C} : a < \operatorname{Re} z < b\}$. Determine a function that conformally maps the domain D onto the half-plane $\Omega = \{z \in \mathbb{C} : \operatorname{Im} z > 0\}$.

Exercise 6.5.14. Let $r > 0$, $a \in \mathbb{C}$, with $|a| = r$, and let us define the domains

$$D = \{z \in \mathbb{C} : |z| < 2r, |z - a| > r\} \quad \text{and} \quad \Omega = \left\{z \in \mathbb{C} : 0 < \operatorname{Re} z < \frac{1}{2}\right\}.$$

Prove that the function $f(z) = \frac{z}{z-2a}$ conformally maps the domain D onto the domain Ω.

Exercise 6.5.15. Determine a function that conformally maps the domain

$$D = \{z \in \mathbb{C} : |z| < 2, |z - i| > 1\}$$

onto the half-plane $\Omega = \{z \in \mathbb{C} : \operatorname{Im} z > 0\}$.

Exercise 6.5.16. Determine a function that conformally maps the domain

$$D = \left\{z \in \mathbb{C} : |z| < r, \left|z + \frac{r}{2}\right| > \frac{r}{2}, \operatorname{Im} z > 0\right\}$$

onto the half-plane $\Omega = \{z \in \mathbb{C} : \operatorname{Im} z > 0\}$.

Exercise 6.5.17. Determine a function that conformally maps the domain

$$D = \left\{ z \in \mathbb{C} : |z| < r, \left| z - \frac{r}{2} \right| > \frac{r}{2}, \right\}$$

onto the half-plane $\Omega = \{z \in \mathbb{C} : \operatorname{Im} z > 0\}$.

Exercise 6.5.18. Determine a function that conformally maps the domain

$$D = \left\{ z \in \mathbb{C} : |z| < 1, 0 < \arg z < \frac{\pi}{2} \right\}$$

onto the disc $\Omega = U(0; 1)$.

Exercise 6.5.19. Let $a, b \in \mathbb{R}$, with $a < b$. Determine a function that conformally maps the domain

$$D = \mathbb{C}_\infty \setminus (\{z \in \mathbb{C} : \operatorname{Im} z = 0, \operatorname{Re} z \leq a\} \cup \{z \in \mathbb{C} : \operatorname{Im} z = 0, \operatorname{Re} z \geq b\})$$

onto the half-plane $\Omega = \{z \in \mathbb{C} : \operatorname{Im} z > 0\}$.

Exercise 6.5.20. Determine a function that conformally maps the domain

$$D = \mathbb{C} \setminus \{z \in \mathbb{C} : \operatorname{Im} z = 0, \operatorname{Re} z \geq 1\}$$

onto the disc $\Omega = U(0; 1)$.

Exercise 6.5.21.

1. Determine a function that conformally maps the set $D = \mathbb{C} \setminus T$ onto the set $\Omega = \{z \in \mathbb{C} : \operatorname{Im} z > 0\} \setminus \{i\}$, where $T = [a, b]$ is a closed interval, that connects the points a and b, with $a, b \in \mathbb{C}$, $a \neq b$;
2. Determine a function that conformally maps the set $D = \mathbb{C} \setminus T$ onto the set $\Omega = \{z \in \mathbb{C} : \operatorname{Im} z > 0\} \setminus \{i\}$, where $T = [1 + i, -3 - i]$ is a closed interval, that connects the points $z_0 = 1 + i$ and $z_1 = -3 - i$;
3. Determine a function that conformally maps the set $D = \mathbb{C} \setminus T$ onto the set $\Omega = \{z \in \mathbb{C} : \operatorname{Im} z > 0\} \setminus \{i\}$, where $T = [-i, i]$ is a closed interval, that connects the points $z_0 = -i$ and $z_1 = i$;
4. Determine a function that conformally maps the set $D = \mathbb{C} \setminus [-1, 1]$ onto the set $\Omega = \{z \in \mathbb{C} : \operatorname{Im} z > 0\} \setminus \{i\}$, where $T = [-1, 1]$ is a closed interval, that connects the points $z_0 = -1$ and $z_1 = 1$.

Exercise 6.5.22. Determine a function that conformally maps the domain

$$D = \{z \in \mathbb{C} : |z - i| > 1, \operatorname{Im} z > 0\}$$

onto the half-plane $\Omega = \{z \in \mathbb{C} : \operatorname{Im} z > 0\}$.

Exercise 6.5.23. Determine a function that conformally maps the domain $D = \{z \in \mathbb{C} : 0 < \arg z < \frac{\pi}{3}\}$ onto the domain $\Omega = \{z \in \mathbb{C} : 0 < \operatorname{Im} z < \frac{\pi}{2}\}$.

Exercise 6.5.24. Let $a \in \mathbb{C}$ and $\alpha \in U(a; r)$. Let us denote by α^* the inverse of the point α with respect to the circle $\partial U(a; r)$.
1. Prove that the function $f(z) = \frac{z-\alpha}{z-\alpha^*}$ conformally maps the disc $D = U(a; r)$ onto the disc $\Omega = U(0; \frac{|\alpha-a|}{r})$;
2. Prove that the function $f(z) = \frac{z-\alpha}{z-\alpha^*}$ conformally maps the set $D = \mathbb{C} \setminus (\overline{U}(a; r) \cup \{\alpha^*\})$ onto the set $\Omega = \mathbb{C} \setminus (U(0; \frac{|\alpha-a|}{r}) \cup \{1\})$.

Exercise 6.5.25. Prove that the Joukowski function $f(z) = \frac{1}{2}(z + \frac{1}{z})$ conformally maps the domain $D = U(0; 1) \setminus \{0\}$ onto the domain $\Omega = \mathbb{C} \setminus [-1, 1]$.

Exercise 6.5.26. Determine a function that conformally maps the domain $D = U(0; 1) \setminus (\{0\} \cup [\frac{1}{2}, 1))$ onto the domain $\Omega = \mathbb{C} \setminus [-1, \frac{5}{4}]$.

Exercise 6.5.27. For a given $a \in (0, 1)$, determine a function that conformally maps the domain

$$D = U(0; 1) \setminus ((-1, 0] \cup [a, 1))$$

onto the half-plane $\Omega = \{z \in \mathbb{C} : \operatorname{Im} z > 0\}$.

Exercise 6.5.28. Determine a function that conformally maps the set $D = U(0; 1) \setminus (\{0\} \cup [\frac{1}{2}, 1))$ onto the set $\Omega = \{z \in \mathbb{C} : \operatorname{Im} z > 0\} \setminus \{i\}$.

Exercise 6.5.29. Determine a function that conformally maps the set

$$D = \{z \in \mathbb{C} : |z - 3| > 9, |z - 8| < 16\}$$

onto the circular ring $\Omega = U(0; \frac{2}{3}, 1)$.

Exercise 6.5.30. Determine a function that conformally maps the set

$$D = \left\{z \in \mathbb{C} : |z| > 1, \left|z + \frac{1}{2}\right| < \sqrt{\frac{5}{2}}\right\}$$

onto the circular ring $\Omega = U(0; 1, \frac{4}{5}\sqrt{\frac{5}{2}})$.

Exercise 6.5.31. Determine a function that conformally maps the set $D = \{z \in \mathbb{C} : \operatorname{Im} z > 0\} \setminus (0, i]$ onto the half-plane $\Omega = \{z \in \mathbb{C} : \operatorname{Im} z > 0\}$.

Exercise 6.5.32. Let $\alpha \in \mathbb{R}$, with $0 < \alpha < 1$, and let

$$D = \{z \in \mathbb{C} : |z| < 1, \operatorname{Im} z > 0\} \setminus (0, \alpha i].$$

Determine a function that conformally maps the domain D onto the half-plane $\Omega = \{z \in \mathbb{C} : \operatorname{Im} z > 0\}$.

7 Solutions to the chapterwise exercises

7.1 Solutions to the exercises of Chapter 1

Solution of Exercise 1.6.1

1. We will use the algebraic form of the complex numbers, i. e., $z = x+iy$, where $x, y \in \mathbb{R}$.
Since $z + i = x + (y + 1)i$, it follows that $\operatorname{Im}(z + i) = y + 1$, and thus

$$\{z \in \mathbb{C} : \operatorname{Im}(z + i) = 0\} = \{z = x + iy \in \mathbb{C} : y = -1\}.$$

Hence

$$\{z \in \mathbb{C} : \operatorname{Im}(z + i) = 0\}$$

is the line parallel with Ox that contains the point $M(-i)$.

2. Since $z - 2i = x + (y - 2)i$, it follows that

$$\arg(z - 2i) = \frac{\pi}{4} \Leftrightarrow \frac{y-2}{x} = 1, \quad x > 0 \Leftrightarrow x - y + 2 = 0, \quad x > 0.$$

Hence

$$\left\{z \in \mathbb{C} : \arg(z - 2i) = \frac{\pi}{4}\right\} = \{z = x + iy \in \mathbb{C} : x - y + 2 = 0, x > 0\},$$

that represents an open half-line.

3. If $\{z \in \mathbb{C} : |z - \alpha| = |z - \beta|\}$, where $\alpha = -i$ and $\beta = 4 + i$, then

$$|z - \alpha| = |z - \beta| \Leftrightarrow |x + (y - 1)i| = |x + 4 + (y + 1)i| \Leftrightarrow y + 2x + 4 = 0.$$

Hence

$$\{z \in \mathbb{C} : |z - \alpha| = |z - \beta|\},$$

where $\alpha = -i$ and $\beta = 4+i$, is the orthogonal line on the segment AB, where $A(\alpha)$, $B(\beta)$, and contains the middle point of the segment $[AB]$.

4. Since

$$|z + i| = 2 \Leftrightarrow |x + (y + 1)i| = 2 \Leftrightarrow x^2 + (y + 1)^2 = 4,$$

it follows that

$$\{z \in \mathbb{C} : |z + i| = 2\} = \partial U(-i; 2),$$

i. e., $x^2 + (y + 1)^2 = 4$.

5. The set $\{z \in \mathbb{C} : |z - 2i| + |z + 4i| = 10\}$ is an ellipse, because

https://doi.org/10.1515/9783110657869-007

$$|z-2i|+|z+4i|=10 \Leftrightarrow \left|x+(y-2)i\right|+\left|x+(y+4)i\right|=10 \Leftrightarrow \frac{x^2}{4^2}+\frac{(y+1)^2}{5^2}=1.$$

From here,

$$\{z\in\mathbb{C}:|z-2i|+|z+4i|=10\}=\left\{z=x+iy\in\mathbb{C}:\frac{x^2}{4^2}+\frac{(y+1)^2}{5^2}=1\right\}.$$

6. If $z=x+iy$, $x,y\in\mathbb{R}$, then

$$\frac{z-1}{z+1}=\frac{x-1+yi}{x+1+yi}=\frac{(x^2+y^2-1)+2yi}{(x+1)^2+y^2},$$

and thus

$$\arg\frac{z-1}{z+1}=\frac{\pi}{3}\Rightarrow\frac{2y}{x^2+y^2-1}=\sqrt{3}\Rightarrow x^2+\left(y-\frac{1}{\sqrt{3}}\right)^2=\frac{4}{3}.$$

Since we have $\frac{2y}{(x+1)^2+y^2}>0$, we deduce that $y>0$. Consequently, we obtain

$$\left\{z\in\mathbb{C}:\arg\frac{z-1}{z+1}=\frac{\pi}{3}\right\}=\left\{z=x+iy\in\mathbb{C}:x^2+\left(y-\frac{1}{\sqrt{3}}\right)^2=\frac{4}{3},y>0\right\}$$

that is the arc of the circle $x^2+(y-\frac{1}{\sqrt{3}})^2=\frac{4}{3}$ contained in the upper half-plane.

7. Clearly,

$$\left\{z\in\mathbb{C}:-\frac{\pi}{4}<\arg(z-i)<\frac{\pi}{6}\right\}=S_i\left(-\frac{\pi}{4},\frac{\pi}{6}\right).$$

Since

$$\arg(z-i)=\frac{\pi}{6}\Leftrightarrow\frac{y-1}{x}=\frac{\sqrt{3}}{3},\quad x>0\Leftrightarrow 3y-\sqrt{3}x-3=0,\quad x>0,$$

and

$$\arg(z-i)=-\frac{\pi}{4}\Leftrightarrow\frac{y-1}{x}=-1\quad x>0\Leftrightarrow y+x-1=0,\quad x>0,$$

it follows that $S_i(-\frac{\pi}{4},\frac{\pi}{6})$ is the angular sector with the apex in i, bounded by the half-lines $3y-\sqrt{3}x-3=0$, $x>0$ and $y+x-1=0$, $x>0$.

8. Since $1-\frac{1}{2i}(z-\overline{z})=1-\operatorname{Im}z$, it follows that

$$|z|<1-\operatorname{Im}z\Leftrightarrow x^2+2\left(y-\frac{1}{2}\right)<0,\quad y\leq 1,$$

and thus

$$\left\{z\in\mathbb{C}:|z|<1-\frac{1}{2i}(z-\overline{z})\right\}=\{z=x+iy\in\mathbb{C}:x^2+2y-1<0\}.$$

This is the open set of the complex plane $\mathbb{C}$ bounded by the parabola $x^2 + 2y - 1 = 0$, which contains the origin O.

9. We deduce immediately that

$$\left|\frac{z}{z+3i}\right| < 1 \Leftrightarrow \frac{|z|}{|z+3i|} < 1 \Leftrightarrow |z| < |z+3i| \Leftrightarrow y > -\frac{3}{2},$$

and thus

$$\left\{z \in \mathbb{C} : \left|\frac{z}{z+3i}\right| < 1\right\} = \left\{z \in \mathbb{C} : \operatorname{Im} z > -\frac{3}{2}\right\}.$$

10. Since

$$\left|\frac{1-z}{1+z}\right| > 3 \Leftrightarrow |1-z| > 3|1+z| \Leftrightarrow \left(x + \frac{5}{4}\right)^2 + y^2 < \frac{9}{16},$$

we obtain that

$$\left\{z \in \mathbb{C} : \left|\frac{1-z}{1+z}\right| > 3\right\} = \left\{z = x + iy \in \mathbb{C} : \left(x + \frac{5}{4}\right)^2 + y^2 < \frac{9}{16}\right\}.$$

11. If $z = x + iy$, where $x, y \in \mathbb{R}$, then

$$|z-4| - |z+6| - 9 > 0 \Leftrightarrow \frac{(x+1)^2}{\frac{9}{2}^2} - \frac{y^2}{\frac{19}{4}} > 1, \quad x < -1.$$

From here,

$$\{z \in \mathbb{C} : |z-4| - |z+6| - 9 > 0\} = \left\{z = x + iy \in \mathbb{C} : \frac{(x+1)^2}{\frac{81}{4}} - \frac{y^2}{\frac{19}{4}} > 1, x < -1\right\},$$

that is the exterior of the hyperbola $\frac{(x+1)^2}{\frac{81}{4}} - \frac{y^2}{\frac{19}{4}} = 1$ that lies on the half-plane $x < -1$.

Solution of Exercise 1.6.2

Using the fact that $\operatorname{Im} z = \frac{1}{2i}(z - \overline{z})$, we obtain

$$\begin{aligned}\left|\frac{z-a}{z-\overline{a}}\right|^2 - 1 &= \frac{z-a}{z-\overline{a}}\overline{\left(\frac{z-a}{z-\overline{a}}\right)} - 1 = \frac{z-a}{z-\overline{a}}\frac{\overline{z}-\overline{a}}{\overline{z}-a} - 1 \\ &= \frac{(a-\overline{a})(z-\overline{z})}{|z-\overline{a}|^2} = \frac{2i \operatorname{Im} a \cdot 2i \operatorname{Im} z}{|z-\overline{a}|^2} = \frac{-4 \operatorname{Im} a \operatorname{Im} z}{|z-\overline{a}|^2}.\end{aligned}$$

Since $|z-\overline{a}|^2 > 0$ for all $z \in \mathbb{C} \setminus \{\overline{a}\}$, we get

$$\operatorname{Im} a \operatorname{Im} z > 0 \Leftrightarrow \left|\frac{z-a}{z-\overline{a}}\right|^2 - 1 < 0 \Leftrightarrow \left|\frac{z-a}{z-\overline{a}}\right| < 1.$$

We will use a similar method to prove the points 2 and 3.

Solution of Exercise 1.6.3

Like in the previous proof, we have

$$\left|\frac{z-a}{z+\overline{a}}\right|^2 - 1 = \frac{z-a}{z+\overline{a}}\overline{\left(\frac{z-a}{z+\overline{a}}\right)} - 1 = \frac{z-a}{z+\overline{a}}\frac{\overline{z}-\overline{a}}{\overline{z}+a} - 1 = \\ = \frac{-(a+\overline{a})(z+\overline{z})}{|z+\overline{a}|^2} = \frac{-4\operatorname{Re} a \operatorname{Re} z}{|z+\overline{a}|^2},$$

where we used the relation $\operatorname{Re} z = \frac{1}{2}(z+\overline{z})$.

Since $|z+\overline{a}|^2 > 0$ for all $z \in \mathbb{C} \setminus \{-\overline{a}\}$, we get

$$\operatorname{Re} a \operatorname{Re} z > 0 \Leftrightarrow \left|\frac{z-a}{z+\overline{a}}\right|^2 - 1 < 0 \Leftrightarrow \left|\frac{z-a}{z+\overline{a}}\right| < 1.$$

The proofs of the next two points are similar.

Solution of Exercise 1.6.4

1. Let $D = \{z \in \mathbb{C} : \operatorname{Im} z > 0\}$. Then $\forall z \in D \Rightarrow z \neq \overline{a}$, because $\operatorname{Im} a > 0 \Rightarrow a \in D \Rightarrow \overline{a} \notin D$. Hence from Exercise 1.6.2,

$$\forall z \in D \Leftrightarrow \operatorname{Im} z \operatorname{Im} a > 0 \Leftrightarrow |f(z)| = \left|\frac{z-a}{z-\overline{a}}\right| < 1 \Leftrightarrow f(z) \in U,$$

so the function f is well-defined.

(i) For the proof of the injectivity, we have

$$f(z_1) = f(z_2), \quad \forall z_1, z_2 \in D \Leftrightarrow \frac{z_1-a}{z_1-\overline{a}} = \frac{z_2-a}{z_2-\overline{a}} \\ \Leftrightarrow z_1(a-\overline{a}) - z_2(a-\overline{a}) = 0 \Leftrightarrow (a-\overline{a})(z_1-z_2) = 0 \Leftrightarrow z_1 = z_2,$$

because $a - \overline{a} = 0 \Leftrightarrow a = \overline{a} \Leftrightarrow \operatorname{Im} a = 0$, but from the assumption we have $\operatorname{Im} a > 0$.

(ii) For the proof of the surjectivity, we will show that for all $w \in U \equiv U(0;1)$ there exists $z \in D$, such that $f(z) = w$, i. e.,

$$f(z) = w \Leftrightarrow \frac{z-a}{z-\overline{a}} = w \Leftrightarrow z = \frac{\overline{a}w - a}{w-1}.$$

We need to prove that the values of z belong to D. Since

$$\operatorname{Im} z = \frac{z-\overline{z}}{2i} = \frac{1}{2i}\left(\frac{\overline{a}w-a}{w-1} - \frac{a\overline{w}-\overline{a}}{\overline{w}-1}\right) \\ = \frac{(\overline{a}-a)(|w|^2-1)}{2i|w-1|^2} = \frac{-\operatorname{Im} a(|w|^2-1)}{|w-1|^2} > 0,$$

because we have from the assumption that $\operatorname{Im} a > 0$, $|w-1|^2 > 0$, and $w \in U \Leftrightarrow |w| < 1 \Leftrightarrow |w|^2 - 1 < 0$, $\forall w \in U$. Hence $z \in D$, and consequently the function f is bijective.

2. We will use the same method like in the previous point.

Using the notation $D = \{z \in \mathbb{C} : \operatorname{Re} z > 0\}$ and $U = U(0,1)$, then $\forall z \in D \Rightarrow z \neq -\overline{a}$, because $\operatorname{Re} a > 0 \Rightarrow a \in D \Rightarrow \overline{a} \in D \Rightarrow -\overline{a} \notin D$. Hence

$$\forall z \in D \Leftrightarrow \operatorname{Re} z \operatorname{Re} a > 0 \overset{\text{Exercise}}{\underset{1.6.3}{\Leftrightarrow}} |f(z)| = \left|\frac{z-a}{z+\overline{a}}\right| < 1 \Leftrightarrow f(z) \in U,$$

so the function f is well-defined.

For the proof of the injectivity, we have

$$\begin{aligned} f(z_1) = f(z_2), \quad \forall z_1, z_2 \in D &\Leftrightarrow \frac{z_1 - a}{z_1 + \overline{a}} = \frac{z_2 - a}{z_2 + \overline{a}} \\ &\Leftrightarrow az_1 - \overline{a}z_2 - az_2 + \overline{a}z_1 = 0 \Leftrightarrow (a + \overline{a})(z_1 - z_2) = 0 \Leftrightarrow z_1 = z_2, \end{aligned}$$

because $a + \overline{a} = 0 \Leftrightarrow \operatorname{Re} a = 0$, but from the assumption $\operatorname{Re} a > 0$.

For the proof of the surjectivity, we will show that for all $w \in U \equiv U(0;1)$ there exists $z \in D$, such that $f(z) = w$, i. e.,

$$f(z) = w \Leftrightarrow z = \frac{\overline{a}w + a}{w - 1}.$$

Now we will prove that $z \in D$. Since

$$\begin{aligned} \operatorname{Re} z = \frac{z + \overline{z}}{2} &= \frac{1}{2}\left(\frac{\overline{a}w + a}{1 - w} + \frac{a\overline{w} + \overline{a}}{1 - \overline{w}}\right) \\ &= \frac{(\overline{a} + a)(1 - |w|^2)}{2|1 - w|^2} = \frac{\operatorname{Re} a(1 - |w|^2)}{|1 - w|^2} > 0, \end{aligned}$$

because from the assumption $\operatorname{Re} a > 0$, $w \in U \Leftrightarrow |w| < 1 \Leftrightarrow 1 - |w|^2 > 0$ and $|1 - w|^2 > 0$, $\forall w \in U$. Hence $z \in D$, and consequently the function f is bijective.

Solution of Exercise 1.6.5

Let $z_k = x_k + y_k$, $k \in \{1,2\}$ and $z = x + iy$. Then

$$\begin{aligned} A \in A_1A_2 &\Leftrightarrow \frac{x - x_1}{x_2 - x_1} = \frac{y - y_1}{y_2 - y_1} \Leftrightarrow \frac{\frac{z+\overline{z}}{2} - \frac{z_1+\overline{z_1}}{2}}{\frac{z_2+\overline{z_2}}{2} - \frac{z_1+\overline{z_1}}{2}} = \frac{\frac{z-\overline{z}}{2i} - \frac{z_1-\overline{z_1}}{2i}}{\frac{z_2-\overline{z_2}}{2i} - \frac{z_1-\overline{z_1}}{2i}} \\ &\Leftrightarrow \frac{z - z_1 + \overline{z - z_1}}{z_2 - z_1 + \overline{z_2 - z_1}} = \frac{z - z_1 - \overline{z - z_1}}{z_2 - z_1 - \overline{z_2 - z_1}} \overset{(7.1)}{=} \frac{z - z_1}{z_2 - z_1} \\ &\Leftrightarrow \frac{z - z_1 + \overline{z - z_1}}{z_2 - z_1 + \overline{z_2 - z_1}} = \frac{z - z_1}{z_2 - z_1} \overset{(7.2)}{=} \frac{\overline{z - z_1}}{\overline{z_2 - z_1}} \Leftrightarrow \frac{z - z_1}{z_2 - z_1} = \overline{\frac{z - z_1}{z_2 - z_1}}, \end{aligned}$$

where we used that

$$x = \operatorname{Re} z = \frac{z + \overline{z}}{2}, \quad y = \operatorname{Im} z = \frac{z - \overline{z}}{2i}$$

and

$$\frac{a}{b} = \frac{c}{d} = \frac{a+c}{b+d} \tag{7.1}$$

$$\frac{a}{b} = \frac{c}{d} = \frac{a-c}{b-d}. \tag{7.2}$$

Solution of Exercise 1.6.6

Using the fact $z = \overline{z} \Leftrightarrow z \in \mathbb{R}$, it follows from Exercise 1.6.5:

$$A_3 \in A_1A_2 \Leftrightarrow \frac{z_3 - z_1}{z_2 - z_1} = \frac{\overline{z_3 - z_1}}{\overline{z_2 - z_1}} = \overline{\left(\frac{z_3 - z_1}{z_2 - z_1}\right)} \Leftrightarrow \frac{z_3 - z_1}{z_2 - z_1} \in \mathbb{R}.$$

Solution of Exercise 1.6.7

It is well known that

$$\begin{aligned} A_3 \in [A_1A_2] &\Leftrightarrow A_3 \in A_1A_2 \quad \text{and} \quad \|A_1A_3\| + \|A_3A_2\| = \|A_1A_2\| \\ &\underset{1.6.6}{\overset{\text{Exercise}}{\Leftrightarrow}} \frac{z_3 - z_1}{z_2 - z_1} = t, \quad t \in \mathbb{R} \quad \text{and} \quad |z_3 - z_1| + |z_2 - z_3| = |z_2 - z_1| \\ &\Leftrightarrow z_3 = (1-t)z_1 + tz_2, \quad t \in \mathbb{R} \quad \text{and} \quad \big|t(z_2 - z_1)\big| + \big|(1-t)(z_2 - z_1)\big| = |z_2 - z_1|, \end{aligned}$$

where we used the relation $\overrightarrow{A_iA_j} = z_j - z_i$.

If $|z_2 - z_1| = 0 \Leftrightarrow z_1 = z_2$, which is a contradiction, so it follows that $|z_2 - z_1| \neq 0$. From the above result, dividing the second equality by $|z_2 - z_1| \neq 0$, we obtain that

$$A_3 \in [A_1A_2] \Leftrightarrow z_3 = (1-t)z_1 + tz_2 \quad \text{and} \quad |t| + |1-t| = 1, t \in \mathbb{R},$$

and using the last relation, we conclude that

$$A_3 \in [A_1A_2] \Leftrightarrow z_3 = (1-t)z_1 + tz_2, \quad t \in [0,1].$$

Solution of Exercise 1.6.8

Since

$$A_1A_2 : \quad \frac{x - x_1}{x_2 - x_1} = \frac{y - y_1}{y_2 - y_1}$$

and

$$A_3A_4: \quad \frac{x - x_3}{x_4 - x_3} = \frac{y - y_3}{y_4 - y_3},$$

using the identities $x = \operatorname{Re} z = \frac{z+\overline{z}}{2}$ and $y = \operatorname{Im} z = \frac{z-\overline{z}}{2i}$ we get that

$$A_1A_2: \quad \frac{x - x_1}{z_2 - z_1 + \overline{(z_2 - z_1)}} = \frac{y - y_1}{-i(z_2 - z_1 - \overline{(z_2 - z_1)})}$$

and

$$A_3A_4: \quad \frac{x - x_3}{z_4 - z_3 + \overline{(z_4 - z_3)}} = \frac{y - y_3}{-i(z_4 - z_3 - \overline{(z_4 - z_3)})}.$$

1. From the above relations,

$$\begin{aligned}
&A_1A_2 \perp A_3A_4 \\
&\quad \Leftrightarrow (z_2 - z_1 + \overline{(z_2 - z_1)})(z_4 - z_3 + \overline{(z_4 - z_3)}) \\
&\qquad + (-i)(z_2 - z_1 - \overline{(z_2 - z_1)})(-i)(z_4 - z_3 - \overline{(z_4 - z_3)}) = 0 \\
&\quad \Leftrightarrow (z_3 - z_4)\overline{(z_1 - z_2)} = -\overline{(z_3 - z_4)}(z_1 - z_2) \\
&\quad \Leftrightarrow \frac{z_1 - z_2}{z_3 - z_4} = -\frac{\overline{z_1 - z_2}}{\overline{z_3 - z_4}} = -\overline{\left(\frac{z_1 - z_2}{z_3 - z_4}\right)} \Leftrightarrow \frac{z_1 - z_2}{z_3 - z_4} \in i\mathbb{R},
\end{aligned}$$

because $z = -\overline{z}$ if and only if $z \in i\mathbb{R}$.

2. Similarly,

$$\begin{aligned}
A_1A_2 \| A_3A_4 &\Leftrightarrow \frac{z_2 - z_1 + \overline{(z_2 - z_1)}}{z_4 - z_3 + \overline{(z_4 - z_3)}} = \frac{-i(z_2 - z_1 - \overline{(z_2 - z_1)})}{-i(z_4 - z_3 - \overline{(z_4 - z_3)})} \\
&\Leftrightarrow (z_3 - z_4)\overline{(z_1 - z_2)} = \overline{(z_3 - z_4)}(z_1 - z_2) \Leftrightarrow \frac{z_1 - z_2}{z_3 - z_4} = \frac{\overline{z_1 - z_2}}{\overline{z_3 - z_4}} = \overline{\left(\frac{z_1 - z_2}{z_3 - z_4}\right)} \\
&\Leftrightarrow \frac{z_1 - z_2}{z_3 - z_4} \in \mathbb{R},
\end{aligned}$$

because $z = \overline{z}$ if and only if $z \in \mathbb{R}$.

Solution of Exercise 1.6.9

Denoting by $\Omega = \Omega(a)$ the center of the circle, we have

$$\begin{aligned}
A(z) \in C(a; r) &\Leftrightarrow \|A\Omega\| = r \Leftrightarrow |z - a| = r \Leftrightarrow |z - a|^2 = r^2 \\
&\Leftrightarrow (z - a)\overline{(z - a)} = r^2 \Leftrightarrow (z - a)(\overline{z} - \overline{a}) = r^2 \Leftrightarrow z\overline{z} - \overline{a}z - a\overline{z} + a\overline{a} = r^2.
\end{aligned}$$

Solution of Exercise 1.6.10

From Exercise 1.6.9, the circle $C(z_1, z_2, z_3)$ has the equation of the form $z\overline{z}-\overline{a}z-a\overline{z}+a\overline{a} = r^2$, or equivalently $z\overline{z} - \overline{a}z - a\overline{z} + |a|^2 - r^2 = 0$.

Denote $u = -\overline{a}$, $v = a$ and $w = |a|^2 - r^2$, then

$$A(z) \in C(z_1, z_2, z_3) \Leftrightarrow A_1(z_1),\ A_2(z_2),\ A_3(z_3) \text{ and } A(z) \text{ are the solutions of the following system:}$$

$$\mathcal{R}: \begin{cases} tz\overline{z} + uz - v\overline{z} + w = 0 \\ tz_1\overline{z_1} + uz_1 - v\overline{z_1} + w = 0 \\ tz_2\overline{z_2} + uz_2 - v\overline{z_2} + w = 0 \\ tz_3\overline{z_3} + uz_3 - v\overline{z_3} + w = 0. \end{cases}$$

This system has the solution

$$(1, -\overline{a}, a, |a|^2 - r^2),$$

and all the other points that satisfy the equation of the circle will satisfy the condition $\det\mathcal{R} = 0$, i. e.,

$$\det\mathcal{R} = 0 \Leftrightarrow \begin{vmatrix} z\overline{z} & z & \overline{z} & 1 \\ z_1\overline{z_1} & z_1 & \overline{z_1} & 1 \\ z_2\overline{z_2} & z_2 & \overline{z_2} & 1 \\ z_3\overline{z_3} & z_3 & \overline{z_3} & 1 \end{vmatrix} = 0.$$

Solution of Exercise 1.6.11

An arbitrary complex number will be real if and only if it coincides with its conjugate, and thus

$$\frac{z_2 - z_1}{z_3 - z_1} : \frac{z_2 - z_4}{z_3 - z_4} \in \mathbb{R} \Leftrightarrow \frac{z_2 - z_1}{z_3 - z_1} \cdot \frac{z_3 - z_4}{z_2 - z_4} = \overline{\frac{z_2 - z_1}{z_3 - z_1} \cdot \frac{z_3 - z_4}{z_2 - z_4}}.$$

A simple computation shows that

$$\frac{z_2 - z_1}{z_3 - z_1} \cdot \frac{z_3 - z_4}{z_2 - z_4} = \overline{\frac{z_2 - z_1}{z_3 - z_1} \cdot \frac{z_3 - z_4}{z_2 - z_4}} \Leftrightarrow \begin{vmatrix} z_1\overline{z_1} & z_1 & \overline{z_1} & 1 \\ z_2\overline{z_2} & z_2 & \overline{z_2} & 1 \\ z_3\overline{z_3} & z_3 & \overline{z_3} & 1 \\ z_4\overline{z_4} & z_4 & \overline{z_4} & 1 \end{vmatrix} = 0$$

$$\overset{\text{Exercise}}{\underset{1.6.10}{\Leftrightarrow}} \text{ there exists a circle } C \text{ that contains the given points.}$$

Solution of Exercise 1.6.12

From (1), it follows that $f(0) = 0$. Since the function f is continuous, from (1) and (2), we deduce that

$$\forall t \in \mathbb{R}, \quad f(tz) = tf(z), \ \forall z \in \mathbb{C}.$$

Let $\forall z' = a + ib$, where $a, b \in \mathbb{R}$, and let $\forall z \in \mathbb{C}$. Then

$$f(az) = af(z), \quad f(ibz) = bf(iz) = bf(i)f(z),$$

and thus

$$f(z')f(z) = f(z'z) = f(az + ibz) = [a + bf(i)]f(z), \quad \forall z \in \mathbb{C}, \ \forall z' = a + ib \in \mathbb{C},$$

or equivalently

$$f(z)[f(z') - (a + bf(i))] = 0, \quad \forall z \in \mathbb{C}, \ \forall z' = a + ib \in \mathbb{C}. \tag{7.3}$$

From the above relation, it follows that

$$\begin{aligned} &\text{(i)} \quad f(z) = 0, \quad \forall z \in \mathbb{C}, \\ &\text{or} \\ &\text{(ii)} \quad \exists z_0 \in \mathbb{C} : f(z_0) \neq 0. \end{aligned}$$

If $\exists z_0 \in \mathbb{C}$ such that $f(z_0) \neq 0$, then the equality (7.3) holds if and only if $f(z') = a+bf(i)$, $\forall z' = a + ib \in \mathbb{C}$. Since $-1 = f(-1) = f(i^2) = f(i)^2$, we obtain that $f(i) = i$ or $f(i) = -i$, and thus:

$$\begin{aligned} &\text{(iii)} \quad \text{if } f(i) = i, \quad \text{then } f(z') = z', \ \forall z' \in \mathbb{C}, \\ &\text{or} \\ &\text{(iv)} \quad \text{if } f(i) = -i, \quad \text{then } f(z') = \overline{z'}, \ \forall z' \in \mathbb{C}. \end{aligned}$$

7.2 Solutions to the exercises of Chapter 2

Solution of Exercise 2.9.1

1. If $f(t) = -3 + i + (1 + i)t$, $t \in \mathbb{R}$, and $f(t) = \alpha(t) + i\beta(t)$, then

$$\begin{cases} x = \alpha(t) \\ y = \beta(t), \quad t \in \mathbb{R} \end{cases} \Leftrightarrow \begin{cases} x = -3 + t \\ y = 1 + t, \quad t \in \mathbb{R} \end{cases} \Leftrightarrow y = x + 4, \quad x \in \mathbb{R}.$$

2. Since $f(t) = 2 - 3i + (1 - 2i)t$, $t \in [1, 3]$, we obtain similarly that

$$\begin{cases} x = 2 + t \\ y = -3 - 2t, \quad t \in [1,3] \end{cases} \Leftrightarrow y = 1 - 2x, \quad x \in [3,5].$$

3. If $f(t) = \frac{16}{3-i+(1+2i)t}$, $t \in \mathbb{R}$, then

$$f(t) = \frac{16(t+3)}{(t+3)^2 + (2t-1)^2} + i\frac{16(1-2t)}{(t+3)^2 + (2t-1)^2}, \quad t \in \mathbb{R},$$

and

$$\begin{cases} x = \frac{16(t+3)}{(t+3)^2+(2t-1)^2} \\ y = \frac{16(1-2t)}{(t+3)^2+(2t-1)^2}, \quad t \in \mathbb{R} \end{cases} \Rightarrow \frac{x}{y} = \frac{t+3}{1-2t} \Leftrightarrow t = \frac{x-3y}{2x+y}.$$

From here, using the first (or the second) equality we obtain that

$$x^2 + y^2 - \frac{32}{7}x - \frac{16}{7}y = 0,$$

i. e., $\partial U(\frac{16}{7} + i\frac{8}{7}; \frac{8\sqrt{5}}{7})$.

4. Since $f(t) = \frac{-12+2i-(1-3i)t}{1+3i-(2-i)t}$, $t \in \mathbb{R}$, then

$$\begin{cases} x = \frac{5t^2+34t-6}{5t^2+2t+10} \\ y = \frac{-5t^2+14t+38}{5t^2+2t+10}, \quad t \in \mathbb{R} \end{cases}$$
$$\Leftrightarrow \begin{cases} x - 1 = \frac{16(2t-1)}{5t^2+2t+10} \\ y + 1 = \frac{16(t+3)}{5t^2+2t+10}, \quad t \in \mathbb{R} \end{cases}$$
$$\Rightarrow \frac{x-1}{y+1} = \frac{2t-1}{t+3} \Leftrightarrow t = \frac{3x+y-2}{-x+2y+3}.$$

If we replace this value into the first or into the second relation, then the required curve has the equation

$$x^2 + y^2 + \frac{2}{7}x - \frac{18}{7}y - \frac{34}{7} = 0,$$

i. e., $\partial U(-\frac{1}{7} + i\frac{9}{7}; \frac{8\sqrt{5}}{7})$.

Solution of Exercise 2.9.2

We will use the *Cauchy–Riemann theorem*. Since $f(z) = \overline{z}$, it follows that $f(x+iy) = x-iy$, $x, y \in \mathbb{R}$. If we will write the function f in the algebraic form $f(x+iy) = u(x,y) + iv(x,y)$, then we get $u(x, y) = x$, $v(x, y) = -y$, hence

$$\frac{\partial u(z)}{\partial x} = 1 \neq \frac{\partial v(z)}{\partial y} = -1, \quad \forall z \in \mathbb{C},$$
$$\frac{\partial u(z)}{\partial y} = 0 = -\frac{\partial v(z)}{\partial x}, \quad \forall z \in \mathbb{C}$$

so there does not exist any point in $\mathbb{C}$ in which the function f is differentiable.

Solution of Exercise 2.9.3

If $f(z) = f(x + iy) = u(x, y) + iv(x, y)$, $x, y \in \mathbb{R}$, then

$$u(x, y) = x^2 + y^2 + x, \quad v(x, y) = 4xy + 3y.$$

Both of the u and v functions have continuously partial derivatives in $\mathbb{C}$, i. e., $f \in C^1(\mathbb{C})$, hence f is $\mathbb{R}$-differentiable on $\mathbb{C}$.

Using the Cauchy–Riemann relations, we will determine those points of $\mathbb{C}$, such that

$$\begin{cases} \frac{\partial u(z)}{\partial x} = \frac{\partial v(z)}{\partial y} \\ \frac{\partial u(z)}{\partial y} = -\frac{\partial v(z)}{\partial x} \end{cases} \Leftrightarrow \begin{cases} 2x + 1 = 4x + 3 \\ 2y = -4y \end{cases} \Leftrightarrow z = -1.$$

So, we get that at the only point $z_0 = -1$ the function f is differentiable. Thus,

$$f'(-1) = \frac{\partial u(-1)}{\partial x} + i\frac{\partial v(-1)}{\partial x} = -1.$$

We can obtain the same result if we will use the equivalent form of the Cauchy–Riemann system, i. e., $\frac{\partial f(z)}{\partial \bar{z}} = 0$, hence

$$\frac{\partial f(z)}{\partial \bar{z}} = 0 \Leftrightarrow z - 2\bar{z} - 1 = 0,$$

and then we will solve this last equation.

Solution of Exercise 2.9.4

We will use a similar method within the previous problem. If $f(z) = u(x, y) + iv(x, y)$, $x, y \in \mathbb{R}$, then

$$u(x, y) = x^2, \quad v(x, y) = xy$$

and the functions u and v belong to the class $C^1(\mathbb{C})$, so the function f is $\mathbb{R}$-differentiable in $\mathbb{C}$.

Since $\frac{\partial u(z)}{\partial x} = 2x$, $\frac{\partial u(z)}{\partial y} = 0$, $\frac{\partial v(z)}{\partial x} = y$ and $\frac{\partial v(z)}{\partial y} = x$, the Cauchy–Riemann system

$$\begin{cases} \frac{\partial u(z)}{\partial x} = \frac{\partial v(z)}{\partial y} \\ \frac{\partial u(z)}{\partial y} = -\frac{\partial v(z)}{\partial x} \end{cases}$$

has the solution $z_0 = 0$. The function is differentiable only at this point, and

$$f'(0) = \frac{\partial u(0)}{\partial x} + i\frac{\partial v(0)}{\partial x} = 0.$$

Solution of Exercise 2.9.5

If $f(z) = u(x,y) + iv(x,y)$, $x, y \in \mathbb{R}$, then

$$u(x,y) = x^2 - y^2 + 2x, \quad v(x,y) = 2xy - 2y.$$

From here, we have that $u, v \in C^1(\mathbb{C})$, hence the function f is $\mathbb{R}$-differentiable in $\mathbb{C}$.

Since $\frac{\partial u(z)}{\partial x} = 2x + 2$, $\frac{\partial u(z)}{\partial y} = -2y$, $\frac{\partial v(z)}{\partial x} = 2y$ and $\frac{\partial v(z)}{\partial y} = 2x - 2$, the Cauchy–Riemann system

$$\begin{cases} \frac{\partial u(z)}{\partial x} = \frac{\partial v(z)}{\partial y} \\ \frac{\partial u(z)}{\partial y} = -\frac{\partial v(z)}{\partial x} \end{cases}$$

has no solutions. It follows that there does not exist any point in $\mathbb{C}$ in which the function f is differentiable.

Solution of the Exercise 2.9.6

We write immediately the real and the imaginary part of f, i. e.,

$$u(x,y) = \ln\sqrt{x^2 + y^2}, \quad v(x,y) = \arctan\frac{y}{x}, \quad z = x + iy \in \mathbb{C}^*.$$

Now we will check the conditions of the *Cauchy–Riemann theorem*:

1. The functions u and v are continuously differentiable in $\mathbb{C}^*$, because

$$\frac{\partial u(z)}{\partial x} = \frac{x}{x^2 + y^2}, \quad \frac{\partial u(z)}{\partial y} = \frac{y}{x^2 + y^2}, \quad \frac{\partial v(z)}{\partial x} = -\frac{y}{x^2 + y^2}, \quad \frac{\partial v(z)}{\partial y} = \frac{x}{x^2 + y^2}.$$

2. Since

$$\frac{\partial u(z)}{\partial x} = \frac{\partial v(z)}{\partial y}, \quad \frac{\partial u(z)}{\partial y} = -\frac{\partial v(z)}{\partial x}, \quad \forall z \in \mathbb{C},$$

it follows that f is differentiable in $\mathbb{C}^*$, i. e., $f \in H(\mathbb{C}^*)$, and the derivative will be

$$f'(z) = \frac{\partial u(z)}{\partial x} + i\frac{\partial v(z)}{\partial x} = \frac{1}{z}, \quad \forall z \in \mathbb{C}^*.$$

Solution of Exercise 2.9.7

If $f(z) = u(x,y) + iv(x,y)$, $x, y \in \mathbb{R}$, then

$$u(x,y) = (e^x - e^{-x})\cos y, \quad v(x,y) = (e^x + e^{-x})\sin y.$$

Since $u, v \in C^1(\mathbb{C})$, and

$$\frac{\partial u(z)}{\partial x} = \frac{\partial v(z)}{\partial y} = (e^x + e^{-x})\cos y,$$
$$\frac{\partial u(z)}{\partial y} = -\frac{\partial v(z)}{\partial x} = -(e^x - e^{-x})\sin y, \quad \forall z \in \mathbb{C},$$

it follows that the function satisfies the conditions from the *Cauchy–Riemann theorem* in the whole complex plane $\mathbb{C}$, and thus $f \in H(\mathbb{C})$. Further,

$$f'(z) = \frac{\partial u(z)}{\partial x} + i\frac{\partial v(z)}{\partial x} = (e^x + e^{-x})\cos y + i(e^x - e^{-x})\sin y = e^z + e^{-z}, \quad \forall z \in \mathbb{C}.$$

Solution of Exercise 2.9.8

1. Since $f = u + iv$ is an entire function, the function will satisfy the conditions from the *Cauchy–Riemann theorem*. We have

$$u(x,y) = x + ay \in C^1(\mathbb{C}) \quad \text{and} \quad v(x,y) = bx + cy \in C^1(\mathbb{C}).$$

Using the Cauchy–Riemann system, we will obtain the required values, i. e.,

$$\begin{cases} \frac{\partial u(x,y)}{\partial x} = \frac{\partial v(x,y)}{\partial y} \\ \frac{\partial u(x,y)}{\partial y} = -\frac{\partial v(x,y)}{\partial x}, \quad \forall (x,y) \in \mathbb{R}^2 \end{cases} \Leftrightarrow \begin{cases} c = 1 \\ a = -b, \end{cases}$$

hence $f(z) = x + ay + i(-ax + y)$, $a \in \mathbb{R}$.

The rest of the points will be treated similarly.

2. Since $f = u + iv$, it follows that

$$u(x,y) = x^2 + axy + by^2, \quad v(x,y) = cx^2 + dxy + y^2,$$

hence $u, v \in C^1(\mathbb{C})$.

Further,

$$\begin{cases} \frac{\partial u(x,y)}{\partial x} = \frac{\partial v(x,y)}{\partial y} \\ \frac{\partial u(x,y)}{\partial y} = -\frac{\partial v(x,y)}{\partial x}, \quad \forall (x,y) \in \mathbb{R}^2 \end{cases}$$

$$\Leftrightarrow \begin{cases} x(2 - d) + y(a - 2) = 0 \\ x(2c + a) + y(d + 2b) = 0, \quad \forall (x,y) \in \mathbb{R}^2 \end{cases} \Leftrightarrow \begin{cases} a = 2 \\ b = -1 \\ c = -1 \\ d = 2, \end{cases}$$

hence $f(z) = x^2 + 2xy - y^2 + i(-x^2 + 2xy + y^2)$.

3. Since $f = u + iv$, we have that

$$u(x,y) = \cos x(\cosh y + a\sinh y), \quad v(x,y) = \sin x(\cosh y + b\sinh y),$$

hence $u, v \in C^1(\mathbb{C})$.

From here,

$$\begin{cases} \frac{\partial u(x,y)}{\partial x} = \frac{\partial v(x,y)}{\partial y} \\ \frac{\partial u(x,y)}{\partial y} = -\frac{\partial v(x,y)}{\partial x}, \quad \forall (x,y) \in \mathbb{R}^2 \end{cases} \Leftrightarrow \begin{cases} a = -1 \\ b = -1, \end{cases}$$

and thus $f(z) = \cos x(\cosh y - \sinh y) + +i\sin x(\cosh y - \sinh y)$.

Solution of Exercise 2.9.9

1.

Method 1.

If $z = x + iy = re^{i\theta}$, $x, y \in \mathbb{R}$, then

$$x = x(r,\theta) = r\cos\theta, \quad y = y(r,\theta) = r\sin\theta$$

and

$$f(z) = u\big(x(r,\theta), y(r,\theta)\big) + iv\big(x(r,\theta), y(r,\theta)\big).$$

From here,

$$\begin{cases} \frac{\partial u}{\partial r} = \frac{\partial u}{\partial x}\frac{\partial x}{\partial r} + \frac{\partial u}{\partial y}\frac{\partial y}{\partial r} = \frac{\partial u}{\partial x}\cos\theta + \frac{\partial u}{\partial y}\sin\theta \\ \frac{\partial u}{\partial \theta} = \frac{\partial u}{\partial x}\frac{\partial x}{\partial \theta} + \frac{\partial u}{\partial y}\frac{\partial y}{\partial \theta} = -\frac{\partial u}{\partial x}r\sin\theta + \frac{\partial u}{\partial y}r\cos\theta \end{cases}$$
$$\Leftrightarrow \frac{\partial u}{\partial x} = \frac{\partial u}{\partial r}\cos\theta - \frac{\partial u}{\partial \theta}\frac{\sin\theta}{r} \quad \text{and} \quad \frac{\partial u}{\partial y} = \frac{\partial u}{\partial r}\sin\theta + \frac{\partial u}{\partial \theta}\frac{\cos\theta}{r}. \tag{7.4}$$

Identically, we obtain similar results for the imaginary part v, i. e.,

$$\begin{cases} \frac{\partial v}{\partial r} = \frac{\partial v}{\partial x}\frac{\partial x}{\partial r} + \frac{\partial v}{\partial y}\frac{\partial y}{\partial r} = \frac{\partial v}{\partial x}\cos\theta + \frac{\partial v}{\partial y}\sin\theta \\ \frac{\partial v}{\partial \theta} = \frac{\partial v}{\partial x}\frac{\partial x}{\partial \theta} + \frac{\partial v}{\partial y}\frac{\partial y}{\partial \theta} = -\frac{\partial v}{\partial x}r\sin\theta + \frac{\partial v}{\partial y}r\cos\theta \end{cases}$$
$$\Leftrightarrow \frac{\partial v}{\partial x} = \frac{\partial v}{\partial r}\cos\theta - \frac{\partial v}{\partial \theta}\frac{\sin\theta}{r} \quad \text{and} \quad \frac{\partial v}{\partial y} = \frac{\partial v}{\partial r}\sin\theta + \frac{\partial v}{\partial \theta}\frac{\cos\theta}{r}. \tag{7.5}$$

Denoting

$$A = \frac{\partial u}{\partial r} - \frac{1}{r}\frac{\partial v}{\partial \theta}, \quad B = \frac{\partial v}{\partial r} + \frac{1}{r}\frac{\partial u}{\partial \theta},$$

from (7.4), (7.5) according to the Cauchy–Riemann system it follows that

$$\begin{cases} \frac{\partial u}{\partial x} = \frac{\partial v}{\partial y} \\ \frac{\partial u}{\partial y} = -\frac{\partial v}{\partial x} \end{cases} \Leftrightarrow \begin{cases} (\frac{\partial u}{\partial r} - \frac{1}{r}\frac{\partial v}{\partial \theta})\cos\theta = (\frac{\partial v}{\partial r} + \frac{1}{r}\frac{\partial u}{\partial \theta})\sin\theta \\ (\frac{\partial u}{\partial r} - \frac{1}{r}\frac{\partial v}{\partial \theta})\sin\theta = -(\frac{\partial v}{\partial r} + \frac{1}{r}\frac{\partial u}{\partial \theta})\cos\theta. \end{cases}$$

$$\Leftrightarrow \begin{cases} A\cos\theta - B\sin\theta = 0 \\ A\sin\theta + B\cos\theta = 0 \end{cases} \Leftrightarrow \begin{cases} A = 0 \\ B = 0 \end{cases}$$

$$\Leftrightarrow \frac{\partial u}{\partial r} = \frac{1}{r}\frac{\partial v}{\partial \theta}, \quad \frac{\partial v}{\partial r} = -\frac{1}{r}\frac{\partial u}{\partial \theta}. \tag{7.6}$$

Using the relations (7.4) and (7.5), we conclude that

$$f'(z) = \frac{\partial u(z)}{\partial x} + i\frac{\partial v(z)}{\partial x} \overset{(7.6)}{=} e^{-i\theta}\left(\frac{\partial u}{\partial r} + i\frac{\partial v}{\partial r}\right) \overset{(7.6)}{=} \frac{e^{-i\theta}}{r}\left(\frac{\partial v}{\partial \theta} - i\frac{\partial u}{\partial \theta}\right),$$

hence

$$f'(z) = e^{-i\theta}\left(\frac{\partial u(z)}{\partial r} + i\frac{\partial v(z)}{\partial r}\right) = \frac{e^{-i\theta}}{r}\left(\frac{\partial v(z)}{\partial \theta} - i\frac{\partial u(z)}{\partial \theta}\right), \quad z = re^{i\theta}.$$

Method 2.
If $z = re^{i\theta}$ and $z_0 = r_0e^{i\theta_0}$, then

$$\frac{f(z) - f(z_0)}{z - z_0} = \begin{cases} \frac{f(re^{i\theta_0}) - f(r_0e^{i\theta_0})}{e^{i\theta_0}(r - r_0)}, & \text{if } \theta = \theta_0 \\ \frac{f(r_0e^{i\theta}) - f(r_0e^{i\theta_0})}{r_0(e^{i\theta} - e^{i\theta_0})}, & \text{if } r = r_0 \end{cases}$$

$$= \begin{cases} \frac{f(re^{i\theta_0}) - f(r_0e^{i\theta_0})}{e^{i\theta_0}(r - r_0)}, & \text{if } \theta = \theta_0 \\ \frac{1}{r_0}\frac{\frac{f(r_0e^{i\theta}) - f(r_0e^{i\theta_0})}{\theta - \theta_0}}{\frac{e^{i\theta} - e^{i\theta_0}}{\theta - \theta_0}}, & \text{if } r = r_0. \end{cases}$$

Since f is differentiable at z_0, we get

$$f'(z_0) = e^{-i\theta_0}\left(\frac{\partial u(z_0)}{\partial r} + i\frac{\partial v(z_0)}{\partial r}\right)$$

and

$$f'(z_0) = \frac{1}{r_0}e^{-i\theta_0}\left(\frac{\partial v(z_0)}{\partial \theta} - i\frac{\partial u(z_0)}{\partial \theta}\right),$$

thus

$$\frac{\partial u(z_0)}{\partial r} = \frac{1}{r_0}\frac{\partial v(z_0)}{\partial \theta}, \quad \frac{\partial v(z_0)}{\partial r} = -\frac{1}{r_0}\frac{\partial u(z_0)}{\partial \theta}.$$

2. If $f(z) = u + iv = Re^{i\Phi}$, $z = x + iy$, where $x, y \in \mathbb{R}$, then

$$u = u(R(x,y), \Phi(x,y)) = R\cos\Phi, \quad v = v(R(x,y), \Phi(x,y)) = R\sin\Phi$$

and

$$f(z) = u(R(x,y), \Phi(x,y)) + iv(R(x,y), \Phi(x,y)).$$

Similarly, as in the above, the Cauchy–Riemann system becomes:

$$\begin{cases} \frac{\partial u}{\partial x} = \frac{\partial v}{\partial y} \\ \frac{\partial u}{\partial y} = -\frac{\partial v}{\partial x} \end{cases} \Leftrightarrow \begin{cases} \frac{\partial u}{\partial R}\frac{\partial R}{\partial x} + \frac{\partial u}{\partial \Phi}\frac{\partial \Phi}{\partial x} = \frac{\partial v}{\partial R}\frac{\partial R}{\partial y} + \frac{\partial v}{\partial \Phi}\frac{\partial \Phi}{\partial y} \\ \frac{\partial u}{\partial R}\frac{\partial R}{\partial y} + \frac{\partial u}{\partial \Phi}\frac{\partial \Phi}{\partial y} = -\frac{\partial v}{\partial R}\frac{\partial R}{\partial x} - \frac{\partial v}{\partial \Phi}\frac{\partial \Phi}{\partial x} \end{cases}$$

$$\Leftrightarrow \begin{cases} \cos\Phi\frac{\partial R}{\partial x} - R\sin\Phi\frac{\partial \Phi}{\partial x} = \sin\Phi\frac{\partial R}{\partial y} + R\cos\Phi\frac{\partial \Phi}{\partial y} \\ \cos\Phi\frac{\partial R}{\partial y} - R\sin\Phi\frac{\partial \Phi}{\partial y} = -\sin\Phi\frac{\partial R}{\partial x} - R\cos\Phi\frac{\partial \Phi}{\partial x} \end{cases}$$

$$\Leftrightarrow \begin{cases} \cos\Phi(\frac{\partial R}{\partial x} - R\frac{\partial \Phi}{\partial y}) = \sin\Phi(\frac{\partial R}{\partial y} + R\frac{\partial \Phi}{\partial x}) \\ \sin\Phi(\frac{\partial R}{\partial x} - R\frac{\partial \Phi}{\partial y}) = -\cos\Phi(\frac{\partial R}{\partial y} + R\frac{\partial \Phi}{\partial x}). \end{cases}$$

Denoting

$$A = \frac{\partial R}{\partial x} - R\frac{\partial \Phi}{\partial y}, \quad B = \frac{\partial R}{\partial y} + R\frac{\partial \Phi}{\partial x},$$

then the last equation system is equivalent to the next one:

$$\begin{cases} A\cos\Phi - B\sin\Phi = 0 \\ A\sin\Phi + B\cos\Phi = 0 \end{cases} \Leftrightarrow \begin{cases} A = 0 \\ B = 0 \end{cases}$$

$$\Leftrightarrow \frac{\partial R}{\partial x} = R\frac{\partial \Phi}{\partial y}, \quad \frac{\partial R}{\partial y} = -R\frac{\partial \Phi}{\partial x}. \tag{7.7}$$

The derivative f' will be

$$f'(z) = \frac{\partial u(z)}{\partial x} + i\frac{\partial v(z)}{\partial x} \overset{(7.7)}{=} e^{i\Phi}\left(\frac{\partial R}{\partial x} + iR\frac{\partial \Phi}{\partial x}\right) \overset{(7.7)}{=} e^{i\Phi}\left(R\frac{\partial \Phi}{\partial y} - i\frac{\partial R}{\partial y}\right),$$

i. e.,

$$f'(z) = e^{i\Phi(z)}\left(\frac{\partial R(z)}{\partial x} + iR\frac{\partial \Phi(z)}{\partial x}\right) = e^{i\Phi(z)}\left(R(z)\frac{\partial \Phi(z)}{\partial y} - i\frac{\partial R(z)}{\partial y}\right).$$

3. If $f(z) = u + iv = Re^{i\Phi}$, then

$$u = u(R(r,\theta), \Phi(r,\theta)) = R\cos\Phi, \quad v = v(R(r,\theta), \Phi(r,\theta)) = R\sin\Phi$$

and

$$f(z) = u(R(r,\theta), \Phi(r,\theta)) + iv(R(r,\theta), \Phi(r,\theta)).$$

Using the results obtained in point 1 of this problem, we get

$$\begin{cases} \frac{\partial u}{\partial r} = \frac{1}{r}\frac{\partial v}{\partial \theta} \\ \frac{\partial v}{\partial r} = -\frac{1}{r}\frac{\partial u}{\partial \theta} \end{cases} \Leftrightarrow \begin{cases} \cos\Phi\frac{\partial R}{\partial r} - R\sin\Phi\frac{\partial \Phi}{\partial r} = \frac{1}{r}(\sin\Phi\frac{\partial R}{\partial \theta} + R\cos\Phi\frac{\partial \Phi}{\partial \theta}) \\ \sin\Phi\frac{\partial R}{\partial r} + R\cos\Phi\frac{\partial \Phi}{\partial r} = -\frac{1}{r}(\cos\Phi\frac{\partial R}{\partial \theta} - R\sin\Phi\frac{\partial \Phi}{\partial \theta}) \end{cases}$$

$$\Leftrightarrow \begin{cases} \cos\Phi(\frac{\partial R}{\partial r} - \frac{R}{r}\frac{\partial \Phi}{\partial \theta}) = \sin\Phi(\frac{1}{r}\frac{\partial R}{\partial \theta} + R\frac{\partial \Phi}{\partial r}) \\ -\sin\Phi(\frac{\partial R}{\partial r} - \frac{R}{r}\frac{\partial \Phi}{\partial \theta}) = \cos\Phi(\frac{1}{r}\frac{\partial R}{\partial \theta} + R\frac{\partial \Phi}{\partial r}). \end{cases}$$

Denoting

$$A = \frac{\partial R}{\partial r} - \frac{R}{r}\frac{\partial \phi}{\partial \theta}, \quad B = \frac{1}{r}\frac{\partial R}{\partial \theta} + R\frac{\partial \phi}{\partial r},$$

then the last system is equivalent to the next one:

$$\begin{cases} A\cos\phi - B\sin\phi = 0 \\ A\sin\phi + B\cos\phi = 0 \end{cases} \Leftrightarrow \begin{cases} A = 0 \\ B = 0 \end{cases}$$
$$\Leftrightarrow \frac{\partial R}{\partial r} = \frac{R}{r}\frac{\partial \Phi}{\partial \theta}, \quad \frac{\partial R}{\partial \theta} = -rR\frac{\partial \Phi}{\partial r}. \tag{7.8}$$

Using the result obtained in point 1 of the problem, we deduce that

$$f'(z) = e^{-i\theta}\left(\frac{\partial u}{\partial r} + i\frac{\partial v}{\partial r}\right) = e^{-i\theta}e^{i\Phi}\left(\frac{\partial R}{\partial r} + iR\frac{\partial \Phi}{\partial r}\right)$$
$$\overset{(7.8)}{=} \frac{e^{-i\theta}e^{i\Phi}}{r}\left(R\frac{\partial \Phi}{\partial \theta} - i\frac{\partial R}{\partial \theta}\right),$$

thus

$$f'(z) = e^{-i\theta}e^{i\Phi(z)}\left(\frac{\partial R(z)}{\partial r} + iR(z)\frac{\partial \Phi(z)}{\partial r}\right)$$
$$= \frac{e^{-i\theta}e^{i\Phi(z)}}{r}\left(R(z)\frac{\partial \Phi(z)}{\partial \theta} - i\frac{\partial R(z)}{\partial \theta}\right), \quad z = re^{i\theta}.$$

Solution of Exercise 2.9.10

We will use the result obtained in point 1 of Exercise 2.9.9 for the Cauchy–Riemann system. For the function f given in the algebraic form, we will write the exponential form

$$f(z) = r^n \cos n\theta + ir^n \sin n\theta = u(r,\theta) + iv(r,\theta).$$

We get that $u = u(r,\theta) = r^n \cos n\theta$, $v = v(r,\theta) = r^n \sin n\theta$ and $u, v \in C^1(\mathbb{C})$. The relations (7.6) of the Cauchy–Riemann system are fulfilled in $\mathbb{C}$, because

$$\frac{\partial u}{\partial r} = \frac{1}{r}\frac{\partial v}{\partial \theta} = nr^{n-1}\cos n\theta, \quad \frac{\partial v}{\partial r} = -\frac{1}{r}\frac{\partial u}{\partial \theta} = nr^{n-1}\sin n\theta,$$

hence $f \in H(\mathbb{C})$.

The derivative of the function will be

$$f'(z) = e^{-i\theta}\left(\frac{\partial u(z)}{\partial r} + i\frac{\partial v(z)}{\partial r}\right) = nz^{n-1}, \quad \forall z \in \mathbb{C}.$$

Solution of Exercise 2.9.11

Suppose that $\exists a, b \in \mathbb{C}$, such that $f(D) = [a, b]$. Let us define the function $f_1(z) = e^{-i\theta} f(z)$, where θ is the angle between $A(a)$ and $B(b)$ line and the real axis, i. e., $\theta = \arg(b - a)$. Since f is differentiable on D, it follows that f_1 is also differentiable on D.

The function f_1 satisfies the fact that its image $f_1(D) = [a, c]$, where $c = e^{-i\theta} b$. It means that the imaginary part v_1 of the function f_1 is constant on the domain D, because $\operatorname{Im} f_1(z) = \operatorname{Im} a$, $\forall z \in D$. Since $f_1 \in H(D)$, and $D \subset \mathbb{C}$ is a domain, we deduce that f_1 is a constant function on D that contradicts the fact that f is a nonconstant function on D.

Solution of Exercise 2.9.12

Define the function

$$\varphi : G \setminus \{z\} \to \mathbb{C}, \quad \text{as } \varphi(\zeta) = \frac{f(\zeta) - f(z)}{\zeta - z},$$

and let use the notation $\zeta = \zeta_1 + i\zeta_2$, $z = x + iy$.

1. We know that

$$\exists \lim_{\zeta \to z} \operatorname{Re} \varphi(\zeta) = \lim_{\zeta \to z} \operatorname{Re} \frac{u(\zeta) - u(z) + i[v(\zeta) - v(z)]}{\zeta - z}$$
$$= \lim_{\zeta \to z} \frac{[u(\zeta) - u(z)](\zeta_1 - x) + [v(\zeta) - v(z)](\zeta_2 - y)}{(\zeta_1 - x)^2 + (\zeta_2 - y)^2} = l_1.$$

If $\zeta_2 = y \Rightarrow [\zeta \to z \Leftrightarrow \zeta_1 \to x]$, thus $\exists l_1 = \lim_{\zeta_1 \to x} \frac{u(\zeta) - u(z)}{\zeta_1 - x} = \frac{\partial u(z)}{\partial x}$.

If $\zeta_1 = x \Rightarrow [\zeta \to z \Leftrightarrow \zeta_2 \to y]$, thus $\exists l_1 = \lim_{\zeta_2 \to y} \frac{v(\zeta) - v(z)}{\zeta_2 - y} = \frac{\partial v(z)}{\partial y}$, hence we deduce that $\frac{\partial u(z)}{\partial x} = \frac{\partial v(z)}{\partial y}$.

2. Similarly,

$$\exists \lim_{\zeta \to z} \operatorname{Im} \varphi(\zeta) = \lim_{\zeta \to z} \frac{-[u(\zeta) - u(z)](\zeta_2 - y) + [v(\zeta) - v(z)](\zeta_1 - x)}{(\zeta_1 - x)^2 + (\zeta_2 - y)^2} = l_2.$$

If $\zeta_2 = y \Rightarrow [\zeta \to z \Leftrightarrow \zeta_1 \to x]$, thus $\exists l_2 = \lim_{\zeta_1 \to x} \frac{v(\zeta) - v(z)}{\zeta_1 - x} = \frac{\partial v(z)}{\partial x}$.

If $\zeta_1 = x \Rightarrow [\zeta \to z \Leftrightarrow \zeta_2 \to y]$, thus $\exists l_2 = \lim_{\zeta_2 \to y} \frac{u(\zeta) - u(z)}{\zeta_2 - y} = -\frac{\partial u(z)}{\partial y}$, and we get $\frac{\partial u(z)}{\partial x} = -\frac{\partial v(z)}{\partial y}$.

Solution of Exercise 2.9.13

Let $\zeta_n = (1 + \frac{z}{n})^n = r_n(\cos\theta_n + i\sin\theta_n)$, where $r_n = |\zeta_n|$, $\theta_n = \arg \zeta_n$. Then

$$r_n = \left|1 + \frac{x + iy}{n}\right|^n = \left[\left(1 + \frac{x}{n}\right)^2 + \left(\frac{y}{n}\right)^2\right]^{\frac{n}{2}} = \left(1 + \frac{2x}{n} + \frac{x^2 + y^2}{n^2}\right)^{\frac{n}{2}}$$

and

$$\theta_n = n \arctan \frac{\frac{y}{n}}{1 + \frac{x}{n}}.$$

Taking the limits for $n \to \infty$, we have

$$\lim_{n\to\infty} r_n = \lim_{n\to\infty} \left(1 + \frac{2x}{n} + \frac{x^2 + y^2}{n^2}\right)^{\frac{n}{2}} = e^x$$

and

$$\lim_{n\to\infty} \theta_n = \lim_{n\to\infty} n \arctan \frac{\frac{y}{n}}{1 + \frac{x}{n}} = y.$$

From the above relations, we conclude that

$$e^z = \lim_{n\to\infty} \zeta_n = \lim_{n\to\infty} \left(1 + \frac{z}{n}\right)^n = e^x(\cos y + i \sin y).$$

Solution of Exercise 2.9.14

1. Using the well-known definitions,

$$\sin z = \frac{e^{iz} - e^{-iz}}{2i}, \quad \cos z = \frac{e^{iz} + e^{-iz}}{2},$$

we obtain that

$$\begin{aligned}\sin z - \cos z = i &\Leftrightarrow \frac{e^{iz} - e^{-iz}}{2i} - \frac{e^{iz} + e^{-iz}}{2} = i\\ &\Leftrightarrow (1-i)e^{2iz} + 2e^{iz} - 1 - i = 0\\ &\Leftrightarrow (1-i)t^2 + 2t - 1 - i = 0, \quad t \overset{\text{not.}}{=} e^{iz} \Leftrightarrow t_{1,2} = \frac{-1 \pm \sqrt{3}}{1-i}.\end{aligned}$$

Thus,

$$\begin{aligned}e^{iz} &= \frac{-1 \pm \sqrt{3}}{1-i} \Leftrightarrow z \in -i\mathrm{Log}\, \frac{-1 \pm \sqrt{3}}{1-i}\\ &= \left\{\frac{\pi}{4} + 2k\pi - i \ln \frac{\sqrt{3}-1}{\sqrt{2}} : k \in \mathbb{Z}\right\} \cup \left\{-\frac{3\pi}{4} + 2k\pi - i \ln \frac{\sqrt{3}+1}{\sqrt{2}} : k \in \mathbb{Z}\right\}.\end{aligned}$$

2. From the next two definitions,

$$\cosh z = \frac{e^z + e^{-z}}{2}, \quad \sinh z = \frac{e^z - e^{-z}}{2},$$

we have that

$$\cosh z - i\sinh z = 1 \Leftrightarrow \frac{e^z + e^{-z}}{2} - i\frac{e^z - e^{-z}}{2} = 1$$
$$\Leftrightarrow (1-i)e^{2z} - 2e^z + 1 + i = 0 \Leftrightarrow (1-i)t^2 - 2t + 1 + i = 0, \quad t \overset{\text{not.}}{=} e^z \Leftrightarrow t_1 = 1, \quad t_2 = i.$$

Thus, we get

$$z \in \operatorname{Log} 1 \cup \operatorname{Log} i = \{2k\pi i : k \in \mathbb{Z}\} \cup \left\{\left(\frac{\pi}{2} + 2k\pi\right)i : k \in \mathbb{Z}\right\}.$$

3. Since

$$\sin z = \frac{e^{iz} - e^{-iz}}{2i},$$

it follows that

$$\sin z = \frac{4i}{3} \Leftrightarrow \frac{e^{iz} - e^{-iz}}{2i} = \frac{4i}{3} \Leftrightarrow 3t^2 + 8t - 3 = 0, \quad t \overset{\text{not.}}{=} e^{iz} \Leftrightarrow t_1 = \frac{1}{3}, \quad t_2 = -3$$
$$\Leftrightarrow z \in \left(-i\operatorname{Log}\frac{1}{3}\right) \cup (-i\operatorname{Log}(-3))$$
$$= \{2k\pi + i\ln 3 : k \in \mathbb{Z}\} \cup \{(2k+1)\pi - i\ln 3 : k \in \mathbb{Z}\}.$$

4. Since

$$\cos z = \frac{e^{iz} + e^{-iz}}{2},$$

we have

$$\cos z = \frac{3+i}{4} \Leftrightarrow \frac{e^{iz} + e^{-iz}}{2} = \frac{3+i}{4} \Leftrightarrow 2t^2 - (3+i)t + 2 = 0, \quad t \overset{\text{not.}}{=} e^{iz}$$
$$\Leftrightarrow t_1 = i + 1, \quad t_2 = \frac{-i+1}{2} \Leftrightarrow z \in (-i\operatorname{Log}(1+i)) \cup \left(-i\operatorname{Log}\frac{1-i}{2}\right)$$
$$= \left\{\frac{\pi}{4} + 2k\pi - i\ln\sqrt{2} : k \in \mathbb{Z}\right\} \cup \left\{-\frac{\pi}{4} + 2k\pi + i\ln\sqrt{2} : k \in \mathbb{Z}\right\}.$$

5. We easily deduce that

$$\sinh z = \frac{i}{2} \Leftrightarrow \frac{e^z - e^{-z}}{2} = \frac{i}{2} \Leftrightarrow t^2 - it - 1 = 0, \quad t \overset{\text{not.}}{=} e^z \Leftrightarrow t_{1,2} = \frac{i \pm \sqrt{3}}{2}$$
$$\Leftrightarrow z \in \operatorname{Log}\frac{i \pm \sqrt{3}}{2} = \left\{\left(\frac{\pi}{6} + 2k\pi\right)i : k \in \mathbb{Z}\right\} \cup \left\{\left(\frac{5\pi}{6} + 2k\pi\right)i : k \in \mathbb{Z}\right\}.$$

6. Since

$$\tan z = \frac{\sin z}{\cos z} = \frac{e^{iz} - e^{-iz}}{i(e^{iz} + e^{-iz})},$$

we deduce that

$$\tan z = \frac{5i}{3} \Leftrightarrow 4t^2 + 1 = 0, \quad t \overset{\text{not.}}{=} e^{iz} \Leftrightarrow t_{1,2} = \pm\frac{i}{2}$$
$$\Leftrightarrow z \in -i\mathrm{Log}\left(\pm\frac{i}{2}\right) = \left\{\pm\frac{\pi}{2} + 2k\pi + i\ln 2 : k \in \mathbb{Z}\right\}.$$

7. We may easily obtain that

$$e^{i3z} = -1 \Leftrightarrow 3iz \in \mathrm{Log}\,(-1) \Leftrightarrow z \in -\frac{i}{3}\mathrm{Log}\,(-1) = \left\{(2k-1)\frac{\pi}{3} : k \in \mathbb{Z}\right\}.$$

8. A simple computation shows that

$$e^{\frac{1}{z^2}} = 1 \Leftrightarrow \frac{1}{z^2} \in \mathrm{Log}\,1 = \{2k\pi i : k \in \mathbb{Z}\} \Leftrightarrow \frac{1}{z^2} = 2k\pi i, \quad k \in \mathbb{Z}$$
$$\Leftrightarrow z_k = \frac{1}{\sqrt{2k\pi i}}, \quad k \in \mathbb{Z}^*$$
$$\Leftrightarrow z \in \left\{\frac{e^{-i(\frac{\pi}{4}+l\pi)}}{\sqrt{2k\pi}} : k \in \mathbb{Z},\, k > 0,\, l \in \{0,1\}\right\} \cup \left\{\frac{e^{i(\frac{\pi}{4}+l\pi)}}{\sqrt{-2k\pi}} : k \in \mathbb{Z},\, k < 0,\, l \in \{0,1\}\right\}.$$

9. We easily deduce the following equivalence:

$$\tanh z = 2 \Leftrightarrow \frac{e^z - e^{-z}}{e^z + e^{-z}} = 2 \Leftrightarrow e^{2z} = -3$$
$$\Leftrightarrow z \in \frac{1}{2}\mathrm{Log}\,(-3) \Leftrightarrow z \in \left\{\frac{\ln 3}{2} + i(2k+1)\frac{\pi}{2} : k \in \mathbb{Z}\right\}.$$

Solution of Exercise 2.9.15

1. $e^i = \cos 1 + i\sin 1.$
2. $\sinh 2i = \frac{e^{2i}-e^{-2i}}{2} = i\sin 2.$
3. $\cosh(2+3i) = \frac{e^{2+3i}+e^{-2-3i}}{2} = \frac{e^2+e^{-2}}{2}\cos 3 + i\frac{e^2-e^{-2}}{2}\sin 3 = \cosh 2\cos 3 + i\sinh 2\sin 3.$
4. $\cos(1-i) = \frac{e^{1+i}+e^{-1-i}}{2} = \frac{e+e^{-1}}{2}\cos 1 + i\frac{e-e^{-1}}{2}\sin 1 = \cosh 1\cos 1 + i\sinh 1\sin 1.$
5. $\tan(1-2i) = \frac{\sin(1-2i)}{\cos(1-2i)} = \frac{\sin 1\cosh 2 - i\cos 1\sinh 2}{\cos 1\cosh 2 + i\sin 1\sinh 2} = \frac{\sin 1\cos 1 - i\sinh 2\cosh 2}{\cos^2 1\cosh^2 2 + \sin^2 1\sinh^2 2}.$
6. $\log(-2i) = \ln 2 - \frac{\pi}{2}i.$
7. $\log(-3+4i) = \ln 5 + i(\pi - \arctan\frac{4}{3}).$
8. Since

$$\zeta_k = \frac{1-i}{\sqrt{3+i}} = \sqrt[4]{\frac{2}{5}}\, e^{-i(\frac{\pi}{4}+\frac{\theta}{2}+k\pi)}, \quad k \in \{0,1\}, \text{ where } \theta = \arctan\frac{1}{3},$$

it follows that

$$\log\zeta_k = \frac{1}{4}\ln\frac{2}{5} - i\left(\frac{\pi}{4} + \frac{\theta}{2} + k\pi\right), \quad k \in \{0,1\}.$$

Solution of Exercise 2.9.16

1. $i^{1-i} = e^{(1-i)\mathrm{Log}\, i} = e^{(1-i)\{i(\frac{\pi}{2}+2k\pi):k\in\mathbb{Z}\}} = e^{\{(1+i)(\frac{\pi}{2}+2k\pi):k\in\mathbb{Z}\}} = \{e^{(1+i)(\frac{\pi}{2}+2k\pi)} : k \in \mathbb{Z}\} = \{ie^{\frac{\pi}{2}+2k\pi} : k \in \mathbb{Z}\}$.

2. $(1+i\sqrt{3})^i = e^{i\mathrm{Log}\,(1+i\sqrt{3})} = e^{i\{\ln 2+i(\frac{\pi}{3}+2k\pi):k\in\mathbb{Z}\}} = e^{\{-(\frac{\pi}{3}+2k\pi)+i\ln 2:k\in\mathbb{Z}\}} = \{e^{-(\frac{\pi}{3}+2k\pi)+i\ln 2} : k \in \mathbb{Z}\} = \{e^{-(\frac{\pi}{3}+2k\pi)}(\cos\ln 2 + i\sin\ln 2) : k \in \mathbb{Z}\}$.

3. $1^{-i} = e^{-i\mathrm{Log}\, 1} = e^{-i\{2k\pi i:k\in\mathbb{Z}\}} = e^{\{2k\pi:k\in\mathbb{Z}\}} = \{e^{2k\pi} : k \in \mathbb{Z}\}$.

4. $e^{\sqrt{i}} = e^{\{\cos(\frac{\pi}{4}+k\pi)+i\sin(\frac{\pi}{4}+k\pi):k\in\mathbb{Z}\}} = \{e^{\cos(\frac{\pi}{4}+k\pi)+i\sin(\frac{\pi}{4}+k\pi)} : k \in \mathbb{Z}\} = \{e^{\cos(\frac{\pi}{4}+k\pi)} \cdot (\cos\sin(\frac{\pi}{4} + k\pi) + i\sin\sin(\frac{\pi}{4} + k\pi)) : k \in \mathbb{Z}\}$.

Solution of Exercise 2.9.17

1. $-i\sinh(iz) = -i\frac{e^{iz}-e^{-iz}}{2} = \frac{e^{iz}-e^{-iz}}{2i} = \sin z$ and $\sin z = -i\sinh(iz) \Leftrightarrow \sinh(iz) = i\sin z$.

2. $\cos(iz) = \frac{e^{-z}+e^{z}}{2} = \cosh z$ and $\cos(i(iz)) = \cosh(iz) \Leftrightarrow \cos z = \cosh(iz)$.

3. $\sinh z_1 \cosh z_2 + \cosh z_1 \sinh z_2 = \frac{e^{z_1}-e^{-z_1}}{2}\frac{e^{z_2}+e^{-z_2}}{2} + \frac{e^{z_1}+e^{-z_1}}{2}\frac{e^{z_2}-e^{-z_2}}{2} = \frac{e^{z_1+z_2}-e^{-(z_1+z_2)}}{2} = \sinh(z_1 + z_2)$ and $\sinh z_1 \cosh z_2 - \cosh z_1 \sinh z_2 = \frac{e^{z_1}-e^{-z_1}}{2}\frac{e^{z_2}+e^{-z_2}}{2} - \frac{e^{z_1}+e^{-z_1}}{2}\frac{e^{z_2}-e^{-z_2}}{2} = \frac{e^{z_1-z_2}-e^{-(z_1-z_2)}}{2} = \sinh(z_1 - z_2)$.

4. $\cos z_1 \cos z_2 - \sin z_1 \sin z_2 = \frac{e^{iz_1}+e^{-iz_1}}{2}\frac{e^{iz_2}+e^{-iz_2}}{2} - \frac{e^{iz_1}-e^{-iz_1}}{2i}\frac{e^{iz_2}-e^{-iz_2}}{2i} = \frac{e^{i(z_1+z_2)}+e^{-i(z_1+z_2)}}{2} = \cos(z_1 + z_2)$ and $\cos z_1 \cos z_2 + \sin z_1 \sin z_2 = \frac{e^{iz_1}+e^{-iz_1}}{2}\frac{e^{iz_2}+e^{-iz_2}}{2} + \frac{e^{iz_1}-e^{-iz_1}}{2i}\frac{e^{iz_2}-e^{-iz_2}}{2i} = \frac{e^{i(z_1-z_2)}+e^{-i(z_1-z_2)}}{2} = \cos(z_1 - z_2)$.

5. $2\sin z\cos z = 2\frac{e^{iz}-e^{-iz}}{2i}\frac{e^{iz}+e^{-iz}}{2} = \frac{e^{i2z}-e^{-i2z}}{2i} = \sin 2z$.

6. $\frac{2\tan z}{1-\tan^2 z} = \frac{2\frac{e^{iz}-e^{-iz}}{i(e^{iz}+e^{-iz})}}{1-(\frac{e^{iz}-e^{-iz}}{i(e^{iz}-e^{-iz})})^2} = \frac{e^{i2z}-e^{-i2z}}{i(e^{i2z}+e^{-i2z})} = \tan 2z$.

7. $\sinh^2 z = (\frac{e^{z}-e^{-z}}{2})^2 = \frac{e^{2z}+e^{-2z}-2}{4} = \frac{\frac{e^{2z}+e^{-2z}}{2}-1}{2} = \frac{\cosh 2z-1}{2}$ and $\cosh^2 z = (\frac{e^{z}+e^{-z}}{2})^2 = \frac{e^{2z}+e^{-2z}+2}{4} = \frac{\frac{e^{2z}+e^{-2z}}{2}+1}{2} = \frac{\cosh 2z+1}{2}$.

8. Since

$$\sin z = \sin x\cosh y + i\cos x\sinh y, \quad z = x + iy,$$

it follows that

$$\begin{aligned}|\sin z|^2 &= \sin^2 x\cosh^2 y + \cos^2 x\sinh^2 y = (1-\cos^2 x)\cosh^2 y + \cos^2 x\sinh^2 y\\ &= \cosh^2 y - \cos^2 x(-\sinh^2 y + \cosh^2 y) = \cosh^2 y - \cos^2 x\\ &\overset{7.}{=} \frac{\cosh 2y + 1}{2} - \frac{\cos 2x + 1}{2} = \frac{1}{2}(\cosh 2y - \cos 2x),\end{aligned}$$

and thus $|\sin z| = \sqrt{\frac{1}{2}(\cosh 2y - \cos 2x)}$.

Similarly,

$$|\sin z|^2 = \sin^2 x \cosh^2 y + \cos^2 x \sinh^2 y$$
$$= \sin^2 x(1+\sinh^2 y) + (1-\sin^2 x)\sinh^2 y = \sin^2 x + \sinh^2 y,$$

and thus $|\sin z| = \sqrt{\sin^2 x + \sinh^2 y}$.

9. Since

$$\cos z = \cos x \cosh y + i \sin x \sinh y, \quad z = x + iy,$$

it follows that

$$|\cos z|^2 = \cos^2 x \cosh^2 y + \sin^2 x \sinh^2 y = \cos^2 x(1+\sinh^2 y) + \sin^2 x \sinh^2 y$$
$$= \cos^2 x + \sinh^2 y \overset{7.}{=} \frac{\cos 2x + 1}{2} + \frac{\cosh 2y - 1}{2} = \frac{1}{2}(\cosh 2y + \cos 2x),$$

hence $|\cos z| = \sqrt{\cos^2 x + \sinh^2 y}$ and $|\cos z| = \sqrt{\frac{1}{2}(\cosh 2y - \cos 2x)}$.

10. $\tan(iz) = \frac{\sin(iz)}{\cos(iz)} = \frac{e^{-z}-e^{z}}{i(e^{-z}+e^{z})} = i\frac{e^{z}-e^{-z}}{e^{z}+e^{-z}} = i\tanh z$ and $\tan z = -\tan(i(iz)) = -i\tanh(iz)$.

11. Using the result of the point 7, we obtain that $\cosh^2 z + \sinh^2 z = \frac{\cosh 2z+1}{2} + \frac{\cosh 2z-1}{2} = \cosh 2z$ and $\cosh^2 z + \sinh^2 z = \frac{\cosh^2 z(1+\frac{\sinh^2 z}{\cosh^2 z})}{\cosh^2 z-\sinh^2 z} = \frac{\cosh^2 z(1+\frac{\sinh^2 z}{\cosh^2 z})}{\cosh^2 z(1-\frac{\sinh^2 z}{\cosh^2 z})} = \frac{1+\tanh^2 z}{1-\tanh^2 z}$.

Solution of Exercise 2.9.18

Since we know that

$$(z_1, z_2, z_3, z_4) = (T(z_1), T(z_2), T(z_3), T(z_4)),$$

denoting by $z_1 = z$, $T(z) = w$ and $T(z_j) = w_j$, $j = 2, 3, 4$, we obtain

$$(z, 1+i, \infty, 1) = (w, i, -1, \infty) \Leftrightarrow \frac{z-1-i}{z-1} : \frac{\infty}{\infty} = \frac{w-i}{\infty} : \frac{-1-i}{\infty}$$
$$\Leftrightarrow \frac{z-1-i}{z-1} = \frac{w-i}{-1-i} \Leftrightarrow w = \frac{z-i}{-z+1}.$$

From $w'(z) = \frac{1-i}{(-z+1)^2} \neq 0$, $\forall z \in \mathbb{C} \setminus \{1\}$, we obtain that the function w is a conformal mapping.

Further, we have

$$w(1) = \infty, \quad w(i) = 0, \quad w(1+2i) = -\frac{1}{2} + \frac{1}{2}i,$$

where $\{1, i, 1+2i\} \subset \{z = x+iy \in \mathbb{C} : (x-1)^2 + (y-1)^2 = 1\}$, hence

$$w(\{z = x+iy \in \mathbb{C} : (x-1)^2 + (y-1)^2 = 1\}) = \{w = u+iv \in \mathbb{C} : u+v = 0\}.$$

Since

$$1+i \in \{z = x+iy \in \mathbb{C} : (x-1)^2 + (y-1)^2 < 1\}$$

and

$$w(1+i) = i \in \{w = u+iv \in \mathbb{C} : u+v > 0\},$$

it follows that

$$w(\{z = x+iy \in \mathbb{C} : (x-1)^2 + (y-1)^2 < 1\}) = \{w = u+iv \in \mathbb{C} : u+v > 0\}.$$

Solution of Exercise 2.9.19

We will determine the inverse of w, and then we will prove that it maps the domain $\{w \in \mathbb{C} : |w| < 1\}$ onto the domain $\{z \in \mathbb{C} : \operatorname{Re} z > 0\}$, which will prove the required result.

A simple computation shows that $g(z) = w^{-1}(z) = -\frac{z+1}{z-1}$, hence

$$g(e^{i\theta}) = i\frac{\sin\theta}{1-\cos\theta} = u(\theta) + iv(\theta) \Leftrightarrow \begin{cases} u = u(\theta) = 0 \\ v = v(\theta) = \frac{\sin\theta}{1-\cos\theta}, \end{cases} \quad \theta \in (0, 2\pi).$$

From $v'(\theta) = \frac{1}{\cos\theta - 1} < 0$, we have that v is a continuous decreasing function, and since

$$\lim_{\theta \downarrow 0} v(\theta) = +\infty, \quad \lim_{\theta \uparrow 2\pi} v(\theta) = -\infty,$$

we deduce $g(\{w \in \mathbb{C} : |w| = 1\}) = \{z \in \mathbb{C} : \operatorname{Re} z = 0\}$.

Since $g(0) = 1 \in \{z \in \mathbb{C} : \operatorname{Re} z > 0\}$, we conclude that

$$g(\{w \in \mathbb{C} : |w| < 1\}) = \{z \in \mathbb{C} : \operatorname{Re} z > 0\},$$

i. e.,

$$w(\{z \in \mathbb{C} : \operatorname{Re} z > 0\}) = \{w \in \mathbb{C} : |w| < 1\}.$$

From $w'(z) = \frac{2}{(z+1)^2} \neq 0$, $\forall z \in \mathbb{C} \setminus \{-1\}$, the function w is a conformal mapping.

Solution of Exercise 2.9.20

Let us denote by $D = \{z \in \mathbb{C} : \operatorname{Re} z < 1\}$ and by $\Delta = \{w \in \mathbb{C} : |w| < R\}$. It is well known that a bilinear transform maps two inverse points with respect of a circle from $\widehat{\mathbb{C}}$, into two inverse points with respect to the image of the given circle.

From here, since $z_1 = -1$ and $z_2 = 3$ are inverse points with respect to the circle $\partial D = \{z \in \mathbb{C} : \operatorname{Re} z = 1\}$, it follows that $w_1 = f(z_1) = 0$ and $w_2 = f(z_2)$ are inverse points with respect to the circle $\partial\Delta = \{w \in \mathbb{C} : |w| = R\}$, hence we deduce that $w_2 = f(z_2) = \infty$.

In order to determine the required function, we need to know the image of three points.

From the assumptions $f(-1) = 0$ and $f(3) = \infty$, it follows that

$$f(z) = k\frac{z+1}{z-3}, \quad k \in \mathbb{C},$$

and by using the assumption $f'(-1) = 1$ we deduce that $f'(-1) = \frac{-k}{4} = 1 \Leftrightarrow k = -4$. Hence, our function is

$$w(z) = -4\frac{z+1}{z-3}.$$

To determine the value of the radius R, it is sufficient to find the module of the image of a chosen point. Since $1 \in \partial D$, it follows that $f(1) \in \partial\Delta$, hence $|f(1)| = R \Leftrightarrow R = 4$.

Since $f'(z) = \frac{16}{(z-3)^2} \neq 0$, $\forall z \in \mathbb{C} \setminus \{3\}$, the function f is a conformal mapping.

Solution of Exercise 2.9.21

Since f is a bilinear transform, the problem will be solved by using the well-known previous technique: we will choose three distinct points on the boundary of the given domain (which is a circle), and the circle determined by its images will be the boundary of the image.

Thus, $f(1) = \beta$, $f(i) = \beta + i(\beta - 1)$, $f(-i) = \beta + i(1-\beta)$ and $f(0) = 1$. From here, since $f'(z) = \frac{2(\beta-1)}{(z+1)^2} \neq 0$, $\forall z \in \mathbb{C} \setminus \{-1\}$, it follows that the function f conformally maps the disc $\{z \in \mathbb{C} : |z| < 1\}$ onto the half-plane $\{w \in \mathbb{C} : \operatorname{Re} w > \beta\}$.

Solution of Exercise 2.9.22

Suppose that $\exists z_1, z_2 \in \{z \in \mathbb{C} : |z| < \frac{1}{2|a|}\}$ such that $f(z_1) = f(z_2)$. Then

$$f(z_1) = f(z_2) \Leftrightarrow (z_1 - z_2)(1 + az_1 + az_2) = 0, \quad \text{where } z_1, z_2 \in U\left(0; \frac{1}{2|a|}\right).$$

Using the property $\operatorname{Re}\zeta \geq -|\zeta|$, $\zeta \in \mathbb{C}$, we deduce that

$$\operatorname{Re}(1 + az_1 + az_2) \geq 1 - |a||z_1| - |a||z_2| > 1 - \frac{1}{2} - \frac{1}{2} = 0, \quad \forall z_1, z_2 \in U\left(0; \frac{1}{2|a|}\right),$$

thus $1 + az_1 + az_2 \neq 0$, $\forall z_1, z_2 \in U(0; \frac{1}{2|a|})$, hence the function f is injective on the disc $\{z \in \mathbb{C} : |z| < \frac{1}{2|a|}\}$.

Solution of Exercise 2.9.23

We will prove the injectivity of the w function. The function w is injective in the set $D \subset \mathbb{C}$, if and only if

$$w(z_1) = w(z_2), \quad z_1, z_2 \in D \Leftrightarrow z_1 = z_2.$$

Since

$$\begin{aligned} w(z_1) = w(z_2), \quad z_1, z_2 \in D &\Leftrightarrow (z_1 - z_2)(5 + z_1 + z_2) = 0, \quad z_1, z_2 \in D \\ &\Leftrightarrow z_1 - z_2 = 0 \quad \text{or} \quad 5 + z_1 + z_2 = 0. \end{aligned}$$

If $D = \{z \in \mathbb{C} : 0 < \operatorname{Re} z < 1, \operatorname{Im} z > 0\}$, then $\forall z_1, z_2 \in D \Rightarrow \operatorname{Re} z_1 > 0, \operatorname{Re} z_2 > 0 \Rightarrow \operatorname{Re}(5 + z_1 + z_2) > 5$, hence $5 + z_1 + z_2 \neq 0$, $\forall z_1, z_2 \in D$.

If $D = \{z \in \mathbb{C} : \operatorname{Re} z > -\frac{5}{2}\}$, then $\forall z_1, z_2 \in D \Leftrightarrow \operatorname{Re} z_1 > -\frac{5}{2}, \operatorname{Re} z_2 > -\frac{5}{2} \Rightarrow \operatorname{Re}(5 + z_1 + z_2) > 0$, hence $5 + z_1 + z_2 \neq 0$, $\forall z_1, z_2 \in D$.

If $D = \{z \in \mathbb{C} : \operatorname{Re} z < -\frac{5}{2}\}$, then $\forall z_1, z_2 \in D \Leftrightarrow \operatorname{Re} z_1 < -\frac{5}{2}, \operatorname{Re} z_2 < -\frac{5}{2}, \Rightarrow \operatorname{Re}(5 + z_1 + z_2) < 0$, hence $5 + z_1 + z_2 \neq 0$, $\forall z_1, z_2 \in D$.

It follows that the function w is injective in all the three cases.

1. Let $z = x + iy \in \partial D = d_1 \cup d_2 \cup d_3$, where

$$d_1 : \begin{cases} x = 0 \\ y = t, \quad t \in [0, +\infty) \end{cases} \qquad d_2 : \begin{cases} x = 1 \\ y = t, \quad t \in [0, +\infty) \end{cases} \qquad d_3 : \begin{cases} x = t \\ y = 0, \quad t \in [0, 1]. \end{cases}$$

Then

$$\forall z \in d_1 \Leftrightarrow z = it, \quad t \in [0, +\infty) \Rightarrow w(z) = 6 - t^2 + 5it, \quad t \in [0, +\infty).$$

Hence, if $d_1' = w(d_1)$, then

$$d_1' : \begin{cases} u = 6 - t^2 \\ v = 5t, \quad t \in [0, +\infty) \end{cases} \Leftrightarrow u = 6 - \frac{v^2}{25}, \quad v \in [0, +\infty).$$

Now,

$$\forall z \in d_2 \Leftrightarrow z = 1 + it, \quad t \in [0, +\infty) \Rightarrow w(z) = 12 - t^2 + 7it, \quad t \in [0, +\infty).$$

Hence, if $d_2' = w(d_2)$, then

$$d_2' : \begin{cases} u = 12 - t^2 \\ v = 7t, \quad t \in [0, +\infty) \end{cases} \Leftrightarrow u = 12 - \frac{v^2}{49}, \quad v \in [0, +\infty).$$

Similarly, we will determine the image of d_3 as follows:

$$\forall z \in d_3 \Leftrightarrow z = t, \quad t \in [0,1] \Rightarrow w(z) = t^2 + 5t + 6, \quad t \in [0,1].$$

Hence, if $d_3' = w(d_3)$, then

$$d_3' : \begin{cases} u = t^2 + 5t + 6 \\ v = 0, \quad t \in [0,1] \end{cases} \Leftrightarrow u = 6t + 6, \quad t \in [0,1], \quad v = 0.$$

Since $z_0 = \frac{1}{2} + \frac{i}{2} \in D \Rightarrow w(z_0) = \frac{17}{2} + 3i \in w(D)$, we deduce that $w(D)$ is the subset of $\mathbb{C}$ that contains the point $M(\frac{17}{2} + 3i)$ and has the boundary $\partial w(D) = d_1' \cup d_2' \cup d_3'$.

2. Let $z = x + iy \in \partial D = d$, where

$$d : \begin{cases} x = -\frac{5}{2} \\ y = t, \quad t \in \mathbb{R}. \end{cases}$$

We have

$$\forall z \in d \Leftrightarrow z = -\frac{5}{2} + it, \quad t \in \mathbb{R} \Rightarrow w(z) = -t^2 - \frac{1}{4}, \quad t \in \mathbb{R}.$$

Hence, if $d' = w(d)$, then

$$d' : \begin{cases} u = -t^2 - \frac{1}{4} \\ v = 0, \quad t \in \mathbb{R}, \end{cases}$$

thus $w(D) = \mathbb{C} \setminus \{w \in \mathbb{C} : \operatorname{Im} w = 0, \operatorname{Re} w \le -\frac{1}{4}\} = E$.

It follows that the function $w : D \to E$, $w(z) = z^2 + 5z + 6$ is bijective, and then

$$\forall w \in E, \quad w(z) = w \Leftrightarrow z = \sqrt{w + \frac{1}{4}} - \frac{5}{2}.$$

From here, since $w(-2) = 0$, we deduce that

$$w^{-1} : E \to D, \quad w^{-1}(w) = \sqrt{w + \frac{1}{4}} - \frac{5}{2}, \quad \text{where } w^{-1}(0) = -2.$$

3. Similarly, since the function w can be written as $w(z) = (z - \frac{5}{2})^2 - \frac{1}{4}$, all the computations are the same like in the previous point.

Hence $w(D) = \mathbb{C} \setminus \{w \in \mathbb{C} : \operatorname{Im} w = 0, \operatorname{Re} w \le -\frac{1}{4}\} = E$ and

$$w^{-1} : E \to D, \quad w^{-1}(w) = \sqrt{w + \frac{1}{4}} - \frac{5}{2}, \quad \text{where } w^{-1}(0) = -3.$$

Solution of Exercise 2.9.24

Let $D = \{z \in \mathbb{C} : 0 < \operatorname{Re} z < a\}$, $a > 0$, and $\Delta = \{w \in \mathbb{C} : \operatorname{Im} w > 0\}$. It is evident that:
(i) If $g(z) = iz$, then $g(D) = \{z \in \mathbb{C} : 0 < \operatorname{Im} z < a\} = \Delta_1$;
(ii) If $h(z) = \frac{\pi}{a} z$, then $h(\Delta_1) = \{z \in \mathbb{C} : 0 < \operatorname{Im} z < \pi\} = \Delta_2$;
(iii) If $k(z) = e^z$, then $k(\Delta_2) = \Delta$.

From here, it follows that the function will be

$$f(z) = (k \circ h \circ g)(z) = e^{\frac{\pi i}{a} z},$$

because $f(D) = \Delta$.

7.3 Solutions to the exercises of Chapter 3

Solution of Exercise 3.10.1

In each of the above cases, first we will determine the parametric equations of the path, then by applying the definition of the complex integral we will compute the given integral.

1. From the equation of the circle,

$$C(O;1) : |z| = 1 \Leftrightarrow z = e^{i\theta}, \quad \theta \in [0, 2\pi],$$

it follows that

$$\widehat{AB} : z = e^{i\theta}, \quad \theta \in \left[-\frac{\pi}{2}, \frac{\pi}{2}\right].$$

Letting $\theta(t) = a + bt$, $t \in [0,1]$, then

$$\begin{cases} \theta(0) = -\frac{\pi}{2} \\ \theta(1) = \frac{\pi}{2} \end{cases} \Leftrightarrow \begin{cases} a = -\frac{\pi}{2} \\ b = \pi \end{cases} \Leftrightarrow \theta(t) = \frac{\pi}{2}(2t-1), \quad t \in [0,1],$$

and the parametric equation of the path is

$$\widehat{AB} : \gamma(t) = e^{i\frac{\pi}{2}(2t-1)}, \quad t \in [0,1].$$

Then the value of the integral is

$$I = \int_{\gamma} \frac{1}{z} \mathrm{d}z = \int_0^1 \frac{1}{\gamma(t)} \gamma'(t) \mathrm{d}t = \pi i.$$

2. The equation of the arc is

$$\widehat{AB} : z = e^{i\theta}, \quad \theta \in \left[-\frac{\pi}{2}, -\frac{3\pi}{2}\right].$$

Letting $\theta(t) = a + bt$, $t \in [0,1]$, then

$$\begin{cases} \theta(0) = -\frac{\pi}{2} \\ \theta(1) = -\frac{3\pi}{2} \end{cases} \Leftrightarrow \begin{cases} a = -\frac{\pi}{2} \\ b = -\pi \end{cases} \Leftrightarrow \theta(t) = -\frac{\pi}{2}(2t+1), \quad t \in [0,1].$$

Hence the parametric equation of the path is

$$\widehat{AB} : \gamma(t) = e^{-i\frac{\pi}{2}(2t+1)}, \quad t \in [0,1],$$

and the integral is given by

$$I = \int_\gamma \frac{1}{z}\mathrm{d}z = \int_0^1 \frac{1}{\gamma(t)}\gamma'(t)\mathrm{d}t = -\pi i.$$

3. Since

$$C(O; \sqrt{2}) : |z| = \sqrt{2} \Leftrightarrow z = \sqrt{2}e^{i\theta}, \quad \theta \in [0, 2\pi],$$

we deduce that

$$\widehat{CD} : z = \sqrt{2}e^{i\theta}, \quad \theta \in \left[-\frac{\pi}{4}, \frac{\pi}{4}\right].$$

Letting $\theta(t) = a + bt$, $t \in [0,1]$, then

$$\begin{cases} \theta(0) = -\frac{\pi}{4} \\ \theta(1) = \frac{\pi}{4} \end{cases} \Leftrightarrow \begin{cases} a = -\frac{\pi}{4} \\ b = \frac{\pi}{2} \end{cases} \Leftrightarrow \theta(t) = \frac{\pi}{4}(2t-1), \quad t \in [0,1].$$

The parametric equation of the path is

$$\widehat{CD} : \gamma(t) = \sqrt{2}e^{i\frac{\pi}{4}(2t-1)}, \quad t \in [0,1],$$

hence

$$I = \int_\gamma \frac{1}{z}\mathrm{d}z = \int_0^1 \frac{1}{\gamma(t)}\gamma'(t)\mathrm{d}t = \frac{\pi}{2}i.$$

4. Since the equation of the path is

$$\overline{CD} : \gamma(t) = 1 - i + 2it, \quad t \in [0,1],$$

it follows that

$$\begin{aligned} I &= \int_\gamma \frac{1}{z}\mathrm{d}z = \int_0^1 \frac{1}{\gamma(t)}\gamma'(t)\mathrm{d}t = \int_0^1 \frac{1}{t + \frac{1-i}{2i}}dt \\ &= \int_0^1 \frac{t - \frac{1}{2}}{t^2 - t + \frac{1}{2}}\mathrm{d}t + \frac{i}{2}\int_0^1 \frac{1}{t^2 - t + \frac{1}{2}}\mathrm{d}t = \frac{\pi}{2}i. \end{aligned}$$

Solution of Exercise 3.10.2

Since $\gamma = \gamma_1 \cup \gamma_2 \cup \gamma_3$, it follows that

$$I = \int_\gamma \frac{z}{|z|}\mathrm{d}z = \int_{\gamma_1} \frac{z}{|z|}\mathrm{d}z + \int_{\gamma_2} \frac{z}{|z|}\mathrm{d}z + \int_{\gamma_3} \frac{z}{|z|}\mathrm{d}z.$$

Now we will compute the value of each of these integrals.

Since

$$\gamma_1(t) = t - 3, \quad t \in [0,1],$$

then

$$I_1 = \int_{\gamma_1} \frac{z}{|z|}\mathrm{d}z = \int_0^1 \frac{\gamma_1(t)}{|\gamma_1(t)|}\gamma_1'(t)\mathrm{d}t = -1.$$

For the path γ_2, we have

$$C(O;2) : |z| = 2 \Leftrightarrow z = 2e^{i\theta}, \quad \theta \in [0, 2\pi],$$

then

$$\gamma_2 : z = 2e^{i\theta}, \quad \theta \in [\pi, 0].$$

Letting $\theta(t) = a + bt$, $t \in [0,1]$, we get

$$\begin{cases} \theta(0) = \pi \\ \theta(1) = 0 \end{cases} \Leftrightarrow \begin{cases} a = \pi \\ b = -\pi \end{cases} \Leftrightarrow \theta(t) = -\pi(t-1), \quad t \in [0,1],$$

We conclude that the path γ_2 has the parametric equation

$$\gamma_2(t) = 2e^{-i\pi(t-1)}, \quad t \in [0,1],$$

and the second integral is

$$I_2 = \int_{\gamma_2} \frac{z}{|z|}\mathrm{d}z = \int_0^1 \frac{\gamma_2(t)}{|\gamma_2(t)|}\gamma_2'(t)\mathrm{d}t = 0.$$

For the path γ_3, we have

$$\gamma_3(t) = t + 2, \quad t \in [0,1],$$

and the value of the third integral is

$$I_3 = \int_{\gamma 3} \frac{z}{|z|}\mathrm{d}z = \int_0^1 \frac{\gamma_3(t)}{|\gamma_3(t)|}\gamma_3'(t)\mathrm{d}t = 1.$$

Now, we conclude that

$$I = I_1 + I_2 + I_3 = 0.$$

Solution of Exercise 3.10.3

From the definition of the multivalued function Log , i. e.,

$$\operatorname{Log} z = \{\ln|z| + i(\arg z + 2k\pi) : k \in \mathbb{Z}\},$$

it is easy to determine the main branch log, since

$$\log 1 = 0 \Rightarrow \log z = \ln|z| + i\arg z.$$

The integral will be

$$\int_\gamma \log z\mathrm{d}z = \int_0^1 \log\gamma(t)\gamma'(t)\mathrm{d}t = \int_0^1 i(2t-1)\frac{\pi}{2}i\pi e^{i(2t-1)\frac{\pi}{2}}\,\mathrm{d}t = -2i,$$

where we used that

$$\log\gamma(t) = i(2t-1)\frac{\pi}{2}, \quad t \in [0,1].$$

Solution of Exercise 3.10.4

Using the definition of the multivalued function Log , i. e.,

$$\operatorname{Log} z = \{\ln|z| + i(\arg z + 2k\pi) : k \in \mathbb{Z}\},$$

since $\sqrt{z} = e^{\frac{1}{2}\operatorname{Log} z}$, from the condition $\sqrt{1} = 1$ we get $k = 0$, and thus

$$\sqrt{z} = \sqrt{|z|}e^{i\frac{\arg z}{2}}, \quad \arg z \in (-\pi,\pi).$$

The value of the integral is

$$\int_\gamma \frac{1}{\sqrt{z}}\mathrm{d}z = \int_0^1 \frac{1}{\sqrt{\gamma(t)}}\gamma'(t)\mathrm{d}t = \int_0^1 \frac{\frac{\pi i}{2}e^{i\frac{\pi t}{2}}}{e^{i\frac{\pi t}{4}}}\mathrm{d}t = \sqrt{2} - 2 + i\sqrt{2},$$

where we used the fact that

$$\sqrt{\gamma(t)} = e^{i\frac{\pi t}{4}}, \quad t \in [0,1].$$

Solution of Exercise 3.10.5

1. From the definition,

$$\operatorname{Log} z = \{\ln|z| + i(\arg z + 2k\pi) : k \in \mathbb{Z}\},$$

and $\sqrt{z} = e^{\frac{1}{2}\operatorname{Log} z}$, since $\sqrt{1} = 1$ it follows that $k = 0$. Then the main branch of the root function is given by

$$\sqrt{z} = \sqrt{|z|}e^{i\frac{\arg z}{2}}, \quad z \in \mathbb{C}^*.$$

The parametric equation of the path is

$$\gamma(t) = r^2 e^{2\pi i t}, \quad t \in [0,1],$$

and the integral will be computed as follows:

$$\int_\gamma \frac{1}{z\sqrt{z}}\mathrm{d}z = \int_0^1 \frac{1}{\gamma(t)\sqrt{\gamma(t)}}\gamma'(t)\mathrm{d}t = \frac{2}{r}\int_0^1 \frac{\pi i}{e^{\pi i t}}\mathrm{d}t = \frac{4}{r},$$

where we used the fact that

$$\sqrt{\gamma(t)} = re^{i\pi t}, \quad t \in [0,1].$$

2. Since $\sqrt{1} = -1$, it follows that the corresponding branch of the root function has the form

$$\sqrt{z} = \sqrt{|z|}e^{i\frac{\arg z + 2\pi}{2}}, \quad z \in \mathbb{C}^*.$$

The integral is given by

$$\int_\gamma \frac{1}{z\sqrt{z}}\mathrm{d}z = \int_0^1 \frac{1}{\gamma(t)\sqrt{\gamma(t)}}\gamma'(t)\mathrm{d}t = \frac{4}{r}\int_0^1 \frac{\pi i}{e^{\pi i(2t+1)}}\mathrm{d}t = 0,$$

where we used the relation

$$\sqrt{\gamma(t)} = re^{i\pi(2t+1)}, \quad t \in [0,1].$$

Solution of Exercise 3.10.6

Like in the previous problem, since the parametric equations of the path are given, we will use the definition of the complex integral to compute the given integral.

1. Since $\gamma(t) = 2t - 1$, $t \in [0,1]$, it follows that

$$\int_\gamma |z|\mathrm{d}z = \int_0^1 2|2t-1|\mathrm{d}t = 1.$$

2. If $\gamma(t) = e^{\pi i t}$, $t \in [0,1]$, then

$$\int_{\gamma} |z| \mathrm{d}z = \int_{0}^{1} |\gamma(t)| \gamma'(t) \mathrm{d}t = \int_{0}^{1} \pi i e^{i\pi t} \mathrm{d}t = -2.$$

3. If $\gamma(t) = e^{\pi i (t-1)}$, $t \in [0,1]$, then

$$\int_{\gamma} |z| \mathrm{d}z = \int_{0}^{1} |\gamma(t)| \gamma'(t) \mathrm{d}t = \int_{0}^{1} \pi i e^{i\pi (t-1)} \mathrm{d}t = 2.$$

Solution of Exercise 3.10.7

We will use the definition of the complex integral to compute the below integrals.

1.

$$\int_{\gamma} \bar{z} \mathrm{d}z = \int_{0}^{1} \overline{\gamma(t)} \gamma'(t) \mathrm{d}t = \int_{0}^{1} 2(2t-1) \mathrm{d}t = 0.$$

2.

$$\int_{\gamma} \bar{z} \mathrm{d}z = \int_{0}^{1} \overline{\gamma(t)} \gamma'(t) \mathrm{d}t = \int_{0}^{1} \pi i e^{-i\frac{(2t-1)\pi}{2}} e^{i\frac{(2t-1)\pi}{2}} \mathrm{d}t = i\pi.$$

3. The parametric equation of the path is

$$\gamma(t) = e^{2\pi i t}, \quad t \in [0,1],$$

and thus

$$\int_{\gamma} \bar{z} \mathrm{d}z = \int_{0}^{1} \overline{\gamma(t)} \gamma'(t) \mathrm{d}t = \int_{0}^{1} 2\pi i e^{-2\pi i t} e^{2\pi i t} \mathrm{d}t = 2\pi i.$$

Solution of Exercise 3.10.8

In all of these three cases, we will use the definition of the complex integral.

1.

$$\begin{aligned}\int_{\gamma} \operatorname{Re} z \mathrm{d}z &= \int_{\gamma_1 \cup \gamma_2} \frac{z+\bar{z}}{2} \mathrm{d}z = \sum_{k=1}^{2} \int_{\gamma_k} \frac{z+\bar{z}}{2} \mathrm{d}z = \sum_{k=1}^{2} \int_{0}^{1} \frac{\gamma_k(t) + \overline{\gamma_k}(t)}{2} \gamma_k'(t) \mathrm{d}t \\ &= \int_{0}^{1} \frac{-2(1-t)}{2} (1+i) \mathrm{d}t + \int_{0}^{1} \frac{2t}{2} (1-i) \mathrm{d}t = -i.\end{aligned}$$

2.

$$\int_{\gamma_3} \operatorname{Re} z\mathrm{d}z = \int_{\gamma_3} \frac{z+\overline{z}}{2}\mathrm{d}z = \int_0^1 \frac{\pi i}{2}\frac{e^{-i\frac{\pi(t-1)}{2}} + e^{i\frac{\pi(t-1)}{2}}}{2}e^{i\frac{\pi(t-1)}{2}}\,\mathrm{d}t$$

$$= \int_0^1 \frac{\pi i}{4}(e^{i\pi(t-1)} + 1)\mathrm{d}t = \frac{2+\pi i}{4}.$$

3. Using results obtained to points 1 and 2, we deduce that

$$\int_{\gamma} \operatorname{Re} z\mathrm{d}z = \int_{\gamma_3\cup(\gamma_1\cup\gamma_2)^-} \operatorname{Re} z\mathrm{d}z = \int_{\gamma_3} \operatorname{Re} z\mathrm{d}z + \int_{(\gamma_1\cup\gamma_2)^-} \operatorname{Re} z\mathrm{d}z$$

$$= \int_{\gamma_3} \operatorname{Re} z\mathrm{d}z - \int_{\gamma_1\cup\gamma_2} \operatorname{Re} z\mathrm{d}z = \frac{2+(\pi+4)i}{4}.$$

Solution of Exercise 3.10.9

1. The parametric equation of the path is

$$C(\alpha;r) : |z-\alpha| = r \Leftrightarrow z = \alpha + re^{i\theta}, \quad \theta \in [0,2\pi],$$

hence

$$\gamma(\theta) = \alpha + re^{i\theta}, \quad \theta \in [0,2\pi], \quad \text{that is} \quad \gamma(t) = \alpha + re^{2\pi i t}, \quad t \in [0,1].$$

We will compute the integral by using the well-known definition of the complex integral, i. e.,

$$I = \int_{\gamma} \frac{1}{z-\alpha}\mathrm{d}z = \int_0^1 \frac{1}{\gamma(t)-\alpha}\gamma'(t)\mathrm{d}t = 2\pi i.$$

2. Since $f(z) = \frac{1}{z-\alpha} \in H(D)$, where $D = \mathbb{C}\setminus\{\alpha\}$, we will discuss two cases. The first one is thus when $\alpha \in \mathbb{C}$ belongs to the interior of the triangle T, and the second one is thus when it belongs to the exterior of the triangle.

Case 1. If $\alpha \in \operatorname{Ext}(T)$, then $\gamma \underset{D}{\sim} 0$, and thus

$$\int_{\gamma} \frac{1}{z-\alpha}\mathrm{d}z = 0.$$

Case 2. If $\alpha \in \operatorname{Int}(T)$, then we will replace the integration path with the homotopic circle $\partial U(\alpha;r)$. Since $\exists r > 0 : \partial U(\alpha;r) \in \operatorname{Int}(T)$, let

$$\gamma_*(t) = \alpha + re^{2\pi i t}, \quad t \in [0,1].$$

Then $\gamma_* \underset{D}{\sim} \gamma$, and thus, according to the *Cauchy theorem* we get

$$\int_{\gamma} \frac{1}{z-\alpha} \mathrm{d}z = \int_{\gamma_*} \frac{1}{z-\alpha} \mathrm{d}z = \int_0^1 \frac{1}{\gamma_*(t)-\alpha} \gamma_*'(t) \mathrm{d}t = 2\pi i.$$

Solution of Exercise 3.10.10

1. Using the definition of the complex integral, we deduce that

$$I_n = \int_{\gamma} (z-a)^n \mathrm{d}z = \int_0^1 (\gamma(t)-a)^n \gamma'(t) \mathrm{d}t = r^{n+1} \int_0^1 \pi i e^{\pi i (n+1)t} \mathrm{d}t. \tag{7.9}$$

Now we will discuss the next two cases according to the values of the parameter $n \in \mathbb{Z}$:

Case 1. If $n \in \mathbb{Z} \setminus \{-1\}$, from the relation (7.9) we have that

$$I_n = \frac{r^{n+1}}{n+1} (e^{\pi i (n+1)} - 1).$$

Case 2. If $n = -1$, from the relation (7.9) it follows that

$$I_{-1} = \int_0^1 \pi i \mathrm{d}t = \pi i.$$

2. The parametric equation of the path γ will be

$$\gamma(t) = a + re^{2\pi i t}, \quad t \in [0,1],$$

hence the integral is given by

$$I_n = \int_{\gamma} (z-a)^n \mathrm{d}z = \int_0^1 (\gamma(t)-a)^n \gamma'(t) \mathrm{d}t = r^{n+1} \int_0^1 2\pi i e^{2\pi i (n+1)t} \mathrm{d}t.$$

Similarly, we will discuss the next two cases according to the values of the parameter $n \in \mathbb{Z}$:

Case 1. If $n \in \mathbb{Z} \setminus \{-1\}$, then

$$I_n = \frac{r^{n+1}}{n+1} (e^{2\pi i (n+1)} - 1).$$

Case 2. If $n = -1$, then

$$I_{-1} = \int_0^1 2\pi i \mathrm{d}t = 2\pi i.$$

3. Since $n \in \mathbb{Z} \setminus \{-1\}$, let

$$g(z) = \frac{(z-a)^{n+1}}{n+1}.$$

This new function satisfies the relation $g'(z) = f(z)$.

Case 1. If $n \in \mathbb{N}$, then $g \in H(\mathbb{C})$ and $g'(z) = f(z)$, $\forall z \in \mathbb{C}$, and thus the function f has a primitive in $\mathbb{C}$. Then

$$\int_\gamma (z-a)^n \mathrm{d}z = g(z_1) - g(z_0) = \frac{(z_1-a)^{n+1} - (z_0-a)^{n+1}}{n+1}, \quad \forall \gamma \in \mathcal{D}(z_0, z_1).$$

Case 2. If $n \in \mathbb{Z}_- \setminus \{-1\}$, and $a \notin \{\gamma\}$, then $g \in H(\mathbb{C}\setminus\{a\})$ and $g'(z) = f(z)$, $\forall z \in \mathbb{C}\setminus\{a\}$, and thus the function f has a primitive in $\mathbb{C} \setminus \{a\}$. It follows that in this case we also have

$$\int_\gamma (z-a)^n \mathrm{d}z = g(z_1) - g(z_0) = \frac{(z_1-a)^{n+1} - (z_0-a)^{n+1}}{n+1}, \quad \forall \gamma \in \mathcal{D}(z_0, z_1) : a \notin \{\gamma\}.$$

4. Since $n = \frac{1}{2}$, then the square root function has two branches: the branch for which

$$\sqrt{1} = 1, \quad \text{and thus} \quad \sqrt{z} = \sqrt{|z|}e^{i\frac{\arg z}{2}}, \quad z \in \mathbb{C}^*,$$

and the branch for which

$$\sqrt{1} = -1, \quad \text{and thus} \quad \sqrt{z} = \sqrt{|z|}e^{i\frac{\arg z + 2\pi}{2}}, \quad z \in \mathbb{C}^*.$$

Case 1. If $\sqrt{1} = 1$, then the value of the integral is

$$\int_\gamma \sqrt{z-a}\mathrm{d}z = \int_0^1 \sqrt{\gamma(t)-a}\gamma'(t)\mathrm{d}t = 2\pi i r\sqrt{r}\int_0^1 e^{3\pi i t}\mathrm{d}t = -\frac{4r\sqrt{r}}{3},$$

because $\sqrt{re^{2\pi i t}} = \sqrt{r}e^{\pi i t}$.

Case 2. If $\sqrt{1} = -1$, then the value of the integral is

$$\int_\gamma \sqrt{z-a}\mathrm{d}z = \int_0^1 \sqrt{\gamma(t)-a}\gamma'(t)\mathrm{d}t = 2\pi i r\sqrt{r}e^{\pi i}\int_0^1 e^{3\pi i t}\mathrm{d}t = \frac{4r\sqrt{r}}{3},$$

because $\sqrt{re^{2\pi i t}} = \sqrt{r}e^{\pi i(t+1)}$.

Solution of Exercise 3.10.11

First, we will determine the poles of the function, i. e.,

$$\sin z + \cos z = 0 \Leftrightarrow \frac{e^{iz} - e^{-iz}}{2i} + \frac{e^{iz} + e^{-iz}}{2} = 0 \Leftrightarrow e^{2iz} = \frac{1-i}{1+i} \Leftrightarrow 2iz \in \operatorname{Log} \frac{1-i}{1+i},$$

and thus the poles of the function are

$$z_k = -\frac{\pi}{4} + k\pi, \quad k \in \mathbb{Z}.$$

It follows that

$$g(z) = \frac{e^z}{\sin z + \cos z} \in H(\mathbb{C} \setminus \{z_k : k \in \mathbb{Z}\}).$$

Since we are interested only about the poles that lie in the disc $U(0, \frac{1}{4})$, we have that

$$z_k \in U\left(0, \frac{1}{4}\right) \Leftrightarrow |z_k| < \frac{1}{4} \Leftrightarrow k \in \emptyset,$$

hence

$$g \in H\left(U\left(0, \frac{1}{4}\right)\right),$$

and thus, according to the *Cauchy theorem* we have that

$$\int_\gamma \frac{e^z}{\sin z + \cos z} \mathrm{d}z = 0.$$

Solution of Exercise 3.10.12

We have $e^z \sin z \in H(\mathbb{C})$, since this function is obtained from the product of two elementary functions, and thus from the *Cauchy theorem* we get

$$\int_\gamma e^z \sin z \mathrm{d}z = 0.$$

The equation of the integrating path γ is $\gamma(t) = 2e^{2\pi i t}$, $t \in [0, 1]$. From the above result, using the definition of the complex integral we deduce that

$$\int_\gamma (\overline{z} + e^z \sin z)\mathrm{d}z = \int_\gamma \overline{z}\mathrm{d}z + \int_\gamma e^z \sin z\mathrm{d}z = \int_\gamma \overline{z}\mathrm{d}z = \int_0^1 8\pi i e^{-2\pi i t} e^{2\pi i t} \mathrm{d}t = 8\pi i.$$

Solution of Exercise 3.10.13

The function

$$g(z) = \frac{e^z}{\cos z \cos iz}$$

is holomorphic in the whole complex plane $\mathbb{C}$, except the zeros of the $\cos z$ and $\cos iz$ functions.

Since

$$\cos z = 0 \Leftrightarrow e^{iz} + e^{-iz} = 0 \Leftrightarrow e^{2iz} = -1 \Leftrightarrow z_k = \frac{\pi}{2} + k\pi, \quad k \in \mathbb{Z},$$

and

$$\cos iz = 0 \Leftrightarrow e^{-z} + e^{z} = 0 \Leftrightarrow e^{2z} = -1 \Leftrightarrow z_k' = i\frac{\pi}{2} + k\pi i, \quad k \in \mathbb{Z}.$$

it follows that $|z_k| \geq \frac{\pi}{2} > 1$, and $|z_k'| \geq \frac{\pi}{2} > 1$, $\forall k \in \mathbb{Z}$, and then $g \in H(U(0,1))$. From the *Cauchy theorem*, we deduce that the value of the integral is zero.

Solution of Exercise 3.10.14

We will determine the zeros of the denominator, i. e.,

$$z^2 + 4 = 0 \Leftrightarrow z = \pm 2i \quad \text{and} \quad z^2 + 16 = 0 \Leftrightarrow z = \pm 4i,$$

and thus

$$g(z) = \frac{1}{(z^2+4)(z^2+16)} \in H(D), \quad D = \mathbb{C} \setminus \{-4i, -2i, 2i, 4i\}.$$

Since any of these zeros does not belong to the interior of the triangle T, then $\gamma \underset{D}{\sim} 0$. Hence, from the *Cauchy theorem* we conclude that the value of the integral is zero.

Solution of Exercise 3.10.15

Since the elementary functions are holomorphic, we have

$$e^{\frac{1}{z-3}} \in H(\mathbb{C} \setminus \{3\}) \quad \text{and} \quad \cos z, (z-2)^n \in H(\mathbb{C}).$$

It follows that

$$g(z) = \frac{e^{\frac{1}{z-3}} \cos z}{(z-2)^n} \in H(D), \quad D = \mathbb{C} \setminus \{2, 3\},$$

has the nth order pole $z_0 = 2$, and has the essential isolated singular point $z_1 = 3$. Both of these singularities does not belong to the disc $U(0;1)$, and thus $\gamma \underset{D}{\sim} 0$. Hence, from the *Cauchy theorem* we obtain that the value of the integral is zero.

Solution of Exercise 3.10.16

Since the function is rational, it will be holomorphic on the whole complex plane except the zeros of the denominator, i. e.,

$$g(z) = \frac{z}{z^2-1} \in H(D), \quad D = \mathbb{C} \setminus \{-1, 1\}.$$

We have to discuss the two cases according to the values of the parameter $a > 0$, with $a \neq 1$.

Case 1. If $0 < a < 1$, then $\gamma \underset{D}{\sim} 0$. Using the *Cauchy theorem*, it follows that the integral is zero.

Case 2. If $a > 1$, then both of the above zeros belong to the disc $U(0; a)$.

Method 1.

We will decompose the given integral into a sum of integrals, where the integration path is disjoint and where the integrated function has only one pole. These kinds of decompositions are useful also for some applications of the *Cauchy formula for the discs*. Thus,

$$\exists r_1, r_2 > 0 : U(-1; r_1) \subset \operatorname{Int} U(0; a), \quad U(1; r_2) \subset \operatorname{Int} U(0; a), \quad U(-1; r_1) \cap U(1; r_2) = \emptyset.$$

If

$$\gamma_1(t) = -1 + r_1 e^{2\pi i t}, \quad t \in [0, 1], \quad \gamma_2(t) = 1 + r_2 e^{2\pi i t}, \quad t \in [0, 1],$$

then $\{\gamma_1\} = \partial U(-1; r_1)$, $\{\gamma_2\} = \partial U(1; r_2)$ and

$$\int_\gamma \frac{z}{z^2-1} dz = \int_{\gamma_1} \frac{z}{z^2-1} dz + \int_{\gamma_2} \frac{z}{z^2-1} dz. \tag{7.10}$$

To compute the integrals from the right-hand side, we see that

$$g(z) = \frac{z}{z^2-1} = \frac{1}{2}\left(\frac{1}{z-1} + \frac{1}{z+1}\right).$$

Thus, according to Exercise 3.10.9 we get

$$\int_{\gamma_1} \frac{z}{z^2-1} dz = \frac{1}{2}\left(\int_{\gamma_1} \frac{1}{z-1} dz + \int_{\gamma_1} \frac{1}{z+1} dz\right) = \frac{1}{2}\int_{\gamma_1} \frac{1}{z+1} dz = \pi i,$$

$$\int_{\gamma_2} \frac{z}{z^2-1} dz = \frac{1}{2}\left(\int_{\gamma_2} \frac{1}{z-1} dz + \int_{\gamma_2} \frac{1}{z+1} dz\right) = \frac{1}{2}\int_{\gamma_2} \frac{1}{z-1} dz = \pi i.$$

From here, using the relation (7.10) we conclude that

$$\int_\gamma \frac{z}{z^2-1} dz = 2\pi i.$$

Method 2.
Similarly,

$$\exists r_1, r_2 > 0 : U(-1; r_1) \subset \operatorname{Int} U(0; a), \quad U(1; r_2) \subset \operatorname{Int} U(0; a), \quad U(-1; r_1) \cap U(1; r_2) = \emptyset.$$

If we introduce the notation

$$\gamma_1(t) = -1 + r_1 e^{2\pi i t}, \quad t \in [0,1], \quad \gamma_2(t) = 1 + r_2 e^{2\pi i t}, \quad t \in [0,1],$$

then $\{\gamma_1\} = \partial U(-1; r_1)$, $\{\gamma_2\} = \partial U(1; r_2)$, and

$$\int_\gamma \frac{z}{z^2-1}\mathrm{d}z = \int_{\gamma_1} \frac{z}{z^2-1}\mathrm{d}z + \int_{\gamma_2} \frac{z}{z^2-1}\mathrm{d}z. \tag{7.11}$$

Now, by using the *Cauchy formula for the discs* we get

$$\int_{\gamma_1} \frac{z}{z^2-1}\mathrm{d}z = \int_{\gamma_1} \frac{\frac{z}{z-1}}{z+1}\mathrm{d}z = 2\pi i\frac{1}{2} = \pi i,$$

and

$$\int_{\gamma_2} \frac{z}{z^2-1}\mathrm{d}z = \int_{\gamma_2} \frac{\frac{z}{z+1}}{z-1}\mathrm{d}z = 2\pi i\frac{1}{2} = \pi i.$$

Replacing these two relations in (7.11), we conclude that

$$\int_\gamma \frac{z}{z^2-1}\mathrm{d}z = 2\pi i.$$

Solution of Exercise 3.10.17

The function under the integral is obtained from the elementary functions, i. e.,

$$\sin z,\ z - \frac{\pi}{2},\ z^2 + 5 \in H(\mathbb{C}),$$

hence

$$g(z) = \frac{\sin z}{(z - \frac{\pi}{2})(z^2+5)} \in H(D), \quad \text{where } D = \mathbb{C} \setminus \left\{\frac{\pi}{2}, -i\sqrt{5}, i\sqrt{5}\right\}.$$

Since only $z_0 = \frac{\pi}{2} \in U(0; 2)$, it follows that $\exists r > 0 : U(\frac{\pi}{2}, r) \subset U(0, 2)$, and

$$\gamma_r(t) = \frac{\pi}{2} + re^{2\pi i t}, \quad t \in [0,1] \Rightarrow \gamma \underset{D}{\sim} \gamma_r.$$

From the *Cauchy integral theorem*, it follows that

$$\int_{\gamma} g(z)\mathrm{d}z = \int_{\gamma_r} g(z)\mathrm{d}z = \int_{\gamma_r} \frac{\frac{\sin z}{z^2+5}}{z-\frac{\pi}{2}}\mathrm{d}z,$$

and using the *Cauchy formula for the disc*, where $f(z) = \frac{\sin z}{z^2+5}$, $z_0 = \frac{\pi}{2}$ and $k = 0$, we get

$$\int_{\gamma} g(z)\mathrm{d}z = \int_{\gamma_r} \frac{\frac{\sin z}{z^2+5}}{z-\frac{\pi}{2}}\mathrm{d}z = \frac{2\pi i}{0!} f\left(\frac{\pi}{2}\right) = \frac{8\pi i}{20+\pi^2}.$$

Solution of Exercise 3.10.18

Since the function under the integral is obtained from the elementary functions, we have

$$g(z) = \frac{z^5}{e^z(z+1)^6} \in H(\mathbb{C}\setminus\{-1\}).$$

Because $z_0 = -1 \in U(0;2)$, from the *Cauchy formula for closed paths* we deduce that

$$\int_{\gamma} g(z)\mathrm{d}z = \int_{\gamma} \frac{\frac{z^5}{e^z}}{(z+1)^6}\mathrm{d}z = \frac{2\pi i}{k!} f^{(k)}(z_0),$$

where $f(z) = \frac{z^5}{e^z}$, $z_0 = -1$ and $k = 5$. Computing the fifth-order derivative of the function f, we have

$$f^{(5)}(z) = 120e^{-z} - 600ze^{-z} + 600z^2e^{-z} - 200z^3e^{-z} + 25z^4e^{-z} - z^5e^{-z},$$

then $f^{(5)}(-1) = 1546e$, and thus

$$\int_{\gamma} g(z)\mathrm{d}z = \frac{2\pi i}{5!} f^{(5)}(-1) = \frac{773}{30}\pi i e.$$

Solution of Exercise 3.10.19

Consider the function

$$g(z) = \frac{ze^z}{(z-2)^n}$$

defined on the maximal domain, where g is holomorphic.

If $n = 0$, then $g \in H(\mathbb{C})$. Since $\gamma \underset{\mathbb{C}}{\sim} 0$, from the *Cauchy integral theorem* it follows that the integral is zero.

If $n \in \mathbb{N}^*$, then $g \in H(D)$, where $D = \mathbb{C} \setminus \{2\}$, thus the point $z_0 = 2$ is a nth order pole for g and $\gamma \underset{D}{\not\sim} 0$.

Since

$$\int_\gamma g(z)\mathrm{d}z = \int_\gamma \frac{ze^z}{(z-2)^n}\mathrm{d}z,$$

from the *Cauchy formula for closed paths* it follows that

$$\int_\gamma \frac{ze^z}{(z-2)^n}\mathrm{d}z = \frac{2\pi i}{k!} f^{(k)}(z_0),$$

where $f(z) = ze^z$, $z_0 = 2$ and $k = n-1$. If we compute the $n-1$th order derivative of the function f, then

$$f^{(n-1)}(z) = (n-1)e^z + ze^z,$$

and using this result we get

$$\int_\gamma \frac{ze^z}{(z-2)^n}\mathrm{d}z = \frac{2\pi i}{(n-1)!}\left((n-1)e^2 + 2e^2\right) = \frac{2(n+1)e^2\pi i}{(n-1)!}.$$

Solution of the Exercise 3.10.20

Case 1. If $a = 0$, in this case the zero of the denominator is a zero for the function $\sin z$. Since

$$\lim_{z\to 0} \frac{e^z \sin z}{z} = 1 \in \mathbb{C},$$

from the *first removability criterion* the point $z_0 = 0$ will be removable. From the *Cauchy integral theorem*, it follows that the required integral is zero, for all $r > 0$.

Case 2. If $a \neq 0$, then

$$g(z) = \frac{e^z \sin z}{z-a} \in H(D), \quad D = \mathbb{C} \setminus \{a\},$$

where $z_0 = a$ is a first-order pole for the function g.

Case 2.1. If $r < |a|$, then $\gamma \underset{D}{\sim} 0$, and from the *Cauchy integral theorem* it follows that

$$\int_{\partial U(0;r)} \frac{e^z \sin z}{z-a}\mathrm{d}z = 0.$$

Case 2.2. If $r > |a|$, then $\gamma \underset{D}{\not\sim} 0$, and from the *Cauchy formula for closed paths* it follows that

$$\int_\gamma g(z)\mathrm{d}z = \int_\gamma \frac{e^z \sin z}{z-a}\mathrm{d}z = \frac{2\pi i}{k!} f^{(k)}(z_0),$$

where $f(z) = e^z \sin z$, $z_0 = a$ and $k = 0$. Thus

$$\int_\gamma \frac{e^z \sin z}{z-a}\mathrm{d}z = 2\pi i f(a) = 2\pi i e^a \sin a.$$

Solution of the Exercise 3.10.21

The function under the integral may be written into the form

$$g(z) = \frac{1}{z^2+1} = \frac{1}{2i}\left(\frac{1}{z-i} - \frac{1}{z+i}\right), \tag{7.12}$$

that is holomorphic on $D = \mathbb{C} \setminus \{-i, i\}$.

1.

Method 1.

If $\{\gamma\} = \partial U(i; 1)$ is a directly oriented path, then the equation of the path is

$$\gamma(t) = i + e^{2\pi i t}, \quad t \in [0, 1].$$

Using the above decomposition, since γ does not turn around the point $-i$, the function $\frac{1}{z+i}$ is holomorphic on the domain bounded by $\{\gamma\}$, hence

$$\int_\gamma \frac{1}{z+i}\mathrm{d}z = 0. \tag{7.13}$$

From the relations (7.12) and (7.13), according to Exercise 3.10.9 we conclude that

$$\int_\gamma \frac{1}{z^2+1}\mathrm{d}z = \frac{1}{2i}\left(\int_\gamma \frac{1}{z-i}\mathrm{d}z - \int_\gamma \frac{1}{z+i}\mathrm{d}z\right) = \frac{1}{2i}\int_\gamma \frac{1}{z-i}\mathrm{d}z = \pi.$$

Method 2.

If the *Cauchy formula for the disc* using then

$$\int_\gamma \frac{1}{z^2+1}\mathrm{d}z = \int_\gamma \frac{\frac{1}{z+i}}{z-i}\mathrm{d}z = 2\pi i \frac{1}{2i} = \pi,$$

the other points of the problem will be solved similarly.

2.

Method 1.

If $\{\gamma\} = \partial U(-i; 1)$ is a directly oriented path, then the equation of the path is

$$\gamma(t) = -i + e^{2\pi it}, \quad t \in [0, 1].$$

The value of the integral is

$$\int_\gamma \frac{1}{z^2+1}\mathrm{d}z = \frac{1}{2i}\left(\int_\gamma \frac{1}{z-i}\mathrm{d}z - \int_\gamma \frac{1}{z+i}\mathrm{d}z\right) = -\frac{1}{2i}\int_\gamma \frac{1}{z+i}\mathrm{d}z = -\pi,$$

where we used the result of Exercise 3.10.9.

Method 2.

If we use the *Cauchy formula for the disc*, then

$$\int_\gamma \frac{1}{z^2+1}\mathrm{d}z = \int_\gamma \frac{\frac{1}{z-i}}{z+i}\mathrm{d}z = 2\pi i\left(-\frac{1}{2i}\right) = -\pi.$$

3.

Method 1.

We decompose our integral into the sum of two integrals, such that the new integration paths are disjoints, and each of them turns around only one pole. Then we will use directly the *Cauchy formula for the disc*. Thus, we have

$$\exists r_1, r_2 > 0 : U(-i; r_1) \subset \operatorname{Int} U(0; 2), \quad U(i; r_2) \subset \operatorname{Int} U(0; 2), \quad U(-i; r_1) \cap U(i; r_2) = \emptyset.$$

If

$$\gamma_1(t) = -i + r_1 e^{2\pi it}, \quad t \in [0, 1], \quad \gamma_2(t) = i + r_2 e^{2\pi it}, \quad t \in [0, 1],$$

then $\{\gamma_1\} = \partial U(-i; r_1)$, $\{\gamma_2\} = \partial U(i; r_2)$.

Since

$$\gamma_1 \underset{D}{\sim} \gamma_1', \quad \text{where } \gamma_1'(t) = -i + e^{2\pi it}, \quad t \in [0, 1],$$

and

$$\gamma_2 \underset{D}{\sim} \gamma_2', \quad \text{where } \gamma_2'(t) = i + e^{2\pi it}, \quad t \in [0, 1],$$

we will use the results from the first two points of the problem to calculate our integral:

$$\begin{aligned}\int_\gamma \frac{1}{z^2+1}\mathrm{d}z &= \int_{\gamma_1} \frac{1}{z^2+1}\mathrm{d}z + \int_{\gamma_2} \frac{1}{z^2+1}\mathrm{d}z \\ &= \int_{\gamma_1'} \frac{1}{z^2+1}\mathrm{d}z + \int_{\gamma_2'} \frac{1}{z^2+1}\mathrm{d}z = -\pi + \pi = 0.\end{aligned}$$

To obtain the above results, we used the fact that

$$\int_{\gamma} \frac{1}{z-a} \mathrm{dz} = 2\pi i,$$

where $\{\gamma\} = \partial U(a;r)$ is a directly oriented path (Exercise 3.10.9).

Method 2.

Similarly,

$$\exists r_1, r_2 > 0 : U(-i;r_1) \subset \operatorname{Int} U(0;2), \quad U(i;r_2) \subset \operatorname{Int} U(0;2), \quad U(-i;r_1) \cap U(i;r_2) = \emptyset.$$

If we use the notation

$$\gamma_1(t) = -i + r_1 e^{2\pi i t}, \quad t \in [0,1], \quad \gamma_2(t) = i + r_2 e^{2\pi i t}, \quad t \in [0,1],$$

then $\{\gamma_1\} = \partial U(-i;r_1)$, $\{\gamma_2\} = \partial U(i;r_2)$.

From the *Cauchy formula for the disc*, we get

$$\begin{aligned}\int_{\gamma} \frac{1}{z^2+1} \mathrm{dz} &= \int_{\gamma_1} \frac{1}{z^2+1} \mathrm{dz} + \int_{\gamma_2} \frac{1}{z^2+1} \mathrm{dz} \\ &= \int_{\gamma_1} \frac{\frac{1}{z-i}}{z+i} \mathrm{dz} + \int_{\gamma_2} \frac{\frac{1}{z+i}}{z-i} \mathrm{dz} = 2\pi i \left(-\frac{1}{2i}\right) + 2\pi i \frac{1}{2i} = -\pi + \pi = 0.\end{aligned}$$

Solution of Exercise 3.10.22

The elementary functions e^z, $\cos z$ and $z^3 + 8$ are holomorphic on the whole complex plane, hence

$$f(z) = \frac{e^z \cos z}{z^3+8} \in H(D), \quad D = \mathbb{C} \setminus A,$$

where

$$A = \{z \in \mathbb{C} : z^3 + 8 = 0\} = \{2e^{i\frac{(2k+1)\pi}{3}} : k \in \{0,1,2\}\}.$$

Since $\forall z \in A$, $|z| = 2 > 1$, the bounded domain with the boundary $\{\gamma\}$ does not contain any pole of the function f. Also, the path $\gamma(t) = e^{2\pi i t}$, $t \in [0,1]$, has the image $\{\gamma\} = \partial U(0;1)$, and thus $\gamma \underset{D}{\sim} 0$. From the *Cauchy integral theorem*, it follows that the integral will be zero, i. e.,

$$\int_{\gamma} \frac{e^z \cos z}{z^3+8} \mathrm{dz} = 0.$$

Solution of Exercise 3.10.23

According to the values of the parameter $k \in \mathbb{Z}$, we will discuss the next cases:

Case 1. If $k \in \mathbb{Z} \cap (-\infty, 0]$, then

$$f(z) = \frac{\sin z}{z^k} \in H(\mathbb{C}),$$

and since a $\gamma(t) = e^{2\pi i t}$, $t \in [0,1]$, the integration path is homotopic with 0 in $\mathbb{C}$. It follows that

$$\int_\gamma \frac{\sin z}{z^k} \mathrm{d}z = 0.$$

Case 2. If $k = 1$, then $f \in H(\mathbb{C}^*)$. Since

$$\lim_{z \to 0} f(z) = 1 \in \mathbb{C},$$

form the *first removability criterion* the point $z_0 = 0$ is removable. Using the *Cauchy integral theorem*, it follows that the integral is equal to zero.

Case 3. If $k \in \mathbb{N} \cap [2, +\infty)$, then $f \in H(\mathbb{C}^*)$. From the *Cauchy formula for the disc* for $f(z) = \sin z$, $z_0 = 0$ and $n = k-1$, we obtain

$$\int_\gamma \frac{\sin z}{z^k} \mathrm{d}z = \frac{2\pi i}{(k-1)!} \sin z^{(k-1)}(0) = \frac{2\pi i}{(k-1)!} u(k),$$

where

$$u(k) = \sin^{(k-1)}(0) = \begin{cases} -1, & \text{if } k = 4p,\ p \in \mathbb{N}^*, \\ 0, & \text{if } k = 4p+1,\ p \in \mathbb{N}^*, \\ 1, & \text{if } k = 4p+2,\ p \in \mathbb{N}, \\ 0, & \text{if } k = 4p+3,\ p \in \mathbb{N}. \end{cases}$$

Solution of Exercise 3.10.24

The function

$$g(z) = \frac{e^z}{z-a}$$

has the unique pole $z_0 = a \in \mathbb{C}^*$, that is a simple pole.

1. If we use the *Cauchy formula for the disc,* we obtain

$$\int_\gamma \frac{e^z}{z-a} \mathrm{d}z = \frac{2\pi i}{k!} f^{(k)}(z_0),$$

where $f(z) = e^z$, $z_0 = a$ and $k = 0$, and thus

$$\int_\gamma \frac{e^z}{z-a} dz = 2\pi i e^a.$$

2. If $\{\gamma\} = \partial U(-a; |a|)$, $a \in \mathbb{C}^*$ is a directly oriented path, then

$$\gamma(t) = -a + |a| e^{2\pi i t}, \quad t \in [0, 1].$$

Since $g \in H(D)$, $D = \mathbb{C} \setminus \{a\}$, it follows that $\gamma \underset{D}{\sim} 0$, and from the *Cauchy integral theorem* we get

$$\int_\gamma \frac{e^z}{z-a} dz = 0.$$

Solution of Exercise 3.10.25

If we use the *Cauchy formula for the disc*, using then

$$\int_\gamma \frac{\sinh z}{z - \frac{\pi i}{2}} dz = \frac{2\pi i}{k!} f^{(k)}(z_0),$$

where $f(z) = \sinh z$, $z_C = \frac{\pi i}{2}$ and $k = 0$, and thus

$$\int_\gamma \frac{\sinh z}{z - \frac{\pi i}{2}} dz = 2\pi i \sinh \frac{\pi i}{2} = -2\pi.$$

Solution of Exercise 3.10.26

According to the values of the parameter $k \in \mathbb{Z}$, we need to discuss a few cases. First, we will denote

$$I_n = \int_\gamma (z - \alpha)^n dz.$$

Case 1. If $n \in \mathbb{N}$, then the function $f(z) = (z - \alpha)^n$ is holomorphic on $\mathbb{C}$, and $\gamma \underset{\mathbb{C}}{\sim} 0$. From the *Cauchy integral theorem*, the integral is equal to zero, i. e.,

$$I_n = 0, \quad \forall n \in \mathbb{N}.$$

Case 2. If $n \in \mathbb{Z} \cap (-\infty, 0)$, then the function f is holomorphic on $\mathbb{C}$ except the point $z_0 = \alpha$, i. e., $f \in H(D)$, where $D = \mathbb{C} \setminus \{\alpha\}$.

Let us see if the integration path turns around this pole.

Case 2.1. If $|\alpha| > r$, the integration path does not turn around the pole $z_0 = \alpha$, and thus $\gamma \underset{D}{\sim} 0$. Now, according to the *Cauchy integral theorem* it follows that the integral is zero.

Case 2.2. If $|\alpha| < r$, the integration path turns around the pole $z_0 = \alpha$. From the *Cauchy formula for the disc,* we obtain that

$$I_n = \int_\gamma \frac{1}{(z-\alpha)^{-n}} \mathrm{d}z = \frac{2\pi i}{k!} f^{(k)}(z_0),$$

where $f(z) = 1$, $z_0 = \alpha$, and $k = -n - 1$.

The function f is constant, thus all its derivatives are zero. The integral will be not zero only in the case $n = -1$, i. e., $I_{-1} = 2\pi i$.

From the above results, we conclude that $I_{-1} = 2\pi i$, and $I_n = 0$ if $n \in \mathbb{Z} \cap (-\infty, -2]$.

Solution of Exercise 3.10.27

The function

$$g(z) = \frac{1}{z(z^2 + z - 2)}$$

has the simple poles $z_1 = 0$, $z_2 = 1$ and $z_3 = -2$, thus $g \in H(D)$, where $D = \mathbb{C} \setminus \{-2, 0, 1\}$.

1. The path $\{\gamma\}$ turns around the points $z_1 = 0$ and $z_2 = 1$. Let

$$\gamma_1(t) = r_1 e^{2\pi i t}, \quad t \in [0, 1]$$

and

$$\gamma_2(t) = 1 + r_2 e^{2\pi i t}, \quad t \in [0, 1],$$

where $0 < r_1 < \frac{1}{2}$, $0 < r_2 < \frac{1}{2}$. Then

$$\int_\gamma g(z)\mathrm{d}z = \int_{\gamma_1} g(z)\mathrm{d}z + \int_{\gamma_2} g(z)\mathrm{d}z. \tag{7.14}$$

Using the *Cauchy formula for the disc,* we have

$$\int_{\gamma_1} \frac{1}{z(z^2 + z - 2)} \mathrm{d}z = \int_{\gamma_1} \frac{\frac{1}{(z^2+z-2)}}{z} \mathrm{d}z = 2\pi i \frac{1}{-2} = -\pi i$$

and

$$\int_{\gamma_2} \frac{1}{z(z^2 + z - 2)} \mathrm{d}z = \int_{\gamma_2} \frac{\frac{1}{z^2+2z}}{z - 1} \mathrm{d}z = 2\pi i \frac{1}{3} = \frac{2\pi i}{3}.$$

From the above results, according to (7.14) we obtain that

$$\int_\gamma \frac{1}{z(z^2+z-2)}\mathrm{d}z = -\pi i + \frac{2\pi i}{3} = -\frac{\pi i}{3}. \tag{7.15}$$

2. In this case, the path $\{\gamma\}$ turns around all the three points $z_1 = 0$, $z_2 = 1$ and $z_3 = -2$. In order to use the relation (7.15), we will decompose our integral in the next sum:

$$\int_\gamma \frac{1}{z(z^2+z-2)}\mathrm{d}z = \int_{\partial U(0;\frac{3}{2})} \frac{1}{z(z^2+z-2)}\mathrm{d}z + \int_{\gamma_3} \frac{1}{z(z^2+z-2)}\mathrm{d}z, \tag{7.16}$$

where $\gamma_3(t) = -2 + r_3 e^{2\pi i t}$, $t \in [0,1]$ and $0 < r_3 < \frac{1}{2}$. To calculate the second integral, we will use the *Cauchy formula for the disc*, i. e.,

$$\int_{\gamma_3} \frac{1}{z(z^2+z-2)}\mathrm{d}z = \int_{\gamma_3} \frac{\frac{1}{z^2-z}}{z+2}\mathrm{d}z = 2\pi i\frac{1}{6} = \frac{\pi i}{3}.$$

Replacing this in (7.16), and using (7.15), we conclude that

$$\int_\gamma \frac{1}{z(z^2+z-2)}\mathrm{d}z = -\frac{\pi i}{3} + \frac{\pi i}{3} = 0.$$

Solution of Exercise 3.10.28

The function has the simple pole $z_1 = 0$, while $z_2 = i$ and $z_3 = -i$ are nth order poles. Hence, the function is holomorphic on $\mathbb{C}$ except at these points, i. e.,

$$g(z) = \frac{f(z)}{z(z^2+1)^n} \in H(D), \quad D = \mathbb{C} \setminus \{0, -i, i\}.$$

1.

Method 1.

The function may be written like

$$\frac{1}{z(z^2+1)} = \frac{1}{z} - \frac{1}{2(z-i)} - \frac{1}{2(z+i)},$$

thus

$$\int_\gamma \frac{1}{z(z^2+1)}\mathrm{d}z = \int_\gamma \frac{1}{z}\mathrm{d}z - \frac{1}{2}\int_\gamma \frac{1}{z-i}\mathrm{d}z - \frac{1}{2}\int_\gamma \frac{1}{z+i}\mathrm{d}z.$$

The integration path is an ellipse that turns around only the point $z_1 = 0$. The fractions $\frac{1}{2(z-i)}$ and $\frac{1}{2(z+i)}$ are holomorphic functions on the bounded domain, which has the boundary this ellipse, and from the *Cauchy formula for closed curves* it follows that these last two integrals are equal to zero.

Using the *Cauchy formula for the disc* for the third integral, we deduce that

$$\int_{\gamma} \frac{1}{z(z^2+1)} dz = \int_{\gamma} \frac{1}{z} dz = 2\pi i.$$

Method 2.

From the *Cauchy formula for closed curves*, it follows that

$$\int_{\gamma} \frac{1}{z(z^2+1)} dz = \int_{\gamma} \frac{\frac{1}{z^2+1}}{z} dz = 2\pi i.$$

2.

Method 1.

Similar to the point 1, the function may be written like

$$\frac{1}{z(z^2+1)} = \frac{1}{z} - \frac{1}{2(z-i)} - \frac{1}{2(z+i)}.$$

The integration path is a circle, and the bounded domain bounded by this circle contains all the poles of the function. We will decompose the integral into the sum of three integrals, i. e.,

$$\int_{\gamma} \frac{1}{z(z^2+1)} dz = \int_{\gamma} \frac{1}{z} dz - \frac{1}{2} \int_{\gamma} \frac{1}{z-i} dz - \frac{1}{2} \int_{\gamma} \frac{1}{z+i} dz.$$

We will use the *Cauchy formula for the disc* for each of the above integrals (or the result of Exercise 3.10.9), and we conclude that

$$\int_{\gamma} \frac{1}{z(z^2+1)} dz = 2\pi i - \pi i - \pi i = 0.$$

Method 2.

Let

$$\gamma_1(t) = r_1 e^{2\pi i t}, \quad \gamma_2(t) = i + r_2 e^{2\pi i t}, \quad \gamma_3(t) = -i + r_3 e^{2\pi i t}, \quad t \in [0,1],$$

where $0 < r_k < \frac{1}{2}$, $k \in \{1,2,3\}$. Now, using the *Cauchy formula for the disc* for each of the integrals we get

$$\begin{aligned}\int_{\gamma} \frac{1}{z(z^2+1)} dz &= \sum_{k=1}^{3} \int_{\gamma_k} \frac{1}{z(z^2+1)} dz \\ &= \int_{\gamma_1} \frac{\frac{1}{z^2+1}}{z} dz + \int_{\gamma_2} \frac{\frac{1}{z(z+i)}}{z-i} dz + \int_{\gamma_3} \frac{\frac{1}{z(z-i)}}{z+i} dz = 2\pi i + 2\pi i\left(-\frac{1}{2}\right) + 2\pi i\left(-\frac{1}{2}\right) = 0.\end{aligned}$$

3.
Method 1.
We will use the same method like to the point 1 of this problem. The function may be decomposed as follows:

$$\frac{e^{iz}}{z(z^2+1)} = \frac{e^{iz}}{z} - \frac{e^{iz}}{2(z-i)} - \frac{e^{iz}}{2(z+i)},$$

and the integration path is an ellipse that turns around only the pole $z_1 = 0$.
If we use the *Cauchy formula for the disc* for $f(z) = e^z$, $z_1 = 0$, and $k = 0$, then

$$\int_\gamma \frac{e^{iz}}{z(z^2+1)} \mathrm{d}z = \int_\gamma \frac{e^{iz}}{z} \mathrm{d}z = 2\pi i. \tag{7.17}$$

Method 2.
Using the *Cauchy formula for closed curves*, we have

$$\int_\gamma \frac{e^{iz}}{z(z^2+1)} \mathrm{d}z = \int_\gamma \frac{\frac{e^{iz}}{z^2+1}}{z} \mathrm{d}z = 2\pi i.$$

4.
Method 1.
This point is similar with the point 2 of this problem. If we write

$$\frac{e^{iz}}{z(z^2+1)} = \frac{e^{iz}}{z} - \frac{e^{iz}}{2(z-i)} - \frac{e^{iz}}{2(z+i)},$$

we obtain that

$$\int_\gamma \frac{e^{iz}}{z(z^2+1)} \mathrm{d}z = \int_\gamma \frac{e^{iz}}{z} \mathrm{d}z - \frac{1}{2}\int_\gamma \frac{e^{iz}}{z-i} \mathrm{d}z - \frac{1}{2}\int_\gamma \frac{e^{iz}}{z+i} \mathrm{d}z.$$

Using the *Cauchy formula for the disc* for each of these integrals, where $f(z) = e^{iz}$, $k = 0$, $z_1 = 0$, $z_2 = i$, and $z_3 = -i$, respectively, we conclude that

$$\int_\gamma \frac{1}{z(z^2+1)} \mathrm{d}z = 2\pi i - \frac{\pi i}{e} - \pi i e = \pi i(2 - e^{-1} - e).$$

Method 2.
Let

$$\gamma_1(t) = r_1 e^{2\pi i t}, \quad \gamma_2(t) = i + r_2 e^{2\pi i t}, \quad \gamma_3(t) = -i + r_3 e^{2\pi i t}, \quad t \in [0,1],$$

where $0 < r_k < \frac{1}{2}$, $k \in \{1,2,3\}$. Now, using the *Cauchy formula for the disc* for each of the integrals we get

$$\begin{aligned}\int_\gamma \frac{e^{iz}}{z(z^2+1)}\mathrm{d}z &= \sum_{k=1}^{3}\int_{\gamma_k}\frac{e^{iz}}{z(z^2+1)}\mathrm{d}z \\ &= \int_{\gamma_1}\frac{\frac{e^{iz}}{z^2+1}}{z}\mathrm{d}z + \int_{\gamma_2}\frac{\frac{e^{iz}}{z(z+i)}}{z-i}\mathrm{d}z + \int_{\gamma_3}\frac{\frac{e^{iz}}{z(z-i)}}{z+i}\mathrm{d}z \\ &= 2\pi i + 2\pi i\left(-\frac{1}{2e}\right) + 2\pi i\left(-\frac{e}{2}\right) = \pi i(2 - e^{-1} - e).\end{aligned}$$

5.

Method 1.

We see that

$$\frac{e^{iz}}{z(z^2+1)^2} = \frac{e^{iz}}{z} - \frac{ze^{iz}}{z^2+1} - \frac{ze^{iz}}{(z^2+1)^2} = \frac{e^{iz}}{z} - \frac{e^{iz}(z^3+2z)}{(z^2+1)^2}.$$

Since the integration path is an ellipse that turns around only the pole $z_1 = 0$, we have

$$\int_\gamma \frac{e^{iz}}{z(z^2+1)^2}\mathrm{d}z = \int_\gamma \frac{e^{iz}}{z}\mathrm{d}z - \int_\gamma \frac{e^{iz}(z^3+2z)}{(z^2+1)^2}\mathrm{d}z = \int_\gamma \frac{e^{iz}}{z}\mathrm{d}z,$$

because the function under the second integral is holomorphic on the bounded domain which is bounded by the ellipse, and from the *Cauchy integral theorem* this integral will be zero. If we use the formula (7.17) to calculate the last integral, we get

$$\int_\gamma \frac{e^{iz}}{z(z^2+1)^2}\mathrm{d}z = 2\pi i.$$

Method 2.

From the *Cauchy formula for closed curves*, we deduce that

$$\int_\gamma \frac{e^{iz}}{z(z^2+1)^2}\mathrm{d}z = \int_\gamma \frac{\frac{e^{iz}}{(z^2+1)^2}}{z}\mathrm{d}z = 2\pi i.$$

6.

Method 1.

If we decompose the function as follows,

$$\frac{e^{iz}}{z(z^2+1)^2} = \frac{e^{iz}}{z} - e^{iz}(z^3+2z)\left(\frac{i}{2(z+i)} - \frac{i}{2(z-i)} + \frac{iz}{4(z-i)^2} - \frac{iz}{4(z+i)^2}\right),$$

the integral can be calculated as the sum of the following three integrals.

The value of the integral

$$I_1 = \int_\gamma \frac{e^{iz}}{z} dz = 2\pi i,$$

is the same with those given by (7.17).

For the integral

$$I_2 = \int_\gamma \frac{-e^{iz} i(z^3 + 2z)}{2(z+i)} dz,$$

we use the *Cauchy formula for the disc*, where $f(z) = \frac{-e^{iz} i(z^3+2z)}{2}$, $z_0 = -i$ and $k = 0$. Thus, we have

$$I_2 = 2\pi i f(-i) = 2\pi i\left(-\frac{e}{2}\right) = -\pi i e.$$

For the integral,

$$I_3 = \int_\gamma \frac{e^{iz} i(z^3 + 2z)}{2(z-i)} dz,$$

using the *Cauchy formula for the disc* where $f(z) = \frac{e^{iz} i(z^3+2z)}{2}$, $z_0 = i$ and $k = 0$, we obtain

$$I_3 = 2\pi i f(i) = 2\pi i\left(-\frac{1}{2e}\right) = -\frac{\pi i}{e}.$$

Using the *Cauchy formula for the disc* for $f(z) = \frac{-e^{iz} iz(z^3+2z)}{4}$, $z_0 = i$ and $k = 1$, we have

$$I_4 = \int_\gamma \frac{-e^{iz} iz(z^3 + 2z)}{4(z-i)^2} dz = 2\pi i f'(i) = 2\pi i\left(-\frac{1}{4e}\right) = -\frac{\pi i}{2e}.$$

If we use the *Cauchy formula for the disc* for $f(z) = \frac{e^{iz} iz(z^3+2z)}{4}$, $z_0 = -i$ and $k = 1$, we obtain that

$$I_5 = \int_\gamma \frac{e^{iz} iz(z^3 + 2z)}{4(z+i)^2} dz = 2\pi i \frac{e}{4} = \frac{\pi i e}{2}.$$

Combining the above results, we conclude that

$$\int_\gamma \frac{e^{iz}}{z(z^2+1)^2} dz = I_1 + I_2 + I_3 + I_4 + I_5 = \pi i\left(2 - \frac{e}{2} - \frac{3}{2e}\right).$$

Method 2.
Using the notation,

$$\gamma_1(t) = r_1 e^{2\pi i t}, \quad \gamma_2(t) = i + r_2 e^{2\pi i t}, \quad \gamma_3(t) = -i + r_3 e^{2\pi i t}, \quad t \in [0,1],$$

where $0 < r_k < \frac{1}{2}$, $k \in \{1, 2, 3\}$. Now, using the *Cauchy formula for the disc* for each of the integrals, we have

$$\begin{aligned}\int_\gamma \frac{e^{iz}}{z(z^2+1)^2}\,\mathrm{d}z &= \sum_{k=1}^{3} \int_{\gamma_k} \frac{e^{iz}}{z(z^2+1)^2}\,\mathrm{d}z \\ &= \int_{\gamma_1} \frac{\frac{e^{iz}}{(z^2+1)^2}}{z}\,\mathrm{d}z + \int_{\gamma_2} \frac{\frac{e^{iz}}{z(z+i)^2}}{(z-i)^2}\,\mathrm{d}z + \int_{\gamma_3} \frac{\frac{e^{iz}}{z(z-i)^2}}{(z+i)^2}\,\mathrm{d}z \\ &= 2\pi i f_1'(0) + 2\pi i f_2'(i) + 2\pi i f_3'(-i),\end{aligned}$$

where $f_1(z) = \frac{e^{iz}}{(z^2+1)^2}$, $f_2(z) = \frac{e^{iz}}{z(z+i)^2}$ and $f_3(z) = \frac{e^{iz}}{z(z-i)^2}$, thus

$$\int_\gamma \frac{e^{iz}}{z(z^2+1)^2}\,\mathrm{d}z = \pi i\left(2 - \frac{e}{2} - \frac{3}{2e}\right).$$

Solution of Exercise 3.10.29

Since the function is obtained from elementary functions, we have that it is holomorphic on the set $D = \mathbb{C} \setminus \{-1, 1\}$.

The integration path is a circle with the radius $|a| > 0$, and the center in $a \in \mathbb{R}^* \setminus \{-\frac{1}{2}, \frac{1}{2}\}$, i. e., $\{\gamma\} = \partial U(a; |a|)$ with direct orientation.

We need to make the following discussion according to the values of the parameter a.

Case 1. If $|a| < \frac{1}{2}$, then the integration path does not turn around any of the poles of the function, hence $\gamma \underset{D}{\sim} 0$. From here, according to the *Cauchy integral theorem*, the value of the integral is zero, i. e.,

$$\int_\gamma \frac{\cos \pi z}{z^2 - 1}\,\mathrm{d}z = 0.$$

Case 2. If $|a| > \frac{1}{2}$, then the integration path turns around only one of the poles of the function, as follows.

Case 2.1. If $a < -\frac{1}{2}$, then $-1 \in U(a; |a|)$ and $1 \notin U(a; |a|)$. Using the *Cauchy formula for closed curves*, where $f(z) = \frac{\cos \pi z}{z-1}$, $z_0 = -1$, and $k = 0$, we have

$$\int_\gamma \frac{\cos \pi z}{z^2 - 1}\,\mathrm{d}z = \int_\gamma \frac{\frac{\cos \pi z}{z-1}}{z+1}\,\mathrm{d}z = 2\pi i f(-1) = \pi i.$$

Case 2.2. If $a > \frac{1}{2}$, then $1 \in U(a; |a|)$ and $-1 \notin U(a; |a|)$. Similarly, using the *Cauchy formula for closed curves*, where $f(z) = \frac{\cos \pi z}{z+1}$, $z_0 = 1$ and $k = 0$, we get

$$\int_\gamma \frac{\cos \pi z}{z^2 - 1} dz = \int_\gamma \frac{\frac{\cos \pi z}{z+1}}{z - 1} dz = 2\pi i f(1) = -\pi i.$$

Solution of Exercise 3.10.30

If $n \in \mathbb{Z} \cap (-\infty, 0]$, then

$$f(z) = \frac{1}{(z^2 + 1)^n} \in H(\mathbb{C}),$$

and since $\gamma \underset{\mathbb{C}}{\sim} 0$, from the *Cauchy integral theorem* it follows that the value of the integral is zero.

If $n \in \mathbb{N}^*$, similarly as in the previous problem we need to determine only the zeros of the denominator. Hence, the function f is holomorphic on the set $D = \mathbb{C} \setminus \{-i, i\}$.

The integration path is a circle with the radius $|\frac{a}{2}|$ and the center $i\frac{a}{2}$, i. e., $\{\gamma\} = \partial U(i\frac{a}{2}; |\frac{a}{2}|)$ with direct orientation.

We need to make the following discussion according to the values of the parameter a.

Case 1. If $|a| < 1$, the function is holomorphic in the disc bounded by the integration circle, hence $\gamma \underset{D}{\sim} 0$. From the *Cauchy integral theorem*, we have that the value of the integral is zero, i. e.,

$$\int_\gamma \frac{1}{(z^2 + 1)^n} dz = 0.$$

Case 2. If $|a| > 1$, then the integration path turns around only one of the poles of the function, as follows.

Case 2.1. If $a < -1$, then $-i \in U(i\frac{a}{2}; |\frac{a}{2}|)$ and $i \notin U(i\frac{a}{2}; |\frac{a}{2}|)$. From the *Cauchy formula for closed curves*, where $f(z) = \frac{1}{(z-i)^n}$, $z_0 = -i$, and $k = n - 1$, we get

$$\int_\gamma \frac{1}{(z^2 + 1)^n} dz = \int_\gamma \frac{\frac{1}{(z-i)^n}}{(z + i)^n} dz = \frac{2\pi i}{(n - 1)!} f^{(n-1)}(-i)$$
$$= -\frac{\pi}{2^{2(n-1)}(n - 1)!} n(n + 1) \cdot \ldots \cdot (2n - 1),$$

because

$$f^{(n-1)}(z) = (-1)^{n-1} n(n + 1)(n + 2) \cdot \ldots \cdot (2n - 1)(z - i)^{-2n+1}.$$

Case 2.2. If $a > 1$, then $i \in U(i\frac{a}{2}; |\frac{a}{2}|)$ and $-i \notin U(i\frac{a}{2}; |\frac{a}{2}|)$. Similarly, from the *Cauchy formula for closed curves*, where $f(z) = \frac{1}{(z+i)^n}$, $z_0 = i$, and $k = n-1$, we have

$$\int_\gamma \frac{1}{(z^2+1)^n} dz = \int_\gamma \frac{\frac{1}{(z+i)^n}}{(z-i)^n} dz = \frac{2\pi i}{(n-1)!} f^{(n-1)}(i)$$
$$= \frac{\pi}{2^{2(n-1)}(n-1)!} n(n+1) \cdot \ldots \cdot (2n-1),$$

because

$$f^{(n-1)}(z) = (-1)^{n-1} n(n+1)(n+2) \cdot \ldots \cdot (2n-1)(z+i)^{-2n+1}.$$

Solution of Exercise 3.10.31

The function is holomorphic on the set $D = \mathbb{C} \setminus \{1\}$.

We need to make the following discussion according to the values of the parameter a.

Case 1. If $0 < a < 1$, then the function is holomorphic in the disc with the center in O and radius a. According to the *Cauchy integral theorem*, the value of the integral is zero, i. e.,

$$\int_\gamma \frac{ze^{\frac{i\pi}{2}z}}{z-1} dz = 0.$$

Case 2. If $a > 1$, then the disc with the center in O and radius a contains the simple pole $z_0 = 1$. If we use the *Cauchy formula for closed curves*, where $f(z) = ze^{\frac{i\pi}{2}z}$, $z_0 = 1$ and $k = 0$, then

$$\int_\gamma \frac{ze^{\frac{i\pi}{2}z}}{z-1} dz = 2\pi i e^{\frac{i\pi}{2}} = -2\pi.$$

Solution of Exercise 3.10.32

The function is holomorphic on the set $D = \mathbb{C} \setminus \{a\}$.

We need to make the following discussion according to the values of the parameter a.

Case 1. If $|a| < 1$, then point a belongs to the unit disc. Using the *Cauchy formula for closed curves*, where $f(z) = z^n$, $z_0 = a$ and $k = 0$, we obtain that

$$\int_\gamma \frac{z^n}{z-a} dz = 2a^n \pi i.$$

Case 2. If $|a| > 1$, then the function is holomorphic in the unit disc, and from the *Cauchy integral theorem* the value of the integral will be zero, i. e.,

$$\int_\gamma \frac{z^n}{z-a} \mathrm{d}z = 0.$$

Solution of Exercise 3.10.33

Since the function is obtained from elementary functions, we need to determine only the zeros of the denominator. Thus, the function f is holomorphic on the set $D = \mathbb{C} \setminus \{0, i\}$.

We need to make the following discussion according to the values of the parameter a.

Case 1. If $0 < a < 1$, then the disc with the center in O and radius $a > 0$ contains only the pole $z_1 = 0$. Using the *Cauchy formula for the disc*, where $f(z) = \frac{e^{\pi z}}{z-i}$, $z_0 = 0$ and $k = 0$, we get

$$\int_\gamma \frac{e^{\pi z}}{z(z-i)} \mathrm{d}z = \int_\gamma \frac{\frac{e^{\pi z}}{z-i}}{z} \mathrm{d}z = 2\pi i\left(-\frac{1}{i}\right) = -2\pi. \tag{7.18}$$

Case 2. If $a > 1$, then both of the poles of the function belong to the disc with the center in O and radius $a > 0$. We decompose the integral into the sum of two integrals, where the integration paths are disjoints and each of them turns around only one of the poles. Then, from a corollary of the *Cauchy integral theorem* we have

$$\int_\gamma \frac{e^{\pi z}}{z(z-i)} \mathrm{d}z = \int_{\gamma_1} \frac{\frac{e^{\pi z}}{z-i}}{z} \mathrm{d}z + \int_{\gamma_2} \frac{\frac{e^{\pi z}}{z}}{z-i} \mathrm{d}z,$$

where

$$\gamma_1(t) = r_1 e^{2\pi i t}, \quad \gamma_2(t) = i + r_2 e^{2\pi i t}, \quad t \in [0,1],$$

and $0 < r_k < \frac{1}{2}$, $k \in \{1, 2, \}$.

Using now the *Cauchy formula for the disc*, where $f(z) = \frac{e^{\pi z}}{z}$, $z_0 = i$ and $k = 0$, together with (7.18), we conclude that

$$\int_\gamma \frac{e^{\pi z}}{z(z-i)} \mathrm{d}z = -2\pi + 2\pi i \frac{e^{\pi i}}{i} = -2\pi - 2\pi = -4\pi.$$

Solution of Exercise 3.10.34

The function is holomorphic on the set $D = \mathbb{C} \setminus \{-\pi, \pi\}$.

The discussion will be similar to that of Exercise 3.10.33.

Case 1. If $0 < a < \pi$, then the function is holomorphic in the disc with the center in O and radius a, and from the *Cauchy integral theorem* the value of the integral is zero, i. e.,

$$\int_{\gamma} \frac{e^{iz}}{z^2 - \pi^2} dz = 0.$$

Case 2. If $a > \pi$, then both of the poles belong to the disc with the center in O and radius a. We decompose the integral into the sum of two integrals, where the integration paths are disjoint and each of them turns around only one of the poles. Then, from a corollary of the *Cauchy integral theorem* we have

$$\int_{\gamma} \frac{e^{iz}}{z^2 - \pi^2} dz = \int_{\gamma_1} \frac{\frac{e^{iz}}{z-\pi}}{z + \pi} dz + \int_{\gamma_2} \frac{\frac{e^{iz}}{z+\pi}}{z - \pi} dz,$$

where

$$\gamma_1(t) = -\pi + r_1 e^{2\pi it}, \quad \gamma_2(t) = \pi + r_2 e^{2\pi it}, \quad t \in [0, 1],$$

and $0 < r_k < a - \pi$, $k \in \{1, 2\}$.

Using the *Cauchy formula for the disc* for the functions $f(z) = \frac{e^{iz}}{z-\pi}$, where $z_0 = -\pi$, $k = 0$ and respectively $f(z) = \frac{e^{iz}}{z+\pi}$, where $z_0 = \pi$, $k = 0$, we have

$$\int_{\gamma} \frac{e^{iz}}{z^2 - \pi^2} dz = 2\pi i\left(-\frac{e^{-\pi i}}{2\pi}\right) + 2\pi i \frac{e^{\pi i}}{2\pi} = i - i = 0.$$

Solution of Exercise 3.10.35

The function is obtained from elementary functions, hence we need to determine only the zeros of the denominator. Thus, the function f is holomorphic on the set $D = \mathbb{C} \setminus \{-2, 2\}$.

The integration path is a circle with radius then $|\frac{a}{2}|$ and the center $\frac{a}{2}$, i. e., $\{\gamma\} = \partial U(\frac{a}{2}; |\frac{a}{2}|)$ with direct orientation.

We need to make the following discussion according to the values of the parameter a.

Case 1. If $|a| < 2$, then the function is holomorphic in the disc with radius the $|\frac{a}{2}|$ and the center $\frac{a}{2}$, and by the *Cauchy integral theorem* the value of the integral is zero, i. e.,

$$\int_{\gamma} \frac{\cos \pi z}{(z^2 - 4)^2} dz = 0.$$

Case 2. If $|a| > 2$, then the integration path turns around only one of the poles of the function, as follows:

Case 2.1. If $a < -2$, then we will use the *Cauchy formula for closed curves*, where $f(z) = \frac{\cos \pi z}{(z-2)^2}$, $z_0 = -2$ and $k = 1$. Since

$$f'(z) = -\frac{\pi z \sin \pi z - 2\pi \sin \pi z + 2\cos \pi z}{(z-2)^3},$$

it follows that

$$\int_\gamma \frac{\cos \pi z}{(z^2-4)^2}\mathrm{d}z = \int_\gamma \frac{\frac{\cos \pi z}{(z-2)^2}}{(z+2)^2}\mathrm{d}z = 2\pi i f'(z_0) = 2\pi i \frac{1}{32} = \frac{\pi i}{16}.$$

Case 2.2. If $a > 2$, then we will use the *Cauchy formula for closed curves*, where $f(z) = \frac{\cos \pi z}{(z+2)^2}$, $z_0 = 2$ and $k = 1$. Since

$$f'(z) = -\frac{\pi z \sin \pi z + 2\pi \sin \pi z + 2\cos \pi z}{(z+2)^3},$$

it follows that

$$\int_\gamma \frac{\cos \pi z}{(z^2-4)^2}\mathrm{d}z = \int_\gamma \frac{\frac{\cos \pi z}{(z+2)^2}}{(z-2)^2}\mathrm{d}z = 2\pi i f'(z_0) = 2\pi i\left(-\frac{1}{32}\right) = -\frac{\pi i}{16}.$$

Solution of Exercise 3.10.36

The function is holomorphic on the set $D = \mathbb{C} \setminus \{-i, i\}$, while the integration part is an ellipse that turns around both of the poles $z_1 = i$, $z_2 = -i$. We decompose the integral into the sum of two integrals, where the integration paths are disjoints and each of them turns around only one of the poles, i. e.,

$$\int_\gamma \frac{z^{100}e^{i\pi z}}{z^2+1}\mathrm{d}z = \int_{\gamma_1} \frac{\frac{z^{100}e^{i\pi z}}{z+i}}{z-i}\mathrm{d}z + \int_{\gamma_2} \frac{\frac{z^{100}e^{i\pi z}}{z-i}}{z+i}\mathrm{d}z;$$

where

$$\gamma_1(t) = i + r_1 e^{2\pi i t}, \quad \gamma_2(t) = -i + r_1 e^{2\pi i t}, \quad t \in [0,1]$$

and $0 < r_k < 1$, $k \in \{1,2\}$. Using the *Cauchy formula for the disc* for $f(z) = \frac{z^{100}e^{i\pi z}}{z+i}$, where $z_0 = i$, $k = 0$, and for $f(z) = \frac{z^{100}e^{i\pi z}}{z-i}$, where $z_0 = -i$, $k = 0$, we deduce that

$$\int_\gamma \frac{z^{100}e^{i\pi z}}{z^2+1}\mathrm{d}z = 2\pi i\frac{i^{100}e^{-\pi}}{2i} + 2\pi i\frac{(-i)^{100}e^{\pi}}{-2i} = \pi e^{-\pi} - \pi e^{\pi} = -2\pi \sinh \pi.$$

Solution of Exercise 3.10.37

The function is holomorphic on $D = \mathbb{C}\setminus\{-i\}$, and the integration path turns around the pole $z = i$. Using the *Cauchy formula for closed curves* for the function $f(z) = \cosh\frac{\pi z}{2}$, where $z_0 = -i$ and $k = 3$, since

$$f^{(3)}(z) = \frac{\pi^3}{8}\sinh\frac{\pi z}{2},$$

it follows that

$$\int_{\gamma} \frac{\cosh\frac{\pi z}{2}}{(z+i)^4}\mathrm{d}z = \frac{2\pi i}{3!}f^{(3)}(-i) = \frac{\pi^4}{24}.$$

Solution of Exercise 3.10.38

If $z = re^{i\frac{\pi(2t-1)}{2}}$, with $t \in [0,1]$, since $\lim_{z\to\infty} zf(z) = 0$ it follows that

$$\lim_{r\to\infty} rf(\gamma_r(t)) = 0, \quad \forall t \in [0,1]. \tag{7.19}$$

Since f is continuous, there exists

$$M(r) = \max\{|f(z)| : z \in \{\gamma_r\}\} = \max\{|f(\gamma_r(t))| : t \in [0,1]\} < +\infty.$$

From (7.19), we obtain that

$$\lim_{r\to\infty} rM(r) = 0. \tag{7.20}$$

From the definition of the integral, we have

$$\begin{aligned}\left|\int_{\gamma_r} e^{az}f(z)\mathrm{d}z\right| &= \left|\int_0^1 e^{a\gamma_r(t)}f(\gamma_r(t))\gamma_r'(t)\mathrm{d}t\right| \le \int_0^1 \left|e^{a\gamma_r(t)}f(\gamma_r(t))\gamma_r'(t)\right|\mathrm{d}t\\ &\le M(r)\int_0^1 \left|e^{a\gamma_r(t)}\gamma_r'(t)\right|\mathrm{d}t = rM(r)\pi\int_0^1 e^{ar\cos\frac{\pi(2t-1)}{2}}\mathrm{d}t,\end{aligned}$$

i. e.,

$$\left|\int_{\gamma_r} e^{az}f(z)\mathrm{d}z\right| \le rM(r)\pi\int_0^1 e^{ar\cos\frac{\pi(2t-1)}{2}}\mathrm{d}t. \tag{7.21}$$

But

$$\int_0^1 e^{ar\cos\frac{\pi(2t-1)}{2}}\mathrm{d}t = \frac{2}{\pi}\int_0^{\frac{\pi}{2}} e^{ar\cos u}\mathrm{d}u,$$

and from $a \le 0$ we have that the function $\varphi(u) = e^{ar\cos u}$ is decreasing and convex in $[0, \frac{\pi}{2}]$, and thus

$$\int_0^{\frac{\pi}{2}} e^{ar\cos u}du \le \frac{(1+e^{ar})\pi}{4},$$

hence

$$\int_0^1 e^{ar\cos\frac{\pi(2t-1)}{2}}dt \le \frac{1+e^{ar}}{2}.$$

From here, using the inequality (7.21) it follows that

$$\left|\int_{\gamma_r} e^{az}f(z)\mathrm{d}z\right| \le rM(r)\frac{(1+e^{ar})\pi}{2}.$$

Since $a \le 0$, from the above inequality and from (7.20) we deduce that

$$\lim_{r\to\infty}\left|\int_{\gamma_r} e^{az}f(z)\mathrm{d}z\right| = 0.$$

Solution of Exercise 3.10.39

The proof is similar with the proof of the previous problem. If $z = re^{i\frac{\pi(2t+1)}{2}}$, with $t \in [0,1]$, since $\lim_{z\to\infty} zf(z) = 0$ it follows that

$$\lim_{r\to\infty} rf(\gamma_r(t)) = 0, \quad \forall t \in [0,1]. \tag{7.22}$$

Since f is continuous, there exists

$$M(r) = \max\{|f(z)| : z \in \{\gamma_r\}\} = \max\{|f(\gamma_r(t))| : t \in [0,1]\} < +\infty.$$

From (7.22), we obtain that

$$\lim_{r\to\infty} rM(r) = 0. \tag{7.23}$$

From the definition of the integral, we have

$$\left|\int_{\gamma_r} e^{az}f(z)\mathrm{d}z\right| = \left|\int_0^1 e^{a\gamma_r(t)}f(\gamma_r(t))\gamma_r'(t)\mathrm{d}t\right| \le \int_0^1 |e^{a\gamma_r(t)}f(\gamma_r(t))\gamma_r'(t)|\mathrm{d}t$$

$$\le M(r)\int_0^1 |e^{a\gamma_r(t)}\gamma_r'(t)|\mathrm{d}t = rM(r)\pi\int_0^1 e^{ar\cos\frac{\pi(2t+1)}{2}}\mathrm{d}t,$$

i. e.,

$$\left|\int_{\gamma_r} e^{az} f(z)\mathrm{d}z\right| \leq rM(r)\pi \int_0^1 e^{ar\cos\frac{\pi(2t+1)}{2}}\mathrm{d}t. \tag{7.24}$$

But

$$\int_0^1 e^{ar\cos\frac{\pi(2t+1)}{2}}\mathrm{d}t = \frac{2}{\pi}\int_0^{\frac{\pi}{2}} e^{-ar\cos u}\mathrm{d}u,$$

and from $a \geq 0$ we have that the function $\varphi(u) = e^{-ar\cos u}$ is decreasing and convex on $[0, \frac{\pi}{2}]$, and thus

$$\int_0^{\frac{\pi}{2}} e^{-ar\cos u}\mathrm{d}u \leq \frac{(1+e^{-ar})\pi}{4},$$

hence

$$\int_0^1 e^{ar\cos\frac{\pi(2t+1)}{2}}\mathrm{d}t \leq \frac{1+e^{-ar}}{2}.$$

From here, using the inequality (7.24) it follows that

$$\left|\int_{\gamma_r} e^{az} f(z)\mathrm{d}z\right| \leq rM(r)\frac{(1+e^{-ar})\pi}{2}.$$

Since $a \geq 0$, from the above inequality and from (7.23) we deduce that

$$\lim_{r\to\infty}\left|\int_{\gamma_r} e^{az} f(z)\mathrm{d}z\right| = 0.$$

Solution of Exercise 3.10.40

If $f = u + iv \in H(G)$, then

$$\frac{\partial^2 u}{\partial x^2} + \frac{\partial^2 u}{\partial y^2} = 0, \quad \frac{\partial^2 v}{\partial x^2} + \frac{\partial^2 v}{\partial y^2} = 0,$$

i. e., u and v are harmonic functions.

If $z = x + iy$ and $z_0 = x_0 + iy_0$, it is well known that

$$u(z) - u(z_0) = \int_{z_0}^{z} du(z) = \int_{z_0}^{z} \frac{\partial u(z)}{\partial x}dx + \frac{\partial u(z)}{\partial y}dy$$

$$= \int_{z_0}^{z} \frac{\partial v(z)}{\partial y}dx - \frac{\partial v(z)}{\partial x}dy = \int_{x_0}^{x} \frac{\partial v(x, y_0)}{\partial y}dx - \int_{y_0}^{y} \frac{\partial v(x, y)}{\partial x}dy$$

and

$$v(z) - v(z_0) = \int_{z_0}^{z} dv(z) = \int_{z_0}^{z} \frac{\partial v(z)}{\partial x}dx + \frac{\partial v(z)}{\partial y}dy$$

$$= \int_{z_0}^{z} -\frac{\partial u(z)}{\partial y}dx + \frac{\partial u(z)}{\partial x}dy = -\int_{x_0}^{x} \frac{\partial u(x, y_0)}{\partial y}dx + \int_{y_0}^{y} \frac{\partial u(x, y)}{\partial x}dy.$$

From here, we deduce that the corresponding real and imaginary parts of the function f are given by

$$u(z) - u(z_0) = \int_{x_0}^{x} \frac{\partial v(x, y_0)}{\partial y}dx - \int_{y_0}^{y} \frac{\partial v(x, y)}{\partial x}dy,$$

$$v(z) - v(z_0) = -\int_{x_0}^{x} \frac{\partial u(x, y_0)}{\partial y}dx + \int_{y_0}^{y} \frac{\partial u(x, y)}{\partial x}dy.$$

In the next special cases, we will use the above methods to determine the real or the imaginary part of the functions.

1. Since $v(x, y) = e^x \sin y$, it follows that

$$\frac{\partial v}{\partial x} = e^x \sin y, \quad \frac{\partial v}{\partial y} = e^x \cos y \Rightarrow \frac{\partial^2 v}{\partial x^2} = e^x \sin y, \quad \frac{\partial^2 v}{\partial y^2} = -e^x \sin y,$$

thus $\frac{\partial^2 v}{\partial x^2} + \frac{\partial^2 v}{\partial y^2} = 0$, hence v is a harmonic function.

Let $z_0 = x_0 + iy_0 = 0$ and $z = x + iy$. We will determine the real part u using the above method:

$$u(z) - u(0, 0) = \int_{0}^{x} \frac{\partial v(x, 0)}{\partial y}dx - \int_{0}^{y} \frac{\partial v(x, y)}{\partial x}dy = e^x \cos y - 1,$$

thus $u(x, y) = e^x \cos y - 1 + u(0, 0) = e^x \cos y + k$, $k \in \mathbb{R}$, hence

$$f(z) = e^x \cos y + k + ie^x \sin y, \quad k \in \mathbb{R}.$$

From the assumption $f(0) = 1$, we will determine the constant $k \in \mathbb{R}$ as follows: $f(0) = 1 \Leftrightarrow k = 0$. Hence, the function will be

$$f(z) = e^x \cos y + ie^x \sin y = e^z.$$

2. Since $v(x,y) = x^2 - y^2 + xy$, it follows that

$$\frac{\partial v}{\partial x} = 2x + y, \quad \frac{\partial v}{\partial y} = -2y + x \Rightarrow \frac{\partial^2 v}{\partial x^2} = 2, \quad \frac{\partial^2 v}{\partial y^2} = -2,$$

thus $\frac{\partial^2 v}{\partial x^2} + \frac{\partial^2 v}{\partial y^2} = 0$, hence v is a harmonic function.

Let $z_0 = x_0 + iy_0 = 0$ and $z = x + iy$. Then

$$u(z) - u(0,0) = \int_0^x \frac{\partial v(x,0)}{\partial y} \mathrm{d}x - \int_0^y \frac{\partial v(x,y)}{\partial x} \mathrm{d}y = \frac{x^2}{2} - 2xy - \frac{y^2}{2},$$

thus $u(x,y) = \frac{x^2}{2} - 2xy - \frac{y^2}{2} + u(0,0) = \frac{x^2}{2} - 2xy - \frac{y^2}{2} + k$, $k \in \mathbb{R}$, hence

$$f(z) = \frac{x^2}{2} - 2xy - \frac{y^2}{2} + k + i(x^2 - y^2 + xy), \quad k \in \mathbb{R}.$$

From the assumption $f(0) = 0$, we will determine the constant $k \in \mathbb{R}$ as follows: $f(0) = 0 \Leftrightarrow k = 0$. Hence, the function will be

$$f(z) = (1 + 2i)\frac{z^2}{2}.$$

3. Since $v(x,y) = 2xy$, it follows that

$$\frac{\partial v}{\partial x} = 2y, \quad \frac{\partial v}{\partial y} = 2x \Rightarrow \frac{\partial^2 v}{\partial x^2} = \frac{\partial^2 v}{\partial y^2} = 0,$$

thus $\frac{\partial^2 v}{\partial x^2} + \frac{\partial^2 v}{\partial y^2} = 0$, hence v is a harmonic function.

Let $z_0 = x_0 + iy_0 = 0$ and $z = x + iy$. Then

$$u(z) - u(0,0) = \int_0^x \frac{\partial v(x,0)}{\partial y} \mathrm{d}x - \int_0^y \frac{\partial v(x,y)}{\partial x} \mathrm{d}y = x^2 - y^2,$$

thus $u(x,y) = x^2 - y^2 + u(0,0) = x^2 - y^2 + k$, $k \in \mathbb{R}$, hence

$$f(z) = x^2 - y^2 + k + i2xy, \quad k \in \mathbb{R}.$$

From the assumption $f(0) = 0$, we will determine the constant $k \in \mathbb{R}$ as follows: $f(0) = 0 \Leftrightarrow k = 0$. Hence, the function will be

$$f(z) = x^2 - y^2 + i2xy = z^2.$$

4. If $u(x,y) = e^x(x\cos y - y\sin y)$, we have that

$$\frac{\partial u}{\partial x} = e^x(x\cos y + \cos y - y\sin y), \quad \frac{\partial u}{\partial y} = -e^x(x\sin y + \sin y + y\cos y)$$

$$\Rightarrow \frac{\partial^2 u}{\partial x^2} = e^x(x\cos y + 2\cos y - y\sin y), \quad \frac{\partial^2 u}{\partial y^2} = -e^x(x\cos y + 2\cos y - y\sin y),$$

hence $\frac{\partial^2 u}{\partial x^2} + \frac{\partial^2 u}{\partial y^2} = 0$, and thus u is a harmonic function.

Let $z_0 = x_0 + iy_0 = 0$ and $z = x + iy$. We will determine the imaginary part v using the above method:

$$v(z) - v(0,0) = -\int_0^x \frac{\partial u(x,0)}{\partial y}\mathrm{d}x + \int_0^y \frac{\partial u(x,y)}{\partial x}\mathrm{d}y = e^x(x\sin y + y\cos y),$$

hence $v(x,y) = e^x(x\sin y + y\cos y) + v(0,0) = e^x(x\sin y + y\cos y) + k$, $k \in \mathbb{R}$, thus

$$f(z) = e^x(x\cos y - y\sin y) + i[e^x(x\sin y + y\cos y) + k], \quad k \in \mathbb{R}.$$

From the assumption $f(0) = 0$, we will determine the constant $k \in \mathbb{R}$ as follows: $f(0) = 0 \Leftrightarrow k = 0$. Now, the function will be

$$f(z) = e^x(x\cos y - y\sin y) + ie^x(x\sin y + y\cos y) = ze^z.$$

Solution of Exercise 3.10.41

The solution is similar to the above problem, with the difference that the *starting point* $z_0 \in \mathbb{C}$ should not be chosen $z_0 = 0$ (because $0 \notin D$), but (for example) it could be $z_0 = 1 + 0i = 1 \in D$.

Since

$$\frac{\partial v}{\partial x} = \frac{-2xy}{(x^2+y^2)^2}, \quad \frac{\partial v}{\partial y} = \frac{x^2-y^2}{(x^2+y^2)^2}$$

$$\Rightarrow \frac{\partial^2 v}{\partial x^2} = \frac{2y(3x^2-y^2)}{(x^2+y^2)^3}, \quad \frac{\partial^2 v}{\partial y^2} = -\frac{2y(3x^2-y^2)}{(x^2+y^2)^3},$$

the function v is harmonic.

Now, the real part u may be found as follows:

$$u(z) - u(1,0) = \int_1^x \frac{\partial v(x,0)}{\partial y}\mathrm{d}x - \int_0^y \frac{\partial v(x,y)}{\partial x}\mathrm{d}y = 1 - \frac{x}{x^2+y^2},$$

hence

$$u(x,y) = 1 - \frac{x}{x^2+y^2} + u(1,0) = -\frac{x}{x^2+y^2} + k, \quad k \in \mathbb{R}.$$

The constant $k \in \mathbb{R}$ can be found by using the assumption $f(1) = 0$, i. e., $f(1) = 0 \Leftrightarrow k = 1$. Finally, we get $f(z) = 1 - \frac{1}{z}$.

Solution of Exercise 3.10.42

1. The function u is well-defined if and only if $(0,0) \notin D$. Since

$$\frac{\partial u}{\partial x} = \frac{-2x(x^2-3y^2)}{(y^2+x^2)^3}, \quad \frac{\partial u}{\partial y} = \frac{-2y(3x^2-y^2)}{(y^2+x^2)^3},$$

it follows that

$$\frac{\partial^2 u}{\partial y^2} = -\frac{\partial^2 u}{\partial x^2} = \frac{-6(x^4+y^4-6y^2x^2)}{(x^2+y^2)^4},$$

thus u is a harmonic function.

If we choose the starting point $z_0 = 1 = 1 + 0i \in D$, then

$$\begin{aligned} v(x,y) - v(1,0) &= -\int_1^x \frac{\partial u(x,0)}{\partial y}\,dx + \int_0^y \frac{\partial u(x,y)}{\partial x}\,dy \\ &= \int_0^y \frac{-2x(x^2-3y^2)}{(y^2+x^2)^3}\,dy = -\frac{2xy}{(x^2+y^2)^2}, \end{aligned}$$

hence

$$f(z) = \frac{1}{z^2} + ki, \quad k \in \mathbb{R},$$

and $f \in H(\mathbb{C}^*)$.

2. From the definition of the function v, we have that $(0,0) \notin D$, and

$$\frac{\partial v}{\partial x} = \frac{2x}{y^2+x^2} - 2x, \quad \frac{\partial v}{\partial y} = \frac{2y}{y^2+x^2} + 2y,$$

then

$$\frac{\partial^2 v}{\partial x^2} = -\frac{\partial^2 v}{\partial y^2} = \frac{2(y^2-x^2)}{(x^2+y^2)^2} - 2,$$

and thus v is a harmonic function.

If we choose the starting point $z_0 = 1 = 1 + 0i \in D$, then

$$\begin{aligned} u(x,y) - u(1,0) &= \int_1^x \frac{\partial v(x,0)}{\partial y}\,dx - \int_0^y \frac{\partial v(x,y)}{\partial x}\,dy \\ &= -\int_0^y \left(\frac{2x}{y^2+x^2} - 2x\right)dy = -2\arctan\frac{y}{x} + 2xy, \end{aligned}$$

and thus

$$f(z) = -2\arctan\frac{y}{x} + 2xy + k + i(\ln(x^2+y^2) - x^2 + y^2)$$
$$= i(2\log z - z^2) + k, \quad k \in \mathbb{R},$$

with $D = \mathbb{C} \setminus \{z \in \mathbb{C} : \operatorname{Re} z \le 0, \operatorname{Im} z = 0\}$.

3. Ha $z = x + iy \in \Delta = \mathbb{C} \setminus \{z \in \mathbb{C} : \operatorname{Re} z = 0\}$, then

$$\frac{\partial u}{\partial x} = -\frac{y}{x^2}e^{\frac{y}{x}}, \quad \frac{\partial u}{\partial y} = e^{\frac{y}{x}}\frac{1}{x},$$

hence

$$\frac{\partial^2 u}{\partial x^2} = e^{\frac{y}{x}}\frac{y(2x+y)}{x^4}, \quad \frac{\partial^2 u}{\partial y^2} = e^{\frac{y}{x}}\frac{1}{x^2}.$$

From here, we have that u is harmonic only if

$$\frac{\partial^2 u}{\partial x^2} + \frac{\partial^2 v}{\partial y^2} = 0 \Leftrightarrow e^{\frac{y}{x}}\frac{y(2x+y)}{x^4} + e^{\frac{y}{x}}\frac{1}{x^2} = 0 \Leftrightarrow (x+y)^2 = 0$$
$$\Leftrightarrow z \in \widetilde{D} = \{z = x + iy \in \mathbb{C} : y = -x, x \neq 0\}.$$

But $\widetilde{D} \subset \mathbb{C}$ is not a domain, thus there does not exist any holomorphic function in a domain $D \subset \mathbb{C}$, of the form $f = u + iv$, such that $u(x,y) = e^{\frac{y}{x}}$.

Solution of Exercise 3.10.43

1. If $v(x,y) = \frac{y}{x^2+y^2}$, first we will check that it is a harmonic function. A simple calculation shows that

$$\frac{\partial v}{\partial x} = -\frac{2xy}{(x^2+y^2)^2}, \quad \frac{\partial v}{\partial y} = \frac{x^2-y^2}{(x^2+y^2)^2} \Rightarrow \frac{\partial^2 v}{\partial x^2} = -\frac{\partial^2 v}{\partial y^2} = \frac{2y(3x^2-y^2)}{(x^2+y^2)^3},$$

hence the function v is harmonic.

Since $z_0 = 0 \notin D$, let choose $z_0 = -1 = -1 + 0i \in D$ to be the starting point. We have

$$u(x,y) - u(-1,0) = \int_{-1}^{x} \frac{\partial v(x,0)}{\partial y}\mathrm{d}x - \int_{0}^{y} \frac{\partial v(x,y)}{\partial x}\mathrm{d}y$$
$$= \int_{-1}^{x} \frac{1}{x^2}\mathrm{d}x - \int_{0}^{y} \frac{2xy}{(x^2+y^2)^2}\mathrm{d}y = -1 - \frac{x}{x^2+y^2},$$

thus $u(x,y) = 1 - \frac{x}{x^2+y^2} + u(-1,0) = -\frac{x}{x^2+y^2} + k$, $k \in \mathbb{R}$. From here, we deduce that

$$f(z) = -\frac{x}{x^2+y^2} + k + i\frac{y}{x^2+y^2}, \quad k \in \mathbb{R}.$$

The constant $k \in \mathbb{R}$ will be determined by using the assumption $f(-2) = 0$, i. e., $k = -\frac{1}{2}$. Finally, we get

$$f(z) = -\frac{x}{x^2+y^2} - \frac{1}{2} + i\frac{y}{x^2+y^2} = -\frac{1}{z} - \frac{1}{2}.$$

2. If $v(x,y) = 3 + x^2 - y^2 - \frac{y}{2(x^2+y^2)}$, similarly

$$\frac{\partial v}{\partial x} = 2x + \frac{xy}{(x^2+y^2)^2}, \quad \frac{\partial v}{\partial y} = -2y - \frac{x^2-y^2}{2(x^2+y^2)^2}$$

$$\Rightarrow \frac{\partial^2 v}{\partial x^2} = -\frac{\partial^2 v}{\partial y^2} = 2 + \frac{y(y^2-3x^2)}{(x^2+y^2)^3},$$

thus v is a harmonic function.

For the starting point $z_0 = 2 = 2 + 0i \in D$, we obtain

$$\begin{aligned} u(x,y) - u(2,0) &= \int_2^x \frac{\partial v(x,0)}{\partial y}\mathrm{d}x - \int_0^y \frac{\partial v(x,y)}{\partial x}\mathrm{d}y \\ &= \int_2^x \left(-\frac{1}{2x^2}\right)\mathrm{d}x - \int_0^y \left(2x + \frac{xy}{(x^2+y^2)^2}\right)\mathrm{d}y = -\frac{1}{4} - 2xy + \frac{x}{2(x^2+y^2)}, \end{aligned}$$

thus $u(x,y) = \frac{x}{2(x^2+y^2)} - 2xy + k$, $k \in \mathbb{R}$. The function f will be of the form

$$f(z) = -2xy + \frac{x}{2(x^2+y^2)} + k + i\left(3 + x^2 - y^2 - \frac{y}{2(x^2+y^2)}\right), \quad k \in \mathbb{R},$$

and since $f(2) = \frac{1}{4} + k + 7i \neq 0$, $\forall k \in \mathbb{R}$, we conclude that there not exists any holomorphic function which satisfies the assumptions.

3. Since $u(x,y) = \frac{1}{2}\ln(x^2+y^2)$, we have

$$\frac{\partial u}{\partial x} = \frac{x}{x^2+y^2}, \quad \frac{\partial u}{\partial y} = \frac{y}{x^2+y^2} \Rightarrow \frac{\partial^2 u}{\partial x^2} = -\frac{\partial^2 u}{\partial y^2} = \frac{y^2-x^2}{(x^2+y^2)^2},$$

thus the function u is harmonic.

Let choose $z_0 = 1 = 1 + 0i \in D$ to be the starting point. Then

$$v(x,y) - v(1,0) = -\int_1^x \frac{\partial u(x,0)}{\partial y}\mathrm{d}x + \int_0^y \frac{\partial u(x,y)}{\partial x}\mathrm{d}y = \int_0^y \frac{x}{x^2+y^2}\mathrm{d}y = \arctan\frac{y}{x},$$

hence $v(x,y) = \arctan\frac{y}{x} + k$, $k \in \mathbb{R}$. From here, we get

$$f(z) = \ln\sqrt{x^2+y^2} + i\left(\arctan\frac{y}{x} + k\right) = \log z + ik, \quad k \in \mathbb{R}.$$

4. If $v(x,y) = \ln(x^2+y^2) + x - 2y$, then

$$\frac{\partial v}{\partial x} = 1 + \frac{2x}{x^2+y^2}, \quad \frac{\partial v}{\partial y} = -2 + \frac{2y}{x^2+y^2} \Rightarrow \frac{\partial^2 v}{\partial x^2} = -\frac{\partial^2 v}{\partial y^2} = \frac{2(y^2-x^2)}{(x^2+y^2)^2},$$

hence the function v is harmonic.

Since $z_0 = 0 \notin D$, choose the starting point to be $z_0 = 1 = 1 + 0i \in D$. Then

$$\begin{aligned} u(x,y) - u(1,0) &= \int_1^x \frac{\partial v(x,0)}{\partial y} dx - \int_0^y \frac{\partial v(x,y)}{\partial x} dy \\ &= \int_1^x (-2) dx - \int_0^y \left(\frac{2x}{x^2+y^2} + 1\right) dy = -2x + 2 - y - 2\arctan\frac{y}{x}, \end{aligned}$$

hence $u(x,y) = -2x - y - 2\arctan\frac{y}{x} + k$, $k \in \mathbb{R}$. The required function will be

$$f(z) = 2i \log z - (2-i)z + k, \quad k \in \mathbb{R}.$$

5. If $v(x,y) = 1 + xy$, then

$$\frac{\partial v}{\partial x} = y, \quad \frac{\partial v}{\partial y} = x \Rightarrow \frac{\partial^2 v}{\partial x^2} = \frac{\partial^2 v}{\partial y^2} = 0,$$

thus v is a harmonic function.

Let $z_0 = 0 \in D$ be the starting point. Then

$$u(x,y) - u(0,0) = \int_0^x \frac{\partial v(x,0)}{\partial y} dx - \int_0^y \frac{\partial v(x,y)}{\partial x} dy = \int_0^x x dx - \int_0^y y dy = \frac{1}{2}(x^2 - y^2),$$

hence

$$f(z) = \frac{z^2}{2} + k + i, \quad k \in \mathbb{R}.$$

6. Since $v(x,y) = y - \frac{y}{x^2+y^2}$, it follows that

$$\frac{\partial v}{\partial x} = \frac{2xy}{(x^2+y^2)^2}, \quad \frac{\partial v}{\partial y} = 1 - \frac{x^2-y^2}{(x^2+y^2)^2} \Rightarrow \frac{\partial^2 v}{\partial x^2} = -\frac{\partial^2 v}{\partial y^2} = \frac{2y(3x^2-y)}{(x^2+y^2)^3},$$

thus v is a harmonic function.

If we choose the starting point $z_0 = i = 0 + i \in D$, we get

$$\begin{aligned} u(x,y) - u(0,1) &= \int_0^x \frac{\partial v(x,1)}{\partial y} dx - \int_1^y \frac{\partial v(x,y)}{\partial x} dy \\ &= \int_0^x \left(1 - \frac{x^2-1}{(x^2+1)^2}\right) dx - \int_1^y \frac{2xy}{(x^2+y^2)^2} dy = x + \frac{x}{x^2+y^2}. \end{aligned}$$

Hence, the function will be of the form

$$f(z) = z + \frac{1}{z} + k, \quad k \in \mathbb{R}.$$

7. If $u(x,y) = \frac{x}{x^2+(y+1)^2} + \frac{x}{x^2+(y-1)^2}$, then

$$\frac{\partial u}{\partial x} = \frac{(y+1)^2 - x^2}{(x^2+(y+1)^2)^2} + \frac{(y-1)^2 - x^2}{(x^2+(y-1)^2)^2},$$
$$\frac{\partial u}{\partial y} = \frac{-2(y+1)x}{(x^2+(y+1)^2)^2} + \frac{-2(y-1)x}{(x^2+(y-1)^2)^2},$$

hence

$$\frac{\partial^2 u}{\partial y^2} = -\frac{\partial^2 u}{\partial x^2} = \frac{2x(3(y+1)^2 - x^2)}{(x^2+(y+1)^2)^3} + \frac{2x(3(y-1)^2 - x^2)}{(x^2+(y-1)^2)^3},$$

and the function u is harmonic.

If we let $z_0 = 0 \in D$ be the starting point, then

$$\begin{aligned} v(x,y) - v(0,0) &= -\int_0^x \frac{\partial u(x,0)}{\partial y}\mathrm{d}x + \int_0^y \frac{\partial u(x,y)}{\partial x}\mathrm{d}y \\ &= \int_0^y \left(\frac{(y+1)^2 - x^2}{(x^2+(y+1)^2)^2} + \frac{(y-1)^2 - x^2}{(x^2+(y-1)^2)^2} \right)\mathrm{d}y \\ &= -\frac{y+1}{(y+1)^2 + x^2} - \frac{y-1}{(y-1)^2 + x^2}, \end{aligned}$$

thus

$$\begin{aligned} f(z) = {} & \frac{x}{x^2+(y+1)^2} + \frac{x}{x^2+(y-1)^2} \\ & + i\left(-\frac{y+1}{(y+1)^2+x^2} - \frac{y-1}{(y-1)^2+x^2} + k \right), \quad k \in \mathbb{R}. \end{aligned}$$

From the assumption $f(1) = 1$, it follows that $k = 0$, and then

$$f(z) = \frac{2z}{z^2+1}.$$

8. From $v(x,y) = -\frac{y}{(x+1)^2+y^2} - \frac{y}{(x-1)^2+y^2}$, we have that

$$\frac{\partial v}{\partial x} = \frac{2(x+1)y}{(y^2+(x+1)^2)^2} + \frac{2(x-1)y}{(y^2+(x-1)^2)^2},$$
$$\frac{\partial v}{\partial y} = -\frac{(x+1)^2 - y^2}{(y^2+(x+1)^2)^2} - \frac{(x-1)^2 - y^2}{(y^2+(x-1)^2)^2},$$

hence

$$\frac{\partial^2 v}{\partial x^2} = -\frac{\partial^2 v}{\partial y^2} = \frac{-2y(3(x+1)^2 - y^2)}{(y^2 + (x+1)^2)^3} + \frac{-2y(3(x-1)^2 - y^2)}{(y^2 + (x-1)^2)^3},$$

thus, the function v is harmonic.

Choosing $z_0 = 0 \in D$, we obtain

$$\begin{aligned} u(x,y) - u(0,0) &= \int_0^x \frac{\partial v(x,0)}{\partial y}\mathrm{d}x - \int_0^y \frac{\partial v(x,y)}{\partial x}\mathrm{d}y \\ &= \int_0^x \left(-\frac{1}{(x+1)^2} - \frac{1}{(x-1)^2}\right)\mathrm{d}x \\ &\quad - \int_0^y \left(\frac{2(x+1)y}{(y^2+(x+1)^2)^2} + \frac{2(x-1)y}{(y^2+(x-1)^2)^2}\right)\mathrm{d}y \\ &= \frac{x+1}{(x+1)^2+y^2} + \frac{x-1}{(x-1)^2+y^2}, \end{aligned}$$

hence

$$\begin{aligned} f(z) = {} & \frac{x+1}{(x+1)^2+y^2} + \frac{x-1}{(x-1)^2+y^2} + k \\ & + i\left(-\frac{y}{y^2+(x+1)^2} - \frac{y}{y^2+(x-1)^2}\right), \quad k \in \mathbb{R}. \end{aligned}$$

The assumption $f(0) = 0$ yields that $k = 0$, and

$$f(z) = \frac{2z}{z^2-1}.$$

9. If $v(x,y) = -\frac{2xy}{(x^2-y^2-1)^2+4x^2y^2}$, then

$$\begin{aligned} \frac{\partial v}{\partial x} &= -\frac{2y((x^2-y^2-1)^2 - 4x^4 + 4x^2)}{((x^2-y^2-1)^2 + 4x^2y^2)^2} \\ \frac{\partial v}{\partial y} &= -\frac{2x((x^2-y^2-1)^2 - 4y^4 + 4y^2)}{((x^2-y^2-1)^2 + 4x^2y^2)^2}. \end{aligned}$$

The relation

$$\begin{aligned} \frac{\partial^2 v}{\partial x^2} &= -\frac{\partial^2 v}{\partial y^2} \\ &= \frac{-8xy(3 - 14x^2y^2 + 3y^2 - 3x^2 - 3x^4 - 3y^4 + 3x^6 - 3y^6 + 3x^4y^2 - 3x^2y^4)}{(x^2-2x+y^2+1)^3(x^2+2x+y^2+1)^3} \end{aligned}$$

shows that v is a harmonic function.

Letting $z_0 = 0 \in D$, we have

$$u(x,y) - u(0,0) = \int_0^x \frac{\partial v(x,0)}{\partial y}\mathrm{d}x - \int_0^y \frac{\partial v(x,y)}{\partial x}\mathrm{d}y$$
$$= \int_0^x \left(-\frac{2x}{(x^2-1)^2}\right)\mathrm{d}x + \int_0^y \frac{2y((x^2-y^2-1)^2 - 4x^4 + 4x^2)}{((x^2-y^2-1)^2 + 4x^2y^2)^2}\mathrm{d}y$$
$$= \frac{x^2-y^2-1}{(x^2-y^2-1)^2+4x^2y^2} + 1,$$

thus

$$f(z) = \frac{x^2-y^2-1}{(x^2-y^2-1)^2+4x^2y^2} + k - i\frac{2xy}{(x^2-y^2-1)^2+4x^2y^2}, \quad k \in \mathbb{R}.$$

From the assumption $f(i) = -\frac{1}{2}$, we obtain $k = 0$, hence

$$f(z) = \frac{1}{z^2-1}.$$

Solution of Exercise 3.10.44

1. We will write the function in the exponential form. If $z = r_1 e^{i\theta_1}$ and $z - 2i = r_2 e^{i\theta_2}$, where $r_1, r_2 > 0$, then $2i - z = r_2 e^{i(\theta_2+\pi)}$. We obtain that

$$w_k(z) = \sqrt[3]{\frac{r_1}{r_2}} e^{i\frac{\theta_1-\theta_2-\pi+2k\pi}{3}}, \quad k \in \{0,1,2\}.$$

If the function w is holomorphic in $D \subset \mathbb{C}$, i. e., $w \in H(D)$, then for any arbitrary closed curve that lies in D, i. e., $\gamma \in \mathcal{D}_D(z_0)$, $\forall z_0 \in D$, the branch of the multivalued function w remains the same whenever the point $z \in \{\gamma\}$ runs over all the path γ.

We will discuss the next three cases:

Case 1. If the closed curve γ turns around only the point 0. If $z \in \{\gamma\}$ runs once over the curve γ, then we obtain the value

$$w_{k+1}(z) = \sqrt[3]{\frac{r_1}{r_2}} e^{i\frac{\theta_1-\theta_2-\pi+2(k+1)\pi}{3}}, \quad k \in \{0,1,2\}.$$

Case 2. If the closed curve γ turns around, we have only the point $2i$. If $z \in \{\gamma\}$ runs once over the curve γ, then we obtain the value

$$w_{k-1}(z) = \sqrt[3]{\frac{r_1}{r_2}} e^{i\frac{\theta_1-\theta_2-\pi+2(k-1)\pi}{3}}, \quad k \in \{0,1,2\}.$$

Case 3. If the closed curve γ turns around both of the points 0 and $2i$. If $z \in \{\gamma\}$ runs once over the curve γ, then we obtain the value

$$w_k(z) = \sqrt[3]{\frac{r_1}{r_2}} e^{i\frac{\theta_1-\theta_2-\pi-2k\pi}{3}}, \quad k \in \{0,1,2\}.$$

It follows that the domain D cannot contain only one of these points, but if D contains one of them, then it is necessary to contain both of these points. Hence

$$D = \mathbb{C} \setminus \{z \in \mathbb{C} : 0 \le \operatorname{Im} z \le 2, \operatorname{Re} z = 0\},$$

or

$$D = \mathbb{C} \setminus \{z \in \mathbb{C} : \operatorname{Im} z \ge 0, \operatorname{Re} z = 0\}, \quad D = \mathbb{C} \setminus \{z \in \mathbb{C} : \operatorname{Im} z \le 2, \operatorname{Re} z = 0\}.$$

The other points of this problem will be solved similarly.

2. The exponential form of the function is

$$w_k(z) = \sqrt[3]{r_1} e^{i\frac{\theta_1+2k\pi}{3}}, \quad k \in \{0,1,2\},$$

if we denote

$$z+1 = z-(-1) = r_1 e^{i\theta_1}, \quad r_1 > 0.$$

Here, we will discuss only one case. If the closed curve γ turns once around the point -1, then we obtain the value

$$w_{k+1}(z) = \sqrt[3]{r_1} e^{i\frac{\theta_1+2(k+1)\pi}{3}}, \quad k \in \{0,1,2\}.$$

It follows that the domain D cannot contain the point -1, hence

$$D = \mathbb{C} \setminus \{z \in \mathbb{C} : \operatorname{Im} z = 0, \operatorname{Re} z \le -1\},$$

or

$$D = \mathbb{C} \setminus \{z \in \mathbb{C} : \operatorname{Im} z = 0, \operatorname{Re} z \ge -1\}.$$

3. We will use the exponential form for the argument. If $z+i = r_1 e^{i\theta_1}$ and $z-i = r_2 e^{i\theta_2}$, where $r_1, r_2 > 0$, then $z^2+1 = r_1 r_2 e^{i(\theta_1+\theta_2)}$. Thus, we have

$$w_k(z) = \ln(r_1 r_2) + i(\theta_1 + \theta_2 + 2k\pi), \quad k \in \mathbb{Z}.$$

Similar to the first point of the problem, we will discuss the next three cases:

Case 1. If the closed curve γ turns around only the point i, and $z \in \{\gamma\}$ runs once over the curve γ, then we obtain the value

$$w_{k+1}(z) = \ln(r_1 r_2) + i(\theta_1 + \theta_2 + 2(k+1)\pi), \quad k \in \mathbb{Z}.$$

Case 2. If the closed curve γ turns around only the point $-i$, and $z \in \{\gamma\}$ runs once over the curve γ, then we obtain the value

$$w_{k+1}(z) = \ln(r_1 r_2) + i(\theta_1 + \theta_2 + 2(k+1)\pi), \quad k \in \mathbb{Z}.$$

Case 3. If the closed curve γ turns around both of the points i and $-i$, and $z \in \{\gamma\}$ runs once over the curve γ, then we obtain the value

$$w_{k+2}(z) = \ln(r_1 r_2) + i(\theta_1 + \theta_2 + 2(k+2)\pi), \quad k \in \mathbb{Z}.$$

It follows that the domain D cannot contain any of these points, hence

$$D = \mathbb{C} \setminus \{z \in \mathbb{C} : |\operatorname{Im} z| \geq 1, \operatorname{Re} z = 0\},$$

or

$$D = \mathbb{C} \setminus \{z \in \mathbb{C} : \operatorname{Im} z \geq -1, \operatorname{Re} z = 0\}, \quad D = \mathbb{C} \setminus \{z \in \mathbb{C} : \operatorname{Im} z \leq 1, \operatorname{Re} z = 0\}.$$

4. We will use the exponential form for the argument, i. e., $z - 1 = r_1 e^{i\theta_1}$ and $z + 1 = r_2 e^{i\theta_2}$, where $r_1, r_2 > 0$. Thus $\frac{z-1}{z+1} = \frac{r_1}{r_2} e^{i(\theta_1 - \theta_2)}$, and we get

$$w_k(z) = \ln \frac{r_1}{r_2} + i(\theta_1 - \theta_2 + 2k\pi), \quad k \in \mathbb{Z}.$$

Since two singular points are involved, we will discuss the next three cases:

Case 1. If the closed curve γ turns around only the point -1, and $z \in \{\gamma\}$ runs once over the curve γ, then we obtain the value

$$w_{k-1}(z) = \ln \frac{r_1}{r_2} + i(\theta_1 - \theta_2 + 2(k-1)\pi), \quad k \in \mathbb{Z}.$$

Case 2. If the closed curve γ turns around only the point 1, and $z \in \{\gamma\}$ runs once over the curve γ, then we obtain the value

$$w_{k+1}(z) = \ln \frac{r_1}{r_2} + i(\theta_1 - \theta_2 + 2(k+1)\pi), \quad k \in \mathbb{Z}.$$

Case 3. If the closed curve γ turns around both of the points -1 and 1, and $z \in \{\gamma\}$ runs once over the curve γ, then we obtain the value

$$w_k(z) = \ln \frac{r_1}{r_2} + i(\theta_1 - \theta_2 + 2k\pi), \quad k \in \mathbb{Z}.$$

It follows that in the domain D we cannot turn around to only one of these points, but we can turn around both of them, hence

$$D = \mathbb{C} \setminus [-1, 1] = \mathbb{C} \setminus \{z \in \mathbb{C} : |\operatorname{Re} z| \leq 1, \operatorname{Im} z = 0\},$$

or

$$D = \mathbb{C} \setminus [-1, +\infty) = \mathbb{C} \setminus \{z \in \mathbb{C} : \operatorname{Re} z \geq -1, \operatorname{Im} z = 0\},$$
$$D = \mathbb{C} \setminus (-\infty, 1] = \mathbb{C} \setminus \{z \in \mathbb{C} : \operatorname{Re} z \leq 1, \operatorname{Im} z = 0\}.$$

5. The exponential form of the function is

$$w_k(z) = \frac{\ln r + i(\theta + 2k\pi)}{re^{i\theta}}, \quad k \in \mathbb{Z},$$

whenever $z = re^{i\theta}$, with $r > 0$.

Here, we will discuss only one case. If the closed curve γ turns once around the point 0, then we obtain the value

$$w_{k+1}(z) = \frac{\ln r + i(\theta + 2(k+1)\pi)}{re^{i\theta}}, \quad k \in \mathbb{Z},$$

because $re^{i(\theta+2\pi)} = re^{i\theta}$. It follows that the domain D cannot contain the point 0, hence

$$D = \mathbb{C} \setminus \{z \in \mathbb{C} : \operatorname{Im} z = 0, \operatorname{Re} z \leq 0\},$$

or

$$D = \mathbb{C} \setminus \{z \in \mathbb{C} : \operatorname{Im} z = 0, \operatorname{Re} z \geq 0\}.$$

6. We will use the exponential form of the function. If $z - i = r_1 e^{i\theta_1}$ and $z + i = r_2 e^{i\theta_2}$, where $r_1, r_2 > 0$, we obtain

$$w_{k,l}(z) = \sqrt{r_1} e^{i\frac{\theta_1 + 2k\pi}{2}} + \sqrt{r_2} e^{i\frac{\theta_2 + 2l\pi}{2}}, \quad k, l \in \{0, 1\}.$$

Since two singular points are involved, we will discuss the next three cases:

Case 1. If the closed curve γ turns around only the point i, and $z \in \{\gamma\}$ runs once over the curve γ, then we obtain the value

$$w_{k+1,l}(z) = \sqrt{r_1} e^{i\frac{\theta_1 + 2(k+1)\pi}{2}} + \sqrt{r_2} e^{i\frac{\theta_2 + 2l\pi}{2}}, \quad k, l \in \{0, 1\}.$$

Case 2. If the closed curve γ turns around only the point $-i$, and $z \in \{\gamma\}$ runs once over the curve γ, then we obtain the value

$$w_{k,l+1}(z) = \sqrt{r_1} e^{i\frac{\theta_1 + 2k\pi}{2}} + \sqrt{r_2} e^{i\frac{\theta_2 + 2(l+1)\pi}{2}}, \quad k, l \in \{0, 1\}.$$

Case 3. If the closed curve γ turns around both of the points i and $-i$, and $z \in \{\gamma\}$ runs once over the curve γ, then we obtain the value

$$w_{k+1,l+1}(z) = \sqrt{r_1} e^{i\frac{\theta_1 + 2(k+1)\pi}{2}} + \sqrt{r_2} e^{i\frac{\theta_2 + 2(l+1)\pi}{2}}, \quad k, l \in \{0, 1\}.$$

It follows that the domain D cannot contain any of these points, hence

$$D = \mathbb{C} \setminus \{z \in \mathbb{C} : |\operatorname{Im} z| \geq 1, \operatorname{Re} z = 0\},$$

or

$$D = \mathbb{C} \setminus \{z \in \mathbb{C} : \operatorname{Im} z \geq -1, \operatorname{Re} z = 0\},$$
$$D = \mathbb{C} \setminus \{z \in \mathbb{C} : \operatorname{Im} z \leq 1, \operatorname{Re} z = 0\}.$$

Solution of Exercise 3.10.45

1. The exponential form of the function is

$$w_k(z) = \sqrt[3]{\frac{r_1}{r_2}} e^{i\frac{\theta_1 - \theta_2 + 2k\pi}{3}}, \quad k \in \{0, 1, 2\},$$

where $z + 1 = r_1 e^{i\theta_1}$, $z - 1 = r_2 e^{i\theta_2}$, and $r_1, r_2 > 0$.

It follows that the domain $\widetilde{D}$ where the function w is holomorphic, cannot contain only one of the points -1 or 1, but necessary both of them. Hence, this domain does not contain any of these points, or both of them, for instance

$$\widetilde{D} = \mathbb{C} \setminus [-1, 1].$$

Since $D \subset \widetilde{D}$, we deduce that there exists a branch of the function w, i. e., holomorphic on the domain D.

2. We will write the exponential form of the function

$$w_k(z) = 2 \ln \frac{r_1}{r_2} + 2i(\theta_1 - \theta_2 + 2k\pi), \quad k \in \mathbb{Z},$$

where $z + 1 = r_1 e^{i\theta_1}$, $z - 1 = r_2 e^{i\theta_2}$ and $r_1, r_2 > 0$.

It follows that the domain $\widetilde{D}$ where the function w is holomorphic, cannot contain only one of the points -1 or 1, but necessary both of them. Hence, this domain does not contain any of these points, or both of them, for instance

$$\widetilde{D} = \mathbb{C} \setminus [-1, 1].$$

Since $D \subset \widetilde{D}$, we deduce that there exists a branch of the function w, i. e., holomorphic on the domain D.

3. The exponential form of the function is

$$w_k(z) = \sqrt{r_1 r_2 r_3 r_4} e^{i\frac{\theta_1 + \theta_2 + \theta_3 + \theta_4 + 2k\pi}{2}}, \quad k \in \{0, 1\},$$

where

$$z+1=r_1e^{i\theta_1},\quad z-1=r_2e^{i\theta_2},\quad z+2=r_3e^{i\theta_3},\quad z-2=r_4e^{i\theta_4}\quad\text{and}\quad r_k>0,\ k\in\{1,2,3,4\}.$$

If $z\in D$, then $\operatorname{Re} z>0$, then the points -1 and -2 does not belong to D. If $\gamma\in\mathcal{D}_D(z_0)$, $z_0\in D$, is a closed curve, and if γ turns around the points 1 and 2, i. e., γ turns around both of these points, then we obtain

$$w_k(z)=w_{k+2}(z)=\sqrt{r_1r_2r_3r_4}e^{i\frac{\theta_1+\theta_2+\theta_3+\theta_4+2(k+2)\pi}{2}},\qquad k\in\{0,1\}.$$

It follows that there exists a branch of the function w that is holomorphic on the domain D.

Solution of Exercise 3.10.46

1. First, we will write the function in the exponential form

$$f_k(z)=\sqrt[3]{r}e^{i\frac{\theta+2k\pi}{3}},\quad k\in\{0,1,2\},\quad\text{where } z+1=re^{i\theta},\ r>0,\ \theta\in[0,2\pi].$$

According to the point 2 of Exercise 3.10.44, the function has an holomorphic branch on those domains that does not contain the point. The given domain satisfies this condition, because $D=\mathbb{C}\setminus[-1,+\infty)$.

Since $-2\in D$, then

$$f_0(-2)=-1\Leftrightarrow e^{i\frac{\pi+2k\pi}{3}}=-1\Leftrightarrow k=1,$$

thus

$$f_0(z)=\sqrt[3]{r}e^{i\frac{\theta+2\pi}{3}},\quad\text{where } z+1=re^{i\theta},\ r>0,\ \theta\in(0,2\pi).$$

2. We will determine the parameters that appeared in the previous exponential form, replacing the variable by i and $-i$, i. e.,

$$z=i\Rightarrow r=\sqrt{2},\quad \theta=\frac{\pi}{4},$$

and

$$z=-i\Rightarrow r=\sqrt{2},\quad \theta=\frac{7\pi}{4}.$$

Then the required values will be

$$f_0(i)=\sqrt[6]{2}e^{i\frac{3\pi}{4}}=\sqrt[6]{2}\Big(-\frac{\sqrt{2}}{2}+i\frac{\sqrt{2}}{2}\Big)$$

and

$$f_0(-i) = \sqrt[6]{2}e^{i\frac{5\pi}{4}} = \sqrt[6]{2}\left(-\frac{\sqrt{2}}{2} - i\frac{\sqrt{2}}{2}\right).$$

The value of the derivative is

$$f_0'(z) = \frac{1}{3(f_0(z))^2}, \quad z \in D,$$

then we get

$$f_0'(i) = \frac{1}{3(\sqrt[6]{2}e^{i\frac{3\pi}{3}})^2} = \frac{i}{3\sqrt[3]{2}} \quad \text{and} \quad f_0'(-i) = \frac{1}{3(\sqrt[6]{2}e^{i\frac{5\pi}{4}})^2} = -\frac{i}{3\sqrt[3]{2}}.$$

3. Since

$$\forall z_f \in T_f \Leftrightarrow z_f + 1 = re^{i0} \Leftrightarrow r = z_f + 1, \quad \theta = 0,$$

the values of the function $\tilde{f}$ on the superior border T_f of T are

$$\tilde{f}(z_f) = \sqrt[3]{z_f + 1}e^{i\frac{2\pi}{3}}, \quad \forall z_f \in T_f.$$

since

$$\forall z_a \in T_a \Leftrightarrow z_a + 1 = re^{i2\pi} \Leftrightarrow r = z_a + 1, \quad \theta = 2\pi,$$

the values of the function $\tilde{f}$ on the border T_a of T are

$$\tilde{f}(z_a) = \sqrt[3]{z_a + 1}e^{i\frac{4\pi}{3}}, \quad \forall z_a \in T_a.$$

Solution of Exercise 3.10.47

1. To the point 6 of Exercise 3.10.44 we already obtained that the function has a holomorphic branch on the given domain, that has the form

$$f_{k,l}(z) = \sqrt{r_1}e^{i\frac{\theta_1+2k\pi}{2}} + \sqrt{r_2}e^{i\frac{\theta_2+2l\pi}{2}}, \quad k,l \in \{0,1\},$$

where

$$z - i = r_1e^{i\theta_1}, \quad r_1 > 0,\ \theta_1 \in \left(-\frac{3\pi}{2}, \frac{\pi}{2}\right) \quad \text{and} \quad z + i = r_2e^{i\theta_2}, \quad r_2 > 0,\ \theta_2 \in \left(-\frac{\pi}{2}, \frac{3\pi}{2}\right).$$

Using the given assumption, we will determine now the required branch. We have

$$z = 0 \Leftrightarrow -i = r_1e^{i\theta_1} \Rightarrow r_1 = 1, \quad \theta_1 = -\frac{\pi}{2},$$

and

$$z = 0 \Leftrightarrow i = r_2 e^{i\theta_2} \Rightarrow r_2 = 1, \quad \theta_2 = \frac{\pi}{2},$$

and since

$$f_0(0) = -i\sqrt{2} \Leftrightarrow e^{i\frac{-\frac{\pi}{2}+2k\pi}{2}} + e^{i\frac{\frac{\pi}{2}+2l\pi}{2}} = -i\sqrt{2} \Leftrightarrow k = 0, \quad l = 1,$$

we obtain

$$f_0(z) = f_{0,1}(z) = \sqrt{r_1}e^{i\frac{\theta_1}{2}} + \sqrt{r_2}e^{i\frac{\theta_2+2\pi}{2}}, \quad \text{where } z - i = r_1 e^{i\theta_1},\ z + i = r_2 e^{i\theta_2}.$$

2. We will determine the parameters that appeared in this last exponential form, replacing the variable by $\frac{1}{\sqrt{3}}$ and 2. Thus

$$z_1 = \frac{1}{\sqrt{3}} \Rightarrow r_1 = \frac{2}{\sqrt{3}}, \quad \theta_1 = -\frac{\pi}{3}, \quad r_2 = \frac{2}{\sqrt{3}}, \quad \theta_2 = \frac{\pi}{3},$$

and

$$z_2 = 2 \Rightarrow r_1 = \sqrt{5}, \quad \theta_1 = -\arctan\frac{1}{2}, \quad r_2 = \sqrt{5}, \quad \theta_2 = \arctan\frac{1}{2}.$$

The required values are

$$f_0\left(\frac{1}{\sqrt{3}}\right) = \sqrt{\frac{2}{\sqrt{3}}}\left(e^{-i\frac{\pi}{6}} + e^{i\frac{7\pi}{6}}\right) = -\frac{i\sqrt{6}\sqrt[4]{3}}{3},$$

and

$$f_0(2) = \sqrt[4]{5}\left(e^{-\frac{i}{2}\arctan\frac{1}{2}} + e^{i\frac{\arctan\frac{1}{2}+2\pi}{2}}\right) = -2i\sqrt[4]{5}\sin\left(\frac{1}{2}\arctan\frac{1}{2}\right).$$

3. Since

$$\forall z \in T_{1j} \Leftrightarrow r_1 = |z - i|, \quad \theta_1 = -\frac{\pi}{2}, \quad r_2 = |z + i|, \quad \theta_2 = -\frac{\pi}{2},$$

the values of the function $\widetilde{f}$ on the right-hand side border T_{1j} of T_1 are given by

$$\widetilde{f}(z) = \sqrt{|z - i|}e^{-i\frac{\pi}{4}} + \sqrt{|z + i|}e^{i\frac{3\pi}{4}}, \quad \forall z \in T_{1j}.$$

Since

$$\forall z \in T_{1b} \Leftrightarrow r_1 = |z - i|, \quad \theta_1 = -\frac{\pi}{2}, \quad r_2 = |z + i|, \quad \theta_2 = \frac{3\pi}{2},$$

the values of the function $\widetilde{f}$ on the left-hand side border T_{1b} of T_1 are given by

$$\widetilde{f}(z) = \sqrt{|z - i|}e^{-i\frac{\pi}{4}} + \sqrt{|z + i|}e^{i\frac{7\pi}{4}}, \quad \forall z \in T_{1b}.$$

Since

$$\forall z \in T_{2j} \Leftrightarrow r_1 = |z - i|, \quad \theta_1 = \frac{\pi}{2}, \quad r_2 = |z + i|, \quad \theta_2 = \frac{\pi}{2},$$

the values of the function $\widetilde{f}$ on the right-hand side border T_{2j} of T_2 are given by

$$\widetilde{f}(z) = \sqrt{|z - i|}e^{i\frac{\pi}{4}} + \sqrt{|z + i|}e^{i\frac{5\pi}{4}}, \quad \forall z \in T_{2j}.$$

Since

$$\forall z \in T_{2b} \Leftrightarrow r_1 = |z - i|, \quad \theta_1 = -\frac{3\pi}{2}, \quad r_2 = |z + i|, \quad \theta_2 = \frac{\pi}{2},$$

the values of the function $\widetilde{f}$ on the left-hand side border T_{2b} of T_2 are given by

$$\widetilde{f}(z) = \sqrt{|z - i|}e^{-i\frac{3\pi}{4}} + \sqrt{|z + i|}e^{i\frac{5\pi}{4}}, \quad \forall z \in T_{2b}.$$

Solution of Exercise 3.10.48

1. To point 3 of Exercise 3.10.44, we have already obtained that the function has a holomorphic branch on the given domain that has the form

$$f_k(z) = \ln(r_1 r_2) + i(\theta_1 + \theta_2 + 2k\pi), \quad k \in \mathbb{Z},$$

where

$$z - i = r_1 e^{i\theta_1}, \quad r_1 > 0,\ \theta_1 \in \left(-\frac{3\pi}{2}, \frac{\pi}{2}\right) \quad \text{and} \quad z + i = r_2 e^{i\theta_2}, \quad r_2 > 0,\ \theta_2 \in \left(-\frac{\pi}{2}, \frac{3\pi}{2}\right).$$

Using the given assumption, we will determine now the required branch. We have

$$z = 0 \Leftrightarrow -i = r_1 e^{i\theta_1} \Rightarrow r_1 = 1, \quad \theta_1 = -\frac{\pi}{2}$$

and

$$z = 0 \Leftrightarrow i = r_2 e^{i\theta_2} \Rightarrow r_2 = 1, \quad \theta_2 = \frac{\pi}{2}.$$

Using the fact

$$f_0(0) = 0 \Leftrightarrow i2k\pi = 0 \Leftrightarrow k = 0,$$

we deduce

$$f_0(z) = \ln(r_1 r_2) + i(\theta_1 + \theta_2), \quad \text{where } z - i = r_1 e^{i\theta_1},\ z + i = r_2 e^{i\theta_2}.$$

2. We will determine the parameters that appeared in this last exponential form, replacing the variable by $2+i$ and $-2-i$. Thus

$$z_1 = 2+i \Rightarrow r_1 = 2, \quad \theta_1 = 0, \quad r_2 = 2\sqrt{2}, \quad \theta_2 = \frac{\pi}{4},$$
$$z_2 = -2-i \Rightarrow r_1 = 2\sqrt{2}, \quad \theta_1 = -\frac{3\pi}{4}, \quad r_2 = 2, \quad \theta_2 = \pi,$$

and

$$z_3 = \frac{i}{2} \Rightarrow r_1 = \frac{1}{2}, \quad \theta_1 = -\frac{\pi}{2}, \quad r_2 = \frac{3}{2}, \quad \theta_2 = \frac{\pi}{2}.$$

The required values are

$$f_0(2+i) = \ln 4\sqrt{2} + i\frac{\pi}{4}, \quad f_0(-2-i) = \ln 4\sqrt{2} - i\frac{\pi}{4} \quad \text{and} \quad f_0\left(\frac{i}{2}\right) = \ln\frac{3}{4}.$$

3. The values of the function $\tilde{f}$ on the right-hand side border T_{1j} of T_1 are given by

$$\tilde{f}(z) = \ln(|z-i||z+i|) - i\pi, \quad \forall z \in T_{1j},$$

because

$$\forall z \in T_{1j} \Leftrightarrow r_1 = |z-i|, \quad \theta_1 = -\frac{\pi}{2}, \quad r_2 = |z+i|, \quad \theta_2 = -\frac{\pi}{2}.$$

The values of the function $\tilde{f}$ on the left-hand side border T_{1b} of T_1 are given by

$$\tilde{f}(z) = \ln(|z-i||z+i|) + i\pi, \quad \forall z \in T_{1b},$$

because

$$\forall z \in T_{1b} \Leftrightarrow r_1 = |z-i|, \quad \theta_1 = -\frac{\pi}{2}, \quad r_2 = |z+i|, \quad \theta_2 = \frac{3\pi}{2}.$$

The values of the function $\tilde{f}$ on the right-hand side border T_{2j} of T_2 are given by

$$\tilde{f}(z) = \ln(|z-i||z+i|) + i\pi, \quad \forall z \in T_{2j},$$

because

$$\forall z \in T_{2j} \Leftrightarrow r_1 = |z-i|, \quad \theta_1 = \frac{\pi}{2}, \quad r_2 = |z+i|, \quad \theta_2 = \frac{\pi}{2}.$$

The values of the function $\tilde{f}$ on the left-hand side border T_{2b} of T_2 are given by

$$\tilde{f}(z) = \ln(|z-i||z+i|) - i\pi, \quad \forall z \in T_{2b},$$

because

$$\forall z \in T_{2b} \Leftrightarrow r_1 = |z-i|, \quad \theta_1 = -\frac{3\pi}{2}, \quad r_2 = |z+i|, \quad \theta_2 = \frac{\pi}{2}.$$

Solution of Exercise 3.10.49

The exponential form of the function is

$$f_k(z) = \sqrt{r_1 r_2} e^{i\frac{\theta_1+\theta_2+2k\pi}{2}}, \quad k \in \{0, 1\},$$

where

$$z - i = r_1 e^{i\theta_1}, \quad z + i = r_2 e^{i\theta_2}, \quad r_1, r_2 > 0.$$

In each of the three cases, the argument θ belongs in the different intervals, as follows:

1.
$$\theta_1 \in \left(-\frac{3\pi}{2}, \frac{\pi}{2}\right), \quad \theta_2 \in \left(-\frac{\pi}{2}, \frac{3\pi}{2}\right).$$

2.
$$\theta_1 \in \left(-\frac{3\pi}{2}, \frac{\pi}{2}\right), \quad \theta_2 \in \left(-\frac{3\pi}{2}, \frac{\pi}{2}\right).$$

3.
$$\theta_1 \in \left[-\frac{\pi}{2}, \frac{3\pi}{2}\right], \quad \theta_2 \in \left[-\frac{\pi}{2}, \frac{3\pi}{2}\right] : (\theta_1, \theta_2) \neq \left(-\frac{\pi}{2}, \frac{\pi}{2}\right), \quad (\theta_1, \theta_2) \neq \left(\frac{3\pi}{2}, \frac{\pi}{2}\right).$$

Solution of Exercise 3.10.50

To point 5 of Exercise 3.10.44, we have already obtained that the function has a holomorphic branch on the given domain that has the form

$$f_k(z) = \frac{\ln r + i(\theta + 2k\pi)}{re^{i\theta}}, \quad k \in \mathbb{Z},$$

where

$$z = re^{i\theta}, \quad r > 0, \quad \theta \in (-\pi, \pi).$$

Using the given assumption, we will determine now the required branch. If $z > 0$, then $r = |z| > 0$ and $\theta = 0$.

Since

$$z \in (0, +\infty) \Rightarrow f_k(z) = \frac{1}{r}(\ln r + i2k\pi) \in \mathbb{R},$$

we get $k = 0$, hence

$$f_0(z) = \frac{1}{r}(\ln r + i\theta)e^{-i\theta}, \quad \text{where } z = re^{i\theta},\ r > 0,\ \theta \in (-\pi, \pi).$$

2. We will determine the parameters that appeared in this last exponential form, replacing the variable by i and $-i$. Since

$$z = i \Leftrightarrow r = 1, \quad \theta = \frac{\pi}{2},$$

and

$$z = -i \Leftrightarrow r = 1, \quad \theta = -\frac{\pi}{2},$$

the required values are

$$f_0(i) = \frac{\pi}{2}, \quad f_0(-i) = \frac{\pi}{2}.$$

Solution of Exercise 3.10.51

Letting $z \in \{\operatorname{Im} z < 0\} \cup \{\operatorname{Im} z \geq 0,\ |z| > r\} = \Delta$, then $\partial\Delta = \{\gamma\}$, and Δ is the unbounded component of the set $\mathbb{C} \setminus \{\gamma\}$, and according to the *index theorem* we have $\mathrm{n}(\gamma, z) = 0$.

Letting $z \in \{\operatorname{Im} z > 0,\ |z| < r\} = D$, then $\partial D = \{\gamma\}$. Then there exists $\rho > 0$ such that $\overline{U}(z;\rho) \subset D$. Denoting by $\widetilde{\gamma}(t) = z + \rho e^{2\pi i t}$, $t \in [0,1]$, then $\{\widetilde{\gamma}\} = \partial U(z;\rho) \subset D$ and $\widetilde{\gamma} \underset{\mathbb{C}\setminus\{z\}}{\sim} \gamma$. From here, we obtain that

$$\mathrm{n}(\gamma, z) = \frac{1}{2\pi i}\int_{\gamma} \frac{d\zeta}{\zeta - z} = \frac{1}{2\pi i}\int_{\widetilde{\gamma}} \frac{d\zeta}{\zeta - z} = \frac{1}{2\pi i}\int_{\partial U(z;\rho)} \frac{d\zeta}{\zeta - z} = 1.$$

Solution of Exercise 3.10.52

Consider $z \in \{z \in \mathbb{C} : r < |z| < \rho,\ \operatorname{Im} z > 0\} = D$. Then $\partial D = \{\gamma\}$, and there exists $r > 0$ such that $\overline{U}(z;r) \subset D$. If $\widetilde{\gamma}(t) = z + re^{2\pi i t}$, $t \in [0,1]$, then $\{\widetilde{\gamma}\} = \partial U(z;r) \subset D$ and $\widetilde{\gamma} \underset{\mathbb{C}\setminus\{z\}}{\sim} \gamma$. It follows that

$$\mathrm{n}(\gamma, z) = \frac{1}{2\pi i}\int_{\gamma} \frac{d\zeta}{\zeta - z} = \frac{1}{2\pi i}\int_{\widetilde{\gamma}} \frac{d\zeta}{\zeta - z} = \frac{1}{2\pi i}\int_{\partial U(z;r)} \frac{d\zeta}{\zeta - z} = 1.$$

Solution of Exercise 3.10.53

We will prove first that the function f has primitives on the set $G \subset \mathbb{C}$, and according to the *Morera theorem* it follows that $f \in H(G)$.

Consider the line $d = AB$, where $A(a)$ and $B(b)$, i. e.,

$$d : d(t) = (1 - t)a + tb, \quad t \in \mathbb{R}.$$

Since $G \setminus d \subset G \setminus [a,b]$, then $f \in C(G) \cap H(G \setminus (G \cap d))$. Let $z_0 \in G$ be a given arbitrary point, and let $r > 0$ such that $U(z_0;r) \subset G$. The disc $U(z_0;r)$ will intersect at most the line d, and according to the *theorem related to the holomorphic functions and the primitive existence* it follows that f has primitives on the disc $U(z_0;r)$. Then, from the *Morera theorem* we get $f \in H(U(z_0;r))$. Since the point $z_0 \in G$ was arbitrary, we deduce that $f \in H(G)$.

Solution of Exercise 3.10.54

According to Exercise 3.10.53, the function f is entire, i. e., $f \in H(\mathbb{C})$. But f is bounded, hence from the *Liouville theorem* it follows that f is constant on $\mathbb{C}$.

Solution of Exercise 3.10.55

It is well known that $g(\{z \in \mathbb{C} : \operatorname{Re} z > 0\}) = U(0;1)$, where $g(z) = \frac{1-z}{1+z}$. The upper half-plane is mapped onto the right half-plane by the rotation with the center at the origin and the rotation angle equal to $-\frac{\pi}{2}$, i. e.,

$$l : \{z \in \mathbb{C} : \operatorname{Im} z > 0\} \to \{w \in \mathbb{C} : \operatorname{Re} w > 0\}, \quad l(z) = -iz.$$

Let us define the function

$$\psi = g \circ l \circ f \in H(\mathbb{C}),$$

which is bounded, because

$$|\psi(z)| = |g(l(f(z))| < 1, \quad \forall z \in \mathbb{C}.$$

Hence ψ is a bounded entire function, and from the *Liouville theorem* it will be constant, i. e.,

$$\psi(z) = c, \quad \forall z \in \mathbb{C}, \quad \text{where } |c| < 1 \Leftrightarrow f(z) = \frac{1}{i}\frac{c-1}{c+1} = k, \quad \forall z \in \mathbb{C}.$$

It follows that there not exists any nonconstant functions that satisfy the required properties.

Solution of Exercise 3.10.56

It is well known that $g(\{z \in \mathbb{C} : \operatorname{Re} z > 0\}) = U(0;1)$, where $g(z) = \frac{1-z}{1+z}$. Let us define the function

$$\psi = g \circ f \in H(\mathbb{C}),$$

which is bounded, because

$$|\psi(z)| = |g(f(z))| < 1, \quad \forall z \in \mathbb{C}.$$

Hence ψ is a bounded entire function, and from the *Liouville theorem* it will be constant, i. e.,

$$\psi(z) = c, \quad \forall z \in \mathbb{C}, \quad \text{where } |c| < 1 \Leftrightarrow f(z) = \frac{1-c}{1+c} = k, \quad \forall z \in \mathbb{C}.$$

It follows that there not exists any nonconstant functions that satisfy the required properties.

Solution of Exercise 3.10.57

For the solution, we will use again the *Liouville theorem*.

It is well known that $g(\{z \in \mathbb{C} : |z| > 1\}) = U(0;1)$, where $g(z) = \frac{1}{z}$. Let us define the function

$$\psi = g \circ f \in H(\mathbb{C}),$$

which is bounded, because

$$|\psi(z)| = |g(f(z))| < 1, \quad \forall z \in \mathbb{C}.$$

Hence ψ is a bounded entire function, and from the *Liouville theorem* it will be constant, i. e.,

$$\psi(z) = c, \quad \forall z \in \mathbb{C}, \quad \text{where } |c| < 1 \Leftrightarrow f(z) = \frac{1}{c} = k, \quad \forall z \in \mathbb{C}.$$

It follows that there not exists any nonconstant functions that satisfy the required properties.

Solution of Exercise 3.10.58

We will use the assumption as follows:

$$\lim_{z\to\infty} f(z) = a \in \mathbb{C} \Leftrightarrow [\forall \varepsilon > 0, \exists \sigma > 0 : \forall z \in \mathbb{C}, |z| > \sigma \Rightarrow |f(z) - a| < \varepsilon].$$

Let $\varepsilon = 1$; hence $\exists \sigma > 0$ such that $\forall z \in \mathbb{C} : |z| > \sigma \Rightarrow |f(z) - a| < 1$. From here, we get

$$f(z) \in U(a;1), \quad \forall z \in \mathbb{C} : |z| > \sigma.$$

If $\forall z \in \overline{U}(0;\sigma)$, that is a compact set, since $f \in C(\overline{U}(0;\sigma))$ it follows that

$$\exists M > 0 : |f(z)| \leq M, \quad \forall z \in \overline{U}(0;\sigma). \tag{7.25}$$

If $\forall z \in \mathbb{C} \setminus \overline{U}(0;\sigma)$, then

$$f(z) \in U(a;1) \Rightarrow \big||f(z)| - |a|\big| \leq |f(z) - a| < 1 \Rightarrow -1 < |f(z)| - |a| < 1,$$

thus

$$|f(z)| < 1 + |a|, \quad \forall z \in \mathbb{C} : |z| > \sigma. \tag{7.26}$$

Letting $N = \max\{M; |a| + 1\}$, from the relations (7.25) and (7.26) we deduce that $|f(z)| \leq N$, $\forall z \in \mathbb{C}$, hence f is a bounded entire function. Now, according to the *Liouville theorem* we conclude that f is a constant function on $\mathbb{C}$.

Solution of Exercise 3.10.59

If $g = \operatorname{Re} f$ is a bounded function, then $\exists M > 0 : |g(z)| < M$, $\forall z \in \mathbb{C}$. Let us define the function

$$h : \mathbb{C} \to \mathbb{C}, \quad h(z) = e^{f(z)}.$$

Since f is holomorphic on $\mathbb{C}$, the function h will be also holomorphic on $\mathbb{C}$. Now, it will be sufficient to prove that h is a bounded function. But

$$|h(z)| = e^{\operatorname{Re} f(z)} = e^{g(z)} < e^{M}, \quad \forall z \in \mathbb{C},$$

and according to the *Liouville theorem* we obtain that

$$h(z) = k, \quad \forall z \in \mathbb{C} \Leftrightarrow e^{f(z)} = k, \quad \forall z \in \mathbb{C}.$$

It follows

$$f(z) \in \{\ln|k| + i(\arg k + 2l\pi) : l \in \mathbb{Z}\} = \{c_l : l \in \mathbb{Z}\},$$

and since f is continuous, then

$$\exists!\, l \in \mathbb{Z} : f(z) = c_l.$$

Hence, the function f is constant on $\mathbb{C}$.

Solution of Exercise 3.10.60

Let denote $K = \partial U(0; r)$, $G = \mathbb{C} \setminus K$, and let

$$g : G \times K \to \mathbb{C}, \quad g(z, \zeta) = \frac{f(\zeta)}{\zeta - z}.$$

Since the function $g_z'(z, \zeta) = -\frac{f(\zeta)}{(\zeta - z)^2}$ is continuous on $G \times K$ (because $f \in C(\mathbb{C})$), then the function

$$F : G \to \mathbb{C}, \quad F(z) = \int\limits_{\partial U(0;r)} \frac{f(\zeta)}{\zeta - z} d\zeta$$

will be holomorphic on $G = \mathbb{C} \setminus K$, i. e., $F \in H(\mathbb{C} \setminus K)$.

Let $a \in \mathbb{C}^*$ an arbitrary fixed point. Then there exists $r > 0$ such that $|a| > r$. Since $F \in H(\mathbb{C} \setminus K)$ and $\forall z \in \mathbb{C} \setminus \overline{U}(0; r)$ we have $f(z) = F(z)$, then it follows that the function f is differentiable at the point $a \in \mathbb{C}^*$, and thus $f \in H(\mathbb{C}^*)$. From here, since $f \in C(\mathbb{C})$, we deduce that $f \in H(\mathbb{C})$.

If $r > 0$ is fixed, since $\forall z \in \mathbb{C} \setminus \overline{U}(0; r)$ we have $f(z) = F(z)$, it follows that

$$\lim_{z \to \infty} f(z) = \lim_{z \to \infty} F(z) = 0,$$

i. e., f is a bounded entire function. Now, from the *Liouville theorem* we obtain that f is a constant function on $\mathbb{C}$, i. e., $f(z) = c$, $\forall z \in \mathbb{C}$. If $\gamma(t) = re^{2\pi i t}$, $t \in [0, 1]$, then

$$c = \int\limits_{\partial U(0;r)} \frac{c}{\zeta - z} d\zeta = c \int\limits_{\gamma} \frac{1}{\zeta - z} d\zeta = cn(\gamma, z) = 0, \quad \forall z \in \mathbb{C} : |z| > r,$$

hence $f(z) = 0$, $\forall z \in \mathbb{C}$.

7.4 Solutions to the exercises of Chapter 4

Solution of Exercise 4.10.1

We will use the formulas from the *Cauchy–Hadamard theorem*, i. e.,

$$R = \frac{1}{\limsup_{n \to +\infty} \sqrt[n]{|a_n|}} \quad \text{or} \quad R = \limsup_{n \to +\infty} \frac{|a_n|}{|a_{n+1}|}.$$

1. $a_n = \frac{1}{n}$, hence

$$R = \frac{1}{\lim_{n \to +\infty} \sqrt[n]{|a_n|}} = \frac{1}{\lim_{n \to +\infty} \sqrt[n]{|\frac{1}{n}|}} = \frac{1}{\lim_{n \to +\infty} \sqrt[n]{\frac{1}{n}}} = \lim_{n \to +\infty} \sqrt[n]{n} = 1.$$

2. $a_n = \frac{1}{n^2}$, hence

$$R = \frac{1}{\lim_{n\to+\infty}\sqrt[n]{|a_n|}} = \frac{1}{\lim_{n\to+\infty}\sqrt[n]{|\frac{1}{n^2}|}} = \frac{1}{\lim_{n\to+\infty}\sqrt[n]{\frac{1}{n^2}}} = \lim_{n\to+\infty}\sqrt[n]{n^2} = 1.$$

3. $a_n = \frac{1}{n!}$, hence

$$R = \lim_{n\to+\infty}\frac{|a_n|}{|a_{n+1}|} = \lim_{n\to+\infty}\frac{\frac{1}{n!}}{\frac{1}{(n+1)!}} = \lim_{n\to+\infty}\frac{(n+1)!}{n!} = \lim_{n\to+\infty}(n+1) = +\infty.$$

4. $a_n = n!$, hence

$$R = \lim_{n\to+\infty}\frac{|a_n|}{|a_{n+1}|} = \lim_{n\to+\infty}\frac{|n!|}{|(n+1)!|} = \lim_{n\to+\infty}\frac{n!}{(n+1)!} = \lim_{n\to+\infty}\frac{1}{n+1} = 0.$$

5. $a_n = \frac{1\cdot 2\cdot\ldots\cdot n}{3\cdot 5\cdot\ldots\cdot(2n+1)}$, hence

$$R = \lim_{n\to+\infty}\frac{|a_n|}{|a_{n+1}|} = \lim_{n\to+\infty}\frac{|\frac{1\cdot 2\cdot\ldots\cdot n}{3\cdot 5\cdot\ldots\cdot(2n+1)}|}{|\frac{1\cdot 2\cdot\ldots\cdot n(n+1)}{3\cdot 5\cdot\ldots\cdot(2n+1)(2n+3)}|} = \lim_{n\to+\infty}\frac{2n+3}{n+1} = 2.$$

6. $a_n = \frac{1}{n\sqrt{n}}$, hence

$$R = \lim_{n\to+\infty}\frac{a_n|}{|a_{n+1}|} = \lim_{n\to+\infty}\frac{|\frac{1}{n\sqrt{n}}|}{|\frac{1}{(n+1)\sqrt{n+1}}|} = \lim_{n\to+\infty}\frac{(n+1)\sqrt{n+1}}{n\sqrt{n}} = 1.$$

7. $a_n = e^{\alpha n}$, $\alpha = a + ib$, hence

$$R = \lim_{n\to+\infty}\frac{|a_n|}{|a_{n+1}|} = \lim_{n\to+\infty}\frac{|e^{\alpha n}|}{|e^{\alpha(n+1)}|} = \lim_{n\to+\infty}\frac{e^{an}}{e^{a(n+1)}} = \lim_{n\to+\infty}e^{-a} = e^{-a}.$$

8. $a_n = \cos n\alpha$, $\alpha = a + ib$, $b \neq 0$, hence

$$\begin{aligned}R &= \lim_{n\to+\infty}\frac{|a_n|}{|a_{n+1}|} = \lim_{n\to+\infty}\frac{|\cos n\alpha|}{|\cos(n+1)\alpha|} = \lim_{n\to+\infty}\frac{|\frac{e^{in\alpha}+e^{-in\alpha}}{2}|}{|\frac{e^{i(n+1)\alpha}+e^{-i(n+1)\alpha}}{2}|}\\ &= \lim_{n\to+\infty}\left|\frac{e^{-in\alpha}(e^{2n\alpha i}+1)}{e^{-i(n+1)\alpha}(e^{2(n+1)\alpha i}+1)}\right| = \lim_{n\to+\infty}\left|\frac{e^{i\alpha}(e^{2n\alpha i}+1)}{e^{2(n+1)\alpha i}+1}\right|\\ &= \lim_{n\to+\infty}|e^{i\alpha}|\left|\frac{e^{2n\alpha i}+1}{e^{2(n+1)\alpha i}+1}\right|\\ &= \lim_{n\to+\infty}e^{-b}\sqrt{\frac{e^{-4nb}+2e^{-2nb}\cos 2na+1}{e^{-4(n+1)b}+2e^{-2(n+1)b}\cos 2(n+1)a+1}}.\end{aligned}$$

If $b > 0$, then

$$R = \lim_{n\to+\infty}e^{-b}\sqrt{\frac{e^{-4nb}+2e^{-2nb}\cos 2na+1}{e^{-4(n+1)b}+2e^{-2(n+1)b}\cos 2(n+1)a+1}} = e^{-b}.$$

If $b < 0$, then

$$R = \lim_{n\to+\infty} e^{-b}\sqrt{\frac{e^{-4nb}(e^{4nb} + 2e^{2nb}\cos 2na + 1)}{e^{-4(n+1)b}(e^{4(n+1)b} + 2e^{2(n+1)b}\cos 2(n+1)a + 1)}}$$

$$= \lim_{n\to+\infty} e^{-b}\sqrt{e^{4b}\frac{e^{4nb} + 2e^{2nb}\cos 2na + 1}{e^{4(n+1)b} + 2e^{2(n+1)b}\cos 2(n+1)a + 1}} = e^{b}.$$

Combining the above two results, we conclude that $R = e^{-|b|}$.

9. $a_n = [3 + (-1)^n]^n$, hence

$$R = \frac{1}{\limsup_{n\to+\infty}\sqrt[n]{|a_n|}} = \frac{1}{\limsup_{n\to+\infty}\sqrt[n]{|[3+(-1)^n]^n|}} = \frac{1}{\limsup_{n\to+\infty}[3+(-1)^n]} = \frac{1}{4},$$

since

$$\lim_{n\to+\infty}[3+(-1)^n] = \begin{cases} 2, & \text{if } n \text{ is odd} \\ 4, & \text{if } n \text{ is even.} \end{cases}$$

10. $a_n = \frac{n}{2} + \frac{1+(-1)^{n+1}}{4}$, hence

$$R = \lim_{n\to+\infty}\frac{|a_n|}{|a_{n+1}|} = \lim_{n\to+\infty}\frac{|\frac{n}{2} + \frac{1+(-1)^{n+1}}{4}|}{|\frac{n+1}{2} + \frac{1+(-1)^{n+2}}{4}|}$$

$$= \begin{cases} \lim_{n\to+\infty}\frac{|\frac{n+1}{2}|}{|\frac{n+1}{2}|}, & \text{if } n \text{ is odd} \\ \lim_{n\to+\infty}\frac{|\frac{n}{2}|}{|\frac{n}{2}+1|}, & \text{if } n \text{ is even} \end{cases} = 1.$$

11. $a_n = \frac{1+(-1)^n}{2}n + \frac{1-(-1)^n}{2}\ln n$, hence

$$R = \frac{1}{\limsup_{n\to+\infty}\sqrt[n]{|a_n|}} = \frac{1}{1} = 1,$$

because

$$\lim_{n\to+\infty}\sqrt[n]{|a_n|} = \begin{cases} \lim_{n\to+\infty}\sqrt[n]{|\ln n|}, & \text{if } n \text{ is odd} \\ \lim_{n\to+\infty}\sqrt[n]{|n|}, & \text{if } n \text{ is even} \end{cases} = \begin{cases} 1, & \text{if } n \text{ is odd} \\ 1, & \text{if } n \text{ is even,} \end{cases}$$

and thus

$$\limsup_{n\to+\infty}\sqrt[n]{|a_n|} = 1.$$

12. If $a_n = \sum_{k=1}^{n}\frac{1}{k}$, then

$$R = \frac{1}{\lim_{n\to+\infty}\sqrt[n]{|a_n|}} = 1.$$

Since

$$1 \le a_n \le n \Rightarrow 1 \le \sqrt[n]{a_n} \le \sqrt[n]{n},$$

taking to the limits in the above double inequality, we get

$$1 = \lim_{n\to+\infty} 1 \le \lim_{n\to+\infty} \sqrt[n]{a_n} \le \lim_{n\to+\infty} \sqrt[n]{n} = 1 \Rightarrow \lim_{n\to+\infty} \sqrt[n]{a_n} = 1.$$

13. $a_n = n + \alpha^n$, $\alpha = a + ib$, hence

$$R = \lim_{n\to+\infty} \frac{|a_n|}{|a_{n+1}|} = \lim_{n\to+\infty} \frac{|n+\alpha^n|}{|n+1+\alpha^{n+1}|}.$$

If $|\alpha| > 1$, then

$$R = \lim_{n\to+\infty} \frac{1}{|\alpha|} \frac{|1+\frac{n}{\alpha^n}|}{|1+\frac{n+1}{\alpha^{n+1}}|} = \frac{1}{|\alpha|},$$

because

$$\lim_{n\to+\infty} \frac{n}{\alpha^n} = 0, \quad \text{if } |\alpha| > 1.$$

If $|\alpha| \le 1$, then

$$R = \lim_{n\to+\infty} \frac{n}{n+1} \frac{|1+\frac{\alpha^n}{n}|}{|1+\frac{\alpha^{n+1}}{n+1}|} = 1,$$

because

$$\lim_{n\to+\infty} \frac{\alpha^n}{n} = 0, \quad \text{if } |\alpha| \le 1.$$

From the above two relations, we conclude that

$$R = \lim_{n\to+\infty} \frac{|n+\alpha^n|}{|n+1+\alpha^{n+1}|} = \begin{cases} 1, & \text{if } |\alpha| \le 1 \\ \frac{1}{|\alpha|}, & \text{if } |\alpha| > 1. \end{cases}$$

14. $a_n = P(n)$, where P is a polynomial with $\deg P = p$, i. e.,

$$P(n) = b_0 + b_1 n + b_2 n^2 + \cdots + b_p n^p, \quad b_p \ne 0.$$

Then

$$\begin{aligned} R &= \lim_{n\to+\infty} \frac{|a_n|}{|a_{n+1}|} \\ &= \lim_{n\to+\infty} \frac{|b_0 + b_1 n + b_2 n^2 + \cdots + b_p n^p|}{|b_0 + b_1(n+1) + b_2(n+1)^2 + \cdots + b_p(n+1)^p|} = 1. \end{aligned}$$

15. $a_n = \frac{(n!)^2}{(3n)!}$, hence

$$R = \frac{1}{\lim_{n\to+\infty}\frac{|a_{n+1}|}{|a_n|}} = \frac{1}{\lim_{n\to+\infty}\frac{(3n)!}{(n!)^2}\frac{[(n+1)!]^2}{[3(n+1)]!}}$$
$$= \frac{1}{\lim_{n\to+\infty}\frac{(n+1)^2}{(3n+1)(3n+2)(3n+3)}} = +\infty.$$

16. $a_n = (\frac{n+1}{n})^n$, hence

$$R = \frac{1}{\lim_{n\to+\infty}\sqrt[n]{|a_n|}} = \frac{1}{\lim_{n\to+\infty}\sqrt[n]{(\frac{n+1}{n})^n}} = \frac{1}{\lim_{n\to+\infty}\frac{n+1}{n}} = 1.$$

17. $a_n = \frac{3^n+(-2)^n}{n}$, hence

$$R = \lim_{n\to+\infty}\frac{|a_n|}{|a_{n+1}|} = \lim_{n\to+\infty}\frac{\frac{3^n+(-2)^n}{n}}{\frac{3^{n+1}+(-2)^{n+1}}{n+1}} = \lim_{n\to+\infty}\frac{n+1}{n}\frac{1+(-\frac{2}{3})^n}{3-2(-\frac{2}{3})^n} = \frac{1}{3}.$$

18. $a_n = 3 + 2(-1)^n$, hence

$$R = \frac{1}{\limsup_{n\to+\infty}\sqrt[n]{|a_n|}} = \frac{1}{1} = 1,$$

because

$$\lim_{n\to+\infty}\sqrt[n]{|a_n|} = \begin{cases}\lim_{n\to+\infty}\sqrt[n]{1}, & \text{if } n \text{ is odd}\\ \lim_{n\to+\infty}\sqrt[n]{5}, & \text{if } n \text{ is even}\end{cases} = 1,$$

and thus

$$\limsup_{n\to+\infty}\sqrt[n]{|a_n|} = 1.$$

19.

$$a_n = \begin{cases}a^n, & n = 2m,\ m \in \mathbb{N},\\ b^n, & n = 2m+1,\ m \in \mathbb{N}\end{cases}$$

hence

$$R = \frac{1}{\limsup_{n\to+\infty}\sqrt[n]{|a_n|}} = \min\left\{\frac{1}{|a|}, \frac{1}{|b|}\right\},$$

because

$$\lim_{n\to+\infty}\sqrt[n]{|a_n|} = \begin{cases}|b|, & \text{if } n \text{ is odd}\\ |a|, & \text{if } n \text{ is even,}\end{cases}$$

hence

$$\limsup_{n\to+\infty} \sqrt[n]{|a_n|} = \max\{|a|, |b|\}.$$

20.

$$a_n = \begin{cases} \frac{1}{n}, & n = 3m,\ m \neq 0, \\ (1 - \frac{1}{n})^n, & n = 3m+1,\ m \in \mathbb{N}, \\ 2^n, & n = 3m+2,\ m \in \mathbb{N} \end{cases}$$

hence

$$R = \frac{1}{\limsup_{n\to+\infty} \sqrt[n]{|a_n|}} = \frac{1}{2},$$

because

$$\lim_{n\to+\infty} \sqrt[n]{|a_n|} = \begin{cases} \lim_{n\to+\infty} \sqrt[n]{\frac{1}{n}}, & n = 3m, \\ & m \neq 0, \\ \lim_{n\to+\infty}(1 - \frac{1}{n}), & n = 3m+1, \\ & m \in \mathbb{N}, \\ \lim_{n\to+\infty} 2, & n = 3m+2, \\ & m \in \mathbb{N}. \end{cases} = \begin{cases} 1, & n = 3m, \\ & m \neq 0, \\ 1, & n = 3m+1, \\ & m \in \mathbb{N}, \\ 2, & n = 3m+2, \\ & m \in \mathbb{N}, \end{cases}$$

and thus

$$\limsup_{n\to+\infty} \sqrt[n]{|a_n|} = \max\{1, 1, 2\} = 2.$$

Solution of Exercise 4.10.2

1. According to point 1 of Exercise 4.10.1, the radius of convergence is $R = 1$, hence the set of convergence will be $U(0; 1)$. Using the well-known relation,

$$\log(1 + z) = \sum_{n=1}^{\infty} (-1)^{n+1} \frac{z^n}{n}, \quad z \in U(0; 1), \quad \text{where } \log 1 = 0,$$

we get

$$\log(1 - z) = -\sum_{n=1}^{\infty} \frac{z^n}{n}, \quad z \in U(0; 1), \quad \text{where } \log 1 = 0. \tag{7.27}$$

Finally, we conclude that

$$1 + \sum_{n=1}^{\infty} \frac{z^n}{n} = 1 - \log(1 - z), \quad z \in U(0; 1).$$

2. The radius of convergence is

$$R = \frac{1}{\lim_{n\to+\infty} \sqrt[n]{|a_n|}} = \frac{1}{\lim_{n\to+\infty} \sqrt[n]{n}} = 1,$$

hence the set of convergence will be $U(0;1)$.

Since

$$\frac{1}{1-z} = \sum_{n=0}^{\infty} z^n, \quad z \in U(0;1),$$

differentiating the both sides of this relation we have

$$\frac{1}{(1-z)^2} = \sum_{n=1}^{\infty} nz^{n-1}, \quad z \in U(0;1), \tag{7.28}$$

and multiplying by z this last identity it follows that

$$\sum_{n=1}^{\infty} nz^n = \frac{z}{(1-z)^2}, \quad z \in U(0;1). \tag{7.29}$$

3. The radius of convergence is

$$R = \frac{1}{\lim_{n\to+\infty} \sqrt[n]{|a_n|}} = \frac{1}{\lim_{n\to+\infty} \sqrt[n]{n^2}} = 1,$$

hence the set of convergence will be $U(0;1)$. Now we will differentiate the relation (7.29), i. e.,

$$\sum_{n=1}^{\infty} nz^n = \frac{z}{(1-z)^2}, \quad z \in U(0;1),$$

and we obtain

$$\sum_{n=1}^{\infty} n^2 z^{n-1} = \frac{1+z}{(1-z)^3}, \quad z \in U(0;1).$$

Multiplying this last identity by z, we obtain the required result

$$\sum_{n=1}^{\infty} n^2 z^n = \frac{z(1+z)}{(1-z)^3}, \quad z \in U(0;1).$$

4. Since

$$\sum_{n=2}^{\infty} n(n-1)z^{n-2} = \sum_{m=0}^{\infty} (m+2)(m+1)z^m,$$

the radius of convergence is given by

$$R = \lim_{m\to+\infty} \frac{|a_m|}{|a_{m+1}|} = \lim_{m\to+\infty} \frac{|(m+2)(m+1)|}{|(m+3)(m+2)|} = 1,$$

hence the set of convergence will be $U(0;1)$. Now we will differentiate the relation (7.28), i. e.,

$$\frac{1}{(1-z)^2} = \sum_{n=1}^{\infty} nz^{n-1}, \quad z \in U(0;1),$$

and we conclude that

$$\sum_{n=2}^{\infty} n(n-1)z^{n-2} = \frac{2}{(1-z)^3}, \quad z \in U(0;1).$$

5. The radius of convergence is

$$\begin{aligned} R &= \lim_{n\to+\infty} \frac{|a_n|}{|a_{n+1}|} = \lim_{n\to+\infty} \frac{|(-1)^n \frac{n}{n^2-1}|}{|(-1)^{n+1} \frac{n+1}{(n+1)^2-1}|} \\ &= \lim_{n\to+\infty} \left| \frac{(-1)^n n((n+1)^2-1)}{(-1)^{n+1}(n^2-1)(n+1)} \right| = 1, \end{aligned}$$

hence the set of convergence will be $U(0;1)$. The sum may be written as

$$\sum_{n=2}^{\infty} (-1)^n \frac{n}{n^2-1} z^n = \frac{1}{2} \sum_{n=2}^{\infty} (-1)^n \frac{z^n}{n-1} + \frac{1}{2} \sum_{n=2}^{\infty} (-1)^n \frac{z^n}{n+1}, \tag{7.30}$$

so we need to determine the sums of both of the terms from the right-hand side. It is well known that

$$\log(1+z) = \sum_{n=1}^{\infty} (-1)^{n+1} \frac{z^n}{n}, \quad z \in U(0;1). \tag{7.31}$$

Multiplying this relation by z, we get

$$z\log(1+z) = \sum_{n=1}^{\infty} (-1)^{n+1} \frac{z^{n+1}}{n}, \quad z \in U(0;1),$$

and the above identity is equivalent to

$$z\log(1+z) = \sum_{n=2}^{\infty} (-1)^n \frac{z^n}{n-1}, \quad z \in U(0;1). \tag{7.32}$$

Dividing the relation (7.31) by z, we have

$$\frac{\log(1+z)}{z} = \sum_{n=1}^{\infty} (-1)^{n+1} \frac{z^{n-1}}{n}, \quad z \in U(0;1),$$

which is equivalent to

$$\frac{\log(1+z)}{z} = 1 - \frac{z}{2} + \sum_{n=2}^{\infty}(-1)^n \frac{z^n}{n+1}, \quad z \in U(0;1). \tag{7.33}$$

Using the identities (7.30), (7.32) and (7.33), we conclude that

$$\sum_{n=2}^{\infty}(-1)^n \frac{n}{n^2-1} z^n = \frac{z}{2}\log(1+z) + \frac{1}{2}\left(\frac{\log(1+z)}{z} - 1 + \frac{z}{2}\right), \quad z \in U(0;1).$$

6. Since

$$\sum_{n=1}^{\infty} nz^{n-1} = \sum_{m=0}^{\infty}(m+1)z^m,$$

the radius of convergence is given by

$$R = \frac{1}{\lim_{m\to+\infty}\sqrt[m]{|a_m|}} = \frac{1}{\lim_{m\to+\infty}\sqrt[m]{m+1}} = 1,$$

hence the set of convergence will be $U(0;1)$. The relation (7.28) obtained to point 2 shows that

$$\frac{1}{(1-z)^2} = \sum_{n=1}^{\infty} nz^{n-1}, \quad z \in U(0;1).$$

7. Since

$$a_n = \begin{cases} 0, & \text{if } n = 2m,\ m \in \mathbb{N}, \\ \frac{1}{2m+1}, & \text{if } n = 2m+1,\ m \in \mathbb{N}, \end{cases}$$

the radius of convergence is

$$R = \frac{1}{\limsup_{n\to+\infty}\sqrt[n]{|a_n|}} = \frac{1}{\lim_{m\to+\infty}\sqrt[2m+1]{\frac{1}{2m+1}}} = 1,$$

hence the set of convergence will be $U(0;1)$. Now we will use the relation (7.27) obtained to point 1, i. e.,

$$\log(1+z) = \sum_{n=1}^{\infty}(-1)^{n+1}\frac{z^n}{n}, \quad z \in U(0;1)$$

$$-\log(1-z) = \sum_{n=1}^{\infty}\frac{z^n}{n}, \quad z \in U(0;1).$$

Adding these two identities, the even terms vanishes, hence we deduce that

$$\sum_{n=0}^{\infty}\frac{z^{2n+1}}{2n+1} = \frac{1}{2}[\log(1+z) - \log(1-z)], \quad z \in U(0;1),$$

and thus

$$\sum_{n=1}^{\infty} \frac{z^{2n+1}}{2n+1} = \frac{1}{2}[\log(1+z) - \log(1-z)] - z, \quad z \in U(0;1).$$

8. The radius of convergence is

$$R = \frac{1}{\limsup_{n\to+\infty} \sqrt[n]{|a_n|}} = \frac{1}{\limsup_{n\to+\infty} \sqrt[n]{|\frac{(-1)^n}{n}|}} = 1,$$

hence the set of convergence will be $U(0;1)$. The required sum is given by the well-known formula

$$\log(1+z) = \sum_{n=1}^{\infty} (-1)^{n+1} \frac{z^n}{n}, \quad z \in U(0;1).$$

9. Substituting $\frac{z}{z+1} = \zeta$, we get

$$\sum_{n=0}^{\infty} \left(\frac{z}{z+1}\right)^n = \sum_{n=0}^{\infty} \zeta^n,$$

and the power series of the right-hand side converges if and only if

$$|\zeta| < 1 \Leftrightarrow \left|\frac{z}{z+1}\right| < 1 \Leftrightarrow \operatorname{Re} z > -\frac{1}{2}.$$

It follows that the convergence set of the given series of functions will be the half-plane

$$D = \left\{z \in \mathbb{C} : \operatorname{Re} z > -\frac{1}{2}\right\}.$$

Starting from the relation

$$\frac{1}{1-\zeta} = \sum_{n=0}^{\infty} \zeta^n, \quad \zeta \in U(0;1),$$

and using the substitution $\zeta = \frac{z}{z+1}$, we deduce that

$$\sum_{n=0}^{\infty} \left(\frac{z}{z+1}\right)^n = \frac{1}{1 - \frac{z}{z+1}} = z+1, \quad \operatorname{Re} z > -\frac{1}{2}.$$

Solution of Exercise 4.10.3

1. A simple computation shows that

$$\begin{aligned} \frac{1}{z^2+a^2} &= \frac{1}{a^2} \frac{1}{1+(\frac{z}{a})^2} \\ &= \frac{1}{a^2} \sum_{n=0}^{\infty} (-1)^n \left(\frac{z}{a}\right)^{2n} = \sum_{n=0}^{\infty} (-1)^n a^{-2n-2} z^{2n}, \quad \text{if } |z| < a,\ a \neq 0, \end{aligned}$$

because if

$$|z| < a, \quad a \neq 0 \Leftrightarrow \left|\frac{z}{a}\right| < 1 \Leftrightarrow \left|\left(\frac{z}{a}\right)^2\right| < 1,$$

we can use the well-known relation

$$\frac{1}{1-z} = \sum_{n=0}^{\infty} z^n, \quad z \in U(0;1).$$

2. We will use the relation obtained to point 3 of Exercise 4.10.2, i. e.,

$$\sum_{n=1}^{\infty} n^2 z^n = \frac{z(1+z)}{(1-z)^3}, \quad z \in U(0;1).$$

Since

$$|z| < a, \quad a \neq 0 \Leftrightarrow \left|\frac{z}{a}\right| < 1,$$

using the substitution $z := \frac{z}{a}$ we get

$$\frac{az(z+a)}{(a-z)^3} = \sum_{n=1}^{\infty} \frac{n^2 z^n}{a^n}, \quad \text{if } |z| < a,\ a \neq 0.$$

Now, by dividing with a the above identity we conclude that

$$\frac{z(z+a)}{(a-z)^3} = \sum_{n=1}^{\infty} \frac{n^2 z^n}{a^{n+1}}, \quad \text{if } |z| < a,\ a \neq 0.$$

3. First, we will prove that

$$\operatorname{Re} z > -\frac{1}{2} \Leftrightarrow \left|\frac{z}{z+1}\right| < 1.$$

This equivalence holds, since if we denote $z = x + iy$, $x, y \in \mathbb{R}$, then

$$\left|\frac{z}{z+1}\right| < 1 \Leftrightarrow |z| < |z+1| \Leftrightarrow x^2 + y^2 < (x+1)^2 + y^2 \Leftrightarrow x = \operatorname{Re} z > -\frac{1}{2}.$$

Using the result from point 9 of Exercise 4.10.2, we deduce that

$$1 + z = \sum_{n=0}^{\infty} \left(\frac{z}{z+1}\right)^n, \quad \text{if } \operatorname{Re} z > -\frac{1}{2}.$$

Solution of Exercise 4.10.4

1. Let

$$|z| = 1 \Leftrightarrow z = e^{i\varphi}, \quad \varphi \in [0, 2\pi].$$

We easily see that

$$\sum_{n=1}^{\infty} \frac{e^{in\varphi}}{n} = \sum_{n=1}^{\infty} \frac{\cos n\varphi}{n} + i \sum_{n=1}^{\infty} \frac{\sin n\varphi}{n},$$

hence it is sufficient to obtain the sum $\sum_{n=1}^{\infty} \frac{e^{in\varphi}}{n}$. For this, we will use the relation

$$-\log(1-z) = \sum_{n=1}^{\infty} \frac{z^n}{n}, \quad z \in U(0;1).$$

and we get

$$\sum_{n=1}^{\infty} \frac{e^{in\varphi}}{n} = -\log(1-e^{i\varphi}) = -\ln\left(2\sin\frac{\varphi}{2}\right) - i\frac{\varphi-\pi}{2}, \quad 0 < \varphi < 2\pi, \tag{7.34}$$

because

$$1 - e^{i\varphi} = 2\sin\frac{\varphi}{2} e^{i\frac{\varphi-\pi}{2}}.$$

Identifying the real parts of the both sides of (7.34), we conclude that

$$\sum_{n=1}^{\infty} \frac{\cos n\varphi}{n} = -\ln\left(2\sin\frac{\varphi}{2}\right), \quad 0 < \varphi < 2\pi.$$

2. Identifying the imaginary parts of the both sides of (7.34), we obtain the required sum

$$\sum_{n=1}^{\infty} \frac{\sin n\varphi}{n} = \frac{\pi-\varphi}{2}, \quad 0 < \varphi < 2\pi.$$

3. We will use a similar proof with in point 1 of this problem. If we denoted

$$|z| = 1 \Leftrightarrow z = e^{i\varphi}, \quad \varphi \in [0, 2\pi],$$

it is easy to see that

$$\sum_{n=0}^{\infty} \frac{e^{i(2n+1)\varphi}}{2n+1} = \sum_{n=0}^{\infty} \frac{\cos(2n+1)\varphi}{2n+1} + i \sum_{n=0}^{\infty} \frac{\sin(2n+1)\varphi}{2n+1}.$$

Hence it is sufficient to compute the sum $\sum_{n=0}^{\infty}\frac{e^{i(2n+1)\varphi}}{2n+1}$. For this, we will use the relation

$$\sum_{n=0}^{\infty}\frac{z^{2n+1}}{2n+1}=\frac{1}{2}[\log(1+z)-\log(1-z)],\quad z\in U(0;1).$$

Thus

$$\begin{aligned}\sum_{n=0}^{\infty}\frac{e^{i(2n+1)\varphi}}{2n+1}&=\frac{1}{2}[\log(1+e^{i\varphi})-\log(1-e^{i\varphi})]\\&=\frac{1}{2}\left[\ln\left(2\cos\frac{\varphi}{2}\right)+i\frac{\varphi}{2}-\ln\left(2\sin\frac{\varphi}{2}\right)-i\frac{\varphi-\pi}{2}\right]\\&=\frac{1}{2}\left[\ln\cot\frac{\varphi}{2}+i\frac{\pi}{2}\right],\quad \text{if } 0<\varphi<\pi,\end{aligned}\tag{7.35}$$

because

$$1-e^{i\varphi}=2\sin\frac{\varphi}{2}e^{i\frac{\varphi-\pi}{2}}\quad\text{and}\quad 1+e^{i\varphi}=2\cos\frac{\varphi}{2}e^{i\frac{\varphi}{2}}.$$

Identifying the real parts of the both sides of (7.35), we deduce that

$$\sum_{n=0}^{\infty}\frac{\cos(2n+1)\varphi}{2n+1}=\frac{1}{2}\ln\left(\cot\frac{\varphi}{2}\right),\quad 0<\varphi<\pi.$$

4. If we identify the imaginary parts of the both sides of the relation (7.35), it follows that

$$\sum_{n=0}^{\infty}\frac{\sin(2n+1)\varphi}{2n+1}=\frac{\pi}{4},\quad 0<\varphi<\pi.$$

Solution of Exercise 4.10.5

1. We will write the function in the form

$$f(z)=\sum_{n=0}^{\infty}a_n(z-z_0)^n,\quad \text{where } z_0=1.$$

In order to determine the coefficients from the above formula, we will decompose the given function by using functions with well-known Taylor series expansions. Thus, we have

$$f(z)=\frac{z}{z+1}=\frac{z+1-1}{z+1}=1-\frac{1}{z+1}.$$

Since the Taylor series expansion is about the point $z_0=1$, we need to emphasize the expression $z-1$, i. e.,

$$f(z)=1-\frac{1}{z+1}=1-\frac{1}{z-1+2}=1-\frac{1}{2(1+\frac{z-1}{2})}=1-\frac{1}{2}\frac{1}{1+\frac{z-1}{2}}.$$

Using the well-known formula,

$$\frac{1}{1+\zeta} = \sum_{n=0}^{\infty}(-1)^n\zeta^n, \quad \zeta \in U(0;1), \tag{7.36}$$

we get

$$f(z) = 1 - \frac{1}{2}\frac{1}{1+\frac{z-1}{2}} = 1 - \frac{1}{2}\sum_{n=0}^{\infty}(-1)^n\left(\frac{z-1}{2}\right)^n, \quad |z-1| < 2,$$

where $|z-1| < 2$, since the function is not defined in the point $z_1 = -1$.

Thus

$$f(z) = 1 + \sum_{n=0}^{\infty}\left(-\frac{1}{2}\right)^{n+1}(z-1)^n, \quad \text{where } |z-1| < 2.$$

2. We will use a similar method like to the point 1 of the problem. It is well known that

$$\frac{1}{(1+\zeta)^2} = \sum_{n=1}^{\infty}(-1)^{n-1}n\zeta^{n-1}, \quad \zeta \in U(0;1). \tag{7.37}$$

Since the Taylor series expansion is about the point $z_0 = 1$, we will emphasize the expression $z-1$, i. e.,

$$\begin{aligned} f(z) &= \frac{z^2}{(z+1)^2} = \frac{(z-1+1)^2}{(z-1+2)^2} = \frac{(z-1)^2+2(z-1)+1}{4(1+\frac{z-1}{2})^2} \\ &= \frac{(z-1)^2+2(z-1)+1}{4}\frac{1}{(1+\frac{z-1}{2})^2}. \end{aligned}$$

Using the relation (7.37), we have

$$f(z) = \frac{(z-1)^2+2(z-1)+1}{4}\sum_{n=1}^{\infty}(-1)^{n-1}n\left(\frac{z-1}{2}\right)^{n-1}, \quad |z-1| < 2,$$

and thus

$$\begin{aligned} f(z) &= \sum_{n=1}^{\infty}\left(-\frac{1}{2}\right)^{n+1}n(z-1)^{n+1} - \sum_{n=1}^{\infty}\left(-\frac{1}{2}\right)^{n}n(z-1)^{n} \\ &\quad + \sum_{n=1}^{\infty}\left(-\frac{1}{2}\right)^{n+1}n(z-1)^{n-1}, \quad |z-1| < 2. \end{aligned}$$

A simple computation gives us the final form of the expansion

$$f(z) = \frac{1}{4} + \frac{1}{4}(z-1) + \sum_{n=3}^{\infty}\left(-\frac{1}{2}\right)^{n+1}(9n-12)(z-1)^{n-1}, \quad \text{where } |z-1| < 2.$$

3. The function is not defined in the point $z_1 = 2$. Thus, the disc $U(0;2)$ is the largest disc with the center at the origin where the given function is holomorphic, and

$$f(z) = \frac{z-1}{z-2} = \frac{z-2+1}{z-2} = 1 + \frac{1}{z-2} = 1 + \frac{1}{2}\frac{1}{\frac{z}{2}-1} = 1 - \frac{1}{2}\frac{1}{1-\frac{z}{2}}.$$

Using the formula,

$$\frac{1}{1-\zeta} = \sum_{n=0}^{\infty} \zeta^n, \quad \zeta \in U(0;1), \tag{7.38}$$

we get

$$f(z) = 1 - \frac{1}{2}\frac{1}{1-\frac{z}{2}} = 1 - \frac{1}{2}\sum_{n=0}^{\infty}\left(\frac{z}{2}\right)^n, \quad |z| < 2,$$

and thus

$$f(z) = 1 - \sum_{n=0}^{\infty}\left(\frac{1}{2}\right)^{n+1} z^n, \quad \text{where } |z| < 2.$$

The expansion about the point $z_0 = i$ will be found similarly, but in this case we need to emphasize the expression $z - i$. The function is not defined in the point $z_1 = 2$, and thus $U(i;\sqrt{5})$ is the maximal disc with the center in $z_0 = i$ where the given function is holomorphic, and

$$f(z) = \frac{z-1}{z-2} = 1 + \frac{1}{z-2} = 1 + \frac{1}{z-i+i-2} = 1 + \frac{1}{i-2}\frac{1}{1+\frac{z-i}{i-2}}.$$

Using the relation (7.36), we get

$$f(z) = 1 + \frac{1}{i-2}\frac{1}{1+\frac{z-i}{i-2}} = 1 + \frac{1}{i-2}\sum_{n=0}^{\infty}(-1)^n\left(\frac{z-i}{i-2}\right)^n, \quad |z-i| < \sqrt{5},$$

and thus

$$f(z) = 1 + \sum_{n=0}^{\infty}\frac{(-1)^n}{(i-2)^{n+1}}z^n, \quad \text{where } |z-i| < \sqrt{5}.$$

4. The function is not defined in points ε and $\overline{\varepsilon}$, where $\varepsilon = e^{i\frac{2\pi}{3}}$ is the third root of the unity. The disc $U(1;\rho)$ is the maximal disc with the center in $z_0 = 1$ where the given function is holomorphic, $|z_0 - \varepsilon| = |z_0 - \overline{\varepsilon}| = \rho = \sqrt{3}$, and

$$\begin{aligned} f(z) &= \frac{1}{z^2+z+1} = \frac{1}{(z-\varepsilon)(z-\overline{\varepsilon})} = \frac{i\sqrt{3}}{3}\left(\frac{1}{z-\overline{\varepsilon}} - \frac{1}{z-\varepsilon}\right) \\ &= \frac{i\sqrt{3}}{3}\left(\frac{1}{(z-1)+1-\overline{\varepsilon}} - \frac{1}{(z-1)+1-\varepsilon}\right) \\ &= \frac{i\sqrt{3}}{3}\left(\frac{1}{1-\overline{\varepsilon}}\frac{1}{1+\frac{z-1}{1-\overline{\varepsilon}}} - \frac{1}{1-\varepsilon}\frac{1}{1+\frac{z-1}{1-\varepsilon}}\right). \end{aligned}$$

Using the relation (7.36), we get

$$f(z) = \frac{i\sqrt{3}}{3(1-\overline{\varepsilon})}\sum_{n=0}^{\infty}(-1)^n\left(\frac{z-1}{1-\overline{\varepsilon}}\right)^n - \frac{i\sqrt{3}}{3(1-\varepsilon)}\sum_{n=0}^{\infty}(-1)^n\left(\frac{z-1}{1-\varepsilon}\right)^n, \quad |z-1| < \rho,$$

if

$$\left|\frac{z-1}{1-\overline{\varepsilon}}\right| < 1 \Leftrightarrow |z-1| < |1-\overline{\varepsilon}| = \rho, \quad \left|\frac{z-1}{1-\varepsilon}\right| < 1 \Leftrightarrow |z-1| < |1-\varepsilon| = \rho.$$

From the above relations, we conclude that

$$f(z) = \sum_{n=0}^{\infty}\frac{i\sqrt{3}}{3}\left[\left(-\frac{1}{1-\varepsilon}\right)^{n+1} - \left(-\frac{1}{1-\overline{\varepsilon}}\right)^{n+1}\right](z-1)^n, \quad \text{where } |z-1| < \sqrt{3}.$$

We will use a similar method for the expansion about the point $z_0 = \infty$, and we need to emphasize the expression $\frac{1}{z}$. The circular ring $U(0; 1, \infty)$ is the maximal ring with the center at the origin where the function is well-defined. Since

$$f(z) = \frac{1}{z^2+z+1} = \frac{i\sqrt{3}}{3}\left(\frac{1}{z-\overline{\varepsilon}} - \frac{1}{z-\varepsilon}\right) = \frac{i\sqrt{3}}{3}\left(\frac{1}{z}\frac{1}{1-\frac{\overline{\varepsilon}}{z}} + \frac{1}{z}\frac{1}{1-\frac{\varepsilon}{z}}\right),$$

using the relation (7.38) we get

$$f(z) = \frac{i\sqrt{3}}{3z}\sum_{n=0}^{\infty}\left(\frac{\overline{\varepsilon}}{z}\right)^n - \frac{i\sqrt{3}}{3z}\sum_{n=0}^{\infty}\left(\frac{\varepsilon}{z}\right)^n, \quad |z| > 1,$$

where

$$\left|\frac{\overline{\varepsilon}}{z}\right| < 1 \Leftrightarrow |z| > |\overline{\varepsilon}| = 1, \quad \left|\frac{\varepsilon}{z}\right| < 1 \Leftrightarrow |z| > |\varepsilon| = 1.$$

Thus,

$$f(z) = \sum_{n=0}^{\infty}\frac{i(\overline{\varepsilon}^n - \varepsilon^n)\sqrt{3}}{3}\frac{1}{z^{n+1}}, \quad \text{where } |z| > 1.$$

5. The solution is similar to that of point 4 of the problem. The function is not defined in points $z_1 = 3$ and $z_2 = 5$. The disc $U(4; 1)$ is the maximal disc with the center in $z_0 = 4$ where the function is well-defined, and

$$\begin{aligned} f(z) &= \frac{z+3}{z^2-8z+15} = \frac{z+3}{(z-3)(z-5)} \\ &= \frac{4}{z-5} - \frac{3}{z-3} = \frac{4}{(z-4)-1} - \frac{3}{(z-4)+1}. \end{aligned}$$

Using the relations (7.36) and (7.38), we get

$$f(z) = -4\sum_{n=0}^{\infty}(z-4)^n - 3\sum_{n=0}^{\infty}(-1)^n(z-4)^n, \quad |z-4| < 1.$$

Consequently, we have that

$$f(z) = -\sum_{n=0}^{\infty}[4+3(-1)^n](z-4)^n, \quad \text{where } |z-4|<1.$$

For the expansion about $z_0 = \infty$, we will emphasize the expression $\frac{1}{z}$. The circular ring $U(0;5,\infty)$ is the maximal ring with the center at the origin where the function is well-defined, and

$$f(z) = \frac{z+3}{z^2-8z+15} = \frac{4}{z-5} - \frac{3}{z-3} = \frac{4}{z}\frac{1}{1-\frac{5}{z}} - \frac{3}{z}\frac{1}{1-\frac{3}{z}}.$$

Using the relation (7.38), we get

$$f(z) = \frac{4}{z}\sum_{n=0}^{\infty}\left(\frac{5}{z}\right)^n - \frac{3}{z}\sum_{n=0}^{\infty}\left(\frac{3}{z}\right)^n, \quad |z|>5,$$

and thus

$$f(z) = \sum_{n=0}^{\infty}[4\cdot 5^n - 3^{n+1}]\frac{1}{z^{n+1}}, \quad \text{where } |z|>5.$$

6. The function $f(z) = \sin\frac{z}{1-z}$ is not defined in the point $z_1 = 1$, and $U(0;1)$ is the maximal disc with the center at the origin where the function is well-defined. The coefficients from the expansion will be determined by using the formula

$$f(z) = f(0) + \frac{f'(0)}{1!}z + \frac{f''(0)}{2!}z + \cdots, \quad z \in U(0;1).$$

We easily obtain that

$$\begin{aligned}
f(0) &= 0\\
f'(z) &= \frac{1}{(1-z)^2}\cos\frac{z}{1-z} \Rightarrow f'(0) = 1\\
f''(z) &= -\frac{\sin\frac{z}{1-z} - 2\cos\frac{z}{1-z} + 2z\cos\frac{z}{1-z}}{1-4z+6z^2-4z^3+z^4} \Rightarrow f''(0) = 2\\
f'''(z) &= \frac{5\cos\frac{z}{1-z} - 6\sin\frac{z}{1-z} + 6z\sin\frac{z}{1-z} - 12z\cos\frac{z}{1-z} + 6z^2\cos\frac{z}{1-z}}{1-6z+15z^2-20z^3+15z^4-6z^5+z^6}\\
&\Rightarrow f'''(0) = 5,
\end{aligned}$$

hence

$$f(z) = z + z^2 + \frac{5}{6}z^3 + \cdots, \quad \text{where } |z|<1.$$

7. The function is defined in the whole complex plane. Using the following formulas,

$$\cosh 2z = 2\cosh^2 z - 1 \quad \text{and} \quad \cosh z = \frac{e^z + e^{-z}}{2},$$

we get

$$f(z) = \cosh^2 z = \frac{\cosh 2z + 1}{2} = \frac{1}{2} + \frac{e^{2z} + e^{-2z}}{4}.$$

Since

$$e^z = \sum_{n=0}^{\infty} \frac{z^n}{n!}, \quad z \in \mathbb{C},$$

the function may be written in the following form:

$$f(z) = \frac{1}{2} + \frac{e^{2z} + e^{-2z}}{4} = \frac{1}{2} + \frac{1}{4}\sum_{n=0}^{\infty} \frac{2^n z^n}{n!} + \frac{1}{4}\sum_{n=0}^{\infty} \frac{(-2)^n z^n}{n!}$$
$$= \frac{1}{2} + \frac{1}{4}\sum_{n=0}^{\infty} 2^n \frac{1 + (-1)^n}{n!} z^n, \quad z \in \mathbb{C}.$$

Thus,

$$f(z) = \frac{1}{2} + \sum_{n=0}^{\infty} \frac{2^{2n-1}}{(2n)!} z^{2n}, \quad \text{where } z \in \mathbb{C}.$$

8. Using the following well-known formula,

$$\cos 3z = 4\cos^3 z - 3\cos z \quad \text{and} \quad \cos z = \sum_{n=0}^{\infty} (-1)^n \frac{z^{2n}}{(2n)!}, \quad z \in \mathbb{C},$$

we get

$$f(z) = \cos^3 z = \frac{\cos 3z + 3\cos z}{4}$$
$$= \frac{1}{4}\sum_{n=0}^{\infty} (-1)^n \frac{(3z)^{2n}}{(2n)!} + \frac{3}{4}\sum_{n=0}^{\infty} (-1)^n \frac{z^{2n}}{(2n)!}, \quad z \in \mathbb{C}.$$

Thus,

$$f(z) = \sum_{n=0}^{\infty} (-1)^n \frac{3^{2n} + 3}{4 \cdot (2n)!} z^{2n}, \quad \text{where } z \in \mathbb{C}.$$

9. The function is defined in the whole complex plane. Using the following formulas,

$$\sin 3z = 3\sin z - 4\sin^3 z \quad \text{and} \quad \sin z = \sum_{n=0}^{\infty} (-1)^n \frac{z^{2n+1}}{(2n+1)!}, \quad z \in \mathbb{C},$$

we get

$$f(z) = \sin^3 z = \frac{3\sin z - \sin 3z}{4}$$
$$= \frac{3}{4}\sum_{n=0}^{\infty} (-1)^n \frac{z^{2n+1}}{(2n+1)!} - \frac{1}{4}\sum_{n=0}^{\infty} (-1)^n \frac{(3z)^{2n+1}}{(2n+1)!}, \quad z \in \mathbb{C},$$

and thus

$$f(z) = \sum_{n=0}^{\infty} (-1)^n \frac{3 - 3^{2n+1}}{4 \cdot (2n+1)!} z^{2n+1}, \quad \text{where } z \in \mathbb{C}.$$

10. Using the following formulas,

$$\sin z = \frac{e^{iz} - e^{-iz}}{2i} \quad \text{and} \quad e^z = \sum_{n=0}^{\infty} \frac{z^n}{n!}, \quad z \in \mathbb{C},$$

we get

$$\begin{aligned} f(z) &= e^z \sin z = e^z \frac{e^{iz} - e^{-iz}}{2i} = \frac{e^{(1+i)z} - e^{(1-i)z}}{2i} \\ &= \frac{1}{2i} \sum_{n=0}^{\infty} \frac{(1+i)^n z^n}{n!} + \frac{1}{2i} \sum_{n=0}^{\infty} \frac{(1-i)^n z^n}{n!}, \quad z \in \mathbb{C}. \end{aligned}$$

Thus,

$$f(z) = \sum_{n=0}^{\infty} \frac{(1+i)^n + (1-i)^n}{2i \cdot n!} z^n, \quad \text{where } z \in \mathbb{C}.$$

11. The given function is $f(z) = \sqrt[3]{z}$, where $f(1) = \frac{-1+i\sqrt{3}}{2}$, $z_0 = 1$. We will write the function in the exponential form, i. e.,

$$\forall z = re^{i\theta}, \quad r > 0, \quad \theta \in [0, 2\pi] \Rightarrow f_k(z) = \sqrt[3]{r} e^{i\frac{\theta + 2k\pi}{3}}, \quad k \in \{0, 1, 2\},$$

and thus the function f has three branches. Using the condition $f(1) = \frac{-1+i\sqrt{3}}{2}$, we determine the required branch:

$$z = 1 \Leftrightarrow z = 1e^{i \cdot 0} \Rightarrow f_k(1) = e^{i\frac{2k\pi}{3}}.$$

Since

$$f(1) = \frac{-1 + i\sqrt{3}}{2} \Leftrightarrow f(1) = e^{i\frac{2\pi}{3}} \Rightarrow k = 1,$$

and thus

$$f(z) = f_1(z) = \sqrt[3]{r} e^{i\frac{\theta + 2\pi}{3}}.$$

The function is not defined at the point $z_1 = 0$, and thus $U(1; 1)$ is the maximal disc with the center in $z_0 = 1$ where the function is well-defined, and

$$f(z) = \sqrt[3]{z} = \sqrt[3]{1 + (z-1)} = [1 + (z-1)]^{\frac{1}{3}}.$$

From the relation,

$$(1+\zeta)^\alpha = (1+\zeta)^\alpha\big|_{\zeta=0}\cdot\left[1+\sum_{n=1}^{\infty}\frac{\alpha(\alpha-1)\cdot\ldots\cdot(\alpha-n+1)}{n!}\zeta^n\right],\quad \zeta\in U(0;1), \tag{7.39}$$

we get

$$f(z) = [1+(z-1)]^{\frac{1}{3}} = \frac{-1+i\sqrt{3}}{2}\left[1+\sum_{n=1}^{\infty}\frac{\frac{1}{3}(\frac{1}{3}-1)\cdot\ldots\cdot(\frac{1}{3}-n+1)}{n!}(z-1)^n\right],$$

for $|z-1|<1$.

12. The given function is $f(z)=\sqrt[5]{z-2i}$, where $f(0)=\sqrt[5]{2}e^{i\frac{7\pi}{10}}$, $z_0=0$ and $z_0=2$. Letting

$$z-2i = re^{i\theta},\quad r>0,\quad \theta\in[0,2\pi]\Rightarrow f_k(z)=\sqrt[5]{r}e^{i\frac{\theta+2k\pi}{5}},\quad k\in\{0,1,2,3,4\},$$

and thus the function f has five branches. Using the given condition, we determine the required branch, i. e.,

$$z=0\Leftrightarrow -2i = 2e^{i\frac{3\pi}{2}}\Rightarrow f_k(0)=\sqrt[5]{2}e^{i\frac{\frac{3\pi}{2}+2k\pi}{5}}.$$

Since

$$f(1)=\sqrt[5]{2}e^{i\frac{7\pi}{10}}\Rightarrow k=1,$$

we get

$$f(z)=f_1(z)=\sqrt[5]{r}e^{i\frac{\theta+2\pi}{5}}.$$

The function is not defined at the point $z_1=2i$, then $U(0;2)$ is the maximal disc with the center in $z_0=0$ where the function is well-defined, and thus

$$f(z)=\sqrt[5]{z-2i}=\sqrt[5]{-2i\left(1-\frac{z}{2i}\right)}=\sqrt[5]{-2i}\cdot\sqrt[5]{1-\frac{z}{2i}}=\sqrt[5]{2}e^{i\frac{7\pi}{10}}\sqrt[5]{1-\frac{z}{2i}},$$

where

$$\sqrt[5]{1-\frac{z}{2i}}\bigg|_{z=0}=1.$$

From the relation (7.39) used to point 11, we get

$$\begin{aligned}f(z)&=\sqrt[5]{2}e^{i\frac{7\pi}{10}}\left(1-\frac{z}{2i}\right)^{\frac{1}{5}}\\ &=\sqrt[5]{2}e^{i\frac{7\pi}{10}}\left(1+\sum_{n=1}^{\infty}(-1)^n\frac{\frac{1}{5}(\frac{1}{5}-1)\cdot\ldots\cdot(\frac{1}{5}-n+1)}{(2i)^n n!}z^n\right),\quad |z|<2.\end{aligned}$$

For the expansion about point $z_0 = 2$, we need to write the given function into the form

$$f(z) = \sum_{n=0}^{\infty} a_n(z-2)^n, \quad z \in U(2;\rho).$$

The disc $U(2; 2\sqrt{2})$ is the maximal disc with the center in $z_0 = 2$ where the function is well-defined, i. e., $\rho = 2\sqrt{2}$, and

$$\begin{aligned} f(z) &= \sqrt[5]{z-2i} = \sqrt[5]{(z-2)+2-2i} \\ &= \sqrt[5]{2-2i} \cdot \sqrt[5]{1+\frac{z-2}{2-2i}} = \sqrt[5]{2\sqrt{2}e^{i\frac{3\pi}{4}}}\sqrt[5]{1+\frac{z-2}{2-2i}}, \end{aligned}$$

where

$$\sqrt[5]{1+\frac{z-2}{2-2i}}\bigg|_{z=2} = 1.$$

From the relation (7.39) used to point 11, we get

$$\begin{aligned} f(z) &= \sqrt[5]{2\sqrt{2}e^{i\frac{3\pi}{4}}}\left(1+\frac{z-2}{2-2i}\right)^{\frac{1}{5}} \\ &= 2^{\frac{3}{10}} e^{i\frac{3\pi}{4}}\left(1+\sum_{n=1}^{\infty}\frac{\frac{1}{5}(\frac{1}{5}-1)\cdot\ldots\cdot(\frac{1}{5}-n+1)}{(2-2i)^n n!}(z-2)^n\right), \quad |z-2| < 2\sqrt{2}. \end{aligned}$$

13. The given function is $f(z) = \log z$, where $f(1+i) = \frac{1}{2}\ln 2 - i\frac{7\pi}{4}$, $z_0 = -i$. We will write the function in the exponential form, i. e.,

$$\forall z = re^{i\theta}, \quad r > 0, \quad \theta \in (-\pi, \pi) \Rightarrow f_k(z) = \ln r + i(\theta + 2k\pi), \quad k \in \mathbb{Z}.$$

Using the given condition we determine the required branch, i. e.,

$$z = 1+i \Leftrightarrow z = \sqrt{2}e^{i\frac{\pi}{4}} \Rightarrow f_k(1+i) = \frac{1}{2}\ln 2 + i\left(\frac{\pi}{4} + 2k\pi\right), \quad k \in \mathbb{Z}.$$

Hence

$$f(1+i) = \frac{1}{2}\ln 2 - i\frac{7\pi}{4} \Rightarrow k = -1,$$

and thus

$$f(z) = f_{-1}(z) = \ln r + i(\theta - 2\pi).$$

The function has a holomorphic branch on the domain $\mathbb{C} \setminus \{z \in \mathbb{C} : \operatorname{Re} z \le 0, z = 0\}$, and thus the disc $U(-i; 1)$ is the maximal disc with the center in $z_0 = -i$ where the function is well-defined. A simple computation shows that

$$\begin{aligned} f(z) &= \log z = \log[(z+i) - i] = \log\left[(-i)\left(1 - \frac{z+i}{i}\right)\right] \\ &= \log_{-1}(-i) + \log\left(1 - \frac{z+i}{i}\right) = -\frac{5\pi i}{2} + \log\left(1 - \frac{z+i}{i}\right), \end{aligned}$$

where

$$\log\left(1-\frac{z+i}{i}\right)\bigg|_{z=-i}=0.$$

Using the formula,

$$\log(1+\zeta)=\sum_{n=1}^{\infty}(-1)^{n+1}\frac{\zeta^n}{n},\quad \zeta\in U(0;1), \tag{7.40}$$

we conclude that

$$f(z)=-\frac{5\pi i}{2}-\sum_{n=1}^{\infty}\frac{(z+i)^n}{i^n n},\quad |z+i|<1.$$

14. The given function is $f(z)=\frac{\log(1+z)}{1+z}$, where $f(0)=-4\pi i$ and $z_0=0$. Letting

$$z+1=re^{i\theta},\quad r>0,\quad \theta\in(-\pi,\pi),$$

then

$$f_k(z)=\frac{\ln r+i(\theta+2k\pi)}{re^{i\theta}},\quad k\in\mathbb{Z}.$$

Using the given condition, we determine the required branch. If

$$z=0\Leftrightarrow 1=e^{i\cdot 0}\Rightarrow f_k(0)=2k\pi i,$$

then

$$f(0)=-4\pi i\Rightarrow k=-2,$$

i. e.,

$$f(z)=f_{-2}(z)=\frac{\ln r+i(\theta-4\pi)}{re^{i\theta}}.$$

The disc $U(0;1)$ is the maximal disc with the center at the origin where the function f is well-defined and it is holomorphic. Since

$$f(z)=\frac{\log(1+z)}{1+z}=\frac{1}{z+1}[-4\pi i+\log(1+z)],$$

where

$$\log(1+z)|_{z=0}=1,$$

from the relation (7.40) used to point 13, we get

$$f(z)=\sum_{n=0}^{\infty}(-1)^n z^n\left(-4\pi i+\sum_{n=1}^{\infty}(-1)^{n+1}\frac{z^n}{n}\right),\quad |z|<1.$$

Solution of Exercise 4.10.6

Since we need to expand all these functions in Laurent series about the point $z_0 = 0$, the expansions will have the form

$$f(z) = \sum_{n=-\infty}^{\infty} a_n z^n.$$

1. The function $f(z) = \frac{1}{z^2-5z+6}$ is not defined in the points $z_1 = 2$ and $z_2 = 3$.
(a) If $|z| < 2$, from the relation (7.38) used in Exercise 4.10.5 we get

$$\begin{aligned} f(z) &= \frac{1}{z^2-5z+6} = \frac{1}{(z-3)(z-2)} = \frac{1}{z-3} - \frac{1}{z-2} = \frac{1}{2}\frac{1}{1-\frac{z}{2}} - \frac{1}{3}\frac{1}{1-\frac{z}{3}} \\ &= \frac{1}{2}\sum_{n=0}^{\infty}\left(\frac{z}{2}\right)^n - \frac{1}{3}\sum_{n=0}^{\infty}\left(\frac{z}{3}\right)^n = \sum_{n=0}^{\infty}\left(\frac{1}{2^{n+1}} - \frac{1}{3^{n+1}}\right)z^n, \quad \text{where } |z| < 2. \end{aligned}$$

(b) If $2 < |z| < 3$, from the relation (7.38) used in Exercise 4.10.5 we get

$$\begin{aligned} f(z) &= \frac{1}{z^2-5z+6} = \frac{1}{z-3} - \frac{1}{z-2} = -\frac{1}{z}\frac{1}{1-\frac{2}{z}} - \frac{1}{3}\frac{1}{1-\frac{z}{3}} \\ &= -\frac{1}{z}\sum_{n=0}^{\infty}\left(\frac{2}{z}\right)^n - \frac{1}{3}\sum_{n=0}^{\infty}\left(\frac{z}{3}\right)^n, \quad \text{where } 2 < |z| < 3, \end{aligned}$$

and where we used the fact that

$$|z| > 2 \Leftrightarrow \left|\frac{2}{z}\right| < 1 \quad \text{and} \quad |z| < 3 \Leftrightarrow \left|\frac{z}{3}\right| < 1.$$

(c) If $3 < |z|$, using the relation (7.38) we get

$$\begin{aligned} f(z) &= \frac{1}{z^2-5z+6} = \frac{1}{z-3} - \frac{1}{z-2} = -\frac{1}{z}\frac{1}{1-\frac{2}{z}} + \frac{1}{z}\frac{1}{1-\frac{3}{z}} \\ &= -\frac{1}{z}\sum_{n=0}^{\infty}\left(\frac{2}{z}\right)^n + \frac{1}{z}\sum_{n=0}^{\infty}\left(\frac{3}{z}\right)^n = \sum_{n=0}^{\infty}(3^n - 2^n)\frac{1}{z^{n+1}}, \quad \text{where } 3 < |z|. \end{aligned}$$

2. The function $f(z) = \frac{1}{z(z+1)(z+2)}$ is not defined in the points $z_1 = 0$, $z_2 = -1$ and $z_3 = -2$.
(a) If $0 < |z| < 1$, from the relation (7.36) used in Exercise 4.10.5, we get

$$\begin{aligned} f(z) &= \frac{1}{z(z+1)(z+2)} = \frac{1}{2z} - \frac{1}{z+1} + \frac{1}{2(z+2)} = \frac{1}{2z} - \frac{1}{z+1} + \frac{1}{4(1+\frac{z}{2})} \\ &= \frac{1}{2z} - \sum_{n=0}^{\infty}(-1)^n z^n + \frac{1}{4}\sum_{n=0}^{\infty}(-1)^n\left(\frac{z}{2}\right)^n \\ &= \frac{1}{2z} - \sum_{n=0}^{\infty}(-1)^n\left(1 - \frac{1}{2^{n+2}}\right)z^n, \quad \text{where } 0 < |z| < 1. \end{aligned}$$

(b) If $1 < |z| < 2$, from the relation (7.36) used in Exercise 4.10.5, we get

$$f(z) = \frac{1}{z(z+1)(z+2)} = \frac{1}{2z} - \frac{1}{z+1} + \frac{1}{2(z+2)} = \frac{1}{2z} - \frac{1}{z}\frac{1}{1+\frac{1}{z}} + \frac{1}{4(1+\frac{z}{2})}$$
$$= \frac{1}{2z} - \sum_{n=0}^{\infty}(-1)^n \frac{1}{z^{n+1}} + \sum_{n=0}^{\infty}\frac{(-1)^n}{2^{n+2}}z^n, \quad \text{where } 1 < |z| < 2,$$

and where we used the fact that

$$|z| > 1 \Leftrightarrow \left|\frac{1}{z}\right| < 1 \quad \text{and} \quad |z| < 2 \Leftrightarrow \left|\frac{z}{2}\right| < 1.$$

(c) If $2 < |z|$, from the relation (7.36) we get

$$f(z) = \frac{1}{z(z+1)(z+2)} = \frac{1}{2z} - \frac{1}{z+1} + \frac{1}{2(z+2)} = \frac{1}{2z} - \frac{1}{z}\frac{1}{1+\frac{1}{z}} + \frac{1}{2z}\frac{1}{(1+\frac{2}{z})}$$
$$= \frac{1}{2z} - \sum_{n=0}^{\infty}(-1)^n \frac{1}{z^{n+1}} + \sum_{n=0}^{\infty}(-1)^n \frac{2^{n-1}}{z^{n+1}}$$
$$= \frac{1}{2z} - \sum_{n=1}^{\infty}(-1)^n (2^{n-1} - 1)\frac{1}{z^{n+1}}, \quad \text{where } 2 < |z|.$$

3. The function $f(z) = \frac{2z^2-3z+3}{z^3-2z^2+z-2}$ is not defined in the points $z_1 = 2$ and $z_{2,3} = \pm i$.

(a) If $|z| < 1$, from the relation (7.36) used in Exercise 4.10.5 together with (7.38), we get

$$f(z) = \frac{2z^2 - 3z + 3}{z^3 - 2z^2 + z - 2} = \frac{1}{z-2} + \frac{1+i}{2}\frac{1}{z-i} + \frac{1-i}{2}\frac{1}{z+i}$$
$$= -\frac{1}{2}\frac{1}{1-\frac{z}{2}} - \frac{1+i}{2i}\frac{1}{1-\frac{z}{i}} + \frac{1-i}{2i}\frac{1}{1+\frac{z}{i}}$$
$$= -\frac{1}{2}\sum_{n=0}^{\infty}\left(\frac{z}{2}\right)^n - \frac{1-i}{2}\sum_{n=0}^{\infty}\left(\frac{z}{i}\right)^n - \frac{1+i}{2}\sum_{n=0}^{\infty}(-1)^n\left(\frac{z}{i}\right)^n$$
$$= -\frac{1}{2}\sum_{n=0}^{\infty}\left(\frac{1}{2^n} + \frac{1-i+(1+i)(-1)^n}{i^n}\right)z^n \quad \text{where } |z| < 1.$$

(b) If $1 < |z| < 2$, similarly as above, we have

$$f(z) = \frac{2z^2 - 3z + 3}{z^3 - 2z^2 + z - 2} = \frac{1}{z-2} + \frac{1+i}{2}\frac{1}{z-i} + \frac{1-i}{2}\frac{1}{z+i}$$
$$= -\frac{1}{2}\frac{1}{1-\frac{z}{2}} + \frac{1+i}{2z}\frac{1}{1-\frac{i}{z}} + \frac{1-i}{2z}\frac{1}{1+\frac{i}{z}}$$
$$= -\frac{1}{2}\sum_{n=0}^{\infty}\left(\frac{z}{2}\right)^n + \frac{1+i}{2z}\sum_{n=0}^{\infty}\left(\frac{i}{z}\right)^n + \frac{1-i}{2z}\sum_{n=0}^{\infty}(-1)^n\left(\frac{i}{z}\right)^n$$
$$= -\sum_{n=0}^{\infty}\frac{z^n}{2^{n+1}} + \sum_{n=0}^{\infty}\frac{1+i+(1-i)(-1)^n}{2}\frac{i^n}{z^{n+1}}, \quad \text{where } 1 < |z| < 2,$$

where we used the fact that

$$|z| > 1 \Leftrightarrow \left|\frac{i}{z}\right| < 1 \quad \text{and} \quad |z| < 2 \Leftrightarrow \left|\frac{z}{2}\right| < 1.$$

(c) If $2 < |z|$, using the formulas (7.36) and (7.38), we obtain

$$\begin{aligned} f(z) &= \frac{2z^2 - 3z + 3}{z^3 - 2z^2 + z - 2} = \frac{1}{z-2} + \frac{1+i}{2}\frac{1}{z-i} + \frac{1-i}{2}\frac{1}{z+i} \\ &= \frac{1}{z}\frac{1}{1-\frac{2}{z}} + \frac{1+i}{2z}\frac{1}{1-\frac{i}{z}} + \frac{1-i}{2z}\frac{1}{1+\frac{i}{z}} \\ &= \frac{1}{z}\sum_{n=0}^{\infty}\left(\frac{2}{z}\right)^n + \frac{1+i}{2z}\sum_{n=0}^{\infty}\left(\frac{i}{z}\right)^n + \frac{1-i}{2z}\sum_{n=0}^{\infty}(-1)^n\left(\frac{i}{z}\right)^n \\ &= \sum_{n=0}^{\infty}\frac{2^n}{z^{n+1}} + \sum_{n=0}^{\infty}\frac{1+i+(1-i)(-1)^n}{2}\frac{i^n}{z^{n+1}} \\ &= \sum_{n=0}^{\infty}\left(2^n + \frac{1+i+(1-i)(-1)^n}{2}i^n\right)\frac{1}{z^{n+1}}, \quad \text{where } 2 < |z|. \end{aligned}$$

4. The function $f(z) = \log\frac{1-z}{1+z}$, with $f(0) = 0$ is not defined in the point $z_1 = 1$. Since $f(0) = 0$, the logarithmic function that appears is the main branch, i. e., $\log 1 = 0$.

(a) If $|z| < 1$, from the relation (7.40) used in Exercise 4.10.5 we get

$$f(z) = \log\frac{1-z}{1+z} = \log(1-z) - \log(1+z),$$

where

$$\log(1-z)|_{z=0} = 0, \quad \log(1+z)|_{z=0} = 0,$$

and it follows that

$$\begin{aligned} f(z) = \log(1-z) - \log(1+z) &= -\sum_{n=1}^{\infty}\frac{z^n}{n} - \sum_{n=1}^{\infty}(-1)^{n+1}\frac{z^n}{n} \\ &= -\sum_{n=1}^{\infty}\frac{1+(-1)^{n+1}}{n}z^n, \quad \text{where } |z| < 1. \end{aligned}$$

(b) If $1 < |z|$, from the relation (7.40) used in Exercise 4.10.5 we get

$$f(z) = \log\frac{1-z}{1+z} = \log(-1)\frac{1-\frac{1}{z}}{1+\frac{1}{z}} = \log(-1) + \log\frac{1-\frac{1}{z}}{1+\frac{1}{z}},$$

where

$$\log\left(1-\frac{1}{z}\right)\Big|_{z=\infty} = 0, \quad \log\left(1+\frac{1}{z}\right)\Big|_{z=\infty} = 0,$$

and thus

$$f(z) = \log\frac{1-z}{1+z} = \log(-1)\frac{1-\frac{1}{z}}{1+\frac{1}{z}} = \log(-1) + \log\frac{1-\frac{1}{z}}{1+\frac{1}{z}}$$
$$= i\pi - \sum_{n=1}^{\infty}\frac{1+(-1)^{n+1}}{n}\frac{1}{z^n}, \quad \text{where } |z| > 1.$$

Solution of Exercise 4.10.7

1. Since $f \in H(\mathbb{C} \setminus \{0, 1, \frac{-1\pm i\sqrt{3}}{2}\})$, the function is holomorphic in the given domain. It follows that

$$f(z) = \frac{1}{z^5 - z^2} = \frac{1}{z^5}\frac{1}{1-\frac{1}{z^3}} = \frac{1}{z^5}\sum_{n=0}^{\infty}\frac{1}{z^{3n}}$$
$$= \sum_{n=0}^{\infty}\frac{1}{z^{3n+5}}, \quad \text{where } |z| > 1,$$

and where we used the relation (7.38), and the fact that

$$|z| > 1 \Leftrightarrow \left|\frac{1}{z}\right| < 1 \Leftrightarrow \left|\frac{1}{z^3}\right| < 1.$$

2. Since $f \in H(\mathbb{C} \setminus \{\pm i, \pm 2, \pm 2i\})$, the function is holomorphic in the given domain. We have

$$f(z) = \frac{z^{15}}{(z^2+1)^5(z^4-16)} = \frac{z}{(1+\frac{1}{z^2})^5(1-\frac{16}{z^4})} = z\left(1+\frac{1}{z^2}\right)^{-5}\frac{1}{1-\frac{16}{z^4}}$$
$$= z\left(1 + \sum_{n=1}^{\infty}\frac{(-5)(-5-1)\cdot\ldots\cdot(-5-n+1)}{n!}\frac{1}{z^{2n}}\right)\sum_{n=0}^{\infty}\left(\frac{16}{z^4}\right)^n$$
$$= z - \frac{5}{z} + \frac{31}{z^3} - \frac{115}{z^5} + \cdots, \quad \text{where } |z| > 2,$$

where we used the relations (7.38) and (7.39), and the fact that

$$|z| > 2 \Leftrightarrow \left|\frac{1}{z}\right| < 1 \Leftrightarrow \left|\frac{1}{z^2}\right| < 1, \quad |z| > 2 \Leftrightarrow \left|\frac{2}{z}\right| < 1 \Leftrightarrow \left|\frac{16}{z^4}\right| < 1.$$

3. The given form of the function is the same with the Laurent series expansion about the point $z_0 = \infty$, and thus

$$f(z) = a_0 + a_1 z + \cdots + a_n z^n, \quad z \in \mathbb{C}.$$

4. We have $f \in H(\mathbb{C} \setminus \{0\})$, and

$$f(z) = z^3 \sin \frac{1}{z} = z^3 \sum_{n=0}^{\infty} \frac{(-1)^n}{(2n+1)!} \frac{1}{z^{2n+1}}$$
$$= \sum_{n=0}^{\infty} \frac{(-1)^n}{(2n+1)!} \frac{1}{z^{2(n-1)}}, \quad \text{where } |z| > 0,$$

and where we used the well-known formula:

$$\sin \zeta = \sum_{n=0}^{\infty} (-1)^n \frac{\zeta^{2n+1}}{(2n+1)!}, \quad \zeta \in \mathbb{C}. \tag{7.41}$$

5. First, we will expand in Laurent series $z_0 = \infty$, the derivative of the given function, i. e.,

$$f'(z) = \frac{a-b}{(z-a)(z-b)} = \frac{1}{z}\left(\frac{1}{1-\frac{a}{z}} - \frac{1}{1-\frac{b}{z}}\right)$$
$$= \sum_{n=0}^{\infty} \frac{a^n - b^n}{z^{n+1}} = \sum_{n=1}^{\infty} \frac{a^n - b^n}{z^{n+1}}, \quad |z| > \max\{|a|, |b|\},$$

where we used the relation (7.38).

Integrating the obtained relation, we deduce that

$$f(z) = \log \frac{z-a}{z-b} = -\sum_{n=1}^{\infty} \frac{a^n - b^n}{n} \frac{1}{z^n} + c, \quad |z| > \max\{|a|, |b|\}.$$

The constant $c \in \mathbb{C}$ will be determined by using the given condition $\log 1 = 2k\pi i$, i. e.,

$$\lim_{z \to \infty} f(z) = \log 1 = 2k\pi i \Rightarrow c = 2k\pi i.$$

Thus,

$$f(z) = 2k\pi i - \sum_{n=1}^{\infty} \frac{a^n - b^n}{n} \frac{1}{z^n}, \quad \text{where } |z| > \max\{|a|, |b|\}.$$

6. Since $f \in H(\mathbb{C} \setminus \{-1\})$, the function is holomorphic in the given domain, and

$$f(z) = e^{\frac{2z}{z+1}} = e^{\frac{2}{1+\frac{1}{z}}} = e^{2(1-\frac{1}{z}+\frac{1}{z^2}-\frac{1}{z^3}+\cdots)} = e^2 e^{2(-\frac{1}{z}+\frac{1}{z^2}-\frac{1}{z^3}+\cdots)}$$
$$= e^2 \sum_{n=0}^{\infty} \frac{2^n(-\frac{1}{z}+\frac{1}{z^2}-\frac{1}{z^3}+\cdots)^n}{n!}$$
$$= e^2\left(1 - \frac{2}{z} + \frac{4}{z^2} - \frac{14}{3}\frac{1}{z^3} + \cdots\right), \quad \text{where } |z| > 1,$$

where we used the relation (7.36), and the fact that

$$e^{\zeta} = \sum_{n=0}^{\infty} \frac{\zeta^n}{n!}, \quad \zeta \in \mathbb{C}. \tag{7.42}$$

Solution of Exercise 4.10.8

1. We have $f \in H(\mathbb{C} \setminus \{\pm 1\})$. The expansion regarding point $z_0 = 1$ will be given in the circular ring $U(1; 0, 2)$, i. e.,

$$\begin{aligned} f(z) &= \frac{1}{(z^2-1)^2} = \frac{1}{(z-1)^2(z+1)^2} = \frac{1}{(z-1)^2} \frac{1}{[(z-1)+2]^2} \\ &= \frac{1}{4(z-1)^2} \frac{1}{(1+\frac{z-1}{2})^2} = \frac{1}{4(z-1)^2} \sum_{n=1}^{\infty} \frac{(-1)^{n-1} n}{2^{n-1}} (z-1)^{n-1} \\ &= \frac{1}{4(z-1)^2} - \frac{1}{4(z-1)} + \frac{3}{16} - \frac{1}{8}(z-1) + \cdots, \quad \text{where } |z-1| < 2, \end{aligned}$$

and where we used that

$$\frac{1}{(1+\zeta)^2} = \sum_{n=1}^{\infty} (-1)^{n-1} n \zeta^{n-1}, \quad \zeta \in U(0;1). \tag{7.43}$$

It follows that point $z_0 = 1$ is a second-order pole for f.

For the expansion regarding the point $z_0 = \infty$, we will expand the function in Laurent series on the circular ring $|z| > 1$.

Letting $\zeta = \frac{1}{z}$, then instead of $z_0 = \infty$ we will expand the function in Laurent series regarding the point $\zeta_0 = 0$. Thus, the circular ring of convergence $|z| > 1$ will be replaced by the unit disc $U(0;1)$, while the type of the singular point $z_0 = \infty$ will be given by the type of the point $\zeta_0 = 0$. Thus,

$$f(z) = \frac{1}{(z^2-1)^2} = \frac{1}{z^4} \frac{1}{(1-\frac{1}{z^2})^2} = \frac{1}{z^4} \sum_{n=1}^{\infty} \frac{n}{z^{2(n-1)}} = \sum_{n=1}^{\infty} \frac{n}{z^{2(n+1)}}, \quad \text{where } |z| > 1,$$

hence

$$f\left(\frac{1}{\zeta}\right) = \sum_{n=1}^{\infty} n \zeta^{2(n+1)}, \quad |\zeta| < 1.$$

We conclude that $z_0 = \infty$ is a removable singular point.

2. We have $f \in H(\mathbb{C} \setminus \{1\})$, and we will expand the function in Laurent series about the point $z_0 = 1$ in the circular ring $U(0; 1, \infty)$, i. e.,

$$\begin{aligned} f(z) &= \sin \frac{1}{1-z} = \sum_{n=0}^{\infty} (-1)^{n+1} \frac{1}{(2n+1)!} \frac{1}{(z-1)^{2n+1}} \\ &= -\frac{1}{(z-1)} + \frac{1}{3!(z-1)^3} - \frac{1}{5!(z-1)^5} + \cdots, \quad \text{where } 0 < |z-1| < \infty, \end{aligned}$$

and where we used the relation (7.41). Hence the point $z_0 = 1$ is an essential isolated singular point.

For the expansion about the point $z_0 = \infty$ we will expand the function in Laurent series in the circular ring $|z| > 1$ as follows:

$$
\begin{aligned}
f(z) &= \sin\frac{1}{1-z} = \sin\left(-\frac{1}{z}\frac{1}{1-\frac{1}{z}}\right) = \sin\left(-\frac{1}{z}\sum_{n=0}^{\infty}\frac{1}{z^n}\right)\\
&= -\sin\left(\frac{1}{z}+\frac{1}{z^2}+\frac{1}{z^3}+\cdots\right)\\
&= \sum_{n=0}^{\infty}(-1)^{n+1}\frac{1}{(2n+1)!}\left(\frac{1}{z}+\frac{1}{z^2}+\frac{1}{z^3}+\cdots\right)^{2n+1}\\
&= -\frac{1}{z}-\frac{1}{z^2}-\frac{5}{6}\frac{1}{z^3}-\frac{1}{2}\frac{1}{z^4}+\cdots, \quad \text{where } |z|>1,
\end{aligned}
$$

and where we used the formulas (7.38) and (7.41).

From here, since

$$
f\left(\frac{1}{\zeta}\right) = \sum_{n=0}^{\infty}(-1)^{n+1}\frac{1}{(2n+1)!}(\zeta+\zeta^2+\zeta^3+\cdots)^{2n+1}, \quad |\zeta|<1,
$$

we conclude that $z_0 = \infty$ is a removable singular point.

3. We have $f \in H(U(0;0,\pi))$, and

$$
f(z) = \frac{1}{z\sin z} = \frac{1}{z^2}\frac{z}{\sin z} = \frac{1}{z^2}g(z), \quad 0<|z|<\pi,
$$

where

$$
g(z) = \frac{z}{\sin z} = \frac{1}{\sum_{n=0}^{\infty}(-1)^n\frac{z^{2n}}{(2n+1)!}} \in H(U(0;\pi)),
$$

because $\lim_{z\to 0} g(z) = 1$. If $g(z) = \sum_{n=0}^{\infty} a_n z^n$, then

$$
\sum_{n=0}^{\infty}(-1)^n\frac{z^{2n}}{(2n+1)!}\sum_{n=0}^{\infty}a_n z^n = 1, \quad z \in U(0;\pi),
$$

thus we get

$$
a_0 = 1, \quad a_1 = 0, \quad a_2 = \frac{1}{6}, \quad a_3 = 0, \quad a_4 = \frac{7}{360}, \quad \ldots .
$$

Hence

$$
f(z) = \frac{1}{z^2}+\frac{1}{6}+\frac{7}{360}z^2+\cdots, \quad \text{where } z \in U(0;0,\pi),
$$

where we used the relation (7.41). It follows that the point $z_0 = 0$ is a second- order pole.

For the expansion about the point $z_0 = \pi$, we will expand the function in Laurent series on the circular ring $U(\pi; 0, \pi)$. Substituting $z = t + \pi$, the new function will be expanded for the corresponding point $t_0 = 0$, i. e.,

$$\begin{aligned}
f(t+\pi) &= \frac{1}{(t+\pi)\sin(t+\pi)} = \frac{-1}{(t+\pi)\sin t} = \frac{-1}{\pi(1+\frac{t}{\pi})}\frac{1}{\sin t} \\
&= -\frac{1}{\pi}\sum_{n=0}^{\infty}(-1)^n\frac{t^n}{\pi^n}\left(1+\frac{t^2}{6}+\frac{7t^4}{360}+\cdots\right)\frac{1}{t} \\
&= \left(-\frac{1}{\pi}+\frac{t}{\pi^2}-\frac{t^2}{\pi^3}+\frac{t^3}{\pi^4}-\cdots\right)\left(1+\frac{t^2}{6}+\frac{7t^4}{360}+\cdots\right)\frac{1}{t} \\
&= -\frac{1}{\pi}\frac{1}{t}+\frac{1}{\pi^2}-\frac{\pi^2+6}{6\pi^3}t+\frac{\pi^2+6}{6\pi^4}t^2-\cdots, \quad |t|<\pi,
\end{aligned}$$

where we used the formulas (7.36) and (7.41). Returning to the original variable z, we get

$$f(z) = -\frac{1}{\pi}\frac{1}{z-\pi}+\frac{1}{\pi^2}-\frac{\pi^2+6}{6\pi^3}(z-\pi)+\frac{\pi^2+6}{6\pi^4}(z-\pi)^2-\cdots, \quad \text{where } z \in U(\pi; 0, \pi),$$

and thus the point $z_0 = \pi$ is a first-order pole.

4. We have $f \in H(\mathbb{C}\backslash\{-1\})$, and the expansion in Laurent series for the point $z_0 = \infty$ will be made in the circular ring $|z| > 1$, i. e.,

$$\begin{aligned}
f(z) &= e^{\frac{2}{1+\frac{1}{z}}} = e^{2\sum_{n=0}^{\infty}\frac{(-1)^n}{z^n}} = e^2 e^{2(-\frac{1}{z}+\frac{1}{z^2}-\frac{1}{z^3}+\cdots)} \\
&= e^2\sum_{n=0}^{\infty}\frac{2^n(-\frac{1}{z}+\frac{1}{z^2}-\frac{1}{z^3}+\cdots)^n}{n!} \\
&= e^2\left(1-\frac{2}{z}+\frac{4}{z^2}+\cdots\right), \quad \text{where } z \in U(0; 1, \infty),
\end{aligned}$$

and where we used the relations (7.36) and (7.42).

From here, since

$$f\left(\frac{1}{\zeta}\right) = e^2(1-2\zeta+4\zeta^2+\cdots), \quad |\zeta|<1,$$

it follows that $z_0 = \infty$ is a removable singular point.

5. We have $f \in H(\mathbb{C} \setminus \{i\})$, and the expansion in Laurent series for the point $z_0 = i$ will be made in the circular ring $U(i; 0, \infty)$, i. e.,

$$\begin{aligned}
f(z) &= e^{i\pi\frac{z+i}{z-i}} = e^{i\pi(1+\frac{2i}{z-i})} = e^{i\pi}e^{-\frac{2\pi}{z-i}} = -e^{-\frac{2\pi}{z-i}} \\
&= -\sum_{n=0}^{\infty}\frac{(-1)^n}{n!}\left(\frac{2\pi}{z-i}\right)^n, \quad \text{where } z \in \mathbb{C} \setminus \{i\},
\end{aligned}$$

where we used the formula (7.42). Thus, $z_0 = i$ is an essential isolated singular point.

6. The function is not defined at $z_0 = 0$, and in the points of the form $z_k = k\pi$, $k \in \mathbb{Z}$. We will expand the function in Laurent series for the point $z_0 = 0$ in the disc $U(0;\pi)$. Using the well-known relation,

$$f(z) = \cot z = \frac{\cos z}{\sin z} = \frac{1}{z}\frac{1 - \frac{z^2}{2!} + \frac{z^4}{4!} - \cdots}{1 - \frac{z^2}{3!} + \frac{z^4}{5!} - \cdots}, \quad 0 < |z| < \pi,$$

and using the undetermined coefficients method, we get

$$f(z) = \frac{1}{z} - \frac{1}{3}z - \frac{1}{45}z^3 - \frac{2}{945}z^5 - \cdots, \quad 0 < |z| < \pi.$$

We conclude that $z_0 = 0$ is a first-order pole.

For the points of the form $z_k = k\pi$, $k \in \mathbb{Z}$, using the above results we get

$$\begin{aligned} f(z) &= \cot(z - k\pi + k\pi) = \cot(z - k\pi) \\ &= \frac{1}{z - k\pi} - \frac{1}{3}(z - k\pi) - \frac{1}{45}(z - k\pi)^3 - \frac{2}{945}(z - k\pi)^5 - \cdots, \\ &\quad 0 < |z - k\pi| < \pi. \end{aligned}$$

Thus, the points $z_k = k\pi$, $k \in \mathbb{Z}$, are first-order poles.

7. We have $f \in H(\mathbb{C} \setminus \{0\})$, and

$$\begin{aligned} f(z) &= \frac{\sin^2 z}{z^5} = \frac{1 - \cos 2z}{z^5} = \frac{1}{z^5}\left(1 - \sum_{n=0}^{\infty}(-1)^n\frac{2^{2n}z^{2n}}{(2n)!}\right) \\ &= \frac{2}{z^3} - \frac{2}{3}\frac{1}{z} + \frac{4}{45}z - \cdots, \quad \text{where } z \in \mathbb{C}^*. \end{aligned}$$

From here, we get that $z_0 = 0$ is a third-order pole for f.

Solution of Exercise 4.10.9

1. The function is not defined in the points $z_1 = 1$ and $z_2 = -2$. Decomposing the function into the simple fractions, we get

$$\begin{aligned} f(z) &= \frac{1}{(z-1)^2(z+2)} = \frac{1}{9}\left(\frac{3}{(z-1)^2} - \frac{1}{z-1} + \frac{1}{z+2}\right) \\ &= \frac{1}{9}\left(\frac{1}{z^2}\frac{3}{(1-\frac{1}{z})^2} - \frac{1}{z}\frac{1}{1-\frac{1}{z}} + \frac{1}{2}\frac{1}{1+\frac{z}{2}}\right). \end{aligned}$$

Using the well-known Taylor series expansions (7.43), (7.38) and (7.36), we obtain

$$\begin{aligned} f(z) &= \frac{1}{9}\left(\frac{3}{z^2}\sum_{n=1}^{\infty}\frac{n}{z^{n-1}} - \frac{1}{z}\sum_{n=0}^{\infty}\frac{1}{z^n} + \frac{1}{2}\sum_{n=0}^{\infty}(-1)^n\frac{z^n}{2^n}\right) \\ &= \frac{1}{9}\left(\sum_{n=1}^{\infty}\frac{3n-1}{z^{n+1}} - \frac{1}{z} + \sum_{n=0}^{\infty}(-1)^n\frac{z^n}{2^{n+1}}\right), \quad 1 < |z| < 2, \end{aligned}$$

where we used that

$$|z| > 1 \Leftrightarrow \left|\frac{1}{z}\right| < 1, \quad |z| < 2 \Leftrightarrow \left|\frac{z}{2}\right| < 1.$$

2. Similarly, for point 1, we have $f \in H(\mathbb{C} \setminus \{-1, -2\})$ and

$$\begin{aligned} f(z) = \frac{z}{(z^2+1)^2(z+2)} &= \frac{2}{25}\frac{z-2}{z^2+1} + \frac{1}{5}\frac{2z+1}{(z^2+1)^2} - \frac{2}{25}\frac{1}{z+2} \\ &= \frac{2z-4}{25z^2}\frac{1}{1+\frac{1}{z^2}} + \frac{2z+1}{5z^4}\frac{1}{(1+\frac{1}{z^2})^2} - \frac{1}{25}\frac{1}{1+\frac{z}{2}} \end{aligned}$$

Using the well-known Taylor series expansions (7.43) and (7.36), we get

$$\begin{aligned} f(z) &= \frac{2z-4}{25z^2}\sum_{n=0}^{\infty}(-1)^n\frac{1}{z^{2n}} + \frac{2z+1}{5z^4}\sum_{n=1}^{\infty}(-1)^{n-1}\frac{n}{z^{2n-2}} - \frac{1}{25}\sum_{n=0}^{\infty}(-1)^n\frac{z^n}{2^n} \\ &= \frac{2z-4}{25z^2} + \sum_{n=1}^{\infty}\frac{2(n-5)z+n+20}{25}\frac{(-1)^{n+1}}{z^{2(n+1)}} - \frac{1}{25}\sum_{n=0}^{\infty}(-1)^n\frac{z^n}{2^n}, \quad 1 < |z| < 2, \end{aligned}$$

where we used that

$$|z| > 1 \Leftrightarrow \left|\frac{1}{z^2}\right| < 1, \quad |z| < 2 \Leftrightarrow \left|\frac{z}{2}\right| < 1.$$

Solution of Exercise 4.10.10

1. Decomposing the function into simple fractions, and using the well-known expansion formulas in Laurent series, we get

$$\begin{aligned} f(z) = \frac{1}{z(z-3)^2} &= \frac{1}{9}\left(\frac{1}{z} - \frac{1}{z-3} + \frac{3}{(z-3)^2}\right) \\ &= \frac{1}{9}\left(\frac{1}{z-1}\frac{1}{1+\frac{1}{z-1}} + \frac{1}{2}\frac{1}{1-\frac{z-1}{2}} + \frac{3}{4}\frac{1}{(1-\frac{z-1}{2})^2}\right) \\ &= \frac{1}{9}\left(\frac{1}{z-1}\sum_{n=0}^{\infty}(-1)^n\frac{1}{(z-1)^n} + \frac{1}{2}\sum_{n=0}^{\infty}\frac{(z-1)^n}{2^n} + \frac{3}{4}\sum_{n=1}^{\infty}n\frac{(z-1)^{n-1}}{2^{n-1}}\right) \\ &= \frac{1}{9}\left(\sum_{n=0}^{\infty}(-1)^n\frac{1}{(z-1)^{n+1}} + \sum_{n=1}^{\infty}\frac{3n+2}{2^{n+1}}(z-1)^{n-1}\right), \quad 1 < |z-1| < 2, \end{aligned}$$

where we used that

$$|z-1| > 1 \Leftrightarrow \left|\frac{1}{z-1}\right| < 1, \quad |z-1| < 2 \Rightarrow \left|\frac{z-1}{2}\right| < 1.$$

2. Similarly, for point 1, we obtain

$$\begin{aligned}f(z) &= \frac{1}{(z^2-1)(z^2+4)^2} = \frac{1}{25}\left(\frac{1}{z^2-1} - \frac{1}{z^2+4} - \frac{5}{(z^2+4)^2}\right)\\ &= \frac{1}{25}\left(\frac{1}{z^2}\frac{1}{1-\frac{1}{z^2}} - \frac{1}{z^2}\frac{1}{1+\frac{4}{z^2}} - \frac{5}{z^4}\frac{1}{(1+\frac{4}{z^2})^2}\right)\\ &= \frac{1}{25}\left(\frac{1}{z^2}\sum_{n=0}^{\infty}\frac{1}{z^{2n}} - \frac{1}{z^2}\sum_{n=0}^{\infty}(-1)^n\frac{4^n}{z^{2n}} - \frac{5}{z^4}\sum_{n=1}^{\infty}(-1)^{n-1}n\frac{4^{n-1}}{z^{2n-2}}\right)\\ &= \sum_{n=1}^{\infty}\frac{1-(-1)^n4^n - 5n(-1)^{n-1}4^{n-1}}{25}\frac{1}{z^{2n+2}}, \quad |z|>2,\end{aligned}$$

where we used that

$$|z|>2 \Leftrightarrow |z|>1 \Leftrightarrow \left|\frac{1}{z^2}\right|<1, \quad |z|>2 \Leftrightarrow \left|\frac{4}{z^2}\right|<1.$$

3. From the relation (7.42), we have

$$f(z) = z^3e^{\frac{1}{z}} = z^3\sum_{n=0}^{\infty}\frac{1}{n!z^n} = \sum_{n=0}^{\infty}\frac{1}{n!z^{n+3}}, \quad z\in\mathbb{C}^*.$$

4. Using the well-known expansion formula in Taylor series for the cosine function, we obtain

$$\begin{aligned}f(z) &= z^3\cos\frac{1}{z-2} = [(z-2)+2]^3\sum_{n=0}^{\infty}(-1)^n\frac{1}{(2n)!(z-2)^{2n}}\\ &= [(z-2)^3+6(z-2)^2+12(z-2)+8]\sum_{n=0}^{\infty}(-1)^n\frac{1}{(2n)!(z-2)^{2n}},\\ &\quad z\in\mathbb{C}\setminus\{2\}.\end{aligned}$$

5. We will use the relation (7.36) and the undetermined coefficients method, hence

$$\frac{1}{\sin z} = \frac{1}{z}\left(1+\frac{1}{6}z^2+\frac{7}{360}z^4+\cdots\right), \quad 0<|z|<\pi.$$

Thus

$$\begin{aligned}f(z) &= \frac{1}{z^2(1-z)\sin z} = \frac{1}{z^3}(1+z+z^2+\cdots)\left(1+\frac{1}{6}z^2+\frac{7}{360}z^4+\cdots\right)\\ &= \frac{1}{z^3}+\frac{1}{z^2}+\frac{7}{6z}+\frac{7}{6}+\frac{427}{360}z+\cdots, \quad 0<|z|<1.\end{aligned}$$

6. Similarly, for point 4, we deduce that

$$f(z) = \cos^2\frac{1}{z} = \frac{1+\cos\frac{2}{z}}{2} = \frac{1}{2}+\frac{1}{2}\cos\frac{2}{z} = \frac{1}{2}+\sum_{n=0}^{\infty}(-1)^n\frac{2^{2n-1}}{(2n)!z^{2n}}, \quad z\in\mathbb{C}^*.$$

Solution of Exercise 4.10.11

In both of these cases, we will use the undetermined coefficient method[1]

1. Using the well-known relation,

$$\sin z = \sum_{n=0}^{\infty} (-1)^n \frac{z^{2n+1}}{(2n+1)!} = z - \frac{z^3}{3!} + \frac{z^5}{5!} - \frac{z^7}{7!} + \cdots, \quad z \in \mathbb{C},$$

we get

$$f(z) = \frac{1}{\sin z} = \frac{1}{z} \frac{z}{\sin z} = \frac{1}{z} \frac{1}{1 - \frac{z^2}{3!} + \frac{z^4}{5!} - \cdots}, \quad z \in \mathbb{C} \setminus \{k\pi : k \in \mathbb{Z}\}.$$

From the undetermined coefficients method,

$$\frac{1}{1 - \frac{z^2}{3!} + \frac{z^4}{5!} - \cdots} = a_0 + a_1 z + a_2 z^2 + \cdots$$

we get

$$a_0 = 1, \quad a_1 = 0, \quad a_2 = \frac{1}{6}, \quad a_3 = 0, \quad a_4 = \frac{7}{360}, \quad \ldots,$$

and thus

$$f(z) = \frac{1}{\sin z} = \frac{1}{z} + \frac{z}{6} + \frac{7z^3}{360} + \cdots, \quad z \in U(0; 0, \pi).$$

2. Using the well-known relation,

$$e^z = \sum_{n=0}^{\infty} \frac{z^n}{n!} = 1 + z + \frac{z^2}{2!} + \frac{z^3}{3!} + \cdots, \quad z \in \mathbb{C},$$

we get

$$f(z) = \frac{1}{e^z - 1} = \frac{1}{z} \frac{z}{e^z - 1} = \frac{1}{z} \frac{1}{1 + \frac{z}{2!} + \frac{z^2}{3!} + \cdots}, \quad z \in \mathbb{C} \setminus \{2k\pi i : k \in \mathbb{Z}\}.$$

According to the undetermined coefficients method,

$$\frac{1}{1 + \frac{z}{2!} + \frac{z^2}{3!} + \cdots} = a_0 + a_1 z + a_2 z^2 + \cdots \Rightarrow a_0 = 1, \quad a_1 = -\frac{1}{2}, \quad a_2 = \frac{1}{12}, \quad \ldots,$$

we conclude that

$$f(z) = \frac{1}{e^z - 1} = \frac{1}{z} - \frac{1}{2} + \frac{z}{12} - \cdots, \quad z \in U(0; 0, 2\pi).$$

---

1 The method of an undetermined coefficient is an approach to find a particular solution for certain problems.

Solution of Exercise 4.10.12

Let denote

$$z - i = r_1 e^{i\theta_1}, \quad z + i = r_2 e^{i\theta_2}, \quad \text{where } \theta_1, \theta_2 \in \left[-\frac{\pi}{2}, \frac{3\pi}{2}\right]$$

and

$$(\theta_1, \theta_2) \neq \left(-\frac{\pi}{2}, \frac{\pi}{2}\right), \quad (\theta_1, \theta_2) \neq \left(\frac{3\pi}{2}, \frac{\pi}{2}\right).$$

Using the above notation, the function f will have the form

$$f(z) = \sqrt{r_1 r_2} e^{i\frac{\theta_1+\theta_2+2k\pi}{2}}, \quad k \in \{0, 1\}.$$

Using the given assumption, we will determine the branch which satisfies the condition

$$f\left(\frac{3}{4}\right) = \frac{5}{4} \Rightarrow k = 0.$$

Hence we deduce that

$$f(z) = \sqrt{z^2\left(1 + \frac{1}{z^2}\right)} = z\left(1 + \frac{1}{z^2}\right)^{\frac{1}{2}}, \quad \text{where } \left(1 + \frac{1}{z^2}\right)^{\frac{1}{2}}\Big|_{z=1} = \sqrt{2},$$

thus

$$\begin{aligned} f(z) = z\left(1 + \frac{1}{z^2}\right)^{\frac{1}{2}} &= z\left(1 + \sum_{n=1}^{\infty} \frac{\frac{1}{2}(\frac{1}{2} - 1)\cdot \ldots \cdot (\frac{1}{2} - n + 1)}{n! z^{2n}}\right) \\ &= z + \sum_{n=1}^{\infty} \frac{\frac{1}{2}(\frac{1}{2} - 1)\cdot \ldots \cdot (\frac{1}{2} - n + 1)}{n! z^{2n-1}}, \quad |z| > 1, \end{aligned}$$

where we used the formula of the binomial expansion, and the fact that

$$|z| > 1 \Leftrightarrow \left|\frac{1}{z^2}\right| < 1.$$

Solution of Exercise 4.10.13

We will determine the required branch similarly as in Exercise 4.10.12 If

$$z + 1 + r_1 e^{i\theta_1}, \quad z - 1 = r_2 e^{i\theta_2}, \quad \text{where } \theta_1, \theta_2 \in [0, 2\pi], \ (\theta_1, \theta_2) \neq (0, \pi),$$

then

$$f(z) = \sqrt{r_1 r_2} e^{i\frac{\theta_1+\theta_2+2k\pi}{2}}, \quad k \in \{0,1\},$$

and

$$f(\sqrt{5}) = 2 \Rightarrow k = 0.$$

It follows that

$$\frac{1}{(z^2-1)^{\frac{1}{2}}} = \frac{1}{z(1-\frac{1}{z^2})^{\frac{1}{2}}} = \frac{1}{z}\left(1-\frac{1}{z^2}\right)^{-\frac{1}{2}}, \quad \text{where } \left(1-\frac{1}{z^2}\right)^{\frac{1}{2}}\Big|_{z=\sqrt{5}} = \frac{2}{\sqrt{5}},$$

hence

$$\begin{aligned}\frac{1}{(z^2-1)^{\frac{1}{2}}} &= \frac{1}{z(1-\frac{1}{z^2})^{\frac{1}{2}}} = \frac{1}{z}\left(1-\frac{1}{z^2}\right)^{-\frac{1}{2}}\\ &= \frac{1}{z} + \sum_{k=1}^{\infty}(-1)^k \frac{-\frac{1}{2}(-\frac{1}{2}-1)\cdot\ldots\cdot(-\frac{1}{2}-k+1)}{k!z^{2k+1}}, \quad |z|>1,\end{aligned}$$

where we used the formula of the binomial expansion, and the fact that

$$|z| > 1 \Rightarrow \left|\frac{1}{z^2}\right| < 1.$$

Substituting $n = -k$, we deduce the required formula

$$\frac{1}{\sqrt{z^2-1}} = \frac{1}{z} + \sum_{n=-\infty}^{-1} \frac{\frac{1}{2}\cdot\frac{3}{2}\cdot\ldots\cdot(\frac{1}{2}-n-1)}{(-n)!} z^{2n-1}, \quad |z| > 1.$$

Solution of Exercise 4.10.14

First, we will decompose the function f into the simple fractions, i. e.,

$$f(z) = \frac{1}{1-z-z^2} = \frac{1}{\sqrt{5}}\left(\frac{1}{z-\frac{-1-\sqrt{5}}{2}} - \frac{1}{z-\frac{-1+\sqrt{5}}{2}}\right).$$

The expansion in Taylor series is

$$f(z) = \sum_{n=0}^{\infty} a_n z^n, \quad z \in U\left(0; \frac{1+\sqrt{5}}{2}\right), \quad \text{where } a_n = \frac{1}{n!}f^{(n)}(0),\ n \in \mathbb{N}.$$

Since

$$\left(\frac{c}{az+b}\right)^{(n)} = (-1)^n c \frac{n!a^n}{(az+b)^{n+1}},$$

we get

$$f^{(n)}(z) = \frac{(-1)^n n!}{\sqrt{5}}\left[\frac{1}{(z-\frac{-1-\sqrt{5}}{2})^{n+1}} - \frac{1}{(z-\frac{-1+\sqrt{5}}{2})^{n+1}}\right],$$

and thus

$$a_n = \frac{1}{\sqrt{5}}\left[\left(\frac{1+\sqrt{5}}{2}\right)^{n+1} - \left(\frac{1-\sqrt{5}}{2}\right)^{n+1}\right].$$

Solution of Exercise 4.10.15

1. We have

$$f(z) = \frac{z}{(z+2)^2} = \frac{z+2-2}{(z+2)^2} = \frac{1}{z+2} - \frac{2}{(z+2)^2}, \quad z \in \mathbb{C}\setminus\{-2\}.$$

2. The function f has the poles $z_k = 2k\pi i$, where $k \in \mathbb{Z}$.

For $z_0 = 0$, the circular ring $U(0;0,2\pi)$ is the maximal ring with the center at the origin where the function f is holomorphic. Then

$$f(z) = \frac{e^z+1}{e^z-1} = \frac{e^z-1+2}{e^z-1} = 1 + \frac{2}{e^z-1}, \quad z \in U(0;0,2\pi),$$

and using point 2 of Exercise 4.10.11 we get

$$f(z) = 1 + 2\left(\frac{1}{z} - \frac{1}{2} + \frac{z}{12} - \cdots\right), \quad z \in U(0;0,2\pi).$$

Thus, the main part of the Laurent series expansion regarding the point $z_0 = -2$ for the function f is $\frac{2}{z}$.

For the points of the form $z_0 = 2\pi i$, substituting $z = t + 2\pi i$ we get

$$f(t+2\pi i) = 1 + \frac{2}{e^{2\pi i+t}-1} = 1 + \frac{2}{e^t-1} = 1 + 2\left(\frac{1}{t} - \frac{1}{2} + \frac{t}{12} - \cdots\right), \quad 0 < |t| < 2\pi,$$

and thus

$$f(z) = 1 + 2\left(\frac{1}{z-2\pi i} - \frac{1}{2} + \frac{z-2\pi i}{12} - \cdots\right), \quad z \in U(2\pi i;0,2\pi).$$

Thus, the main part of the Laurent series expansion regarding the point $z_0 = 2\pi i$ for the function f is $\frac{2}{z-2\pi i}$.

3. Using the result of point 1 of Exercise 4.10.11, we have

$$\begin{aligned} f(z) &= \frac{z-1}{\sin^2 z} = (z-1)\left(\frac{1}{\sin z}\right)^2 \\ &= (z-1)\left(\frac{1}{z} + \frac{z}{6} + \frac{7z^3}{360} + \cdots\right)^2 = (z-1)\left(\frac{1}{z^2} + \frac{1}{3} + \cdots\right) \\ &= -\frac{1}{z^2} + \frac{1}{z} - \frac{1}{3} + \cdots, \quad z \in U(0;0,\pi), \end{aligned}$$

hence the main part of the Laurent series expansion is $-\frac{1}{z^2} + \frac{1}{z}$.

Solution of Exercise 4.10.16

We will use the first removability criterion.

1. Since

$$\lim_{z\to 1} f(z) = \lim_{z\to 1} \frac{z^2-1}{z-1} = 2 \in \mathbb{C},$$

it follows that $z_0 = 1$ is a removable singular point.

2. The point $z_0 = 0$ is a removable singular point, because

$$\lim_{z\to 0} f(z) = \lim_{z\to 0} \frac{\sin z}{z} = 1 \in \mathbb{C}.$$

3. Since

$$\lim_{z\to 0} f(z) = \lim_{z\to 0} \frac{z}{\tan z} = \lim_{z\to 0} \frac{z}{\frac{\sin z}{\cos z}} = \lim_{z\to 0} \frac{\cos z}{\frac{\sin z}{z}} = 1 \in \mathbb{C},$$

it follows that $z_0 = 0$ is a removable singular point.

4. To determine the type of the point $z_0 = \frac{\pi}{2}$, we will expand the given function in Laurent series for the point z_0. Since

$$f(z) = \frac{1}{\cos^2 z} - \frac{1}{(z-\frac{\pi}{2})^2} = \frac{1}{\sin^2(z-\frac{\pi}{2})} - \frac{1}{(z-\frac{\pi}{2})^2},$$

we will use the result obtained in point 1 of Exercise 4.10.11, for $t = z - \frac{\pi}{2}$, i. e.,

$$\frac{1}{\sin(z-\frac{\pi}{2})} = \frac{1}{z-\frac{\pi}{2}} + \frac{z-\frac{\pi}{2}}{6} + \frac{7(z-\frac{\pi}{2})^3}{360} + \cdots, \quad 0 < \left|z - \frac{\pi}{2}\right| < \pi.$$

Then the main part of the expansion of $\frac{1}{\sin^2(z-\frac{\pi}{2})}$ will be $\frac{1}{(z-\frac{\pi}{2})^2}$, hence the main part for the Laurent series expansion of the function f about z_0 is 0. From the second removability criterion, it follows that $z_0 = \frac{\pi}{2}$ is a removable singular point.

5. In this case, we will also use the second removability criterion. Since

$$\begin{aligned} f(z) &= \frac{1-\cos z}{z^2} = \frac{1}{z^2} - \frac{1}{z^2}\sum_{n=0}^{\infty}(-1)^n \frac{z^{2n}}{(2n)!} = \frac{1}{z^2} - \frac{1}{z^2}\left(1 - \frac{z^2}{2!} + \frac{z^4}{4!} - \cdots\right) \\ &= \frac{1}{2} - \frac{z^2}{4!} + \cdots, \quad z \in \mathbb{C}^*, \end{aligned}$$

we conclude that $z_0 = 0$ is a removable singular point.

6. Using the results of Exercise 4.10.11, we have

$$\begin{aligned} f(z) &= \frac{1}{e^z - 1} - \frac{1}{\sin z} = \left(\frac{1}{z} - \frac{1}{2} + \frac{z}{12} - \cdots\right) - \left(\frac{1}{z} + \frac{z}{6} + \frac{7z^3}{360} + \cdots\right) \\ &= -\frac{1}{2} - \frac{z}{12} - \cdots, \quad 0 < |z| < \pi. \end{aligned}$$

The main part of the Laurent series expansion of f for z_0 is 0; then from the second removability criterion, it follows that $z_0 = 0$ is a removable singular point.

7. Since

$$\lim_{z\to-1} f(z) = \lim_{z\to-1} \frac{z^2-1}{z^3+1} = \lim_{z\to-1} \frac{z-1}{z^2-z+1} = -\frac{2}{3} \in \mathbb{C},$$

from the first removability criterion it follows that a $z_0 = -1$ is a removable singular point.

Solution of Exercise 4.10.17

We will prove the above results by using the *theorem of characterization of the poles.*

1. Since

$$\lim_{z\to 0} f(z) = \lim_{z\to 0} \frac{1}{z} = \infty,$$

it follows that $z_0 = 0$ is a first-order pole for f.

2. The point $z_0 = i$ is a double pole for f, because

$$\lim_{z\to i} f(z) = \lim_{z\to i} \frac{1}{(z^2+1)^2} = \infty.$$

3. Since

$$\lim_{z\to-1} f(z) = \lim_{z\to-1} \frac{z^2+1}{z+1} = \infty,$$

the point $z_0 = -1$ is a first-order pole for f.

4. We will determine the main part of the Laurent series about the point z_0. Since

$$f(z) = \frac{1}{1-\cos z} = \frac{1}{2\sin^2\frac{z}{2}},$$

using point 1 of Exercise 4.10.11 we get

$$\frac{1}{\sin t} = \frac{1}{t} + \frac{t}{6} + \frac{7t^3}{360} + \cdots, \quad 0 < |t| < \pi,$$

hence it follows that

$$\frac{1}{\sin^2 t} = \frac{1}{t^2} + \frac{1}{3} + \frac{t^2}{15} + \cdots, \quad 0 < |t| < \pi. \tag{7.44}$$

Substituting $t = \frac{z}{2}$, we have

$$f(z) = \frac{1}{2\sin^2\frac{z}{2}} = \frac{2}{z^2} + \frac{1}{6} + \frac{z^2}{120} + \cdots, \quad 0 < |z| < 2\pi,$$

and it follows that $z_0 = 0$ is a double pole for f.

5. Since

$$\lim_{z\to -i} f(z) = \lim_{z\to -i} \frac{1}{(z^2+1)^2} = \infty,$$

the point $z_0 = -i$ double pole for f.

6. We will determine the main part of the Laurent-series about the point z_0. A simple calculation shows that

$$f(z) = \frac{z}{1-\cos z} = \frac{z-2\pi+2\pi}{2\sin^2\frac{z}{2}} = \frac{z-2\pi}{2\sin^2\frac{z-2\pi}{2}} + \frac{2\pi}{2\sin^2\frac{z-2\pi}{2}}.$$

Using the relation (7.44) from the point 4, we get

$$\frac{1}{\sin^2 t} = \frac{1}{t^2} + \frac{1}{3} + \frac{t^2}{15} + \cdots, \quad 0 < |t| < \pi,$$

Letting $t = \frac{z-2\pi}{2}$, we have

$$\begin{aligned} f(z) &= \frac{z-2\pi}{2\sin^2\frac{z-2\pi}{2}} + \frac{2\pi}{2\sin^2\frac{z-2\pi}{2}} = \frac{z-2\pi}{2}\left(\frac{4}{(z-2\pi)^2} + \frac{1}{3} + \frac{(z-2\pi)^2}{60} + \cdots\right) \\ &\quad + \pi\left(\frac{4}{(z-2\pi)^2} + \frac{1}{3} + \frac{(z-2\pi)^2}{60} + \cdots\right) \\ &= \frac{4\pi}{(z-2\pi)^2} + \frac{2}{z-2\pi} + \frac{\pi}{3} + \frac{z-2\pi}{6} + \frac{\pi(z-2\pi)^2}{60} + \cdots, \quad 0 < |z-2\pi| < 2\pi, \end{aligned}$$

then it follows that $z_0 = 2\pi$ is a double pole for f.

7. We will determine the main part of the Laurent series about the point z_0. Since

$$f(z) = \frac{z}{e^z+1} = \frac{z}{e^{\pi i}e^{z-\pi i}+1} = -\frac{z-\pi i+\pi i}{e^{z-\pi i}-1} = -\frac{z-\pi i}{e^{z-\pi i}-1} - \frac{\pi i}{e^{z-\pi i}-1},$$

using the result from point 2 of Exercise 4.10.11 we have

$$\frac{1}{e^t-1} = \frac{1}{t} - \frac{1}{2} + \frac{t}{12} - \cdots, \quad 0 < |t| < 2\pi i,$$

and substituting $t = z - \pi i$ it follows that

$$\begin{aligned} f(z) &= -\frac{z-\pi i}{e^{z-\pi i}-1} - \frac{\pi i}{e^{z-\pi i}-1} = -(z-\pi i)\left(\frac{1}{z-\pi i} - \frac{1}{2} + \frac{z-\pi i}{12} - \cdots\right) \\ &\quad - \pi i\left(\frac{1}{z-\pi i} - \frac{1}{2} + \frac{z-\pi i}{12} - \cdots\right) \\ &= \frac{-\pi i}{z-\pi i} + \frac{\pi i-2}{2} + \frac{6-\pi i}{12}(z-\pi i) + \cdots, \quad 0 < |z-\pi i| < 2\pi. \end{aligned}$$

Thus, $z_0 = \pi i$ is a first-order pole for f.

Solution of Exercise 4.10.18

1. We will determine the main part of the Laurent series for the point z_0, by using the formula (7.41), i. e.,

$$f(z) = \sin\frac{\pi}{z^2} = \sum_{n=0}^{\infty}(-1)^n\frac{\pi^{2n+1}}{(2n+1)!z^{4n+2}}, \quad z \in \mathbb{C}^*.$$

Since the main part of the Laurent series expansion has an infinite number of terms, it follows that $z_0 = 0$ is an essential isolated singular for f.

2. The Laurent series expansion of f for z_0 is

$$f(z) = (z-1)^2\cos\frac{\pi}{z-1} = \sum_{n=0}^{\infty}(-1)^n\frac{\pi^{2n}}{(2n)!(z-1)^{2(n-1)}}, \quad z \in \mathbb{C}\setminus\{1\}.$$

The main part of the Laurent series expansion has an infinite number of terms, hence $z_0 = 1$ is an essential isolated singular for f.

3. The Laurent series expansion of f about z_0 is

$$f(z) = e^{\frac{1}{z+1}} = \sum_{n=0}^{\infty}\frac{1}{n!(z+1)^n}, \quad z \in \mathbb{C}\setminus\{-1\},$$

thus $z_0 = -1$ is an essential isolated singular for f, because the main part of the Laurent series expansion has an infinite number of terms.

4. The Laurent series expansion for z_0 may be determined as follows:

$$\begin{aligned} f(z) &= \cos\frac{z}{z+1} = \cos\left(1-\frac{1}{z+1}\right) = \cos 1\cos\frac{1}{z+1} + \sin 1\sin\frac{1}{z+1} \\ &= \cos 1\sum_{n=0}^{\infty}\frac{(-1)^n}{(2n)!(z+1)^{2n}} + \sin 1\sum_{n=0}^{\infty}\frac{(-1)^n}{(2n+1)!(z+1)^{2n+1}}, \quad z \in \mathbb{C}\setminus\{-1\}. \end{aligned}$$

Since the main part of the Laurent series expansion has an infinite number of terms, it follows that $z_0 = -1$ is an essential isolated singular for f.

5. The function is not defined in the points $z_{1,2} = \pm i$, thus $f \in H(U(i;0,2))$. Since

$$\begin{aligned} &\frac{\pi}{z^2+1} = -\frac{i\pi}{2(z-i)}\frac{1}{1+\frac{z-i}{2i}} = -\frac{i\pi}{2(z-i)}\sum_{m=0}^{\infty}(-1)^m\left(\frac{z-i}{2i}\right)^m, \\ &0 < |z-i| < 2, \end{aligned}$$

where

$$|z-i| < 2 \Leftrightarrow \left|\frac{z-i}{2i}\right| < 1,$$

by using the expansion of the sine function in Taylor series, it follows that

$$f(z) = \sin\frac{\pi}{z^2+1} = \sum_{n=0}^{\infty}\frac{(-1)^n}{(2n+1)!}\left[-\frac{i\pi}{2(z-i)}\sum_{m=0}^{\infty}(-1)^m\left(\frac{z-i}{2i}\right)^m\right]^{2n+1},$$
$$0 < |z-i| < 2.$$

For $m = 0$, the main part of the Laurent series expansion for z_0 has an infinite number of terms, hence $z_0 = i$ is an essential isolated singular for f.

6. We will consequently use the expansion formulas in the Taylor series of the sine and of the exponential functions, i. e.,

$$f(z) = \sin e^{\frac{1}{z}} = \sum_{n=0}^{\infty}\frac{(-1)^n}{(2n+1)!}e^{\frac{2n+1}{z}} = \sum_{n=0}^{\infty}\frac{(-1)^n}{(2n+1)!}\left(\sum_{m=0}^{\infty}\frac{(2n+1)^m}{m!z^m}\right), \quad z \in \mathbb{C}^*.$$

We deduce that $z_0 = 0$ is an essential isolated singular for f, because the main part of the Laurent series expansion for z_0 has an infinite number of terms.

Solution of Exercise 4.10.19

Since $a \in D$, $g \in H(D)$ and $g(a) = 0$, according the *properties of holomorphic functions of zeros*, it follows that there exists $r > 0$, such that $g(z) \neq 0$, $\forall z \in \dot{U}(a;r)$. Hence the point $a \in D$ is an isolated singular point for the function h. Then

$$\lim_{z\to a}\frac{f(z)}{g(z)} = \lim_{z\to a}\frac{\frac{f(z)-f(a)}{z-a}}{\frac{g(z)-g(a)}{z-a}} = \frac{f'(a)}{g'(a)} \in \mathbb{C},$$

because $g'(a) \neq 0$, and using the *first removability criterion* we deduce that the point $a \in D$ is a removable isolated singular point for the function h.

Solution of Exercise 4.10.20

Since $f : U(a;r) \setminus \{a\} \to \mathbb{C}^*$, the function f has no zeros, thus the function h is well-defined, and

$$f \in H(U(a;r) \setminus \{a\}) \Rightarrow \frac{1}{f} \in H(U(a;r) \setminus \{a\}) \Rightarrow h \in H(U(a;r) \setminus \{a\}).$$

We need to prove that the function h is differentiable at the point a, whenever a is a pole for f. Using the *theorem of characterization of the poles*, the point $z_0 = a$ is a pole for f if and only if

$$\exists!\, n \in \mathbb{N}^*,\ \exists\rho > 0,\ \exists g \in H(U(a;\rho)) : g(a) \neq 0, \quad f(z) = \frac{g(z)}{(z-a)^n}, \quad \forall z \in \dot{U}(a;\rho).$$

Using these

$$\lim_{z\to a}\frac{h(z)-h(a)}{z-a}=\lim_{z\to a}\frac{\frac{1}{f(z)}}{z-a}$$
$$=\lim_{z\to a}\frac{\frac{(z-a)^n}{g(z)}}{z-a}=\lim_{z\to a}\frac{(z-a)^{n-1}}{g(z)}\in\begin{cases}0, & \text{if } n>1\\ \frac{1}{g(a)}, & \text{if } n=1,\end{cases}$$

i. e., $\exists \lim_{z\to a}\frac{h(z)-h(a)}{z-a}\in\mathbb{C}$, hence the function h is holomorphic in the disc $U(a;r)$.

Solution of Exercise 4.10.21

Since $f\in H(D)$, the function $\varphi(z)=e^{f(z)}$ is holomorphic on the domain D, with $\varphi(z)\neq 0$, $\forall z\in D$. From here, according to the *theorem of the module maximum of the holomorphic functions*, we have that $|\varphi|$ has no locally maximum or minimum points in the domain D. But

$$|\varphi(z)|=\left|e^{f(z)}\right|=e^{\mathrm{Re}\,f(z)},$$

hence the function $g=\mathrm{Re}\,f$ has no locally maximum or minimum points in the domain D.

Similarly, if we denote $h=\mathrm{Im}\,f$, and using the fact $\mathrm{Re}(-if)=\mathrm{Im}\,f$, it follows that $h=\mathrm{Im}\,f$ has no locally maximum or minimum points in the domain D.

Solution of Exercise 4.10.22

From Exercise 4.10.21, it is known that the function $\mathrm{Re}\,f$ has no locally minimum point in the disc $U(0;r)$, whenever f is a nonconstant function. Then

$$\mathrm{Re}\,f(z)<A,\quad \forall z\in U(0;r), \tag{7.45}$$

hence, from $\mathrm{Re}\,f(0)=0$ it follows that $A>0$.

Let

$$\varphi: U(0;r)\to\mathbb{C},\quad \varphi(z)=\frac{f(z)}{2A-f(z)},$$

that is also a holomorphic function on the disc $U(0;r)$, because $\mathrm{Re}\,f(z)\leq A$, $\forall z\in U(0;r)$. Then $\varphi(0)=0$, and since $A>0$, from the inequality (7.45), we deduce that

$$|\varphi(z)|^2=\frac{|f(z)|^2}{|2A-f(z)|^2}=\frac{|f(z)|^2}{4A[A-\mathrm{Re}\,f(z)]+|f(z)|^2}<1,\quad \forall z\in U(0;r),$$

i. e.,

$$|\varphi(z)|<1,\quad \forall z\in U(0;r).$$

Using the *generalized Schwarz lemma*, we get

$$|\varphi(z)| \le \frac{|z|}{r}, \quad z \in U(0;r).$$

From here, since

$$\varphi(z) = \frac{f(z)}{2A - f(z)} \Leftrightarrow f(z) = \frac{2A\varphi(z)}{1+\varphi(z)},$$

we deduce the required inequality

$$|f(z)| = \left|\frac{2A\varphi(z)}{1+\varphi(z)}\right| \le \frac{2A|\varphi(z)|}{1-|\varphi(z)|} \le \frac{2A|z|}{r-|z|}, \quad \forall z \in U(0;r).$$

Solution of Exercise 4.10.23

Since $|a| < 1$, the function

$$h : U(0;1) \to U(0;1), \quad h(z) = \frac{z+a}{1+\overline{a}z}$$

is bijective and its inverse will be

$$h^{-1} : U(0;1) \to U(0;1), \quad h^{-1}(z) = \frac{z-a}{1-\overline{a}z}.$$

Thus $g = h^{-1} \circ f \circ h \in H(U(0;1))$, $g(0) = h^{-1}(f(a)) = h^{-1}(a) = 0$ and $|g(z)| < 1$, $\forall z \in U(0;1)$. From here, according to the *Schwarz lemma*, we get

$$|g(z)| \le |z|, \quad \forall z \in U(0;1),$$

and since

$$g\left(\frac{b-a}{1-\overline{a}b}\right) = h^{-1}(f(b)) = h^{-1}(b) = \frac{b-a}{1-\overline{a}b},$$

using again the *Schwarz lemma* we deduce that

$$g(z) = cz, \quad z \in U(0;1) \quad \text{where } |c| = 1.$$

since there exists $z_0 = \frac{b-a}{1-\overline{a}b} \in U(0;1)$, such that $g(z_0) = z_0$, it follows $c = 1$. Hence $g(z) = z$, $\forall z \in U(0;1)$.

From here, we have $f(h(z)) = h(z)$, $\forall z \in U(0;1)$ and since $h : U(0;1) \to U(0;1)$ is a bijection we conclude that

$$f(z) = z, \quad \forall z \in U(0;1).$$

Solution of Exercise 4.10.24

Since $|\alpha| < 1$, the function

$$\varphi : U(0;1) \to U(0;1), \quad \varphi(z) = \frac{z-\alpha}{1-\overline{\alpha}z}$$

is bijective and its inverse will be

$$\varphi^{-1} : U(0;1) \to U(0;1), \quad \varphi^{-1}(z) = \frac{z+\alpha}{1+\overline{\alpha}z}.$$

Then $g = f \circ \varphi^{-1} \in H(U(0;1))$, $g(0) = f(\alpha) = 0$ and $|g(z)| = |f(\varphi^{-1}(z))| < 1$, $\forall z \in U(0;1)$, because from the *theorem of the module maximum of the holomorphic functions* there does not exist points $z_0 \in U(0;1)$ such that $|f(z_0)| = 1$.

Using the *Schwarz lemma* for the function g, we have

$$|f(\varphi^{-1}(z))| \le |z|, \quad \forall z \in U(0;1).$$

From here, since the function $\varphi : U(0;1) \to U(0;1)$ is bijective, it follows that

$$|f(z)| \le \left|\frac{z-\alpha}{1-\overline{\alpha}z}\right|, \quad \forall z \in U(0;1),$$

Solution of Exercise 4.10.25

Since

$$f\left(\frac{1}{n}\right) = \frac{1}{n+1}, \quad \forall n \in \mathbb{N}^* \Leftrightarrow f\left(\frac{1}{n}\right) = \frac{\frac{1}{n}}{1+\frac{1}{n}}, \quad \forall n \in \mathbb{N}^*,$$

it follows that the function

$$f(z) = \frac{z}{z+1} \in H(\mathbb{C} \setminus \{-1\})$$

satisfies the assumption of the problem.

Since $\lim_{n\to+\infty} \frac{1}{n} = 0 \in \mathbb{C} \setminus \{-1\}$, according to the *theorem of the uniqueness of holomorphic functions* there does not exist another function that satisfies these properties.

Solution of Exercise 4.10.26

1. Since

$$f\left(\frac{1}{n}\right) = \begin{cases} 0, & \text{if } n = 2k+1,\ k \in \mathbb{N} \\ \frac{2}{n}, & \text{if } n = 2k,\ k \in \mathbb{N}^* \end{cases} \quad \text{and} \quad \lim_{k\to\infty} \frac{1}{2k+1} = 0 \in U\left(\frac{1}{2}; \frac{5}{6}\right),$$

from the *theorem of the uniqueness of holomorphic functions*, since $f \in H(U(\frac{1}{2}; \frac{5}{6}))$, it follows that $f(z) = 0$, $\forall z \in U(\frac{1}{2}; \frac{5}{6})$. But this function does not satisfy the assumption $f(\frac{1}{n}) = \frac{2}{n}$, $n = 2k$ and $k \in \mathbb{N}^*$.

2. We will use the same method as in point 1. Since

$$f\left(\frac{1}{n}\right) = \begin{cases} \frac{1}{k+2}, & \text{if } n = 2k+1,\ k \in \mathbb{N} \\ \frac{1}{k+1}, & \text{if } n = 2k,\ k \in \mathbb{N}^* \end{cases}$$
$$= \begin{cases} \frac{\frac{2}{n}}{1+\frac{3}{n}}, & \text{if } n = 2k+1,\ k \in \mathbb{N} \\ \frac{\frac{2}{n}}{1+\frac{2}{n}}, & \text{if } n = 2k,\ k \in \mathbb{N}^* \end{cases}$$

and $\lim_{k\to\infty} \frac{1}{2k} = 0 \in U(\frac{1}{2}; \frac{5}{6})$, from the *theorem of the uniqueness of holomorphic functions*, since $f \in H(U(\frac{1}{2}; \frac{5}{6}))$, it follows that $f(z) = \frac{2z}{2z+1}$, $\forall z \in U(\frac{1}{2}; \frac{5}{6})$. But this function does not satisfy the assumptions $f(\frac{1}{n}) = \frac{2}{n+3}$, $n = 2k+1$ and $k \in \mathbb{N}$.

Solution of Exercise 4.10.27

1. The function $f(z) = z^2$, $\forall z \in U(0;1)$, is holomorphic in the disc $U(0;1)$ and it satisfies the assumption. From the *theorem of the uniqueness of holomorphic functions*, this function is the only one from $H(U(0;1))$ that satisfies the given assumptions.

2. Since

$$f\left(\frac{1}{n}\right) = \frac{1}{n^3}, \quad n \in \mathbb{N}^* \text{ and } \lim_{n\to+\infty} \frac{1}{n} = 0 \in U(0;1),$$

from the *theorem of the uniqueness of holomorphic functions*, since $f \in H(U(0;1))$, it follows that $f(z) = z^3$, $\forall z \in U(0;1)$. But this function does not satisfy the assumption $f(-\frac{1}{n}) = \frac{1}{n^3}$, $n \in \mathbb{N}^*$.

7.5 Solutions to the exercises of Chapter 5

Solution of Exercise 5.4.1

1. From the formula of *the residues value at the pole*, we get

$$\operatorname{Res}(f;2) = \lim_{z\to2} f(z)(z-2) = \lim_{z\to2} \frac{\sin z}{z-2}(z-2) = \sin 2,$$

because $z_0 = 2$ is a simple pole for the function f.

2. The point $z_0 = 1$ is a simple pole for the function, hence

$$\operatorname{Res}(f;1) = \lim_{z\to1} f(z)(z-1) = \lim_{z\to1} \frac{z}{(z-1)(z-2)^2}(z-1) = 1.$$

Since the point $z_1 = 2$ is a second-order pole, we have

$$\operatorname{Res}(f;2) = \frac{1}{1!}\lim_{z\to 2}[f(z)(z-2)^2]' = \lim_{z\to 2}\left(\frac{z}{z-1}\right)' = \lim_{z\to 2}\frac{-1}{(z-1)^2} = -1.$$

3. The residue of the function in the point z_1 is

$$\begin{aligned}\operatorname{Res}(f;z_1) &= \frac{1}{(p-1)!}\lim_{z\to 1}\left[\frac{z}{(z-z_1)^p(z-z_2)}(z-z_1)^p\right]^{(p-1)}\\ &= \frac{1}{(p-1)!}\lim_{z\to 1}\left(\frac{z}{z-z_2}\right)^{(p-1)} = \frac{1}{(p-1)!}\lim_{z\to 1}(-1)^{p-1}\frac{(p-1)!z_2}{(z-z_2)^p}\\ &= (-1)^{p-1}\frac{z_2}{(z_1-z_2)^p},\end{aligned}$$

because z_1 is a pth order pole.

The point z_2 is a simple pole for the function, hence

$$\operatorname{Res}(f;z_2) = \lim_{z\to z_2}\frac{z}{(z-z_1)^p(z-z_2)}(z-z_2) = \frac{z_2}{(z_2-z_1)^p}.$$

4. The point $z_0 = -1$ is a first-order pole, hence

$$\operatorname{Res}(f;-1) = \lim_{z\to -1}\frac{1}{z^3-z^5}(z+1) = \lim_{z\to -1}\frac{1}{z^3(1-z)} = -\frac{1}{2}.$$

The point $z_1 = 0$ is a third-order pole, thus

$$\begin{aligned}\operatorname{Res}(f;0) &= \frac{1}{2!}\lim_{z\to 0}\left(\frac{1}{z^3-z^5}z^3\right)'' = \frac{1}{2!}\lim_{z\to 0}\left(\frac{1}{1-z^2}\right)''\\ &= \frac{1}{2}\lim_{z\to 0}\frac{2+6z^2}{(1-z^2)^3} = 1.\end{aligned}$$

The residue in the point $z_2 = 1$ is

$$\operatorname{Res}(f;1) = \lim_{z\to 1}\frac{1}{z^3-z^5}(z-1) = \lim_{z\to 1}\frac{-1}{z^3(1+z)} = -\frac{1}{2},$$

since z_2 is a simple pole.

5. Both of the points are simple poles, hence the residues are

$$\operatorname{Res}(f;-i) = \lim_{z\to -i}\frac{z^2}{z^2+1}(z+i) = \lim_{z\to -i}\frac{z^2}{z-i} = -\frac{i}{2}$$

and

$$\operatorname{Res}(f;i) = \lim_{z\to i}\frac{z^2}{z^2+1}(z-i) = \lim_{z\to i}\frac{z^2}{z+i} = \frac{i}{2}.$$

6. The point $z_0 = -1$ is an nth order pole, hence

$$\operatorname{Res}(f;-1) = \frac{1}{(n-1)!}\lim_{z\to-1}\left[\frac{z^{2n}}{(z+1)^n}(z+1)^n\right]^{(n-1)} = \frac{1}{(n-1)!}\lim_{z\to-1}\left(z^{2n}\right)^{(n-1)}$$
$$= \frac{1}{(n-1)!}\frac{(2n)!}{(n+1)!}(-1)^{n+1} = (-1)^{n+1}\,C_{2n}^{n-1}\,.$$

Since the sum of the residues of all the poles is zero, we deduce that the residue in $z_1 = \infty$ is

$$\operatorname{Res}(f;\infty) = -\operatorname{Res}(f;-1) = (-1)^n\,C_{2n}^{n-1}\,.$$

7. The points $z_k = 2k\pi i$ are first-order poles, for all $k \in \mathbb{Z}$, hence

$$\operatorname{Res}(f;z_k) = \lim_{z\to 2k\pi i}\frac{z-2k\pi i}{1-e^z} = \lim_{\zeta\to0}\frac{\zeta}{1-e^{\zeta+2k\pi i}} = \lim_{\zeta\to0}\frac{\zeta}{1-e^{\zeta}}$$
$$= \lim_{\zeta\to0}\frac{\zeta}{1-\sum_{n=0}^{\infty}\frac{\zeta^n}{n!}} = \lim_{\zeta\to0}\frac{1}{-\sum_{n=1}^{\infty}\frac{\zeta^{n-1}}{n!}} = -1.$$

Solution of Exercise 5.4.2

1. The function f has the simple poles $z_k = e^{i\frac{(2k+1)\pi}{n}}$, where $k \in \{0,1,\dots,n-1\}$, while $z_n = \infty$ is an nth order pole.

The residues in the points $z_k = e^{i\frac{(2k+1)\pi}{n}}$ are

$$\operatorname{Res}(f;z_k) = \left.\frac{z^{2n}}{nz^{n-1}}\right|_{z=z_k} = \left.\frac{z^n z}{n}\right|_{z=z_k} = -\frac{z_k}{n} = -\frac{e^{i\frac{(2k+1)\pi}{n}}}{n}.$$

To calculate the residue in the pole $z_n = \infty$, we will use the definition of the residue at the infinity, i. e.,

$$\operatorname{Res}(f;\infty) = -a_{-1},$$

where

$$f(z) = \sum_{n=-\infty}^{\infty} a_n z^n,\quad z \in U(0;r,\infty).$$

In a neighborhood of the infinity, the expansion of the function in Laurent series is

$$f(z) = \frac{z^{2n}}{1+z^n} = z^n\frac{1}{1+\frac{1}{z^n}} = z^n\sum_{k=0}^{\infty}(-1)^k\left(\frac{1}{z^n}\right)^k$$
$$= z^n\left(1-\frac{1}{z^n}+\frac{1}{z^{2n}}-\frac{1}{z^{3n}}+\cdots\right) = z^n - 1 + \frac{1}{z^n} - \frac{1}{z^{2n}} + \cdots,$$
where $|z| > 1$,

hence

$$\operatorname{Res}(f;\infty) = \begin{cases} 0, & \text{if } n \neq 1 \\ -1, & \text{if } n = 1. \end{cases}$$

2. The points of the form $z_k = i(2k+1)2\pi$, $k \in \mathbb{Z}$, are second-order poles. The residues in the points z_k are

$$\operatorname{Res}(f;z_k) = \lim_{z\to z_k}\left[\frac{e^{az}}{(1+e^{\frac{z}{2}})^2}(z-z_k)^2\right]' = e^{az_k}\lim_{t\to 0}\left[\frac{t^2e^{at}}{(1-e^{\frac{t}{2}})^2}\right]',$$

where we used the substitution $z = t + z_k$.

The expansion in Taylor series of the function e^z is well known, hence

$$\begin{aligned}(1-e^{\frac{t}{2}})^2 &= \left(1-\sum_{n=0}^{\infty}\frac{t^n}{2^n n!}\right)^2 = \left(\sum_{n=1}^{\infty}\frac{t^n}{2^n n!}\right)^2 \\ &= \frac{t^2}{4}\left(1+\frac{t}{4}+\frac{t^2}{24}+\cdots\right)^2, \quad \text{where } t \in \mathbb{C},\end{aligned}$$

and thus

$$\begin{aligned}\operatorname{Res}(f;z_k) &= 4e^{az_k}\lim_{t\to 0}\left[\frac{e^{at}}{(1+\frac{t}{4}+\frac{t^2}{24}+\cdots)^2}\right]' \\ &= 4e^{az_k}\lim_{t\to 0}\left[e^{at}\frac{a-\frac{1}{2}+(\frac{a}{4}-\frac{1}{6})t+\cdots}{(1+\frac{t}{4}+\frac{t^2}{24}+\cdots)^3}\right] \\ &= 2(2a-1)e^{az_k} = 2(2a-1)e^{i(2k+1)2\pi a}, \quad k \in \mathbb{Z}.\end{aligned}$$

3. The function has the essential singular points $z_0 = 0$ and $z_1 = \infty$.

To determine the residue in the point $z_0 = 0$, we will expand the function in Taylor series, i. e.,

$$\begin{aligned}f(z) = e^{z-\frac{1}{z}} = e^z e^{-\frac{1}{z}} &= \sum_{k=0}^{\infty}\frac{z^k}{k!}\cdot\sum_{j=0}^{\infty}(-1)^j\frac{1}{j!z^j} \\ &= \sum_{k=0}^{\infty}\sum_{j=0}^{\infty}(-1)^j\frac{1}{k!j!}z^{k-j}, \quad \text{where } z \in \mathbb{C}^*,\end{aligned}$$

hence

$$\operatorname{Res}(f;0) = -\operatorname{Res}(f;\infty) = \sum_{\substack{k=0 \\ k-j=-1}}^{\infty}\sum_{j=0}^{\infty}(-1)^j\frac{1}{k!j!} = \sum_{k=0}^{\infty}\frac{(-1)^{k+1}}{k!(k+1)!}.$$

4. The point $z_0 = 0$ is an essential singular point, $z_1 = -1$ is a simple pole while $z_2 = \infty$ is a $n-1$-th order pole.

To determine the residue of f in the point $z_0 = 0$, we will expand the function in Laurent series. Using the following formulas,

$$\frac{1}{1+z} = \sum_{n=0}^{\infty}(-1)^n z^n, \quad \text{if } |z|<1, \quad e^{\frac{1}{z}} = \sum_{n=0}^{\infty}\frac{1}{n!z^n}, \quad \text{if } z \in \mathbb{C}^*,$$

we get

$$f(z) = \frac{z^n e^{\frac{1}{z}}}{1+z} = z^n(1-z+z^2-z^3+\cdots)\left(1+\frac{1}{z}+\frac{1}{2!z^2}+\frac{1}{3!z^3}+\cdots\right),$$
$$0<|z|<1,$$

hence

$$\begin{aligned}\operatorname{Res}(f;0) = a_{-1} &= \frac{1}{(n+1)!} - \frac{1}{(n+2)!} + \frac{1}{(n+3)!} - \frac{1}{(n+4)!} + \cdots \\ &= \begin{cases} -\frac{1}{e} + \frac{1}{2!} - \frac{1}{3!} + \cdots + \frac{1}{n!}, & \text{if } n \text{ is even} \\ \frac{1}{e} - \frac{1}{2!} + \frac{1}{3!} - \cdots + \frac{1}{n!}, & \text{if } n \text{ is odd.} \end{cases}\end{aligned}$$

From the formulas of *the residues value at the pole*, for the point $z_1 = -1$ we have

$$\operatorname{Res}(f;-1) = \lim_{z\to -1}\left[\frac{z^n e^{\frac{1}{z}}}{1+z}(1+z)\right] = \lim_{z\to -1} z^n e^{\frac{1}{z}} = \frac{(-1)^n}{e}.$$

To calculate the residue at the infinity, we will use the definition of the residue at the infinity.

Case 1. If $n = 1$, we will use the well-known Taylor series expansion formula,

$$\begin{aligned} f(z) &= \frac{e^{\frac{1}{z}}}{1+\frac{1}{z}} = \sum_{k=0}^{\infty}\frac{1}{k!z^k}\sum_{j=0}^{\infty}\frac{(-1)^j}{z^j} \\ &= \left(1+\frac{1}{1!z}+\frac{1}{2!z^2}+\frac{1}{3!z^3}+\cdots\right)\left(1-\frac{1}{z}+\frac{1}{z^2}-\frac{1}{z^3}+\cdots\right), \quad |z|>1, \end{aligned}$$

hence

$$\operatorname{Res}(f;\infty) = -a_{-1} = 0.$$

Case 2. If $n > 1$, we get similarly

$$\begin{aligned} f(z) &= \frac{z^{n-1}e^{\frac{1}{z}}}{1+\frac{1}{z}} = z^{n-1}\sum_{k=0}^{\infty}\frac{1}{k!z^k}\sum_{j=0}^{\infty}\frac{(-1)^j}{z^j} \\ &= z^{n-1}\left(1+\frac{1}{1!z}+\frac{1}{2!z^2}+\frac{1}{3!z^3}+\cdots\right)\left(1-\frac{1}{z}+\frac{1}{z^2}-\frac{1}{z^3}+\cdots\right), \\ & |z|>1, \end{aligned}$$

and thus

$$\operatorname{Res}(f;\infty) = -a_{-1} = -\frac{1}{n!} + \frac{1}{(n-1)!} - \frac{1}{(n-2)!} + \cdots + (-1)^{n-1}\frac{1}{2!}.$$

Solution of Exercise 5.4.3

1. The function has the simple poles $z_k = k\pi i$, where $k \in \mathbb{Z}$, hence

$$\begin{aligned}\operatorname{Res}(f;z_k) &= \lim_{z\to z_k}\left[\frac{e^{iaz}}{\sinh z}(z-z_k)\right] = \lim_{t\to 0}\frac{te^{ia(z_k+t)}}{\sinh(z_k+t)} = e^{iaz_k}\lim_{t\to 0}\frac{te^{iat}}{(-1)^k\sinh t}\\ &= (-1)^k e^{iaz_k}\lim_{t\to 0}\frac{te^{iat}}{\sum_{k=0}^{\infty}\frac{t^{2k+1}}{(2k+1)!}} = (-1)^k e^{iaz_k}\lim_{t\to 0}\frac{e^{iat}}{\sum_{k=0}^{\infty}\frac{t^{2k}}{(2k+1)!}}\\ &= (-1)^k e^{-ak\pi}.\end{aligned}$$

2. The function f has the second-order pole $z_0 = 1$, hence

$$\operatorname{Res}(f;1) = \lim_{z\to 1}\left[\frac{e^{\frac{1}{z}}}{(z-1)^2}(z-1)^2\right]' = \lim_{z\to 1}\frac{-e^{\frac{1}{z}}}{z^2} = -e.$$

The point $z_1 = 0$ is an isolated essential singular point, and

$$\begin{aligned}f(z) &= e^{\frac{1}{z}}\frac{1}{(1-z)^2} = \sum_{k=0}^{\infty}\frac{1}{k!z^k}\sum_{j=1}^{\infty}jz^{j-1}\\ &= \left(1+\frac{1}{1!z}+\frac{1}{2!z^2}+\frac{1}{3!z^3}+\cdots\right)(1+2z+3z^2+4z^3+\cdots),\quad 0<|z|<1,\end{aligned}$$

hence

$$\operatorname{Res}(f;0) = 1+\frac{1}{1!}+\frac{1}{2!}+\frac{1}{3!}+\cdots = e.$$

3. For the given function, the points $z_0 = i$ and $z_1 = -i$ are nth order poles, hence

$$\begin{aligned}\operatorname{Res}(f;i) &= \frac{1}{(n-1)!}\lim_{z\to i}\left[\frac{1}{(z^2+1)^n}(z-i)^n\right]^{(n-1)}\\ &= \frac{1}{(n-1)!}\lim_{z\to i}\left[\frac{1}{(z+i)^n}\right]^{(n-1)} = -\frac{i}{(n-1)!}\frac{n(n+1)\cdot\ldots\cdot(2n-2)}{2^{2n-1}}\end{aligned}$$

and

$$\begin{aligned}\operatorname{Res}(f;-i) &= \frac{1}{(n-1)!}\lim_{z\to -i}\left[\frac{1}{(z^2+1)^n}(z+i)^n\right]^{(n-1)}\\ &= \frac{1}{(n-1)!}\lim_{z\to i}\left[\frac{1}{(z-i)^n}\right]^{(n-1)} = \frac{i}{(n-1)!}\frac{n(n+1)\cdot\ldots\cdot(2n-2)}{2^{2n-1}}.\end{aligned}$$

4. The function has the singular points $z_k = k\pi$, where $k \in \mathbb{Z}^*$ that are simple poles, and $z_0 = 0$ is a third-order pole. The point $z_* = \infty$ is an accumulation point for the poles of the function f, hence it is not an isolated singular point of the function f.

We have

$$\operatorname{Res}(f;k\pi) = \frac{1}{(z^2\sin z)'}\bigg|_{z=z_k} = \frac{1}{2z\sin z + z^2\cos z}\bigg|_{z=z_k} = \frac{(-1)^k}{k^2\pi^2}, \quad \text{where } k \in \mathbb{Z}^*.$$

Since

$$\operatorname{Res}(f;0) = \frac{1}{2!}\lim_{z\to 0}\left[\frac{1}{z^2\sin z}z^3\right] = \frac{1}{2!}\lim_{z\to 0}\left[\frac{z}{\sin z}\right]'',$$

we define the function

$$g(z) = \frac{z}{\sin z} = \frac{1}{\sum_{n=0}^{\infty}(-1)^n\frac{z^{2n}}{(2n+1)!}}.$$

Then $g \in H(U(0;\pi))$, because $\lim_{z\to 0} g(z) = 1$. If $g(z) = \sum_{n=0}^{\infty} a_n z^n$, then

$$\sum_{n=0}^{\infty}(-1)^n\frac{z^{2n}}{(2n+1)!}\sum_{n=0}^{\infty} a_n z^n = 1, \quad z \in U(0;\pi),$$

hence it follows that

$$a_0 = 1, \quad a_1 = 0, \quad a_2 = \frac{1}{6}, \quad a_3 = 0, \quad a_4 = \frac{7}{360}, \quad \dots$$

and thus

$$g(z) = 1 + \frac{1}{6}z^2 + \frac{7}{360}z^4 + \cdots, \quad z \in U(0;\pi).$$

From here, we deduce that

$$\operatorname{Res}(f;0) = \frac{1}{2!}\lim_{z\to 0} g''(z) = \frac{1}{2}\frac{1}{3} = \frac{1}{6},$$

since

$$g'(z) = \left(\frac{z}{\sin z}\right)' = \frac{z}{3} + \frac{7z^3}{90} + \cdots, \quad 0 < |z| < \pi,$$

and

$$g''(z) = \left(\frac{z}{\sin z}\right)'' = \frac{1}{3} + \frac{7z^2}{30} + \cdots, \quad 0 < |z| < \pi.$$

5. The point $z_0 = 1$ is an essential singular point for the function, while $z_1 = \infty$ is a third-order pole.

The function has the expansion about the point $z_0 = 1$ of the form

$$f(z) = z^3 e^{\frac{1}{1-z}} = \left[(z-1)^3 + 3(z-1)^2 + 3(z-1) + 1\right]\sum_{n=0}^{\infty}\frac{(-1)^n}{n!(z-1)^n},$$

$z \in U(1;0,\infty)$.

Hence

$$\mathrm{Res}(f;1) = a_{-1} = \frac{1}{4!} - \frac{3}{3!} + \frac{3}{2!} - 1 = \frac{1}{24},$$

and from the sum of the residues it follows that

$$\mathrm{Res}(f;\infty) = -\,\mathrm{Res}(f;1) = -\frac{1}{24}.$$

6. The function has only the second-order poles $z_k = i(2k+1)\frac{\pi}{2}$, where $k \in \mathbb{Z}$. From here, we get

$$\begin{aligned}\mathrm{Res}(f;z_k) &= \lim_{z\to z_k}\left[\frac{e^{iaz}}{\cosh^2 z}(z-z_k)^2\right]' = -e^{iaz_k}\lim_{t\to 0}\left[\frac{t^2 e^{iat}}{\sinh^2 t}\right]'\\ &= -e^{iaz_k}\lim_{t\to 0}\left[ie^{iat}\frac{t}{\sinh t}\left(a\frac{t}{\sinh t} + 2i\frac{t\cosh t - \sinh t}{\sinh^2 t}\right)\right] = -iae^{-a(2k+1)\frac{\pi}{2}},\end{aligned}$$

where we used the substitution $z = t + z_k$ and the following relations:

$$\cosh(t+z_k) = i(-1)^k \sinh t, \quad \lim_{t\to 0}\frac{\sinh t}{t} = 1, \quad \lim_{t\to 0}\frac{t\cosh t - \sinh t}{\sinh^2 t} = 0.$$

Solution of Exercise 5.4.4

1. The function has singular points $z_0 = 0$ that is a first-order pole, and $z_1 = 2i$, $z_2 = -2i$ that are second-order poles. Then

$$\begin{aligned}\mathrm{Res}(f;0) &= \lim_{z\to 0}\frac{z+2}{(z^2+4)^2} = \frac{1}{8},\\ \mathrm{Res}(f;2i) &= \lim_{z\to 2i}\left[\frac{z+2}{z(z+2i)^2}\right]' = \lim_{z\to 2i} -2\frac{z^2+3z+2i}{z^2(z+2i)^3} = -\frac{1}{16} - \frac{1}{32}i\end{aligned}$$

and

$$\mathrm{Res}(f;-2i) = \lim_{z\to -2i}\left[\frac{z+2}{z(z-2i)^2}\right]' = \lim_{z\to -2i} -2\frac{z^2+3z-2i}{z^2(z-2i)^3} = -\frac{1}{16} + \frac{1}{32}i.$$

(a) The path $\{\gamma\} = \partial U(0;1)$ that is directly oriented, turns around only the point z_0, hence from the *residues theorem* we have

$$\int_\gamma \frac{z+2}{z(z^2+4)^2}\mathrm{d}z = 2\pi i\,\mathrm{Res}(f;0) = \frac{\pi i}{4}.$$

(b) The path $\{\gamma\} = \partial U(0;5)$ turns around all of the singular points of f, and from the *residues theorem* it follows that

$$\int_\gamma \frac{z+2}{z(z^2+4)^2}\mathrm{d}z = 2\pi i[\mathrm{Res}(f;0) + \mathrm{Res}(f;2i) + \mathrm{Res}(f;-2i)] = 0.$$

2. The point $z_0 = 1$ is a second-order pole for the function, while $z_1 = i$ and $z_2 = -i$ are simple poles. The disc $U(1+i;2)$ contains the points z_0 and z_1. Since

$$\operatorname{Res}(f;1) = \lim_{z\to 1}\left[\frac{1}{z^2+1}\right]' = \lim_{z\to 1}\frac{-2z}{(z^2+1)^2} = -\frac{1}{2}$$

and

$$\operatorname{Res}(f;i) = \lim_{z\to i}\frac{1}{(z-1)^2(z+i)} = \frac{1}{4},$$

from the *residues theorem*, it follows that

$$\int_\gamma \frac{1}{(z-1)^2(z^2+1)}\mathrm{d}z = 2\pi i[\operatorname{Res}(f;1)+\operatorname{Res}(f;i)] = -\frac{\pi i}{2}.$$

3. Case 1. If $n = 1$, then $f(z) = \frac{1}{z+2} \in H(\mathbb{C}\setminus\{-2\})$. From the *residues theorem*, it follows that

$$\int_\gamma \frac{1}{z+2}\mathrm{d}z = 2\pi i\operatorname{Res}(f;-2) = 2\pi i.$$

Case 2. If $n > 1$, then the function f has the singular points of the form $z_k = \sqrt[n]{2}e^{i\frac{\pi+2k\pi}{n}}$, $k \in \{0,1,\dots,n-1\}$, that are simple poles and $|z_k| = \sqrt[n]{2}$ for all $k \in \{0,1,\dots,n-1\}$.

The path $\gamma(t) = (3+\varepsilon)e^{2\pi i t}+1$, $t \in [0,1]$, with $\varepsilon > 0$ turns around all of these poles, hence

$$\int_\gamma \frac{1}{z^n+2}\mathrm{d}z = 2\pi i\sum_{k=0}^{n-1}\operatorname{Res}(f;z_k) = -2\pi i\operatorname{Res}(f;\infty).$$

Since

$$\frac{1}{z^n+1} = \frac{1}{z^n}\frac{1}{1+\frac{2}{z^n}} = \frac{1}{z^n}\sum_{k=0}^{\infty}(-1)^k\frac{2^k}{z^{kn}} = \frac{1}{z^n}\left(1-\frac{2}{z^n}+\frac{2^2}{z^{2n}}-\frac{2^3}{z^{3n}}+\cdots\right),$$
$$|z| > \sqrt[n]{2},$$

we have $\operatorname{Res}(f;\infty) = 0$, hence

$$\int_\gamma \frac{1}{z^n+2}\mathrm{d}z = 0.$$

4. The function has singular points all the nth order roots of the unity, and also all of the third-order roots of the unity. But all of these points do not belong to the $U(0;r)$

disc, whenever $r < 1$, and from the *Cauchy integral theorem* the value of the integral will be zero, i. e.,

$$\int_{\gamma} \frac{1}{(z^n - 1)(z^3 - 1)} \mathrm{d}z = 0.$$

5. The points $z_0 = 1$, $z_1 = i$ and $z_2 = -i$ are simple poles for the function, and the path $\gamma(t) = \sqrt{2}e^{2\pi i t} + (1 + i)$, $t \in [0, 1]$, turns around only the points z_0 and z_1. Since

$$\operatorname{Res}(f; 1) = \lim_{z \to 1} \frac{1}{z^2 + 1} = \frac{1}{2}$$

and

$$\operatorname{Res}(f; i) = \lim_{z \to i} \frac{1}{(z - 1)(z + i)} = -\frac{1 - i}{4},$$

it follows that

$$\int_{\gamma} \frac{1}{(z - 1)(z^2 + 1)} \mathrm{d}z = 2\pi i[\operatorname{Res}(f; 1) + \operatorname{Res}(f; i)] = \frac{\pi(i - 1)}{2}.$$

6. The points $z_k = e^{\frac{k\pi i}{5}}$, $k \in \{0, 1, \ldots, 9\}$, are simple poles, while $z_{10} = 0$ is a third-order pole for f.

All these singular points belong to the disc $U(0; 2)$, because $|z_k| = 1$, for all $k \in \{0, 1, \ldots, 9\}$, and $|z_{10}| = 0$. Using the *theorem related to the sum of the residues in all of the singular points*, we get

$$\int_{\gamma} \frac{1}{z^3(z^{10} - 1)} \mathrm{d}z = 2\pi i \sum_{k=0}^{10} \operatorname{Res}(f; z_k) = -2\pi i \operatorname{Res}(f; \infty).$$

Since

$$\frac{1}{z^3(z^{10} - 1)} = \frac{1}{z^{13}} \frac{1}{1 - \frac{1}{z^{10}}} = \frac{1}{z^{13}} \sum_{n=0}^{\infty} \frac{1}{z^{10n}}, \quad |z| > 1,$$

it follows that $\operatorname{Res}(f; \infty) = 0$, hence

$$\int_{\gamma} \frac{1}{z^3(z^{10} - 1)} \mathrm{d}z = 0.$$

7. The point $z_0 = 1$ is an essential singular point for the function. Since

$$e^{\frac{1}{1-z}} = \sum_{n=0}^{\infty} \frac{(-1)^n}{n!(z - 1)^n} = 1 - \frac{1}{z - 1} + \frac{1}{2!(z - 1)^2} - \frac{1}{3!(z - 1)^3} + \cdots,$$

where $z \in \mathbb{C} \setminus \{1\}$,

it follows that

$$\operatorname{Res}(f;1) = a_{-1} = -1,$$

hence

$$\int_\gamma e^{\frac{1}{1-z}}\,dz = 2\pi i \operatorname{Res}(f;1) = -2\pi i.$$

8. The function f has the first-order poles of the form $z_k = \frac{1}{2} + k$, where $k \in \mathbb{Z}$, and

$$\operatorname{Res}(f;z_k) = \left.\frac{\sin \pi z}{(\cos \pi z)'}\right|_{z=z_k} = \left.\frac{\sin \pi z}{-\pi \sin \pi z}\right|_{z=z_k} = -\frac{1}{\pi}, \quad \forall k \in \mathbb{Z}.$$

The disc $U(0;n)$, $n \in \mathbb{N}^*$, contains the point z_k if and only if

$$|z_k| < n \Leftrightarrow \left|\frac{1}{2} + k\right| < n.$$

Thus,

$$\int_\gamma \tan \pi z\,dz = 2\pi i \sum_{k=-n}^{k=n-1} \operatorname{Res}(f;z_k) = -4ni.$$

9. The function has singular points $z_k = \frac{\pi}{2} + k\pi$, where $k \in \mathbb{Z}$, which are simple poles, and $z_* = 0$ is also a simple pole. The disc $U(0;2)$ contains only the poles $z_* = 0$, $z_1 = \frac{\pi}{2}$ and $z_{-1} = -\frac{\pi}{2}$. Since

$$\operatorname{Res}(f;0) = \lim_{z\to 0}\left[\frac{\tan z}{z^2} z\right] = \lim_{z\to 0} \frac{\tan z}{z} = 1,$$

$$\operatorname{Res}\left(f;\frac{\pi}{2}\right) = \left.\frac{\sin z}{(z^2 \cos z)'}\right|_{z=\frac{\pi}{2}} = \left.\frac{\sin z}{2z\cos z - z^2 \sin z}\right|_{z=\frac{\pi}{2}} = \frac{1}{-\frac{\pi^2}{4}} = -\frac{4}{\pi^2}$$

and

$$\operatorname{Res}\left(f;-\frac{\pi}{2}\right) = \left.\frac{\sin z}{(z^2 \cos z)'}\right|_{z=-\frac{\pi}{2}} = \left.\frac{\sin z}{2z\cos z - z^2 \sin z}\right|_{z=-\frac{\pi}{2}} = \frac{-1}{\frac{\pi^2}{4}} = -\frac{4}{\pi^2},$$

it follows that

$$\int_\gamma \frac{\tan z}{z^2}\,dz = 2\pi i\left[\operatorname{Res}(f;0) + \operatorname{Res}\left(f;\frac{\pi}{2}\right) + \operatorname{Res}\left(f;-\frac{\pi}{2}\right)\right] = 2\pi i\left(1 - \frac{8}{\pi^2}\right).$$

10. The function f has the simple poles of the form $z_k = e^{i\frac{\pi+2k\pi}{4}}$, where $k \in \{0,1,2,3\}$. Since the disc $U(1;1)$ contains only the poles z_0 and z_3, we have

$$\operatorname{Res}(f;z_0) = \left.\frac{1}{(z^4+1)'}\right|_{z=z_0} = \frac{z_0}{4z_0^4} = -\frac{z_0}{4} = -\frac{\sqrt{2}}{8}(1+i)$$

and similarly

$$\operatorname{Res}(f, z_3) = \frac{1}{(z^4+1)'}\bigg|_{z=z_3} = \frac{z_3}{4z_3^4} = -\frac{z_3}{4} = -\frac{\sqrt{2}}{8}(1-i).$$

Thus,

$$\int_\gamma \frac{1}{z^4+1}\,dz = 2\pi i[\operatorname{Res}(f; z_0) + \operatorname{Res}(f; z_3)] = -\frac{\pi i\sqrt{2}}{2}.$$

11. The point $z_0 = 0$ is a simple pole, while $z_1 = 1$ is an essential isolated singular point. The directly oriented path $\{\gamma\} = \partial U(0;2)$ turns around both of these points, hence

$$\operatorname{Res}(f; 0) = \lim_{z\to 0}\left[z\frac{1}{z}\sin\frac{1}{(z-1)^2}\right] = \sin 1,$$

to determine the residue in $z_1 = 1$ we will expand the function in Laurent series about this point. Using the relations,

$$\frac{1}{z} = \frac{1}{(z-1)+1} = \sum_{n=0}^{\infty}(-1)^n(z-1)^n, \quad |z-1| < 1,$$

$$\sin\frac{1}{(z-1)^2} = \sum_{n=0}^{\infty}(-1)^n\frac{1}{(2n+1)!(z-1)^{2(2n+1)}}, \quad z \in \mathbb{C}\setminus\{1\},$$

it follows that

$$f(z) = (1-(z-1)+(z-1)^2-(z-1)^3+\cdots)$$
$$\cdot\left(\frac{1}{(z-1)^2} - \frac{1}{3!(z-1)^6} + \frac{1}{5!(z-1)^{10}} - \cdots\right), \quad 0 < |z-1| < 1,$$

hence the residue in $z_1 = 1$ is

$$\operatorname{Res}(f; 1) = a_{-1} = -1 + \frac{1}{3!} - \frac{1}{5!} + \frac{1}{7!} - \cdots = \sin(-1) = -\sin 1.$$

Thus,

$$\int_\gamma \frac{1}{z}\sin\frac{1}{(z-1)^2}\,dz = 2\pi i[\operatorname{Res}(f; 0) + \operatorname{Res}(f; 1)] = 0.$$

12. The directly oriented path $\{\gamma\} = \partial U(0;2)$ turns around the simple pole $z_0 = 0$ and the essential isolated singular point $z_1 = 1$. Then

$$\operatorname{Res}(f; 0) = \lim_{z\to 0}(z-1)^3 e^{\frac{1}{z-1}} = -\frac{1}{e}.$$

We will expand the function in Laurent series for the point $z_1 = 1$ as follows:

$$\frac{1}{z} = \frac{1}{(z-1)+1} = \sum_{n=0}^{\infty} (-1)^n (z-1)^n, \quad |z-1| < 1,$$

$$e^{\frac{1}{z-1}} = \sum_{n=0}^{\infty} \frac{1}{n!(z-1)^n}, \quad z \in \mathbb{C} \setminus \{1\},$$

and we deduce that

$$f(z) = (1 - (z-1) + (z-1)^2 - (z-1)^3 + \cdots)(z-1)^3 \cdot \left(1 + \frac{1}{z-1} + \frac{1}{2!(z-1)^2} + \frac{1}{3!(z-1)^3} + \cdots\right), \quad 0 < |z-1| < 1.$$

From here, the residue in the point $z_1 = 1$ will be

$$\operatorname{Res}(f;1) = a_{-1} = \frac{1}{4!} - \frac{1}{5!} + \frac{1}{6!} - \frac{1}{7!} + \cdots = e^{-1} - \left(1 - \frac{1}{1!} + \frac{1}{2!} - \frac{1}{3!}\right) = \frac{1}{e} - \frac{1}{3},$$

hence

$$\int_{\gamma} \frac{(z-1)^3}{z} e^{\frac{1}{z-1}} \mathrm{d}z = 2\pi i [\operatorname{Res}(f;0) + \operatorname{Res}(f;1)] = -\frac{2\pi i}{3}.$$

13. Since the directly oriented path $\{\gamma\} = \partial U(0;r)$ has the radius $r > 0$ enough big, such that all the zeros of the function $z^{111} + z^{11} + z + 1$ belong to $U(0;r)$, using the *theorem related to the sum of the residues in all of the singular points* we get

$$\int_{\gamma} \frac{1}{z^{111} + z^{11} + z + 1} \mathrm{d}z = -2\pi i \operatorname{Res}(f;\infty).$$

If

$$f(z) = \frac{1}{z^{111} + z^{11} + z + 1}, \quad |z| > r,$$

letting $\varphi(\zeta) = f(\frac{1}{\zeta})$, we have

$$\varphi(\zeta) = \frac{\zeta^{111}}{1 + \zeta^{100} + \zeta^{110} + \zeta^{111}} = \zeta^{111}(1 - \zeta^{100} + \cdots) = \zeta^{111} - \zeta^{211} + \cdots, \quad |\zeta| < \frac{1}{r},$$

hence

$$f(z) = \frac{1}{z^{111}} - \frac{1}{z^{211}} + \cdots, \quad z \in U(0;r,\infty).$$

From here, it follows that $\operatorname{Res}(f;\infty) = 0$, and then

$$\int_{\gamma} \frac{1}{z^{111} + z^{11} + z + 1} \mathrm{d}z = 0.$$

14. The point $z_0 = 1$ is an essential isolated singular point.

Case 1. If $r < 1$, then the function f is holomorphic in the disc $U(0; r)$. According to the *Cauchy integral theorem*, we have

$$\int_\gamma \frac{1}{z-1} e^{\frac{1}{z-1}}\, dz = 0.$$

Case 2. If $r > 1$, then $z_1 \in U(0; r)$. To determine the residue in the point z_1, we will expand the function in Laurent series about z_1, i. e.,

$$\begin{aligned} f(z) &= \frac{1}{z-1} e^{\frac{1}{z-1}} = \frac{1}{z-1} \sum_{n=0}^{\infty} \frac{1}{n!(z-1)^n} \\ &= \frac{1}{z-1}\left(1 + \frac{1}{z-1} + \frac{1}{2!(z-1)^2} + \frac{1}{3!(z-1)^3} + \cdots\right), \quad z \in \mathbb{C} \setminus \{1\}, \end{aligned}$$

hence

$$\operatorname{Res}(f; 1) = a_{-1} = 1.$$

We conclude that

$$\int_\gamma \frac{1}{z-1} e^{\frac{1}{z-1}}\, dz = 2\pi i \operatorname{Res}(f; 1) = 2\pi i.$$

15. The disc $U(0; 3)$ contains the next singular points of f: the point $z_0 = 0$ that is an essential isolated singular point, and the simple poles $z_1 = 1$ and $z_2 = -1$. Then

$$\operatorname{Res}(f; 1) = \lim_{z\to 1} \frac{z^n e^{\frac{1}{z}}}{z+1} = \frac{e}{2}$$

and

$$\operatorname{Res}(f; -1) = \lim_{z\to -1} \frac{z^n e^{\frac{1}{z}}}{z-1} = \frac{(-1)^{n-1}}{2e}.$$

To obtain the residue in the point z_0, we will expand the function in Laurent series about z_0. If we use the relations,

$$\begin{aligned} \frac{1}{z^2-1} &= \frac{1}{2}\left(\frac{1}{z-1} - \frac{1}{z+1}\right) = -\frac{1}{2}\left(\sum_{n=0}^{\infty} z^n + \sum_{n=0}^{\infty} (-1)^n z^n\right) = -\sum_{n=0}^{\infty} z^{2n}, \quad |z| < 1, \\ e^{\frac{1}{z}} &= \sum_{n=0}^{\infty} \frac{1}{n! z^n}, \quad z \in \mathbb{C}^*, \end{aligned}$$

then

$$f(z) = -z^n(1 + z^2 + z^4 + z^6 + \cdots)\left(1 + \frac{1}{z} + \frac{1}{2!z^2} + \frac{1}{3!z^3} + \cdots\right), \quad 0 < |z| < 1.$$

Since the residue in the origin is

$$\operatorname{Res}(f;0) = a_{-1} = -\frac{1}{(n+1)!} - \frac{1}{(n+3)!} - \frac{1}{(n+5)!} - \cdots,$$

we get

$$\int_\gamma \frac{z^n e^{\frac{1}{z}}}{z^2-1}\mathrm{d}z = 2\pi i[\operatorname{Res}(f;0) + \operatorname{Res}(f;1) + \operatorname{Res}(f;-1)]$$

$$= 2\pi i\left[\frac{e}{2} + \frac{(-1)^{n-1}}{2e} + \operatorname{Res}(f;0)\right]$$

$$= \begin{cases} 2\pi i[\frac{1}{1!} + \frac{1}{3!} + \frac{1}{5!} + \cdots + \frac{1}{(n-1)!}], & \text{if } n \text{ is even} \\ 2\pi i[1 + \frac{1}{2!} + \frac{1}{4!} + \cdots + \frac{1}{(n-1)!}], & \text{if } n \text{ is odd,} \end{cases}$$

where we used the expansion in Taylor series of the function e^z, and we replaced the variable z by 1, and respectively by -1.

16. The function has an essential isolated singular point $z_0 = 0$, and the directly oriented path $\{\gamma\} = \partial U(0;r)$, $r > 0$, turns around the point z_0. We will expand the function in Laurent series for the point z_0. Since we know that

$$\sin\frac{1}{z} = \frac{1}{z} - \frac{1}{3!z^3} + \frac{1}{5!z^5} - \cdots, \quad z \in \mathbb{C}^*,$$

it follows

$$f(z) = \sin^n\frac{1}{z} = \left(\frac{1}{z} - \frac{1}{3!z^3} + \frac{1}{5!z^5} - \cdots\right)^n, \quad z \in \mathbb{C}^*.$$

Case 1. If $n = 1$, then $\operatorname{Res}(f;0) = 1$, hence we get

$$\int_\gamma \sin^n\frac{1}{z}\mathrm{d}z = 2\pi i \operatorname{Res}(f;0) = 2\pi i.$$

Case 2. If $n > 1$, then $\operatorname{Res}(f;0) = 0$, hence we get

$$\int_\gamma \sin^n\frac{1}{z}\mathrm{d}z = 2\pi i \operatorname{Res}(f;0) = 0.$$

17. The directly oriented path $\{\gamma\} = \partial U(0;3)$ turns around the points $z_0 = 0$, $z_1 = 1$ and $z_2 = 2$, that are essential singular points. Let consider the directly oriented paths γ_0, γ_1 and γ_2, where $\{\gamma_k\} = \partial U(z_k;\frac{1}{2})$, with $k \in \{0,1,2\}$. From a consequence of the *Cauchy integral theorem*, it is well known that

$$\int_\gamma f(z)\mathrm{d}z = \int_{\gamma_0} f(z)\mathrm{d}z + \int_{\gamma_1} f(z)\mathrm{d}z + \int_{\gamma_2} f(z)\mathrm{d}z.$$

The integral on the path γ_0 is given by

$$\int_{\gamma_0} f(z)\mathrm{d}z = \int_{\gamma_0} (1+z+z^2)e^{\frac{1}{z}}\mathrm{d}z + \int_{\gamma_0} (1+z+z^2)(e^{\frac{1}{z-1}} + e^{\frac{1}{z-2}})\mathrm{d}z,$$

and since $(1+z+z^2)(e^{\frac{1}{z-1}} + e^{\frac{1}{z-2}})$ is a holomorphic function in the disc $U(0;\frac{1}{2})$ we get

$$\int_{\gamma_0} (1+z+z^2)(e^{\frac{1}{z-1}} + e^{\frac{1}{z-2}})\mathrm{d}z = 0.$$

To calculate the integral $\int_{\gamma_0}(1+z+z^2)e^{\frac{1}{z}}\mathrm{d}z$, we need to expand in Laurent series for $z_0 = 0$ the function $f_0(z) = (1+z+z^2)e^{\frac{1}{z}}$, i. e.,

$$\begin{aligned} f_0(z) = (1+z+z^2)e^{\frac{1}{z}} &= (1+z+z^2)\sum_{n=0}^{\infty}\frac{1}{n!z^n} \\ &= (1+z+z^2)\left(1+\frac{1}{z}+\frac{1}{2!z^2}+\frac{1}{3!z^3}+\cdots\right), \quad z\in\mathbb{C}^*. \end{aligned}$$

It follows that

$$\operatorname{Res}(f_0;0) = a_{-1} = 1 + \frac{1}{2!} + \frac{1}{3!} = \frac{5}{3},$$

hence

$$\int_{\gamma_0} f_0(z)\mathrm{d}z = \frac{10\pi i}{3}. \tag{7.46}$$

The other integrals can be calculated similarly, i. e.,

$$\int_{\gamma_1} f(z)\mathrm{d}z = \int_{\gamma_1} (1+z+z^2)e^{\frac{1}{z-1}}\mathrm{d}z + \int_{\gamma_1} (1+z+z^2)(e^{\frac{1}{z}} + e^{\frac{1}{z-2}})\mathrm{d}z,$$

where

$$\int_{\gamma_1} (1+z+z^2)(e^{\frac{1}{z}} + e^{\frac{1}{z-2}})\mathrm{d}z = 0.$$

Since

$$\begin{aligned} f_1(z) = (1+z+z^2)e^{\frac{1}{z-1}} &= ((z-1)^2 + 3(z-1) + 3) \\ &\cdot\left(1+\frac{1}{1!(z-1)}+\frac{1}{2!(z-1)^2}+\frac{1}{3!(z-1)^3}+\cdots\right), \quad z\in\mathbb{C}\setminus\{1\}, \end{aligned}$$

it follows that

$$\operatorname{Res}(f_1,1) = a_{-1} = 3 + \frac{3}{2!} + \frac{1}{3!} = \frac{14}{3},$$

hence

$$\int_{\gamma_1} f_1(z)\mathrm{d}z = \frac{28\pi i}{3}. \tag{7.47}$$

The value of the integral on the path γ_2 is given by

$$\int_{\gamma_2} f(z)\mathrm{d}z = \int_{\gamma_2} (1+z+z^2)e^{\frac{1}{z-2}}\mathrm{d}z + \int_{\gamma_2} (1+z+z^2)(e^{\frac{1}{z}} + e^{\frac{1}{z-1}})\mathrm{d}z,$$

where

$$\int_{\gamma_2} (1+z+z^2)(e^{\frac{1}{z}} + e^{\frac{1}{z-1}})\mathrm{d}z = 0.$$

Since

$$\begin{aligned} f_2(z) &= (1+z+z^2)e^{\frac{1}{z-2}}((z-2)^2 + 5(z-2) + 7) \\ &\quad \cdot \left(1 + \frac{1}{1!(z-2)} + \frac{1}{2!(z-2)^2} + \frac{1}{3!(z-2)^3} + \cdots\right), \quad z \in \mathbb{C} \setminus \{2\}, \end{aligned}$$

it follows that

$$\operatorname{Res}(f_2, 2) = a_{-1} = 7 + \frac{5}{2!} + \frac{1}{3!} = \frac{29}{3},$$

hence

$$\int_{\gamma_2} f_2(z)\mathrm{d}z = \frac{58\pi i}{3}. \tag{7.48}$$

Using the relations (7.46), (7.47) and (7.48), we conclude that

$$\int_{\gamma} (1+z+z^2)(e^{\frac{1}{z}} + e^{\frac{1}{z-1}} + e^{\frac{1}{z-2}})\mathrm{d}z = 32\pi i.$$

18. The points $z_k = (2k+1)\pi i$, with $k \in \mathbb{Z}$ are second-order poles for the function. The directly oriented path $\{\gamma\} = \partial U(0; r)$, with $(2n-1)\pi < r < (2n+1)\pi$ and $n \in \mathbb{N}^*$, turns around the point z_k, whenever $k \in \{-n, -n+1, -n+2, \ldots, n-2, n-1\}$.

The residues in the points z_k are

$$\operatorname{Res}(f; z_k) = \lim_{z \to z_k} \left[\frac{1}{(1+e^z)^2}(z-z_k)^2\right]' = \lim_{t \to 0} \left[\frac{t^2}{(1-e^t)^2}\right]',$$

where we used the substitution $z = t + z_k$.

Using the expansion in Taylor series of the function e^z, we deduce that

$$(1-e^t)^2 = \left(1 - \sum_{n=0}^{\infty} \frac{t^n}{n!}\right)^2 = \left(\sum_{n=1}^{\infty} \frac{t^n}{n!}\right)^2 = t^2\left(1 + \frac{t}{2} + \frac{t^2}{6} + \cdots\right)^2, \quad z \in \mathbb{C}.$$

Hence

$$\operatorname{Res}(f;z_k) = \lim_{t\to 0}\left[\frac{t^2}{t^2(1+\frac{t}{2}+\frac{t^2}{6}+\cdots)^2}\right]' = \lim_{t\to 0}\frac{-1-\frac{7}{6}t+\cdots}{(1+\frac{t}{2}+\frac{t^2}{6}+\cdots)^3} = -1,$$

and the value of the given integral is

$$\int_\gamma \frac{1}{(1+e^z)^2}dz = 2\pi i\sum_{k=-n}^{n-1}\operatorname{Res}(f;z_k) = -4n\pi i.$$

Solution of Exercise 5.4.5

1. The point $z_1 = -i$ is a fourth-order pole, and the path $\{\gamma\} = \{z \in \mathbb{C} : |z+2i| = 2\}$ turns around this pole. The residue of the function f in this pole is

$$\operatorname{Res}(f;-i) = \frac{1}{3!}\lim_{z\to -i}\left[\cosh\frac{\pi z}{2}\right]''' = -\frac{\pi^3 i}{48},$$

where we used that

$$f'(z) = \frac{\pi}{2}\sinh\frac{\pi z}{2},$$
$$f''(z) = \frac{\pi^2}{4}\cosh\frac{\pi z}{2},$$
$$f'''(z) = \frac{\pi^3}{8}\sinh\frac{\pi z}{2}.$$

From the *residues theorem*, it follows that

$$\int_\gamma \frac{\cosh\frac{\pi z}{2}}{(z+i)^4}dz = 2\pi i\operatorname{Res}(f;-i) = \frac{\pi^4}{4!}.$$

2. The function $f(z) = \frac{z^{100}e^{i\pi z}}{z^2+1}$ has the simple poles $z_1 = i$ and $z_2 = -i$, and the ellipse $\{\gamma\} = \{z = x+iy \in \mathbb{C} : 4x^2+y^2-4 = 0\}$ turns around these poles. The residues of the given function in these poles are

$$\operatorname{Res}(f;i) = \lim_{z\to i}\frac{z^{100}e^{i\pi z}}{z+i} = \frac{e^{-\pi}}{2i},$$
$$\operatorname{Res}(f;-i) = \lim_{z\to -i}\frac{z^{100}e^{i\pi z}}{z-i} = -\frac{e^{\pi}}{2i}.$$

The value of the integral is

$$\int_\gamma \frac{z^{100}e^{i\pi z}}{z^2-1}dz = 2\pi i[\operatorname{Res}(f;i)+\operatorname{Res}(f;-i)] = -2\pi\sinh\pi.$$

3. For the function $f(z) = \frac{\sqrt[5]{\frac{z+3i}{3-z}}}{z^3}$, the points $z_1 = -3i$ and $z_2 = 3$ are essential isolated singular points, while $z_3 = 0$ is a third-order pole. We will determine the branch of the function involved in this problem. Denoting

$$z + 3i = r_1 e^{i\theta_1}, \quad 3 - z = r_2 e^{i\theta_2},$$

it follows that

$$g_k(z) = \sqrt[5]{\frac{z+3i}{3-z}} = \sqrt[5]{\frac{r_1}{r_2}} e^{i\frac{\theta_1-\theta_2+2k\pi}{5}}, \quad k \in \{0, 1, 2, 3, 4\},$$

and for $k = 2$, we get

$$g(z) = g_2(z) = \sqrt[5]{\frac{z+3i}{3-z}}\bigg|_{z=3i} = \sqrt[10]{2} e^{i\frac{19\pi}{20}}.$$

Thus, the given function f is holomorphic on the circular ring $\dot{U}(0; r)$, $r < 3$. From the *residues theorem*, we need to compute only the residue in the origin, i. e.,

$$\operatorname{Res}(f; 0) = \frac{1}{2!} \lim_{z\to 0} \left[\sqrt[5]{\frac{z+3i}{3-z}} \right]'' = \frac{1}{2!} \lim_{z\to 0} g''(z) = \frac{5-i}{225} e^{i\frac{9\pi}{10}},$$

because

$$g'(z) = \left(\frac{3}{5} + \frac{3}{5}i\right) \frac{\sqrt[5]{\frac{z+3i}{3-z}}}{(3-z)(z+3i)},$$

$$g''(z) = \left(\frac{6}{25} + \frac{6}{25}i\right)(-6 + 9i + 5z) \frac{\sqrt[5]{\frac{z+3i}{3-z}}}{(z+3i)^2(3-z)^2},$$

and then

$$g(z) = g_2(z) = \sqrt[5]{\frac{r_1}{r_2}} e^{i\frac{\theta_1-\theta_2+4\pi}{5}}.$$

From here, it follows

$$\int_\gamma \frac{\sqrt[5]{\frac{z+3i}{3-z}}}{z^3} \mathrm{d}z = 2\pi i \operatorname{Res}(f; 0) = \frac{2\pi(1+5i)}{225} e^{i\frac{9\pi}{10}}.$$

4. The function has an infinite number of singular points, and the path $\{\gamma\} = \{z = x + iy \in \mathbb{C} : x^2 + y^2 - 2y - 3 = 0\}$ turns around only the points $z_1 = 0$, $z_2 = -\sqrt{\frac{\pi}{2}}$, $z_3 = \sqrt{\frac{\pi}{2}}$, $z_4 = i\sqrt{\frac{\pi}{2}}$, $z_5 = i\sqrt{\frac{3\pi}{2}}$, and $z_6 = i\sqrt{\frac{5\pi}{2}}$. The residues in these points are

$$\begin{aligned} \operatorname{Res}(f; z_k) &= \lim_{z\to z_k} \frac{1}{(z\cos z^2)'} = \frac{1}{\cos z_k^2 - 2z_k^2 \sin z_k^2} \\ &= \begin{cases} 1, & \text{if } k = 1, \\ -\frac{1}{2z_k^2}, & \text{if } k \in \{2, 3, 5\}, \\ \frac{1}{2z_k^2}, & \text{if } k \in \{4, 6\}. \end{cases} \end{aligned}$$

Using the *residues theorem*, it follows that

$$\int_{\gamma} \frac{1}{z\cos z^2}\,dz = 2\pi i \sum_{k=1}^{6} \operatorname{Res}(f; z_k) = 2\pi i\left(1 - \frac{58}{15\pi}\right).$$

5. The function $f(z) = z^2 e^{\frac{2z}{z+1}}$ has an essential singular point $z_0 = -1$, and the path $\{\gamma\} = \{z = x + iy \in \mathbb{C} : x^2 + y^2 + 2x = 0\}$ turns around this point. Expanding the function in Laurent series for the point $z_0 = -1$, we have

$$\begin{aligned} f(z) = z^2 e^{\frac{2z}{z+1}} &= (z+1-1)^2 e^{2-\frac{2}{z+1}} \\ &= \left[(z+1)^2 - 2(z+1) + 1\right] e^2 \sum_{n=0}^{\infty} \frac{(-2)^n}{n!(z+1)^n} \\ &= \cdots - \frac{22e^2}{3}\frac{1}{z+1} + \cdots, \quad z \in \mathbb{C}\setminus\{-1\}, \end{aligned}$$

hence

$$\operatorname{Res}(f; -1) = -\frac{22}{3}e^2.$$

Thus,

$$\int_{\gamma} z^2 e^{\frac{2z}{z+1}}\,dz = -\frac{44i}{3}\pi e^2.$$

6. The path $\{\gamma\} = \{z = x + iy \in \mathbb{C} : 4x^2 + 9y^2 - 36 = 0\}$ turns around to all of the isolated singular points of the function $f(z) = \frac{z^{13}}{(z-2)^4(z^5+3)^2}$, that are: the fourth-order pole $z_5 = 2$, and second-order poles $z_k = \sqrt[5]{3}e^{i\frac{(2k+1)\pi}{5}}$, with $k \in \{0, 1, 2, 3, 4\}$. Since the sum of all of the residues is zero, it is easier to calculate the integral by computing the residue at the infinity. Thus, we will expand the function in Laurent series for the infinity, i. e.,

$$\begin{aligned} f(z) = \frac{z^{13}}{(z-2)^4(z^5+3)^2} &= \frac{1}{z}\frac{1}{(1-\frac{2}{z})^4(1+\frac{3}{z^5})^2} \\ &= \frac{1}{z}\left(1 + 4\frac{2}{z} + \frac{4\cdot 5}{2}\frac{2^2}{z^2} + \cdots\right)\left(1 - 2\frac{3}{z^5} + 3\frac{3^2}{z^{10}} + \cdots\right) \\ &= \cdots + \frac{1}{z} + \cdots, \quad |z| > 2. \end{aligned}$$

Thus

$$\operatorname{Res}(f; \infty) = -1,$$

hence it follows that

$$\int_{\gamma} \frac{z^{13}}{(z-2)^4(z^5+3)^2}\,dz = -2\pi i \operatorname{Res}(f; \infty) = 2\pi i.$$

7. The function $f(z) = \frac{1}{z\sqrt{4z^2+12z+13}}$ has isolated essential singular points $z_1 = -\frac{3}{2}+i$ and $z_2 = -\frac{3}{2} - i$, and the first-order pole $z_3 = 0$.

If $0 < r < \frac{\sqrt{13}}{2}$, from the *residues theorem* we have

$$\int_\gamma \frac{1}{\sqrt{4z^2+12z+13}} \mathrm{d}z = 2\pi i \operatorname{Res}(f;0) = 2\pi i \lim_{z\to 0} \frac{1}{\sqrt{4z^2+12z+13}} = \frac{2\pi i}{\sqrt{13}}.$$

8. Since the function $f(z) = \frac{\log(z-a)}{z^2}$ is not defined in the point $z = a$, we will replace the path $\{\gamma\} = \{z \in \mathbb{C} : z = Re^{i\theta}, 0 \le \theta < 2\pi\}$, $R > a > 0$, with the path $\Gamma = \gamma_R \cup \gamma_{[R,a+\varepsilon]} \cup \gamma_\varepsilon^- \cup \gamma_{[a+\varepsilon,R]}$, where

$$\begin{aligned}
\gamma_R(t) &= Re^{2\pi i t}, \quad t \in [0,1],\\
\gamma_{[R,a+\varepsilon]}(t) &= (1-t)R + t(a+\varepsilon), \quad t \in [0,1],\\
\gamma_{a,\varepsilon}(t) &= \varepsilon e^{2\pi i t}, \quad t \in [0,1],\\
\gamma_{[a+\varepsilon,R]}(t) &= (1-t)(a+\varepsilon) + tR, \quad t \in [0,1].
\end{aligned}$$

The point $z = 0$ is a second-order pole for the function, the residue in this point is

$$\operatorname{Res}(f;0) = \lim_{z\to 0}[\log(z-a)]' = -\frac{1}{a},$$

hence

$$\int_\Gamma \frac{\log(z-a)}{z^2} \mathrm{d}z = -\frac{2\pi i}{a}.$$

Also, we have

$$\int_{\gamma_R} \frac{\log(z-a)}{z^2} \mathrm{d}z + \int_R^{a+\varepsilon} \frac{\log(x-a) + 2\pi i}{x^2} \mathrm{d}x$$
$$+ \int_{\gamma_{a,\varepsilon}} \frac{\log(z-a)}{z^2} \mathrm{d}z + \int_{a+\varepsilon}^R \frac{\log(x-a)}{x^2} \mathrm{d}x = -\frac{2\pi i}{a}.$$

If $z - a = \varepsilon e^{i\theta}$, then $\mathrm{d}z = \varepsilon \mathrm{d}\theta$ and

$$\begin{aligned}
|\log(z-a)| &= \sqrt{(\ln\varepsilon)^2 + \theta} \le \sqrt{(\ln\varepsilon)^2 + 2\pi},\\
\frac{1}{|z|^2} &= \frac{1}{|z-a+a|^2} \le \frac{1}{(|z-a| - |a|)^2} = \frac{1}{a^2 - 2a\varepsilon + \varepsilon^2}.
\end{aligned}$$

Hence, if $|z-a| = \varepsilon \to 0$ then the integral $\int_{\gamma_{a,\varepsilon}} \frac{\log(z-a)}{z^2} \mathrm{d}z$ tends to zero, because

$$\left|\int_{\gamma_{a,\varepsilon}} \frac{\log(z-a)}{z^2} \mathrm{d}z\right| \le \int_0^{2\pi} \frac{\sqrt{(\ln\varepsilon)^2 + 2\pi}}{a^2 - 2a\varepsilon + \varepsilon^2} \varepsilon \mathrm{d}\theta = \frac{2\pi\varepsilon\sqrt{(\ln\varepsilon)^2 + 2\pi}}{a^2 - 2a\varepsilon + \varepsilon^2} \to 0, \quad \text{if } \varepsilon \to 0.$$

Thus, if $\varepsilon \to 0$ it follows that

$$\int_{\gamma_R} \frac{\log(z-a)}{z^2}\mathrm{d}z + 2\pi i \int_R^a \frac{1}{x^2}\mathrm{d}x = -\frac{2\pi i}{a},$$

hence

$$\int_{\gamma} \frac{\log(z-a)}{z^2}\mathrm{d}z = -\frac{2\pi i}{R}.$$

Solution of Exercise 5.4.6

1. The function $f(z) = \frac{1}{z\sin z}$ has the following isolated singular points: the second-order pole $z_0 = 0$, and the simple poles $z_k = k\pi$, where $k \in \mathbb{Z}^*$. The residues of the function in these points are

$$\operatorname{Res}(f;0) = \lim_{z\to 0}\left[\frac{z^2}{z\sin z}\right]' = \lim_{z\to 0}\frac{\sin z - z\cos z}{\sin^2 z} = 0,$$

and

$$\begin{aligned}\operatorname{Res}(f;k\pi) &= \lim_{z\to k\pi}\frac{z-k\pi}{z\sin z}\\ &= \lim_{\varepsilon\to 0}\frac{\varepsilon}{(k\pi+\varepsilon)\sin(k\pi+\varepsilon)} = \lim_{\varepsilon\to 0}\frac{(-1)^k\varepsilon}{(k\pi+\varepsilon)\sin\varepsilon} = \frac{(-1)^k\varepsilon}{k\pi},\end{aligned}$$

where $k \in \mathbb{Z}^*$. From here, we see that

$$\operatorname{Res}(f;k\pi) + \operatorname{Res}(f;-k\pi) = 0.$$

If the path $\{\gamma\} = \{z \in \mathbb{C} : |z| = r\}$, $r \neq k\pi$, $k \in \mathbb{Z}$, turns around the points z_k, then it turns around the points z_{-k}, hence

$$\int_{\gamma} \frac{1}{z\sin z}\mathrm{d}z = 2\pi i\left[\operatorname{Res}(f;0) + \sum_{k=1}^{n}[\operatorname{Res}(f;k\pi) + \operatorname{Res}(f;-k\pi)]\right] = 0,$$

where $n\pi < r < (n+1)\pi$.

2. The function $f(z) = \frac{1}{3\sin z - \sin 3z}$ may be written in the form

$$f(z) = \frac{1}{4\sin^3 z},$$

where we used that $\sin 3z = 3\sin z - 4\sin^3 z$. Hence the function f has the third- order poles $z_k = k\pi$, with $k \in \mathbb{Z}$. The circle $\{\gamma\} = \{z \in \mathbb{C} : |z| = 4\}$ turns around the points

$z_0 = 0, z_1 = \pi$ and $z_{-1} = -\pi$. To determine the residues of f, we will expand the function in Laurent series for these points. Using the relation,

$$\frac{1}{\sin z} = \frac{1}{z}\left(1 + \frac{z^2}{6} + \frac{7z^4}{360} + \cdots\right), \quad z \in U(0; 0, \pi),$$

we get

$$f(z) = \frac{1}{4\sin^3 z} = \frac{1}{4z^3}\left(1 + \frac{z^2}{2} + \frac{17z^4}{120} + \cdots\right), \quad z \in U(0; 0, \pi).$$

From here, it follows that

$$\operatorname{Res}(f; 0) = \frac{1}{8}.$$

For the point $z_1 = \pi$, substituting $z = t + \pi$ and using the expansion about the point $z_* = 0$ we get

$$f(t + \pi) = -\frac{1}{4\sin^3 t} = -\frac{1}{4t^3}\left(1 + \frac{t^2}{2} + \frac{17t^4}{120} + \cdots\right), \quad t \in U(0; 0, \pi),$$

or

$$f(z) = -\frac{1}{4(z - \pi)^3}\left(1 + \frac{(z - \pi)^2}{2} + \frac{17(z - \pi)^4}{120} + \cdots\right), \quad z \in U(\pi; 0, \pi),$$

hence

$$\operatorname{Res}(f; \pi) = -\frac{1}{8}.$$

For the point $z_{-1} = -\pi$, substituting $z = t - \pi$ and using the expansion for the point $z_* = 0$ we get

$$f(z) = -\frac{1}{4(z + \pi)^3}\left(1 + \frac{(z + \pi)^2}{2} + \frac{17(z + \pi)^4}{120} + \cdots\right), \quad z \in U(-\pi; 0, \pi),$$

hence

$$\operatorname{Res}(f; -\pi) = -\frac{1}{8}.$$

Combining the above results, it follows that

$$\int_\gamma \frac{1}{3\sin z - \sin 3z} dz = 2\pi i[\operatorname{Res}(f; 0) + \operatorname{Res}(f; \pi) + \operatorname{Res}(f; -\pi)] = -\frac{\pi i}{4}.$$

3. For the function $f(z) = \frac{e^{\frac{1}{z}}}{(z-1)^2}$, the point $z_0 = 0$ is an isolated essential singular point, while $z_1 = 1$ is a second-order pole. Since the origin is an isolated essential

singular point, we will expand the function in Laurent series for the origin, using also the relations

$$e^{\frac{1}{z}} = \sum_{n=0}^{\infty} \frac{1}{n!z^n}, \quad z \in \mathbb{C}^*,$$

and

$$\frac{1}{(z-1)^2} = \sum_{n=1}^{\infty} nz^{n-1}, \quad |z| < 1.$$

Thus,

$$f(z) = (1 + 2z + 3z^2 + \cdots)\left(1 + \frac{1}{z} + \frac{1}{2!z^2} + \frac{1}{3!z^3} + \cdots\right), \quad 0 < |z| < 1,$$

hence it follows that

$$\operatorname{Res}(f;0) = 1 + \frac{1}{1!} + \frac{1}{2!} + \frac{1}{3!} + \frac{1}{4!} + \cdots = \sum_{n=0}^{\infty} \frac{1}{n!} = e.$$

The residue in $z_1 = 1$ is

$$\operatorname{Res}(f;1) = \lim_{z\to 1}(e^{\frac{1}{z}})' = \lim_{z\to 1}\left(-\frac{1}{z^2}e^{\frac{1}{z}}\right) = -e.$$

(a) If $\{\gamma\} = \{z \in \mathbb{C} : |z| = \frac{\sqrt{2}}{2}\}$, using the *residues theorem* it follows that

$$\int_\gamma \frac{e^{\frac{1}{z}}}{(z-1)^2}\mathrm{d}z = 2\pi i \operatorname{Res}(f;0) = 2\pi i e.$$

(b) If $\{\gamma\} = \{z \in \mathbb{C} : |z| = 2\}$, we have

$$\int_\gamma \frac{e^{\frac{1}{z}}}{(z-1)^2}\mathrm{d}z = 2\pi i[\operatorname{Res}(f;0) + \operatorname{Res}(f;1)] = 0,$$

because the path of integration turns around all the singular points of f.

4. The function $f(z) = \frac{\log\frac{z-1}{z+1}}{(z^2+1)(z^2-4)}$ has the simple poles $z_{1,2} = \pm i$, $z_{3,4} = \pm 2$, and the isolated essential singular points $z_{5,6} = \pm 1$. Using the given relation $\log\frac{z-1}{z+1}|_{z=0} = \pi i$, we deduce that $k = 0$ because

$$\log\frac{z-1}{z+1}\bigg|_{z=0} = \pi i \Leftrightarrow \log(-1) = i(\pi + 2k\pi) = \pi i \Leftrightarrow k = 0.$$

(a) If $\{\gamma\} = \{z \in \mathbb{C} : |z| = \frac{1}{2}\}$, then the path does not turn around any singular points, and since $f \in H(D)$ where $D = \mathbb{C} \setminus ((-\infty, -1] \cup [1, +\infty))$, it follows that

$$\int_\gamma \frac{\log\frac{z-1}{z+1}}{(z^2+1)(z^2-4)}\mathrm{d}z = 0.$$

(b) If $\{\gamma\} = \{z \in \mathbb{C} : |z| = \sqrt{2}\}$, then the path turns around the points $z_{1,2}$. Since $f \in H(D)$, where $D = \mathbb{C} \setminus [-1, 1]$, we need to calculate the next residues:

$$\operatorname{Res}(f; i) = \lim_{z \to i} \frac{\log \frac{z-1}{z+1}}{(z+i)(z^2-4)} = -\frac{\pi}{20},$$

and

$$\operatorname{Res}(f; -i) = \lim_{z \to -i} \frac{\log \frac{z-1}{z+1}}{(z-i)(z^2-4)} = -\frac{\pi}{20}.$$

From here, it follows that

$$\int_\gamma \frac{\log \frac{z-1}{z+1}}{(z^2+1)(z^2-4)} dz = 2\pi i(\operatorname{Res}(f; i) + \operatorname{Res}(f; -i)) = -\frac{\pi^2 i}{5}.$$

(c) The path $\{\gamma\} = \{z = x + iy \in \mathbb{C} : 3x^2 + y^2 - 2 = 0\}$ turns around only the simple poles $z_{1,2} = \pm i$, and $f \in H(D)$, where $D = \mathbb{C} \setminus ((-\infty, -1] \cup [1, +\infty))$. From the *residues theorem*, it follows that

$$\int_\gamma \frac{\log \frac{z-1}{z+1}}{(z^2+1)(z^2-4)} dz = 2\pi i(\operatorname{Res}(f; i) + \operatorname{Res}(f; -i)) = -\frac{\pi^2 i}{5}.$$

(d) The path $\{\gamma\} = \{z \in \mathbb{C} : |z| = 3\}$ turns around the simple poles $z_{1,2} = \pm i$ and a $z_{3,4} = \pm 2$, and $f \in H(D)$ where $D = \mathbb{C} \setminus [-1, 1]$. Since

$$\operatorname{Res}(f; 2) = \lim_{z \to 2} \frac{\log \frac{z-1}{z+1}}{(z^2+1)(z+2)} = -\frac{\ln 3}{20},$$

and

$$\operatorname{Res}(f; -2) = \lim_{z \to -2} \frac{\log \frac{z-1}{z+1}}{(z^2+1)(z-2)} = -\frac{\ln 3}{20},$$

we have

$$\int_\gamma \frac{\log \frac{z-1}{z+1}}{(z^2+1)(z^2-4)} dz = 2\pi i(\operatorname{Res}(f; i) + \operatorname{Res}(f; -i) + \operatorname{Res}(f; 2) + \operatorname{Res}(f; -2))$$
$$= -\frac{\pi i}{5}(\pi + \ln 3).$$

5. The function $f(z) = \frac{\sin z}{z^2(z^4+1)}$ has the following isolated singular points: $z_0 = 0$, $z_1 = \frac{\sqrt{2}}{2}(1+i)$, $z_2 = \frac{\sqrt{2}}{2}(1-i)$, $z_3 = -\frac{\sqrt{2}}{2}(1+i)$ and $z_4 = -\frac{\sqrt{2}}{2}(1-i)$. Since $\sin z \in H(\mathbb{C})$ and

$$\sin z = z - \frac{z^3}{3!} + \frac{z^5}{5!} - \cdots, \quad z \in \mathbb{C},$$

all of these singular points are simple poles. The circle $\{\gamma\} = \{z \in \mathbb{C} : |z| = 2\}$ turns around all these points, and the residues are

$$\operatorname{Res}(f;0) = \lim_{z\to 0} \frac{\sin z}{z(z^4+1)} = \lim_{z\to 0} \frac{\sin z}{z} \frac{1}{z^4+1} = 1,$$

and

$$\operatorname{Res}(f;z_k) = \lim_{z\to z_k} \frac{(z-z_k)\sin z}{z^2(z-z_1)(z-z_2)(z-z_3)(z-z_4)}.$$

It follows that

$$\begin{aligned}\operatorname{Res}(f;z_1) &= \operatorname{Res}(f;z_3)\\ &= -\frac{1}{8\sqrt{2}}\left[(1-i)(e^{\frac{\sqrt{2}}{2}} + e^{-\frac{\sqrt{2}}{2}})\sin\frac{\sqrt{2}}{2} + (1+i)(e^{\frac{\sqrt{2}}{2}} - e^{-\frac{\sqrt{2}}{2}})\cos\frac{\sqrt{2}}{2}\right],\end{aligned}$$

and

$$\begin{aligned}\operatorname{Res}(f;z_2) &= \operatorname{Res}(f;z_4)\\ &= -\frac{1}{8\sqrt{2}}\left[(1+i)(e^{\frac{\sqrt{2}}{2}} + e^{-\frac{\sqrt{2}}{2}})\sin\frac{\sqrt{2}}{2} + (1-i)(e^{\frac{\sqrt{2}}{2}} - e^{-\frac{\sqrt{2}}{2}})\cos\frac{\sqrt{2}}{2}\right],\end{aligned}$$

where we used that

$$\begin{aligned}\sin\left[\frac{\sqrt{2}}{2}(1+i)\right] &= \sin\frac{\sqrt{2}}{2}\cos\frac{i\sqrt{2}}{2} + \sin\frac{i\sqrt{2}}{2}\cos\frac{\sqrt{2}}{2}\\ &= \frac{1}{2}\left[\sin\frac{\sqrt{2}}{2}(e^{\frac{\sqrt{2}}{2}} + e^{-\frac{\sqrt{2}}{2}}) + i\cos\frac{\sqrt{2}}{2}(e^{\frac{\sqrt{2}}{2}} - e^{-\frac{\sqrt{2}}{2}})\right].\end{aligned}$$

Using the *residues theorem*, we conclude

$$\begin{aligned}\int_\gamma \frac{\sin z}{z^2(z^4+1)}dz &= 2\pi i(\operatorname{Res}(f;0) + 2\operatorname{Res}(f;z_1) + 2\operatorname{Res}(f;z_2))\\ &= 2\pi i - \frac{\pi\sqrt{2}}{2}\left[(e^{\frac{\sqrt{2}}{2}} + e^{-\frac{\sqrt{2}}{2}})\sin\frac{\sqrt{2}}{2} + (e^{\frac{\sqrt{2}}{2}} - e^{-\frac{\sqrt{2}}{2}})\cos\frac{\sqrt{2}}{2}\right]i.\end{aligned}$$

6. The function $f(z) = \frac{1}{z^2(z-1)\sin z}$ has the third-order pole $z_0 = 0$, and the simple poles $z_* = 1$ and $z_k = k\pi$, with $k \in \mathbb{Z}^*$. The path $\{\gamma\} = \{z \in \mathbb{C} : |z| = r\}$, $n\pi < r < (n+1)\pi$, $n \in \mathbb{N}$, turns around the points z_0, z_* and z_k, with $k \in \{\pm 1, \pm 2, \dots, \pm n\}$. The residue of the function in these points is

$$\operatorname{Res}(f;1) = \lim_{z\to 1} \frac{1}{z^2\sin z} = \frac{1}{\sin 1},$$

and

$$\begin{aligned}\operatorname{Res}(f;k\pi) &= \lim_{z\to k\pi} \frac{z-k\pi}{z^2(z-1)\sin z}\\ &= \frac{1}{k^2\pi^2(k\pi-1)} \lim_{\varepsilon\to 0} \frac{\varepsilon}{\sin(k\pi+\varepsilon)} = \frac{(-1)^k}{k^2\pi^2(k\pi-1)},\end{aligned}$$

whenever $k \in \{\pm 1, \pm 2, \dots, \pm n\}$. To determine the residue in the origin, we will expand the function in Laurent series for the origin, i. e.,

$$\begin{aligned} f(z) &= \frac{1}{z^2}\frac{1}{z-1}\frac{1}{\sin z} = -\frac{1}{z^3}(1+z+z^2+z^3+\cdots)\left(1+\frac{z^2}{6}+\frac{7z^4}{360}+\cdots\right) \\ &= -\left(\frac{1}{z^3}+\frac{1}{z^2}+\frac{7}{6}\frac{1}{z}+\frac{7}{6}+\cdots\right), \quad z \in U(0;0,1), \end{aligned}$$

hence

$$\operatorname{Res}(f;0) = -\frac{7}{6}.$$

Combining the above results, by using the *residues theorem*, we get

$$\begin{aligned} \int_\gamma \frac{1}{z^2(z-1)\sin z}dz &= 2\pi i\left(\operatorname{Res}(f;0)+\operatorname{Res}(f;1)+\sum_{k=-n}^{n}\operatorname{Res}(f;k\pi)\right) \\ &= 2\pi i\left(\frac{1}{\sin 1}-\frac{7}{6}+\sum_{k=-n}^{n}\frac{(-1)^k}{k^2\pi^2(k\pi-1)}\right). \end{aligned}$$

7. The function $f(z) = \frac{z}{\sin z(1-\cos z)}$ has the isolated singular points $z_k = k\pi$, where $k \in \mathbb{Z}$. The point $z_0 = 0$ is a second-order pole, while $z_k = k\pi$, $k = 2k'+1$, $k' \in \mathbb{Z}$, are simple poles, and $z_k = k\pi$, $k = 2k'$, $k' \in \mathbb{Z}^*$, are third-order poles. The circle $\{\gamma\} = \{z \in \mathbb{C} : |z| = 4\}$ turns around the points $z_0 = 0$, $z_{-1} = -\pi$ and $z_1 = \pi$. The residues in z_1 and z_{-1} are given by

$$\operatorname{Res}(f;\pi) = \lim_{z\to\pi}\frac{z(z-\pi)}{\sin z(1-\cos z)} = \lim_{z\to\pi}\frac{z}{1-\cos z}\lim_{z\to\pi}\frac{z-\pi}{\sin z} = -\frac{\pi}{2},$$

and similarly for the point $z_{-1} = -\pi$, we have

$$\operatorname{Res}(f;-\pi) = \lim_{z\to-\pi}\frac{z(z+\pi)}{\sin z(1-\cos z)} = \frac{\pi}{2}.$$

To determine the residue in $z_0 = 0$, we will expand the function in Laurent series for the origin. Using the facts that

$$\begin{aligned} \sin z &= z - \frac{z^3}{3!} + \frac{z^5}{5!} - \cdots, \quad z \in \mathbb{C}, \\ 1-\cos z &= \frac{z^2}{2!} - \frac{z^4}{4!} + \cdots, \quad z \in \mathbb{C}, \end{aligned}$$

we deduce

$$f(z) = \frac{1}{z^2}\frac{1}{1-\frac{z^2}{3!}+\frac{z^4}{5!}-\cdots}\frac{1}{\frac{1}{2!}-\frac{z^2}{4!}+\cdots} = \frac{1}{z^2}\left(\frac{1}{2}-\frac{1}{12}z^2+\cdots\right), \quad z \in U(0;0,\pi),$$

hence

$$\operatorname{Res}(f;0) = 0.$$

Finally, we conclude that

$$\int_{\gamma} \frac{z}{\sin z(1-\cos z)}\,dz = 2\pi i(\operatorname{Res}(f;\pi) + \operatorname{Res}(f;-\pi) + \operatorname{Res}(f;0)) = 0.$$

8. The function $f(z) = \frac{z^3 e^{\frac{1}{z}}}{z+1}$ has the isolated essential singular point $z_0 = 0$, and the simple pole $z_1 = -1$.

(a) The paths $\{\gamma\} = \{z \in \mathbb{C} : |z| = \frac{\sqrt{2}}{2}\}$ turns around only the point $z_0 = 0$. Since the origin is an isolated essential singular point, we will expand the function in Laurent series for the origin. Using the relations,

$$\begin{aligned} \frac{1}{z+1} &= 1 - z + z^2 - \cdots, \quad z \in U(0;1), \\ e^{\frac{1}{z}} &= 1 + \frac{1}{z} + \frac{1}{2!z^2} + \cdots, \quad z \in \mathbb{C}^*, \end{aligned}$$

it follows that

$$f(z) = z^3(1 - z + z^2 - \cdots)\left(1 + \frac{1}{1!z} + \frac{1}{2!z^2} + \cdots\right), \quad z \in U(0;0,1),$$

hence the coefficient of the term $\frac{1}{z}$ is given by

$$\operatorname{Res}(f;0) = \frac{1}{4!} - \frac{1}{5!} + \frac{1}{6!} - \cdots = \frac{1}{e} - \frac{1}{3}.$$

Thus,

$$\int_{\gamma} \frac{z^3 e^{\frac{1}{z}}}{z+1}\,dz = 2\pi i \operatorname{Res}(f;0) = \frac{2\pi i(3-e)}{3e}.$$

(b) The path $\{\gamma\} = \{z \in \mathbb{C} : |z| = \sqrt{2}\}$ turns around both of the singular points of f. The residue in the simple pole $z_1 = -1$ is

$$\operatorname{Res}(f;-1) = \lim_{z\to-1} z^3 e^{\frac{1}{z}} = -\frac{1}{e},$$

and using the *residues theorem*, it follows that

$$\int_{\gamma} \frac{z^3 e^{\frac{1}{z}}}{z+1}\,dz = 2\pi i(\operatorname{Res}(f;0) + \operatorname{Res}(f;-1)) = -\frac{2\pi i}{3}.$$

9. The function $f(z) = \frac{z^n e^{\frac{1}{z}}}{z^2-1}$ has the following isolated singular points: the isolated essential singular point $z_0 = 0$, and the simple poles $z_1 = 1$ and $z_2 = -1$. The residues in these points are

$$\operatorname{Res}(f;1) = \lim_{z\to 1} \frac{z^n e^{\frac{1}{z}}}{z+1} = \frac{e}{2}$$

and

$$\operatorname{Res}(f;-1) = \lim_{z\to -1} \frac{z^n e^{\frac{1}{z}}}{z-1} = \frac{(-1)^{n+1}}{2e}.$$

The expansion of the function in Laurent series about the origin is

$$\begin{aligned} f(z) &= -z^n \frac{e^{\frac{1}{z}}}{1-z^2} \\ &= -z^n(1+z^2+z^4+\cdots)\left(1+\frac{1}{1!z}+\frac{1}{2!z^2}+\cdots\right), \quad z \in U(0;0,1), \end{aligned}$$

hence it follows that

$$\operatorname{Res}(f;0) = -\left[\frac{1}{(n+1)!}+\frac{1}{(n+3)!}+\frac{1}{(n+5)!}+\cdots\right].$$

Using the *residues theorem*, it follows that:
(a) If $\{\gamma\} = \{z \in \mathbb{C} : |z| = \frac{1}{2}\}$, then

$$\int_\gamma \frac{z^n e^{\frac{1}{z}}}{z^2-1} dz = 2\pi i \operatorname{Res}(f;0) = -2\pi i\left[\frac{1}{(n+1)!}+\frac{1}{(n+3)!}+\frac{1}{(n+5)!}+\cdots\right].$$

(b) If $\{\gamma\} = \{z \in \mathbb{C} : |z| = 2\}$, then

$$\begin{aligned} \int_\gamma \frac{z^n e^{\frac{1}{z}}}{z^2-1} dz &= 2\pi i(\operatorname{Res}(f;0)+\operatorname{Res}(f;1)+\operatorname{Res}(f;-1)) \\ &= -2\pi i\left[\frac{1}{(n+1)!}+\frac{1}{(n+3)!}+\frac{1}{(n+5)!}+\cdots\right]+\frac{\pi i(e^2-(-1)^n)}{e}. \end{aligned}$$

(c) If $\{\gamma\} = \{z = x+iy \in \mathbb{C} : x^2+y^2-2x-\frac{5}{4} = 0\}$, then

$$\begin{aligned} \int_\gamma \frac{z^n e^{\frac{1}{z}}}{z^2-1} dz &= 2\pi i(\operatorname{Res}(f;0)+\operatorname{Res}(f;1)) \\ &= -2\pi i\left[\frac{1}{(n+1)!}+\frac{1}{(n+3)!}+\frac{1}{(n+5)!}+\cdots\right]+\pi i e. \end{aligned}$$

10. The function $f(z) = \frac{e^{\frac{\pi}{z-i}}}{z^2+1}$ has the essential singular point $z_1 = i$, and the simple pole $z_2 = -i$. The given integration paths $\{\gamma\} = \{z \in \mathbb{C} : |z| = 2\}$ turns around both of these singular points, and the first residue is

$$\operatorname{Res}(f; -i) = \lim_{z \to -i} \frac{e^{\frac{\pi}{z-i}}}{z-i} = \frac{e^{-\frac{\pi}{2i}}}{-2i} = \frac{e^{\frac{\pi i}{2}}}{-2i} = -\frac{1}{2}.$$

To determine the residue in $z_1 = i$, we will expand the function in Laurent series for this point, using the substitution $z = t + i$, i. e.,

$$\begin{aligned} f(t+i) &= \frac{e^{\frac{\pi}{t}}}{t^2 + 2it} = \frac{1}{2it} \frac{1}{1 - \frac{ti}{2}} e^{\frac{\pi}{t}} \\ &= \frac{1}{2it}\left(1 + \frac{ti}{2} + \frac{(ti)^2}{2^2} + \frac{(ti)^3}{2^3} + \cdots\right)\left(1 + \frac{\pi}{1!t} + \frac{\pi^2}{2!t^2} + \frac{\pi^3}{3!t^3} + \cdots\right), \\ &t \in U(0; 0, 2), \end{aligned}$$

hence

$$\begin{aligned} f(z) = {}& \frac{1}{2i(z-i)}\left(1 + \frac{(z-i)i}{2} + \frac{((z-i)i)^2}{2^2} + \frac{((z-i)i)^3}{2^3} + \cdots\right) \\ & \cdot\left(1 + \frac{\pi}{1!(z-i)} + \frac{\pi^2}{2!(z-i)^2} + \frac{\pi^3}{3!(z-i)^3} + \cdots\right), \quad z \in U(i; 0, 2). \end{aligned}$$

It follows that

$$\operatorname{Res}(f; i) = \frac{1}{2i}\left(1 + \frac{\pi i}{1!2} + \frac{(\frac{\pi i}{2})^2}{2!} + \frac{(\frac{\pi i}{2})^3}{3!} + \cdots\right) = \frac{1}{2i} e^{\frac{\pi i}{2}} = \frac{1}{2}.$$

From here, we deduce that

$$\int_{\gamma} \frac{e^{\frac{\pi}{z-i}}}{z^2+1} dz = 2\pi i(\operatorname{Res}(f; i) + \operatorname{Res}(f; -i)) = 0.$$

11. The function $f(z) = \frac{1}{z(z^2+a^2)^2}$ has the simple pole $z_0 = 0$, while $z_1 = ai$ and $z_2 = -ai$ are second-order poles. Thus,

$$\begin{aligned} \operatorname{Res}(f; 0) &= \lim_{z \to 0} \frac{1}{(z^2 + a^2)^2} = \frac{1}{a^4}, \\ \operatorname{Res}(f; ai) &= \lim_{z \to ai} \left(\frac{1}{z(z + ai)^2}\right)' = \lim_{z \to ai} \frac{-3z - ai}{z^2(z + ai)^3} = -\frac{1}{2a^4}, \end{aligned}$$

and

$$\operatorname{Res}(f; -ai) = \lim_{z \to -ai} \left(\frac{1}{z(z - ai)^2}\right)' = \lim_{z \to -ai} \frac{-3z + ai}{z^2(z - ai)^3} = -\frac{1}{2a^4}.$$

From the *residues theorem*, it follows that

$$\int_\gamma \frac{1}{z(z^2+a^2)^2}\mathrm{d}z = 2\pi i(\operatorname{Res}(f;0) + \operatorname{Res}(f;ai) + \operatorname{Res}(f;-ai)) = 0.$$

12. The function $f(z) = \frac{1}{z^n+1}$ has the nth order roots of the unit $z_k = e^{i\frac{\pi+2k\pi}{n}}$, with $k \in \{0, 1, 2, \ldots, n-1\}$, as simple poles. We have the following discussion with respect to the radius $r > 0$, and $r \neq 1$ of the integration circle $\{\gamma\} = \{z \in \mathbb{C} : |z| = r\}$.

Case 1. If $r < 1$, the circle does not turn around any of the above poles. Using the *Cauchy integral theorem*, the value of the integral is zero, i. e.,

$$\int_\gamma \frac{1}{z^n+1}\mathrm{d}z = 0.$$

Case 2. If $r > 1$, the circle turns around all the above poles. Using the *theorem related to the sum of the residues in all of the singular points*, we need to determine the residue of the function in the point $z = \infty$. Hence we will expand the function in Laurent series for this point, i. e.,

$$f(z) = \frac{1}{z^n+1} = \frac{1}{z^n}\frac{1}{1+\frac{1}{z^n}} = \frac{1}{z^n}\left(1 - \frac{1}{z^n} + \frac{1}{z^{2n}} - \cdots\right) = \frac{1}{z^n} - \frac{1}{z^{2n}} + \cdots,$$
$$\left|\frac{1}{z}\right| < 1,$$

and it follows the two cases:

2. Subcase 1. If $n = 1$, then

$$\operatorname{Res}(f;\infty) = -1,$$

hence

$$\int_\gamma \frac{1}{z^n+1}\mathrm{d}z = -2\pi i \operatorname{Res}(f;\infty) = 2\pi i.$$

2. Subcase 2. If $n > 1$, then

$$\operatorname{Res}(f;\infty) = 0,$$

hence we have

$$\int_\gamma \frac{1}{z^n+1}\mathrm{d}z = -2\pi i \operatorname{Res}(f;\infty) = 0.$$

13. The singular points of the functions $f(z) = \frac{1}{z^2 \sin z}$ have been determined to point 4 of Exercise 5.4.3. Thus, the points of the form $z_k = k\pi$, with $k \in \mathbb{Z}^*$, are simple

poles, while $z_0 = 0$ is a third-order pole. The circle $\{\gamma\} = \{z \in \mathbb{C} : |z| = r\}$, with $n\pi < r < (n+1)\pi$, and $n \in \mathbb{N}^*$, turns around all these z_k points, whenever $|k| < n+1$. We easily deduce that

$$\mathrm{Res}(f; 0) = \frac{1}{6},$$

and

$$\mathrm{Res}(f; k\pi) = \frac{(-1)^k}{k^2\pi^2}, \quad k \in \mathbb{Z}^*.$$

Thus,

$$\int_\gamma \frac{1}{z^2 \sin z} dz$$

$$= 2\pi i \sum_{k=-n}^{n} \mathrm{Res}(f; z_k) = 2\pi i\left(\frac{1}{6} + \frac{2}{\pi^2}\sum_{k=1}^{n}\frac{(-1)^k}{k^2}\right) = \frac{\pi i}{3} + \frac{4i}{\pi}\sum_{k=1}^{n}\frac{(-1)^k}{k^2}.$$

14. The function $f(z) = \frac{\sqrt{1+z^2}}{(1-z^2)(z^2-4)}$ has the simple poles $z_1 = 1$, $z_2 = -1$, $z_3 = 2$ and $z_4 = -2$.

If

$$z - i = r_1 e^{i\theta_1}, \quad z + i = r_2 e^{i\theta_2}, \quad \text{where } r_1 > 0,\ r_2 > 0,\ \theta_1, \theta_2 \in \left[-\frac{\pi}{2}, \frac{3\pi}{2}\right],$$

then

$$\sqrt{z^2+1} = \sqrt{r_1 r_2}\, e^{i\frac{\theta_1+\theta_2+2k\pi}{2}}, \quad k \in \{0, 1\}.$$

The branch of the given function is obtained for $k = 0$, because of the given condition $\sqrt{1+z^2}|_{z=0} = 1$. Then $f \in H(\mathbb{C} \setminus (T_1 \cup T_2))$, where

$$T_1 = \{z = x + iy \in \mathbb{C} : x = 0, y \le -1\}, \quad T_2 = \{z = x + iy \in \mathbb{C} : x = 0, y \ge 1\}.$$

(a) The circle $\{\gamma\} = \{z \in \mathbb{C} : |z| = \frac{\sqrt{2}}{2}\}$ does not turn around any of the above poles, and from the *Cauchy integral theorem* we have

$$\int_\gamma \frac{\sqrt{1+z^2}}{(1-z^2)(z^2-4)} dz = 0.$$

(b) The path $\{\gamma\} = \{z = x + iy \in \mathbb{C} : \frac{x^2}{2} + 4y^2 - 1 = 0, x \ge 0\} \cup \{z = x + iy \in \mathbb{C} : x^2 + y^2 - \frac{1}{4} = 0, x < 0\}$ turns around only the point $z_1 = 1$. The residue of the function at this point is

$$\mathrm{Res}(f; 1) = \lim_{z\to 1} \frac{\sqrt{1+z^2}}{-(z+1)(z^2-4)} = \frac{\sqrt{2}}{6},$$

hence

$$\int_\gamma \frac{\sqrt{1+z^2}}{(1-z^2)(z^2-4)}\mathrm{d}z = 2\pi i \operatorname{Res}(f;1) = \frac{\sqrt{2}\pi i}{3}.$$

15. The function $f(z) = \frac{\sqrt{z^2-1}\log\frac{z-1}{z+1}}{(z^2+1)(z^2+4)}$ has the simple poles $z_1 = i$, $z_2 = -i$, $z_3 = 2i$ and $z_4 = -2i$.

If

$$z-1 = r_1 e^{i\theta_1}, \quad z+1 = r_2 e^{i\theta_2}, \quad \text{where } r_1, r_2 \in [0, 2\pi],$$

then

$$\sqrt{z^2-1} = \sqrt{r_1 r_2} e^{i\frac{\theta_1+\theta_2+2k\pi}{2}}, \quad k \in \{0,1\}$$

and

$$\log\frac{z-1}{z+1} = \ln\frac{r_1}{r_2} + i(\theta_1 - \theta_2 + 2l\pi), \quad l \in \mathbb{Z}.$$

From the conditions $\sqrt{z^2-1}|_{z=0} = i$ and $\log\frac{z-1}{z+1}|_{z=0} = \pi i$, we obtain the values $k = 0$ and $l = 0$. Then $f \in H(\mathbb{C}\setminus(T_1 \cup T_2))$, where

$$T_1 = \{z = x+iy \in \mathbb{C} : x \le -1, y = 0\}, \quad T_2 = \{z = x+iy \in \mathbb{C} : x \ge 1, y = 0\}.$$

The residues in the simple poles are:

$$\operatorname{Res}(f;i) = \lim_{z\to i}\frac{\sqrt{z^2-1}\log\frac{z-1}{z+1}}{(z+i)(z^2+4)} = \frac{\pi i\sqrt{2}}{12},$$
$$\operatorname{Res}(f;-i) = \lim_{z\to -i}\frac{\sqrt{z^2-1}\log\frac{z-1}{z+1}}{(z-i)(z^2+4)} = -\frac{\pi i\sqrt{2}}{4},$$
$$\operatorname{Res}(f;2i) = \lim_{z\to 2i}\frac{\sqrt{z^2-1}\log\frac{z-1}{z+1}}{(z^2+1)(z+2i)} = -\frac{i\sqrt{5}}{12}(\pi - 2\arctan 2),$$

and

$$\operatorname{Res}(f;-2i) = \lim_{z\to -2i}\frac{\sqrt{z^2-1}\log\frac{z-1}{z+1}}{(z^2+1)(z-2i)} = \frac{i\sqrt{5}}{12}(\pi + 2\arctan 2).$$

(a) The path $\{\gamma\} = \{z \in \mathbb{C} : |z| = \frac{\sqrt{2}}{2}\}$ does not turn around any of these singular points, hence

$$\int_\gamma \frac{\sqrt{z^2-1}\log\frac{z-1}{z+1}}{(z^2+1)(z^2+4)}\mathrm{d}z = 0.$$

(b) The ellipse $\{\gamma\} = \{z = x + iy \in \mathbb{C} : 36x^2 + 4y^2 - 9 = 0\}$ turns around only the points $z_1 = i$ and $z_2 = -i$. From the *residues theorem*, it follows that

$$\int_\gamma \frac{\sqrt{z^2-1}\log\frac{z-1}{z+1}}{(z^2+1)(z^2+4)}dz = 2\pi i(\mathrm{Res}(f;i) + \mathrm{Res}(f;-i)) = \frac{\pi^2\sqrt{2}}{3}.$$

(c) The ellipse $\{\gamma\} = \{z = x + iy \in \mathbb{C} : 10x^2 + y^2 - 5 = 0\}$ turns around to all of the simple poles, and thus

$$\int_\gamma \frac{\sqrt{z^2-1}\log\frac{z-1}{z+1}}{(z^2+1)(z^2+4)}dz = 2\pi i(\mathrm{Res}(f;i) + \mathrm{Res}(f;-i) + \mathrm{Res}(f;2i) + \mathrm{Res}(f;-2i))$$

$$= 2\pi\left(\frac{\pi\sqrt{2}}{6} - \frac{\sqrt{5}}{3}\arctan 2\right).$$

(d) The circle $\{\gamma\} = \{z = x + iy \in \mathbb{C} : x^2 + y^2 - 5y = 0\}$ turns around only the simple poles $z_1 = i$ and $z_3 = 2i$, and thus

$$\int_\gamma \frac{\sqrt{z^2-1}\log\frac{z-1}{z+1}}{(z^2+1)(z^2+4)}dz = 2\pi i(\mathrm{Res}(f;i) + \mathrm{Res}(f;2i))$$

$$= \frac{\pi}{6}(\sqrt{5}(\pi - \arctan 2) - \pi\sqrt{2}).$$

Solution of Exercise 5.4.7

All the integrals are of the form

$$I = \int_0^{2\pi} R(\sin\theta, \cos\theta)d\theta,$$

that can be computed as follows:

$$I = 2\pi \sum_{|z|<1} \mathrm{Res}(g;z), \quad \text{where } g(z) = \frac{1}{z}R\left(\frac{z - z^{-1}}{2i}, \frac{z + z^{-1}}{2}\right).$$

We will use the above relation to calculate all the above integrals.

1. Since $f(\theta) = \frac{1}{2-\sin^2\theta}$, then

$$g(z) = \frac{1}{z}\frac{1}{2 - (\frac{z-z^{-1}}{2i})^2} = \frac{4z}{z^4 + 6z^2 + 1}.$$

The function g has the simple poles $z_{1,2} = \pm i(\sqrt{2}-1)$ and $z_{3,4} = \pm i(\sqrt{2}+1)$. But $|z_{1,2}| < 1$ and $|z_{3,4}| > 1$. From

$$\mathrm{Res}(g;z_1) = \lim_{z\to z_1}(z - z_1)g(z) = \frac{\sqrt{2}}{4}$$

and

$$\operatorname{Res}(g;z_2) = \lim_{z\to z_2}(z-z_2)g(z) = \frac{\sqrt{2}}{4},$$

we have that

$$\int_0^{2\pi} \frac{1}{2-\sin^2\theta}\mathrm{d}\theta = 2\pi\sum_{|z|<1}\operatorname{Res}(g;z) = 2\pi(\operatorname{Res}(g;z_1)+\operatorname{Res}(g;z_2)) = \pi\sqrt{2}.$$

2. Since $f(\theta) = \frac{1+\sin\theta}{2+\cos\theta}$, then the corresponding function g will be

$$g(z) = \frac{1}{z}\frac{1+\frac{z-z^{-1}}{2i}}{2+\frac{z+z^{-1}}{2}} = \frac{z^2+2iz-1}{iz(z^2+4z+1)}.$$

The function g has the singular points $z_0 = 0$, $z_1 = -2+\sqrt{3}$ and $z_2 = -2-\sqrt{3}$ that are simple poles. The unit circle turns around only the points z_0 and z_1. For these points, the residues are

$$\operatorname{Res}(g;0) = \lim_{z\to 0}\frac{z^2+2iz-1}{i(z^2+4z+1)} = i,$$

and

$$\operatorname{Res}(g;-2+\sqrt{3}) = \lim_{z\to -2+\sqrt{3}}\frac{z^2+2iz-1}{z(z+2+\sqrt{3})} = \frac{\sqrt{3}}{3} - i.$$

Thus

$$\int_0^{2\pi}\frac{1+\sin\theta}{2+\cos\theta}\mathrm{d}\theta = 2\pi(\operatorname{Res}(g;0)+\operatorname{Res}(g;-2+\sqrt{3})) = \frac{2\pi\sqrt{3}}{3}.$$

3. Since $f(\theta) = \frac{2+\sin\theta}{2+\cos\theta}$, the function g will be

$$g(z) = \frac{1}{z}\frac{2+\frac{z-z^{-1}}{2i}}{2+\frac{z+z^{-1}}{2}} = \frac{z^2+4iz-1}{iz(z^2+4z+1)}.$$

The function g has the singular points $z_0 = 0$, $z_1 = -2+\sqrt{3}$ and $z_2 = -2-\sqrt{3}$ that are simple poles. The unit circle turns around only the points z_0 and z_1, and the corresponding residues are

$$\operatorname{Res}(g;0) = \lim_{z\to 0}\frac{z^2+4iz-1}{i(z^2+4z+1)} = i,$$

and

$$\operatorname{Res}(g;-2+\sqrt{3}) = \lim_{z\to -2+\sqrt{3}} \frac{z^2+4iz-1}{iz(z+2+\sqrt{3})} = \frac{2\sqrt{3}}{3} - i.$$

Thus

$$\int_0^{2\pi} \frac{2+\sin\theta}{2+\cos\theta}\,\mathrm{d}\theta = 2\pi(\operatorname{Res}(g;0)+\operatorname{Res}(g;-2+\sqrt{3})) = \frac{4\pi\sqrt{3}}{3}.$$

4. Since $f(\theta)=\frac{1}{a+\cos\theta}$, the function g will be given by

$$g(z) = \frac{1}{z}\frac{1}{a+\frac{z+z^{-1}}{2}} = \frac{2}{z^2+2az+1}.$$

The function g has the simple poles $z_1 = -a+\sqrt{a^2-1}$ and $z_2 = -a-\sqrt{a^2-1}$. Since $a>1$, the unit disc contains only the point z_1, with the residue

$$\operatorname{Res}(g;-a+\sqrt{a^2-1}) = \lim_{z\to -a+\sqrt{a^2-1}} \frac{2}{z+a+\sqrt{a^2-1}} = \frac{1}{\sqrt{a^2-1}}.$$

It follows that

$$\int_0^{2\pi} \frac{1}{a-\cos\theta}\,\mathrm{d}\theta = 2\pi\operatorname{Res}(g;-a+\sqrt{a^2-1}) = \frac{2\pi}{\sqrt{a^2-1}}.$$

5. We have $f(\theta)=\frac{1}{1+a\sin\theta}$, and g will be

$$g(z) = \frac{1}{z}\frac{1}{1+a\frac{z-z^{-1}}{2i}} = \frac{2i}{az^2+2iz-a}.$$

The function g has the simple poles $z_1 = i\frac{-1+\sqrt{1-a^2}}{a}$ and $z_2 = i\frac{-1-\sqrt{1-a^2}}{a}$. From $|a|<1$, the unit circle turns around only the point z_1. The residue is

$$\operatorname{Res}\left(g;i\frac{-1+\sqrt{1-a^2}}{a}\right) = \lim_{z\to i\frac{-1+\sqrt{1-a^2}}{a}} \frac{2i}{a(z+i\frac{1+\sqrt{1-a^2}}{a})} = \frac{1}{\sqrt{1-a^2}},$$

hence

$$\int_0^{2\pi} \frac{1}{1+a\sin\theta}\,\mathrm{d}\theta = 2\pi\operatorname{Res}\left(g;i\frac{-1+\sqrt{1-a^2}}{a}\right) = \frac{2\pi}{\sqrt{1-a^2}}.$$

6. Since $f(\theta)=\frac{1}{(a+b\cos\theta)^2}$, the correspondent function g is

$$g(z) = \frac{1}{z}\frac{1}{(a+b\frac{z+z^{-1}}{2})^2} = \frac{4z}{(bz^2+2az+b)^2}.$$

The function g has the singular points $z_1 = \frac{-a+\sqrt{a^2-b^2}}{b}$ and $z_2 = \frac{-a-\sqrt{a^2-b^2}}{b}$ that are second-order poles. The unit disc contains only the point z_1, whenever $a > b > 0$. The residue is

$$\operatorname{Res}(g; z_1) = \lim_{z\to z_1}\left(\frac{4z}{b^2(z + \frac{a+\sqrt{a^2-b^2}}{b})^2}\right)' = \frac{a}{(a^2-b^2)^{\frac{3}{2}}},$$

hence we deduce that

$$\int_0^{2\pi} \frac{1}{(a+b\cos\theta)^2}\,\mathrm{d}\theta = 2\pi\operatorname{Res}(g; z_1) = \frac{2\pi a}{(a^2-b^2)^{\frac{3}{2}}}.$$

7. Since $f(\theta) = \frac{1+\cos\theta}{(13-5\cos\theta)^2}$, we have that g will be

$$g(z) = \frac{1}{z}\frac{1+\frac{z+z^{-1}}{2}}{(13-5\frac{z+z^{-1}}{2})^2} = \frac{2(z+1)^2}{25(z-5)^2(z-\frac{1}{5})^2}.$$

The function g has the second-order poles $z_1 = 5$ and $z_2 = \frac{1}{5}$, and the unit disc contains only the pole z_2. The residue in this point is

$$\operatorname{Res}\left(g; \frac{1}{5}\right) = \lim_{z\to\frac{1}{5}}\left(\frac{2(z+1)^2}{25(z-5)^2}\right)' = \lim_{z\to\frac{1}{5}}\frac{-24(z+1)}{25(z-5)^3} = \frac{1}{96},$$

hence we get

$$\int_0^{2\pi}\frac{1+\cos\theta}{(13-5\cos\theta)^2}\,\mathrm{d}\theta = 2\pi\operatorname{Res}\left(g; \frac{1}{5}\right) = \frac{\pi}{48}.$$

8. The function $f(\theta) = \frac{1}{(17+8\cos\theta)^2}$ is a special case of the point 6, where $a = 17$ and $b = 8$. Hence

$$\int_0^{2\pi}\frac{1}{(17+8\cos\theta)^2}\,\mathrm{d}\theta = \frac{2\pi\cdot 17}{(17^2-8^2)^{\frac{3}{2}}} = \frac{34\pi}{15^3}.$$

9. First, we will use the exponential form of the function $\cos\theta$, and then we will use the following substitutions:

$$\cos\theta = \frac{e^{i\theta}+e^{-i\theta}}{2}, \quad \theta = 2\pi t, \quad z = e^{i2\pi t}.$$

Thus

$$\int_0^{2\pi} \frac{1}{1-2p\cos\theta+p^2}\,d\theta = \int_0^{2\pi} \frac{1}{1-p(e^{i\theta}+e^{-i\theta})+p^2}\,d\theta$$
$$= 2\pi \int_0^1 \frac{1}{1-p(e^{i2\pi t}+e^{-i2\pi t})+p^2}\,dt$$
$$= \int_0^1 \frac{1}{(1+p^2)e^{i2\pi t}-p(e^{i4\pi t}+1)} 2\pi i e^{i2\pi t}\,dt$$
$$= \frac{1}{i}\int_{\partial U(0;1)} \frac{1}{(1+p^2)z-p(z^2+1)}\,dz = \frac{1}{i}\int_{\partial U(0;1)} \frac{1}{(z-p)(1-pz)}\,dz.$$

The function $g(z) = \frac{1}{(z-p)(1-pz)}$, where $0 < p < 1$, has the only one singular point $z_0 = p$ with the module less than 1, and then

$$\mathrm{Res}(g;p) = \lim_{z\to p} \frac{1}{1-pz} = \frac{1}{1-p^2}.$$

It follows that

$$\int_0^{2\pi} \frac{1}{1-2p\cos\theta+p^2}\,d\theta = \frac{1}{i}\int_{\partial U(0;1)} \frac{1}{(z-p)(1-pz)}\,dz = 2\pi\,\mathrm{Res}(g;p) = \frac{2\pi}{1-p^2}.$$

10. We will use the same method as in point 9. Using the substitution $z = e^{i\theta}$, we get

$$\int_0^{2\pi} \frac{\cos^2 2\theta}{1-2p\cos\theta+p^2}\,d\theta = \frac{1}{2}\int_0^{2\pi} \frac{1+\cos 4\theta}{1-2p\cos\theta+p^2}\,d\theta$$
$$= \frac{1}{4}\int_0^{2\pi} \frac{2+e^{i4\theta}+e^{-i4\theta}}{1-p(e^{i\theta}+e^{-i\theta})+p^2}\,d\theta = \frac{1}{4}\int_{\partial U(0;1)} \frac{2+z^4+z^{-4}}{1-p(z+z^{-1})+p^2}\frac{1}{iz}\,dz$$
$$= \frac{1}{4i}\int_{\partial U(0;1)} \frac{z^8+2z^4+1}{(p^2z-p-pz^2+z)z^4}\,dz$$
$$= \frac{1}{4i}\int_{\partial U(0;1)} \frac{(z^4+1)^2}{-pz^4(z-p)(z-\frac{1}{p})}\,dz = \frac{\pi}{2}\sum_{|z|<1} \mathrm{Res}(g;z),$$

where $g(z) = \frac{(z^4+1)^2}{-pz^4(z-p)(z-\frac{1}{p})}$. From the assumption $0 < p < 1$, the points $z_0 = 0$ and $z_1 = p$ are being the only poles in the unit disc. The residues at these points are

$$\mathrm{Res}(g;p) = \lim_{z\to p} \frac{(z^4+1)^2}{-pz^4(z-\frac{1}{p})} = -\frac{(p^4+1)^2}{p^4(p^2-1)}$$

and

$$\operatorname{Res}(g;0)=\frac{1}{3!}\lim_{z\to 0}\left(\frac{(z^4+1)^2}{-p(z-p)(z-\frac{1}{p})}\right)^{(3)}=-\frac{1+p^2+p^4+p^6}{p^4}.$$

We conclude that

$$\int_0^{2\pi}\frac{\cos^2 2\theta}{1-2p\cos\theta+p^2}\mathrm{d}\theta=\frac{\pi}{2}(\operatorname{Res}(g;p)+\operatorname{Res}(g;0))=-\pi\frac{p^4+1}{p^2-1}.$$

11. We will use a similar method as in the previous two points of the problem. Using the substitution $z=e^{2i\theta}$ together with the fact that variable of the integral runs twice over the curve $\partial U(0;1)$, we need to compute the integral on the curve $2\times\partial U(0;1)$. Thus

$$\begin{aligned}
&\int_0^{2\pi}\frac{\cos^2 3\theta}{1-2p\cos 2\theta+p^2}\mathrm{d}\theta\\
&=\frac{1}{2}\int_0^{2\pi}\frac{1+\cos 6\theta}{1-2p\cos 2\theta+p^2}\mathrm{d}\theta\\
&=\frac{1}{4}\int_0^{2\pi}\frac{2+e^{i6\theta}+e^{-i6\theta}}{1-p(e^{i2\theta}+e^{-i2\theta})+p^2}\mathrm{d}\theta=\frac{1}{4}\int_{2\times\partial U(0;1)}\frac{2+z^3+z^{-3}}{1-p(z+z^{-1})+p^2}\frac{1}{2iz}\mathrm{d}z\\
&=\frac{1}{8i}\int_{2\times\partial U(0;1)}\frac{(z^3+1)^2}{-pz^3(z-p)(z-\frac{1}{p})}\mathrm{d}z\\
&=\frac{\pi}{4}[\mathrm{n}(2\times\partial U(0;1),p)\operatorname{Res}(g;p)+\mathrm{n}(2\times\partial U(0;1),0)\operatorname{Res}(g;0)]\\
&=\frac{\pi}{4}2(\operatorname{Res}(g;p)+\operatorname{Res}(g;0))=\frac{\pi}{2}\left(-(p^2-p+1)\frac{p^3+1}{(p-1)p^3}-\frac{1+p^2+p^4}{p^3}\right)\\
&=\frac{\pi(p^2-p+1)}{1-p},
\end{aligned}$$

where $g(z)=\frac{(z^3+1)^2}{-pz^3(z-p)(z-\frac{1}{p})}$, because

$$\operatorname{Res}(g;p)=\lim_{z\to p}\frac{(z^3+1)^2}{-pz^3(z-\frac{1}{p})}=-(p^2-p+1)\frac{p^3+1}{(p-1)p^3}$$

and

$$\operatorname{Res}(g;0)=\frac{1}{2!}\lim_{z\to 0}\left(\frac{(z^3+1)^2}{-p(z-p)(z-\frac{1}{p})}\right)''=-\frac{1+p^2+p^4}{p^3}.$$

Solution of Exercise 5.4.8

1. Let us use the notation

$$I = \int_0^{2\pi} \frac{\cos n\theta}{1 - 2a\cos\theta + a^2} d\theta, \quad J = \int_0^{2\pi} \frac{\sin n\theta}{1 - 2a\cos\theta + a^2} d\theta.$$

We will compute the integral $I + iJ$, and the real part of it will be the integral I, while the imaginary part will be J. Thus

$$I + iJ = \int_0^{2\pi} \frac{e^{in\theta}}{1 - 2a\cos\theta + a^2} d\theta = \int_{\partial U(0;1)} \frac{z^n}{1 - a(z + z^{-1}) + a^2} \frac{1}{iz} dz$$
$$= \int_{\partial U(0;1)} \frac{z^n}{-ai(z - a)(z - \frac{1}{a})} dz = 2\pi \operatorname{Res}\left(g; \frac{1}{a}\right) = \frac{2\pi}{a^n(a^2 - 1)},$$

where we used the substitution $z = e^{i\theta}$, and we denoted by g the function $g(z) = \frac{z^n}{-a(z-a)(z-\frac{1}{a})}$. Since $a > 1$, the unit disc contains only the simple pole $z_1 = \frac{1}{a}$. From the above result,

$$I = \int_0^{2\pi} \frac{\cos n\theta}{1 - 2a\cos\theta + a^2} d\theta = \frac{2\pi}{a^n(a^2 - 1)}, \quad J = \int_0^{2\pi} \frac{\sin n\theta}{1 - 2a\cos\theta + a^2} d\theta = 0.$$

If we use the notation,

$$I = \int_0^{2\pi} \frac{\cos n\theta}{1 - 2a\sin\theta + a^2} d\theta, \quad J = \int_0^{2\pi} \frac{\sin n\theta}{1 - 2a\sin\theta + a^2} d\theta,$$

then

$$I + iJ = \int_0^{2\pi} \frac{e^{in\theta}}{1 - 2a\sin\theta + a^2} d\theta = \int_{\partial U(0;1)} \frac{z^n}{1 - a\frac{z - z^{-1}}{i} + a^2} \frac{1}{iz} dz$$
$$= \int_{\partial U(0;1)} \frac{z^n}{-a(z - ia)(z - \frac{i}{a})} dz = 2\pi i \operatorname{Res}\left(g; \frac{i}{a}\right) = \frac{2\pi i^n}{a^n(a^2 - 1)},$$

where we used the substitution $z = e^{i\theta}$, and we denoted by g the function $g(z) = \frac{z^n}{-a(z-ia)(z-\frac{i}{a})}$. Since $a > 1$, the unit disc contains only the simple pole $z_1 = \frac{i}{a}$. From here, we get

$$I = \int_0^{2\pi} \frac{\cos n\theta}{1 - 2a\sin\theta + a^2} d\theta = \frac{2\pi}{a^n(a^2 - 1)} \operatorname{Re} i^n,$$

$$J = \int_0^{2\pi} \frac{\sin n\theta}{1 - 2a\sin\theta + a^2} d\theta = \frac{2\pi}{a^n(a^2 - 1)} \operatorname{Im} i^n.$$

2. We will use a similar method with in point 1. Let

$$I = \int_0^{2\pi} \frac{\sin\theta \sin n\theta}{5 - 4\cos\theta} \mathrm{d}\theta, \quad J = \int_0^{2\pi} \frac{\sin\theta \cos n\theta}{5 - 4\cos\theta} \mathrm{d}\theta.$$

Then

$$\begin{aligned} J + iI &= \int_0^{2\pi} \frac{\sin\theta e^{in\theta}}{5 - 4\cos\theta} \mathrm{d}\theta = -\frac{1}{2} \int_{\partial U(0;1)} \frac{(z - z^{-1}) z^n}{5 - 2(z + z^{-1})} \frac{1}{z} \mathrm{d}z \\ &= \frac{1}{2} \int_{\partial U(0;1)} \frac{(z^2 - 1) z^{n-1}}{2z^2 - 5z + 2} \mathrm{d}z = \frac{1}{2} \int_{\partial U(0;1)} \frac{(z^2 - 1) z^{n-1}}{2(z - 2)(z - \frac{1}{2})} \mathrm{d}z \\ &= \pi i \operatorname{Res}\left(g; \frac{1}{2}\right) = \frac{\pi i}{2^{n+1}}, \end{aligned}$$

where $g(z) = \frac{(z^2-1)z^{n-1}}{2(z-2)(z-\frac{1}{2})}$. Only the pole $z_1 = \frac{1}{2}$ lies in the unit disc, and the residue in this point is

$$\operatorname{Res}\left(g; \frac{1}{2}\right) = \lim_{z \to \frac{1}{2}} \frac{(z^2 - 1) z^{n-1}}{2(z - 2)} = \frac{1}{2^{n+1}}.$$

The integrals are

$$I = \int_0^{2\pi} \frac{\sin\theta \sin n\theta}{5 - 4\cos\theta} \mathrm{d}\theta = \frac{\pi}{2^{n+1}}, \quad J = \int_0^{2\pi} \frac{\sin\theta \cos n\theta}{5 - 4\cos\theta} \mathrm{d}\theta = 0.$$

Solution of Exercise 5.4.9

1. Since $f(\theta) = \frac{\cos n\theta}{b - ia\cos\theta}$ is an even function, we have

$$I = \int_0^{\pi} \frac{\cos n\theta}{b - ia\cos\theta} \mathrm{d}\theta = \frac{1}{2} \int_{-\pi}^{\pi} \frac{\cos n\theta}{b - ia\cos\theta} \mathrm{d}\theta.$$

Using this property, we will define the integral

$$J = \frac{1}{2} \int_{-\pi}^{\pi} \frac{\sin n\theta}{b - ia\cos\theta} \mathrm{d}\theta,$$

and from the fact that $h(z) = \frac{\sin n\theta}{b - ia\cos\theta}$ is an odd function, it follows that $J = 0$.

Using the substitution $z = e^{i\theta}$, we get $d\theta = \frac{1}{iz}dz$, hence

$$I + iJ = \frac{1}{2}\int_{-\pi}^{\pi} \frac{e^{in\theta}}{b - ia\cos\theta}d\theta = \int_{\partial U(0;1)} \frac{z^n}{2b - ia(z + z^{-1})}\frac{1}{iz}dz$$

$$= \int_{\partial U(0;1)} \frac{z^n}{az^2 + 2ibz + a}dz.$$

The function $g(z) = \frac{z^n}{az^2+2ibz+a}$ has the simple poles $z_1 = i\frac{-b+\sqrt{a^2+b^2}}{a}$ and $z_2 = i\frac{-b-\sqrt{a^2+b^2}}{a}$, and only z_1 lies in the unit disc, whenever $a > 0$, $b > 0$. The residue will be

$$\operatorname{Res}(g; z_1) = \lim_{z\to z_1} \frac{z^n}{2az + 2bi} = \frac{i^n}{2i\sqrt{a^2 + b^2}}\frac{a^n}{(\sqrt{a^2 + b^2} + b)^n},$$

and since

$$I + iJ = 2\pi i \operatorname{Res}(g; z_1)$$

we obtain that

$$\int_0^{\pi} \frac{\cos n\theta}{b - ia\cos\theta}d\theta = \frac{\pi i^n}{\sqrt{a^2 + b^2}}\frac{a^n}{(\sqrt{a^2 + b^2} + b)^n}.$$

2. We will substitute $z = e^{i\theta}$, hence

$$I = \int_0^{2\pi} \frac{1}{a + b\cos^2\theta}d\theta = \frac{4}{i}\int_{\partial U(0;1)} \frac{z}{bz^4 + 2(2a + b)z^2 + b}dz.$$

If we denote $\zeta = z^2$, then we get

$$I = \frac{2}{i}\int_{2\times\partial U(0;1)} \frac{1}{b\zeta^2 + 2(2a + b)\zeta + b}d\zeta = \frac{4}{i}\int_{\partial U(0;1)} \frac{1}{b\zeta^2 + 2(2a + b)\zeta + b}d\zeta.$$

The function $g(z) = \frac{1}{b\zeta^2+2(2a+b)\zeta+b}$ has the simple poles

$$\zeta_1 = \frac{-(2a + b) + 2\sqrt{a(a + b)}}{b} \quad \text{and} \quad \zeta_2 = \frac{-(2a + b) - 2\sqrt{a(a + b)}}{b}.$$

Only the point ζ_1 lies in the unit disc, whenever $a > 0$, $b > 0$, and the residue will be

$$\operatorname{Res}(g; \zeta_1) = \lim_{z\to\zeta_1} \frac{1}{b(\zeta - \zeta_2)} = \frac{1}{4\sqrt{a(a + b)}}.$$

From the *residues theorem*, we have

$$\int_0^{2\pi} \frac{1}{a + b\cos^2\theta}d\theta = 8\pi \operatorname{Res}(g; \zeta_1) = \frac{2\pi}{\sqrt{a(a + b)}}.$$

Solution of Exercise 5.4.10

1. To calculate this integral, we will use the next well-known result: if P and Q are real polynomials, such that Q has no real roots, and $\operatorname{gr} Q \geq \operatorname{gr} P$, then

$$\int_0^{+\infty} \left[\frac{P(x)}{Q(x)} e^{ix} - \frac{P(-x)}{Q(-x)} e^{-ix} \right] \frac{dx}{x} = 2\pi i \left[\frac{1}{2} \frac{P(0)}{Q(0)} + \sum_{\operatorname{Im} z_k > 0} \operatorname{Res}\left(\frac{P(z)}{zQ(z)} e^{iz}; z_k \right) \right].$$

We will write the required integral in a such form, to be able to use the above formula, i. e.,

$$\int_0^{+\infty} \frac{\sin x}{x} dx = \frac{1}{2i} \int_0^{+\infty} \frac{e^{ix} - e^{-ix}}{x} dx = \frac{2\pi i}{2i} \frac{1}{2} \frac{P(0)}{Q(0)} = \frac{\pi}{2},$$

where $P(x) = 1$ and $Q(x) = 1$. Hence

$$\int_0^{+\infty} \frac{\sin x}{x} dx = \frac{\pi}{2}.$$

2. In this case, we will use another well-known result: if P and Q are real polynomials, such that Q has no real roots, and $\operatorname{gr} Q \geq \operatorname{gr} P + 2$, then

$$\int_{-\infty}^{+\infty} \frac{P(x)}{Q(x)} dx = 2\pi i \sum_{\operatorname{Im} z_k > 0} \operatorname{Res}\left(\frac{P}{Q}; z_k \right). \tag{7.49}$$

The function $f(z) = \frac{1}{z^4+1}$ has the simple poles $z_k = e^{i\frac{\pi+2k\pi}{4}}$, where $k \in \{0, 1, 2, 3\}$. The condition $\operatorname{Im} z_k > 0$ holds only for the points $z_0 = e^{i\frac{\pi}{4}}$ and $z_1 = e^{i\frac{3\pi}{4}}$. The corresponding residues are

$$\operatorname{Res}(f; z_0) = \lim_{z \to z_0} \frac{1}{4z^3} = -\frac{z_0}{4},$$

and

$$\operatorname{Res}(f; z_1) = \lim_{z \to z_1} \frac{1}{4z^3} = -\frac{z_1}{4},$$

hence

$$\int_{-\infty}^{+\infty} \frac{1}{x^4 + 1} dx = 2\pi i (\operatorname{Res}(f; z_0) + \operatorname{Res}(f; z_1)) = -\frac{\pi i}{2}(z_0 + z_1) = \frac{\pi\sqrt{2}}{2}.$$

3. As in point 2, we will use the relation (7.49). To change the integration limits, we have

$$\begin{aligned}\int_{-\infty}^{+\infty}\frac{1}{(a+bx^2)^n}\mathrm{d}x &= \int_{-\infty}^{0}\frac{1}{(a+bx^2)^n}\mathrm{d}x+\int_{0}^{+\infty}\frac{1}{(a+bx^2)^n}\mathrm{d}x\\ &= \int_{+\infty}^{0}\frac{1}{(a+by^2)^n}(-\mathrm{d}y)+\int_{0}^{+\infty}\frac{1}{(a+bx^2)^n}\mathrm{d}x\\ &= 2\int_{0}^{+\infty}\frac{1}{(a+bx^2)^n}\mathrm{d}x,\end{aligned}\tag{7.50}$$

where we used the $y=-x$ variable changes. Now we will apply the above mentioned relation. The function $f(z)=\frac{1}{(a+bz^2)^n}$ has the nth order poles $z_1=i\sqrt{\frac{a}{b}}$ and $z_2=-i\sqrt{\frac{a}{b}}$. The condition $\operatorname{Im} z_k>0$ holds only for the point $z_1=i\sqrt{\frac{a}{b}}$, and the residue is

$$\begin{aligned}\operatorname{Res}(f;z_1) &= \frac{1}{(n-1)!}\lim_{z\to z_1}\left(\frac{1}{b^n(z-z_2)^n}\right)^{(n-1)}\\ &= \frac{1}{(n-1)!b^n}\lim_{z\to z_1}\frac{(-1)^{n-1}(2n-2)!}{(n-1)!(z-z_2)^{2n-1}}=\frac{(-1)^{n-1}(2n-2)!\sqrt{a}}{(-1)^n[(n-1)!]^2 2^{2n-1}a^n\sqrt{b}}i\\ &= -\frac{(2n-2)!\sqrt{a}}{[(n-1)!]^2 2^{2n-1}a^n\sqrt{b}}i.\end{aligned}$$

Using the relation (7.50), we get

$$\int_{0}^{+\infty}\frac{1}{(a+bx^2)^n}\mathrm{d}x=\frac{1}{2}2\pi i\operatorname{Res}(f;z_1)=\frac{(2n-2)!\pi\sqrt{a}}{[(n-1)!]^2 2^{2n-1}a^n\sqrt{b}}.$$

4. Similarly, as in the previous point, we will use the relation (7.49). The function $f(z)=\frac{1}{1+z^{2n}}$ has the simple poles $z_k=e^{i\frac{\pi+2k\pi}{2n}}$, where $k\in\{0,1,2,\dots,2n-1\}$. The condition $\operatorname{Im} z_k>0$ holds only for the points $z_k=e^{i\frac{\pi+2k\pi}{2n}}$, $k\in\{0,1,\dots,n-1\}$, and the residues in these points are

$$\operatorname{Res}(f;z_k)=\lim_{z\to z_k}\frac{1}{2nz^{2n-1}}=-\frac{z_k}{2n},$$

hence

$$\int_{-\infty}^{+\infty}\frac{1}{1+x^{2n}}\mathrm{d}x=2\pi i\sum_{k=0}^{n-1}\operatorname{Res}(f;z_k)=-\frac{\pi i}{n}\sum_{k=0}^{n-1}z_k=-\frac{\pi i}{n}\frac{i}{\sin\frac{\pi}{2n}}=\frac{\pi}{n\sin\frac{\pi}{2n}}.$$

5. Here, we will use another well-known result: if P and Q are real polynomials, such that Q has no real nonnegative roots, and $\operatorname{gr} Q\ge\operatorname{gr} P+2$, then

$$\int_{0}^{+\infty}\frac{P(x)}{Q(x)}\ln x\mathrm{d}x=-\frac{1}{2}\operatorname{Re}\left[\sum_{z_k\in\mathbb{C}^*}\operatorname{Res}\left(\frac{P(z)}{Q(z)}(\log z)^2;z_k\right)\right],$$

where $\log 1 = 0$. It follows that we need to study the function $f(z) = \frac{(\log z)^2}{(z+a)^2+b^2}$. The numbers $z_1 = -a + bi$ and $z_2 = -a - bi$ are simple poles, and both satisfy the condition $z_k \in \mathbb{C}^*$, because $a, b > 0$. The residues in these points are

$$\operatorname{Res}(f; z_1) = \lim_{z \to z_1} \frac{(\log z)^2}{z + a + bi} = \frac{(\log(-a + bi))^2}{2bi},$$

and

$$\operatorname{Res}(f; z_2) = \lim_{z \to z_2} \frac{(\log z)^2}{z + a - bi} = \frac{(\log(-a - bi))^2}{-2bi}.$$

Thus

$$\begin{aligned}
&\int_0^{+\infty} \frac{\ln x}{(x + a)^2 + b^2} \mathrm{d}x \\
&= -\frac{1}{2} \operatorname{Re}(\operatorname{Res}(f; z_1) + \operatorname{Res}(f; z_2)) \\
&= -\frac{1}{2} \operatorname{Re} \frac{(\log(-a + bi))^2 - (\log(-a - bi))^2}{2bi} \\
&= -\frac{1}{2} \operatorname{Re} \frac{(\ln \sqrt{a^2 + b^2} + i(\pi - \arctan \frac{b}{a}))^2 - (\ln \sqrt{a^2 + b^2} + i(\pi + \arctan \frac{b}{a}))^2}{2bi} \\
&= -\frac{1}{2} \operatorname{Re} \frac{2\pi \arctan \frac{b}{a} - 2i \ln \sqrt{a^2 + b^2} \arctan \frac{b}{a}}{bi} = \frac{\ln \sqrt{a^2 + b^2}}{b} \arctan \frac{b}{a}.
\end{aligned}$$

6. We will change the limits of integration as follows:

$$\begin{aligned}
I = \int_{-\infty}^{+\infty} \frac{\cos ax}{(x^2 + b^2)^2} \mathrm{d}x &= \int_{-\infty}^{0} \frac{\cos ax}{(x^2 + b^2)^2} \mathrm{d}x + \int_{0}^{+\infty} \frac{\cos ax}{(x^2 + b^2)^2} \mathrm{d}x \\
&= \int_{+\infty}^{0} \frac{\cos ay}{(y^2 + b^2)^2} (-\mathrm{d}y) + \int_{0}^{+\infty} \frac{\cos ax}{(x^2 + b^2)^2} \mathrm{d}x = 2 \int_{0}^{+\infty} \frac{\cos ax}{(x^2 + b^2)^2} \mathrm{d}x.
\end{aligned}$$

Let

$$\begin{aligned}
J = \int_{-\infty}^{+\infty} \frac{\sin ax}{(x^2 + b^2)^2} \mathrm{d}x &= \int_{-\infty}^{0} \frac{\sin ax}{(x^2 + b^2)^2} \mathrm{d}x + \int_{0}^{+\infty} \frac{\sin ax}{(x^2 + b^2)^2} \mathrm{d}x \\
&= \int_{+\infty}^{0} \frac{-\sin ay}{(y^2 + b^2)^2} (-\mathrm{d}y) + \int_{0}^{+\infty} \frac{\sin ax}{(x^2 + b^2)^2} \mathrm{d}x = 0.
\end{aligned}$$

We get

$$I = I + iJ = \int_{-\infty}^{+\infty} \frac{e^{ixa}}{(x^2 + b^2)^2} \mathrm{d}x.$$

To compute this last integral, we will use the following well-known result: if P and Q are real polynomials, such that Q has no real roots, and $\operatorname{gr} Q \geq \operatorname{gr} P + 1$, then

$$\int_{-\infty}^{+\infty} \frac{P(x)}{Q(x)} e^{ix} \mathrm{d}x = 2\pi i \sum_{\operatorname{Im} z_k > 0} \operatorname{Res}\left(\frac{P(z)}{Q(z)} e^{iz}; z_k\right). \tag{7.51}$$

The function $f(z) = \frac{e^{iaz}}{(z^2+b^2)^2}$ has the second-order pole $z_1 = bi$, with the corresponding residue

$$\operatorname{Res}(f; z_1) = \lim_{z \to bi} \left(\frac{e^{iaz}}{(z + bi)^2}\right)' = -\frac{ab + 1}{4e^{ab} b^3} i,$$

hence

$$I = \int_{0}^{+\infty} \frac{\cos ax}{(x^2 + b^2)^2} \mathrm{d}x = \frac{1}{2} 2\pi i \operatorname{Res}(f; z_1) = \frac{(ab + 1)\pi}{4e^{ab} b^3}.$$

7. We will integrate the function $f(z) = \frac{e^{iaz} - e^{ibz}}{z^2}$ along the curve $\gamma = \gamma_{[-r,-\varepsilon]} \cup \gamma_\varepsilon^- \cup \gamma_{[\varepsilon,r]} \cup \gamma_r$, where

$$\begin{aligned}
&\gamma_{[-r,-\varepsilon]}(t) = (1 - t)(-r) + t(-\varepsilon),\\
&\gamma_\varepsilon(t) = \varepsilon e^{i\pi t},\\
&\gamma_{[\varepsilon,r]}(t) = (1 - t)\varepsilon + tr,\\
&\gamma_r(t) = re^{i\pi t}, \quad t \in [0, 1],
\end{aligned}$$

and $0 < \varepsilon < r$.

Since $\int_\gamma f(z)\mathrm{d}z = 0$, then

$$\int_{-r}^{-\varepsilon} f(x)\mathrm{d}x - \int_{\gamma_\varepsilon} f(z)\mathrm{d}z + \int_{\varepsilon}^{r} f(x)\mathrm{d}x + \int_{\gamma_r} f(z)\mathrm{d}z = 0,$$

and we deduce that

$$\int_{-\infty}^{0} \frac{e^{iax} - e^{ibx}}{x^2} \mathrm{d}x + \int_{0}^{+\infty} \frac{e^{iax} - e^{ibx}}{x^2} \mathrm{d}x = \lim_{\varepsilon \to 0} \int_{\gamma_\varepsilon} f(z)\mathrm{d}z - \lim_{r \to +\infty} \int_{\gamma_r} f(z)\mathrm{d}z.$$

But

$$\begin{aligned}
&\int_{-\infty}^{0} \frac{e^{iax} - e^{ibx}}{x^2} \mathrm{d}x + \int_{0}^{+\infty} \frac{e^{iax} - e^{ibx}}{x^2} \mathrm{d}x\\
&\quad = \int_{+\infty}^{0} \frac{e^{-iat} - e^{-ibt}}{t^2} \mathrm{d}t + \int_{0}^{+\infty} \frac{e^{iax} - e^{ibx}}{x^2} \mathrm{d}x = 2 \int_{0}^{+\infty} \frac{\cos ax - \cos bx}{x^2} \mathrm{d}x,
\end{aligned}$$

and thus

$$\int_0^{+\infty} \frac{\cos ax - \cos bx}{x^2}\mathrm{d}x = \frac{1}{2}\Big[\lim_{\varepsilon\to 0}\int_{\gamma_\varepsilon} f(z)\mathrm{d}z - \lim_{r\to+\infty}\int_{\gamma_r} f(z)\mathrm{d}z\Big]. \tag{7.52}$$

Since $f \in H(S_0[0,\pi] \setminus \{0\})$ and $z_0 = 0$ is a simple pole for f, using the fact that

$$\operatorname{Res}(f;0) = \lim_{z\to 0}(e^{iaz} - e^{ibz})' = i(a-b),$$

it follows

$$\lim_{\varepsilon\to 0}\int_{\gamma_\varepsilon} f(z)\mathrm{d}z = \pi i \operatorname{Res}(f;0) = \pi(b-a). \tag{7.53}$$

Substituting $az = \zeta$, we have

$$\int_{\gamma_r} \frac{e^{iaz}}{z^2}\mathrm{d}z = \frac{1}{a}\int_{\gamma_r} \varphi(\zeta)e^{i\zeta}\mathrm{d}\zeta, \quad \text{where } \varphi(\zeta) = \frac{a^2}{\zeta^2},$$

hence we get

$$\lim_{r\to+\infty}\varphi(\zeta) = 0 \Rightarrow \lim_{r\to+\infty}\int_{\gamma_r}\varphi(\zeta)e^{i\zeta}\mathrm{d}\zeta = 0.$$

Thus

$$\lim_{r\to+\infty}\int_{\gamma_r}\frac{e^{iaz}}{z^2}\mathrm{d}z = 0,$$

and similarly we obtain

$$\lim_{r\to+\infty}\int_{\gamma_r}\frac{e^{ibz}}{z^2}\mathrm{d}z = 0,$$

thus

$$\lim_{r\to+\infty}\int_{\gamma_r} f(z)\mathrm{d}z = \lim_{r\to+\infty}\int_{\gamma_r}\frac{e^{iaz} - e^{ibz}}{z^2}\mathrm{d}z = 0. \tag{7.54}$$

Combining the relations (7.52), (7.53) and (7.54), we have

$$\int_0^{+\infty}\frac{\cos ax - \cos bx}{x^2}\mathrm{d}x = \frac{\pi(b-a)}{2}.$$

8. Define the integral

$$I = \int_{\partial C} e^{-az^2}\mathrm{d}z,$$

where C is the boundary of the rectangle determined by the points $A(R)$, $B(R + i\frac{b}{2a})$, $C(-R + i\frac{b}{2a})$ and $D(-R)$. The function $f(z) = e^{-az^2}$ is holomorphic in the whole complex plane, thus

$$I = \int_{C} e^{-az^2}\mathrm{d}z = 0,$$

i. e.,

$$\int_{\overline{AB}} f(z)\mathrm{d}z + \int_{\overline{BC}} f(z)\mathrm{d}z + \int_{\overline{CD}} f(z)\mathrm{d}z + \int_{\overline{DA}} f(z)\mathrm{d}z = 0. \tag{7.55}$$

On the real axis, we have

$$\int_{\overline{DA}} f(z)\mathrm{d}z = \int_{-R}^{R} e^{-ax^2}\mathrm{d}x = 2\int_{0}^{R} e^{-ax^2}\mathrm{d}x,$$

and using the substitution $x = \frac{t}{\sqrt{a}}$ we get

$$\int_{\overline{DA}} f(z)\mathrm{d}z = \frac{2}{\sqrt{a}}\int_{0}^{R\sqrt{a}} e^{-t^2}\mathrm{d}t.$$

On the $\overline{BC}$ segment, where $z = x + i\frac{b}{2a}$, we have

$$\int_{\overline{BC}} f(z)\mathrm{d}z = e^{\frac{b^2}{4a}}\int_{R}^{-R} e^{-ax^2}e^{-ibx}\mathrm{d}x.$$

Since

$$\int_{-R}^{0} e^{-ax^2}e^{-ibx}\mathrm{d}x = \int_{0}^{R} e^{-ax^2}e^{ibx}\mathrm{d}x,$$

we obtain that

$$\int_{\overline{BC}} f(z)\mathrm{d}z = -e^{\frac{b^2}{4a}}\int_{0}^{R} e^{-ax^2}(e^{ibx} + e^{-ibx})\mathrm{d}x,$$

or

$$\int_{\overline{BC}} f(z)\mathrm{d}z = -2e^{\frac{b^2}{4a}} \int_0^R e^{-ax^2} \cos bx\mathrm{d}x.$$

Replacing the above results in (7.55), it follows that

$$\frac{2}{\sqrt{a}} \int_0^{R\sqrt{a}} e^{-t^2}\mathrm{d}t - 2e^{\frac{b^2}{4a}} \int_0^R e^{-ax^2} \cos bx\mathrm{d}x = -\int_{\overline{AB}} f(z)\mathrm{d}z - \int_{\overline{CD}} f(z)\mathrm{d}z.$$

Since

$$\int_0^{+\infty} e^{-t^2}\mathrm{d}t = \frac{\sqrt{\pi}}{2},$$

for $R \to +\infty$, we have

$$\sqrt{\frac{\pi}{a}} - 2e^{\frac{b^2}{4a}} \int_0^{+\infty} e^{-ax^2} \cos bx\mathrm{d}x = -\lim_{R\to\infty}\left[\int_{\overline{AB}} f(z)\mathrm{d}z + \int_{\overline{CD}} f(z)\mathrm{d}z\right].$$

The equations of the segments $\overline{AB}$ and $\overline{CD}$ are $z = R + iy$, $y \in [0, \frac{b}{2a}]$ and, respectively, $z = -R + iy$, $y \in [0, \frac{b}{2a}]$, thus

$$|f(z)| = |e^{-a(R^2 \pm 2Riy - y^2)}| = e^{-a(R^2-y^2)} \le e^{\frac{b^2}{4a}} e^{-aR^2}, \quad z \in \overline{AB} \cup \overline{CD}.$$

From here, we get

$$\left|\int_{\overline{AB}} f(z)\mathrm{d}z\right| \le \max\{|f(z)| : z \in \overline{AB}\} V(\overline{AB}) \le e^{\frac{b^2}{4a}} e^{-aR^2} \frac{b}{2a},$$

hence

$$\lim_{R\to\infty} \int_{\overline{AB}} f(z)\mathrm{d}z = 0.$$

We will prove similarly that

$$\lim_{R\to\infty} \int_{\overline{CD}} f(z)\mathrm{d}z = 0,$$

and thus it follows that

$$\int_0^{+\infty} e^{-ax^2} \cos bx\mathrm{d}x = \frac{1}{2}\sqrt{\frac{\pi}{a}} e^{-\frac{b^2}{4a}}.$$

Solution of Exercise 5.4.11

We will determine the number of the zeros that lies in the disc $U(0;1)$ by using the *Rouché theorem*.

1. The equation $z^8 - 4z^5 + z^2 - 1 = 0$ may be written as $f(z) + g(z) = 0$, where

$$f(z) = -4z^5$$

and

$$g(z) = z^8 + z^2 - 1.$$

If $\zeta \in \{\gamma\} = \partial U(0;1)$, then

$$|f(\zeta)| = |-4\zeta^5| = 4,$$

and

$$|g(\zeta)| = |\zeta^8 + \zeta^2 - 1| \leq |\zeta^8| + |\zeta^2| + 1 = 3,$$

i. e.,

$$|g(\zeta)| \leq 3 < 4 = |f(\zeta)|, \quad \forall \zeta \in \{\gamma\}.$$

From the *Rouché theorem*, we obtain that the given equation has five zeros in the unit disc, because

$$\Theta(f + g, U(0;1)) = \Theta(f, U(0;1)) = 5.$$

2. The equation $z^8 - 11z + 1 = 0$ may be written as $f(z) + g(z) = 0$, where

$$f(z) = -11z$$

and

$$g(z) = z^8 + 1.$$

If $\zeta \in \{\gamma\} = \partial U(0;1)$, then

$$|f(\zeta)| = |-11\zeta| = 11,$$

and

$$|g(\zeta)| = |\zeta^8 + 1| \leq |\zeta^8| + 1 = 2,$$

i. e.,

$$|g(\zeta)| \leq 2 < 11 = |f(\zeta)|, \quad \forall \zeta \in \{\gamma\}.$$

From the *Rouché theorem*, we obtain that the given equation has one zero in the unit disc, because

$$\Theta(f+g, U(0;1)) = \Theta(f, U(0;1)) = 1.$$

3. The equation $23z^5 - 10z^3 + 6z - 5 = 0$ may be written as $f(z) + g(z) = 0$, where

$$f(z) = 23z^5$$

and

$$g(z) = -10z^3 + 6z - 5.$$

If $\zeta \in \{\gamma\} = \partial U(0;1)$, then

$$|f(\zeta)| = |23\zeta^5| = 23,$$

and

$$|g(\zeta)| = |-10\zeta^3 + 6\zeta - 5| \leq 21,$$

i. e.,

$$|g(\zeta)| \leq 21 < 23 = |f(\zeta)|, \quad \forall \zeta \in \{\gamma\}.$$

From the *Rouché theorem*, we obtain that the given equation has five zeros in the unit disc.

4. For the equation $e^z - 3iz = 0$, let

$$f(z) = -3iz, \quad g(z) = e^z,$$

where $g(z) + f(z) = 0$. If $\zeta \in \{\gamma\} = \partial U(0;1)$, then

$$|f(\zeta)| = |-3i\zeta| = 3,$$

and

$$|g(\zeta)| = |e^{\zeta}| \leq e.$$

We deduce that

$$|g(\zeta)| \leq e < 3 = |f(\zeta)|, \quad \forall \zeta \in \{\gamma\}.$$

From the *Rouché theorem*, we obtain that the given equation has one zero in the unit disc, because

$$\Theta(f+g, U(0;1)) = \Theta(f, U(0;1)) = 1.$$

5. The equation $z^n + 8z^2 + 1 = 0$, $n \in \mathbb{N}$, $n \geq 3$, may be written as $f(z) + g(z) = 0$, where

$$f(z) = 8z^2, \quad g(z) = z^n + 1.$$

If $\zeta \in \{\gamma\} = \partial U(0;1)$, then

$$|f(\zeta)| = |8\zeta^2| = 8,$$

and

$$|g(\zeta)| = |\zeta^n + 1| \leq |\zeta^n| + 1 = 2,$$

i. e.,

$$|g(\zeta)| \leq 2 < 8 = |f(\zeta)|, \quad \forall \zeta \in \{\gamma\}.$$

From the *Rouché theorem*, we obtain that the given equation has two zeros in the unit disc.

6. The equation $z^n + 3z^2 + 1 = 0$, $n \in \mathbb{N}$, $n \geq 3$, may be written as $f(z) + g(z) = 0$, where

$$f(z) = 3z^2$$

and

$$g(z) = z^n + 1.$$

If $\zeta \in \{\gamma\} = \partial U(0;1)$, then

$$|f(\zeta)| = |3\zeta^2| = 3,$$

and

$$|g(\zeta)| = |\zeta^n + 1| \leq |\zeta^n| + 1 = 2,$$

i. e.,

$$|g(\zeta)| \leq 2 < 3 = |f(\zeta)|, \quad \forall \zeta \in \{\gamma\}.$$

From the *Rouché theorem*, we obtain that the given equation has two zeros in the unit disc, because

$$\Theta(f+g, U(0;1)) = \Theta(f, U(0;1)) = 2.$$

7. Similar to point 4, the equation has the form

$$f(z) + g(z) = 0,$$

where

$$f(z) = -4z^n,$$

and

$$g(z) = e^z + 1.$$

If $\zeta \in \{\gamma\} = \partial U(0;1)$, then

$$|f(\zeta)| = |-4\zeta^n| = 4,$$
$$|g(\zeta)| = |e^\zeta + 1| \le e + 1,$$

i. e.,

$$|g(\zeta)| \le e + 1 < 4 = |f(\zeta)|, \quad \forall \zeta \in \{\gamma\}.$$

From the *Rouché theorem*, the given equation has n zeros in the unit disc.

8. If $z^n + pz^2 + qz + r = 0$, $n \in \mathbb{N}$, $n \ge 3$, where $|p| > |q| + |r| + 1$, let

$$f(z) = pz^2$$

and

$$g(z) = z^n + qz + r.$$

If $\zeta \in \{\gamma\} = \partial U(0;1)$, we deduce that

$$|f(\zeta)| = |p\zeta^2| = |p|,$$

and

$$|g(\zeta)| = |\zeta^n + q\zeta + r| \le 1 + |q| + |r|.$$

Because

$$|g(\zeta)| \le 1 + |q| + |r| < |p| = |f(\zeta)|, \quad \forall \zeta \in \{\gamma\},$$

according to the *Rouché theorem*, the given equation has two zeros in the unit disc.

9. The equation $z^{10} - 9z^6 + 3z^3 + z^2 - 2 = 0$ has the form $f(z) + g(z) = 0$, where

$$f(z) = -9z^6$$

and

$$g(z) = z^{10} + 3z^3 + z^2 - 2.$$

If $\zeta \in \{\gamma\} = \partial U(0; 1)$, we get

$$|f(\zeta)| = |-9\zeta^6| = 9,$$

and

$$|g(\zeta)| = |\zeta^{10} + 3\zeta^3 + \zeta^2 - 2| \leq 7.$$

From here, it follows that

$$|g(\zeta)| \leq 7 < 9 = |f(\zeta)|, \quad \forall \zeta \in \{\gamma\},$$

according to the *Rouché theorem*, the given equation has six zeros in the unit disc, because

$$\Theta(f + g, U(0; 1)) = \Theta(f, U(0; 1)) = 6.$$

Solution of Exercise 5.4.12

To solve this problem, we will use the *Rouché theorem*. Let

$$f(z) = az^n,$$

and

$$g(z) = -e^z.$$

Since we need to determine the number of the zeros from the disc $U(0; r)$, if $\zeta \in \{\gamma\} = \partial U(0; r)$, then

$$|f(\zeta)| = |a\zeta^n| = |a|r^n,$$

and

$$|g(\zeta)| = |-e^\zeta| \leq e^r.$$

From the assumption $|a|r^n > e^r$, we get

$$|g(\zeta)| \leq e^r < |a|r^n = |f(\zeta)|, \quad \forall \zeta \in \{\gamma\},$$

and according to the *Rouché theorem*, the equation has n zeros in the disc $U(0; r)$.

Solution of Exercise 5.4.13

We will again use the *Rouché theorem*. The solution consists of the following two steps: we will determine the number of the roots in the bigger circular ring, then in the smaller circular ring, and the result will be the difference between these numbers.

1. (i) First, we will determine the number of the roots in the disc $U(0;1)$.

The left-hand side of the equation may be written like the sum of the next functions:

$$f(z) = -9z,$$
$$g(z) = z^4 + 1.$$

If $\zeta \in \{\gamma_1\} = \partial U(0;1)$, then

$$|f(\zeta)| = |-9\zeta| = 9,$$

and

$$|g(\zeta)| = |\zeta^4 + 1| \leq 2.$$

It follows that

$$|g(\zeta)| \leq 2 < 9 = |f(\zeta)|, \quad \forall \zeta \in \{\gamma_1\},$$

and using the *Rouché theorem* we obtain that the equation has one root in the disc $U(0;1)$.

(ii) Now we will determine the number of the roots in the disc $U(0;3)$.

The left-hand side of the equation may be written like the sum of the next functions:

$$f(z) = z^4,$$
$$g(z) = -9z + 1.$$

If $\zeta \in \{\gamma_2\} = \partial U(0;3)$, then

$$|f(\zeta)| = |\zeta^4| = 3^4 = 81,$$

and

$$|g(\zeta)| = |-9\zeta + 1| \leq |-9\zeta| + 1 = 28,$$

hence

$$|g(\zeta)| \leq 28 < 81 = |f(\zeta)|, \quad \forall \zeta \in \{\gamma_2\}.$$

Using the *Rouché theorem,* we obtain that the equation has four roots in the disc $U(0;3)$.

In conclusion, the equation has three roots in the circular ring $U(0;1,3)$.

2. (i) We will determine the number of the roots in the disc $U(0;1)$.

The left-hand side of the equation may be written like the sum of the next functions:

$$f(z) = -7z^3,$$
$$g(z) = 3z^5 + z^2 - 2.$$

If $\zeta \in \{\gamma_1\} = \partial U(0;1)$, then

$$|f(\zeta)| = |-7\zeta^3| = 7,$$
$$|g(\zeta)| = |3\zeta^5 + \zeta^2 - 2| \le 6.$$

From here, we get

$$|g(\zeta)| \le 6 < 7 = |f(\zeta)|, \quad \forall \zeta \in \{\gamma_1\},$$

and from the *Rouché theorem* we obtain

$$\Theta(f + g, U(0;1)) = \Theta(f, U(0;1)) = 3.$$

(ii) Next, we will determine the number of the roots in the disc $U(0;2)$.

The left-hand side of the equation may be written like the sum of the next functions:

$$f(z) = 3z^5,$$
$$g(z) = -7z^3 + z^2 - 2.$$

If $\zeta \in \{\gamma_2\} = \partial U(0;2)$, then

$$|f(\zeta)| = |3\zeta^5| = 3 \cdot 2^5 = 96,$$

and

$$|g(\zeta)| = |-7\zeta^3 + \zeta^2 - 2| \le |-7\zeta^3| + |\zeta^2| + 2 = 62,$$

hence

$$|g(\zeta)| \le 62 < 96 = |f(\zeta)|, \quad \forall \zeta \in \{\gamma_2\}.$$

Using the *Rouché theorem,* we obtain

$$\Theta(f + g, U(0;2)) = \Theta(f, U(0;2)) = 5.$$

We conclude that

$$\Theta(f+g, U(0;1,2)) = 5-3 = 2,$$

hence the equation has two roots in the circular ring $U(0;1,2)$.

3. (i) We will determine the number of the roots in the disc $U(0;1)$.

The left-hand side of the equation may be written like the sum of the next functions:

$$f(z) = -4z,$$
$$g(z) = z^4 + z^3 + 1.$$

If $\zeta \in \{\gamma_1\} = \partial U(0;1)$, we have

$$|f(\zeta)| = |-4\zeta| = 4,$$

and

$$|g(\zeta)| = |\zeta^4 + \zeta^3 + 1| \le 3.$$

It follows that

$$|g(\zeta)| \le 3 < 4 = |f(\zeta)|, \quad \forall \zeta \in \{\gamma_1\},$$

and using the *Rouché theorem* we obtain

$$\Theta(f+g, U(0;1)) = \Theta(f, U(0;1)) = 1.$$

(ii) Next, we will determine the number of the roots in the disc $U(0;3)$.

The left-hand side of the equation may be written like the sum of the next functions:

$$f(z) = z^4,$$
$$g(z) = z^3 - 4z + 1.$$

If $\zeta \in \{\gamma_2\} = \partial U(0;3)$, we get

$$|f(\zeta)| = |\zeta^4| = 3^4 = 81,$$

and

$$|g(\zeta)| = |\zeta^3 - 4\zeta + 1| \le |\zeta^3| + |-4\zeta^2| + 1 = 40.$$

Thus

$$|g(\zeta)| \le 40 < 81 = |f(\zeta)|, \quad \forall \zeta \in \{\gamma_2\},$$

and using the *Rouché theorem* we obtain

$$\Theta(f+g, U(0;2)) = \Theta(f, U(0;2)) = 4.$$

We conclude that

$$\Theta(f+g, U(0;1,3)) = 4-1 = 3,$$

hence the equation has three roots in the circular ring $U(0;1,3)$.

Solution of Exercise 5.4.14

1. We will use the *Rouché theorem* for the functions

$$f(z) = z,$$

and

$$g(z) = -ae^z.$$

Since we need to determine the number of the roots in the disc $U(0;1)$, if $\zeta \in \{\gamma\} = \partial U(0;1)$ then

$$\begin{aligned}|f(\zeta)| &= |\zeta| = 1,\\ |g(\zeta)| &= |-ae^\zeta| \le |a|e.\end{aligned}$$

From the assumption $|a|e < 1$, we get

$$|g(\zeta)| \le |a|e < 1 = |f(\zeta)|, \quad \forall \zeta \in \{\gamma\}.$$

It follows that the equation has only one root in the disc $U(0;1)$.

2. The proof is identical to the previous one. The left-hand side of the equation $z^2 = ae^z$ may be written as the sum of

$$f(z) = z^2,$$

and

$$g(z) = -ae^z.$$

Since we need to determine the number of the roots in the disc $U(0;1)$, if $\zeta \in \{\gamma\} = \partial U(0;1)$ then

$$\begin{aligned}|f(\zeta)| &= |\zeta^2| = 1,\\ |g(\zeta)| &= |-ae^\zeta| \le |a|e.\end{aligned}$$

From the assumption $|a|e < 1$, we get

$$|g(\zeta)| \le |a|e < 1 = |f(\zeta)|, \quad \forall \zeta \in \{\gamma\}.$$

It follows that the equation has two roots in the disc $U(0;1)$.

7.6 Solutions to the exercises of Chapter 6

Solution of Exercise 6.5.1

1. It is sufficient to prove that the function f is bijective, $f \in H(D)$ and $f'(z) \neq 0, \forall z \in D$. If $\forall \zeta = u + iv \in \Omega$, then $v > 0$ and

$$f(z) = \zeta \Leftrightarrow iz = \zeta \Leftrightarrow \exists! z = -i\zeta = v - iu \in D.$$

It follows that f is bijective. But $f \in H(D)$ and $f'(z) = i \neq 0, \forall z \in D$, thus the conclusion of the problem holds, i. e., the function f maps conformally the half-plane D onto the half-plane Ω.

2. Let now $g : \Omega \to D, g(z) = -iz$. For $\forall \zeta = u + iv \in D$, then $u > 0$ and

$$g(z) = \zeta \Leftrightarrow -iz = \zeta \Leftrightarrow \exists! z = i\zeta = -v + iu \in \Omega.$$

It follows that g is bijective. But $g \in H(\Omega)$ and $g'(z) = -i \neq 0, \forall z \in \Omega$, thus the conclusion of the problem holds, i. e., the function g maps conformally the half-plane Ω onto the half-plane D.

Solution of Exercise 6.5.2

We know that conformal mappings maps the domains onto the domains, and maps the boundary of the domain onto the boundary of the image domain. These properties will be used in the following solution. The boundary of the domain D is $\partial D = \widehat{AB} \cup [BA]$, where

$$\widehat{AB} : \gamma_{\widehat{AB}}(t) = e^{i\pi\frac{2t-1}{2}}, \quad t \in [0, 1],$$

and

$$[BA] : \gamma_{[BA]}(t) = i(1 - t) + t(-i) = i(1 - 2t), \quad t \in [0, 1].$$

The image of ∂D will be

$$f(\widehat{AB}) = \{f(e^{i\pi\frac{2t-1}{2}}) : t \in [0, 1]\} = \{e^{i\pi(2t-1)} : t \in [0, 1]\} = \partial U(0; 1),$$

that represents the unit circle, and

$$f([BA]) = \{f(i - 2ti) : t \in [0, 1]\} = \{-(2t - 1)^2 : t \in [0, 1]\}.$$

Thus, if

$$t \in \left[0, \frac{1}{2}\right] \Rightarrow f(i - 2ti) = [-1, 0],$$
$$t \in \left[\frac{1}{2}, 1\right] \Rightarrow f(i - 2ti) = [0, -1],$$

hence

$$f([BA]) = [-1,0] \cup [0,-1].$$

We need to determine in which of the domains bounded by $\partial\Omega$, the function f maps the domain D. For this, let us choose an arbitrary point in D, and we will check if its image belongs to Ω:

$$\frac{1}{2} \in D \Rightarrow f\left(\frac{1}{2}\right) = \frac{1}{4} \in \Omega.$$

Thus, the function $f(z) = z^2$ conformally maps the domain D onto the domain Ω.

Solution of Exercise 6.5.3

We will try to determine a circular transform f that conformally maps the half-plane D onto the disc Ω. It is well known that a circular transform maps the circles from $\mathbb{C}_\infty$ onto the circles from $\mathbb{C}_\infty$, hence it is sufficient to determine the image of the three points from the boundary of the domain of definition. Thus

$$\{-1,0,1\} \subset \partial D \Rightarrow \{g(-1) = i, g(-0) = -1, g(1) = -i\} \subset \partial\Omega,$$

and

$$i \in D \Rightarrow g(i) = 0 \in \Omega.$$

It follows that the function $f = g|_D$ conformally maps the half-plane D onto the disc Ω.

Solution of Exercise 6.5.4

Similarly, we will determine a circular transform f that conformally maps the domain D onto the domain Ω. Thus

$$\{-i,-1,i\} \subset \partial D \Rightarrow \{f(-i) = 1, f(-1) = 0, f(i) = -1\} \subset \partial\Omega,$$

where

$$\partial D = \partial U(0;1), \quad \partial\Omega = \{z \in \mathbb{C} : \operatorname{Im} z = 0\},$$

and let consider a point from D:

$$0 \in D \Rightarrow f(-0) = i \in \Omega.$$

It follows that the function f conformally maps the disc D onto the half-plane Ω.

Solution of Exercise 6.5.5

We will determine a circular transform f that conformally maps the domain D onto the domain, but we cannot use the previous method because the boundary of the domain of the definition is not a circle from $\mathbb{C}_\infty$. We will follow the solution of Exercise 6.5.2, since a conformal mappings maps the domains onto the domains, and maps the boundary of the domain onto the boundary of the image domain. Thus

$$\begin{aligned}\partial D &= \{z \in \mathbb{C} : \operatorname{Re} z = 0, \operatorname{Im} z \geq 0\} \cup \{z \in \mathbb{C} : \operatorname{Re} z \leq 0, \operatorname{Im} z = 0\}\\ &= \{x + iy \in \mathbb{C} : x = 0, y \geq 0\} \cup \{x + iy \in \mathbb{C} : x \leq 0, y = 0\}.\end{aligned}$$

Case 1. If $z \in \{x + iy \in \mathbb{C} : x = 0, y \geq 0\}$, then

$$f(z) = f(iy) = \frac{yi - i}{yi + i} = \frac{y-1}{y+1}, \quad y \geq 0.$$

Let

$$f_1 : [0, +\infty) \to \mathbb{R}, \quad f_1(y) = \frac{y-1}{y+1},$$

and we need to determine the image of f_1. Since

$$f_1'(y) = \frac{2}{(y+1)^2} > 0, \quad \forall y \in [0, +\infty)$$

and

$$\lim_{y\to+\infty} f_1(y) = 1, \quad \lim_{y\downarrow 0} f_1(y) = -1,$$

we get

$$f(\{x + iy \in \mathbb{C} : x = 0, y \geq 0\}) = [-1, 1].$$

Case 2. If $z \in \{x + iy \in \mathbb{C} : x \leq 0, y = 0\}$, then the function will be

$$f(z) = f(x) = \frac{x-i}{x+i} = \frac{x^2-1}{x^2+1} - \frac{2x}{x^2+1}i, \quad x \leq 0,$$

hence

$$|f(z)| = \sqrt{\left(\frac{x^2-1}{x^2+1}\right)^2 + \left(\frac{2x}{x^2+1}\right)^2} = 1.$$

We have obtained that the images $f(z)$ that belong to the unit circle. We need to prove that these images are only on the required half-circle.

Since $\operatorname{Im} f(z) = -\frac{2x}{x^2+1}$, let

$$f_2 : (-\infty, 0] \to \mathbb{R}, \quad f_2(x) = -\frac{2x}{x^2-1}.$$

Then

$$f_2'(x) = 2\frac{x^2-1}{(x^2+1)^2}, \quad x \in (-\infty, 0]$$

and

$$\lim_{x\to-\infty} f_2(x) = 0, \quad \lim_{x\uparrow 0} f_2(x) = 0, \quad f_2(-1) = 1,$$

hence we have that

$$f(\{x+iy \in \mathbb{C} : x \le 0, y = 0\}) = \{z \in \mathbb{C} : |z| = 1, \operatorname{Im} z \ge 0\}.$$

From both of these cases, we deduce that

$$f(\partial D) = \partial \Omega.$$

We will choose a point from the domain D, i. e.,

$$-1+i \in D \Rightarrow f(-1+i) = \frac{1}{5} + \frac{2}{5}i \in \Omega,$$

which completes our proof.

Solution of Exercise 6.5.6

We will use a similar method as in Exercise 6.5.5. We will determine the image of the boundary of the domain a D, i. e.,

$$\partial D = \{e^{i\theta} \in \mathbb{C} : \theta \in [0, \pi]\} \cup \{x \in \mathbb{R} : x \in [-1, 1]\}.$$

Case 1. If $z \in \{e^{i\theta} \in \mathbb{C} : \theta \in [0, \pi]\}$, the image is

$$f(z) = f(e^{i\theta}) = \frac{1}{2}\left(e^{i\theta} + \frac{1}{e^{i\theta}}\right) = \cos\theta, \quad \theta \in [0, \pi],$$

i. e.,

$$\{f(e^{i\theta}) : \theta \in [0, \pi]\} = [-1, 1].$$

Case 2. if $z \in \{x \in \mathbb{R} : x \in [-1, 1]\}$, then

$$f(z) = f(x) = \frac{1}{2}\left(x + \frac{1}{x}\right) = \frac{1}{2}\frac{x^2+1}{x}, \quad x \in [-1, 1].$$

Let

$$f_1 : [-1,1] \setminus \{0\} \to \mathbb{R}, \quad f_1(x) = \frac{1}{2}\frac{x^2+1}{x},$$

hence we need to find the image of f_1. But

$$f_1'(x) = \frac{1}{2}\frac{x^2-1}{x^2} \Rightarrow f_1 \quad \text{is decreasing on } [-1,0) \text{ and } (0,1],$$

and

$$\lim_{x\downarrow -1} f_1(x) = -1, \quad \lim_{x\uparrow 0} f_1(x) = -\infty,$$
$$\lim_{x\downarrow 0} f_1(x) = +\infty, \quad \lim_{x\uparrow 1} f_1(x) = 1,$$

thus

$$f([-1,1] \setminus \{0\}) = \mathbb{R} \setminus [-1,1].$$

From the above results, we get

$$f(\partial D) = \partial\Omega.$$

The image of an arbitrary point from D is

$$\frac{i}{2} \in D \Rightarrow f\left(\frac{i}{2}\right) = -\frac{3i}{4} \in \Omega,$$

hence the Joukowski function conformally maps the domain D onto the half-plane Ω.

Solution of Exercise 6.5.7

We will use a similar proof as in Exercise 6.5.6, by determining the image of the boundary of the domain D, where

$$\partial D = (\mathbb{R} \setminus [-1,1]) \cup \{e^{i\theta} \in \mathbb{C} : \theta \in [0,\pi]\}.$$

Case 1. If $z \in \mathbb{R} \setminus [-1,1]$, then

$$g(z) = g(x) = x + \frac{1}{x} = \frac{x^2+1}{x}, \quad x \in \mathbb{R} \setminus [-1,1].$$

If

$$g_1 : \mathbb{R} \setminus [-1,1] \to \mathbb{R}, \quad g_1(x) = \frac{x^2+1}{x},$$

we will determine the image of the function g_1. Since

$$g_1'(x) = \frac{x^2 - 1}{x^2} \Rightarrow g_1 \quad \text{is increasing in } (-\infty, -1) \text{ and } (1, +\infty),$$

and

$$\lim_{x \to -\infty} g_1(x) = -\infty, \quad \lim_{x \to -1} g_1(x) = -2,$$
$$\lim_{x \to 1} g_1(x) = 2, \quad \lim_{x \to +\infty} g_1(x) = +\infty,$$

we get

$$g(\mathbb{R} \setminus [-1, 1]) = \mathbb{R} \setminus [-2, 2].$$

Case 2. If $z \in \{e^{i\theta} \in \mathbb{C} : \theta \in [0, \pi]\}$, the image of the function will be

$$g(z) = g(e^{i\theta}) = e^{i\theta} + \frac{1}{e^{i\theta}} = 2\cos\theta, \quad \theta \in [0, \pi],$$

i. e.,

$$g(\{e^{i\theta} \in \mathbb{C} : \theta \in [0, \pi]\}) = [-2, 2].$$

Combining the results of the both cases, we deduce that the function g maps the boundary of the domain D onto the boundary of the domain Ω. The image of a point that belongs to the domain D is

$$2i \in D \Rightarrow f(2i) = \frac{3i}{2} \in \Omega,$$

thus the function $f = g|_D$ conformally maps the domain D onto the half-plane Ω.

Solution of Exercise 6.5.8

We will determine the required function by using the images of some well-known elementary functions. It is easy to prove that

$$g(D) = \{z \in \mathbb{C} : \operatorname{Im} z > 0\}, \quad \text{where } g(z) = \left(\frac{z-1}{z+1}\right)^2.$$

According to Exercise 6.5.3, we have

$$h(\{z \in \mathbb{C} : \operatorname{Im} z > 0\}) = \Omega, \quad \text{where } h(z) = \frac{z - i}{z + i},$$

and composing both of the above functions we get $f = h \circ g$, i. e.,

$$f(z) = \frac{(z-1)^2 - i(z+1)^2}{(z-1)^2 + i(z+1)^2}.$$

Solution of Exercise 6.5.9

We will use a similar proof as in Exercise 6.5.6. If

$$\partial D = \{z = e^{i\theta} \in \mathbb{C} : \theta \in [0, 2\pi]\},$$

then

$$z \in \partial D \Rightarrow f(z) = f(e^{i\theta}) = \cos\theta, \quad \theta \in [0, 2\pi].$$

For the function $f_1(\theta) = \cos\theta$, $\theta \in [0, 2\pi]$, we have

$$\lim_{\theta \downarrow 0} f_1(\theta) = 1, \quad \lim_{\theta \to \pi} f_1(\theta) = -1, \quad \lim_{\theta \to 2\pi} f_1(\theta) = 1,$$

thus

$$f(\partial U(0;1)) = [-1, 1].$$

The image of a point that belongs to the domain D is

$$2 \in D \Rightarrow f(2) = \frac{5}{4} \in \Omega.$$

But

$$f'(z) = \frac{z^2 - 1}{2z^2} \neq 0, \quad \forall z \in D,$$

hence the Joukowski function conformally maps the domain D onto the domain Ω.

Solution of Exercise 6.5.10

It is well known that a conformal mappings maps the domains onto the domains, and maps the boundary of the domain onto the boundary of the image domain. Thus, the boundary of the domain D is

$$\partial D = \{z \in \mathbb{C} : \operatorname{Re} z = 0\}.$$

The image of a point that belongs to this boundary is

$$z = iy \in \partial D \Rightarrow f(iy) = (iy)^2 = -y^2, \quad y \in \mathbb{R}.$$

If we let $f_1(y) = -y^2$, then

$$\lim_{y \to -\infty} f_1(y) = -\infty, \quad \lim_{y \to 0} f_1(y) = 0, \quad \lim_{y \to +\infty} f_1(y) = +\infty.$$

From here, we get

$$f(\partial D) = (-\infty, 0] = \partial\Omega,$$

and the image of a point that belongs to the domain D will be

$$1 \in D \Rightarrow f(1) = 1 \in \Omega.$$

Hence the function f conformally maps the half-plane D onto the domain Ω.

Solution of Exercise 6.5.11

Similar to the proof of Exercise 6.5.8, the required function will be determined by using the images of some known functions. Using the result of Exercise 6.5.6, we know that

$$f_1(D) = \{z \in \mathbb{C} : \operatorname{Im} z < 0\} = \Delta_1, \quad \text{where } f_1(z) = \frac{1}{2}\left(z + \frac{1}{z}\right).$$

Rotating the domain Δ_1 with the angle π we get

$$f_2(\Delta_1) = \{z \in \mathbb{C} : \operatorname{Im} z > 0\} = \Delta_2, \quad \text{where } f_2(z) = e^{i\pi} z = -z.$$

Finally, we will use the result of Exercise 6.5.3, i. e.,

$$f_3(\Delta_2) = \Omega, \quad \text{where } f_3(z) = \frac{z - i}{z + i}.$$

Composing the above functions, we get $f = f_3 \circ f_2 \circ f_1$, thus

$$f(z) = \frac{z^2 + 2iz + 1}{z^2 - 2iz + 1}.$$

Solution of Exercise 6.5.12

We will use a similar solution as in the previous problem. Multiplying the argument by 3, we have

$$f_1(D) = \{z \in \mathbb{C} : |z| < 2, \operatorname{Im} z > 0\} = \Delta_1, \quad \text{where } f_1(z) = z^3.$$

The function

$$f_2 : \Delta_1 \to \{z \in \mathbb{C} : |z| < 1, \operatorname{Im} z > 0\} = \Delta_2, \quad f_2(z) = \frac{z}{2}$$

divide the modules by 2 and $f_2(\Delta_1) = \Delta_2$. According to Exercise 6.5.6 we get

$$f_3(\Delta_2) = \{z \in \mathbb{C} : \operatorname{Im} z < 0\} = \Delta_3, \quad f_3(z) = \frac{1}{2}\left(z + \frac{1}{z}\right).$$

Rotating the domain Δ_3 with the angle π, we have

$$f_4(\Delta_3) = \Omega, \quad f_4(z) = e^{i\pi} z = -z.$$

From the above results, we conclude that $f = f_4 \circ f_3 \circ f_2 \circ f_1$, hence

$$f(z) = -\frac{z^6 + 4}{4z^3}.$$

Solution of Exercise 6.5.13

The function will be determined by using the images of some known functions. First, if

$$f_1 : D \to \{z \in \mathbb{C} : 0 < \operatorname{Re} z < 1\} = \Delta_1, \quad f_1(z) = \frac{z-a}{b-a},$$

we easily prove that $f_1(D) = \Delta_1$.

The domain Δ_1 is independent on the parameters a and b. Rotating the domain Δ_1 with the angle $\frac{\pi}{2}$ and multiplying the modules by π, we get

$$f_2 : \Delta_1 \to \{z \in \mathbb{C} : 0 < \operatorname{Im} z < \pi\} = \Delta_2, \quad f_2(z) = \pi e^{i\frac{\pi}{2}} z = \pi i z,$$

and $f_2(\Delta_1) = \Delta_2$.

Finally, using the function

$$f_3 : \Delta_2 \to \Omega, \quad \text{where } f_3(z) = e^z,$$

we get the required domain, because $f_3(\Delta_2) = \Omega$. The function will be $f = f_3 \circ f_2 \circ f_1$, i. e.,

$$f(z) = e^{\frac{\pi i}{b-a}(z-a)}.$$

Solution of Exercise 6.5.14

We will determine a circular transform f that conformally maps the half-plane D onto the disc Ω, using the fact that a circular transform maps the circles from $\mathbb{C}_\infty$ onto the circles from $\mathbb{C}_\infty$. It is sufficient to determine the image of the three points from the boundary of the domain of definition. The boundary of the domain D consists in the union of two circles, i. e.,

$$\partial D = \partial U(0; 2r) \cup \partial U(a; r).$$

To find the image of one of these circles, it is sufficient to calculate the images of the three points from the circle:

$$\{2r, 2ri, -2r\} \subset \partial U(0; 2r)$$

$$\Rightarrow f(2r) = \frac{r}{r-a}, \quad f(2ri) = \frac{r^2}{a^2+r^2} - i\frac{ar}{a^2+r^2}, \quad f(-2r) = \frac{r}{r+a},$$

where $|a| = r$, i. e., $a = re^{i\theta}$.

Thus

$$f(2r) = \frac{r}{r-a} = \frac{1}{2} + i\frac{\sin\theta}{2 - 2\cos\theta} \Rightarrow \operatorname{Re} f(2ri) = \frac{1}{2}.$$

We get similarly

$$\operatorname{Re} f(2ri) = \frac{1}{2}, \quad \operatorname{Re} f(-2r) = \frac{1}{2},$$

and thus, we deduce

$$f(\partial U(0; 2r)) = \left\{z \in \mathbb{C} : \operatorname{Re} z = \frac{1}{2}\right\}.$$

Now we will determine the image of the circle $\partial U(a; r)$:

$$\{0, r-a, -r-a\} \subset \partial U(a; r) \Rightarrow f(0) = 0, \quad f(r-a) = \frac{r-a}{r-3a}, \quad f(-r-a) = \frac{r+a}{r+3a},$$

where $a = re^{i\theta}$, hence

$$\operatorname{Re} f(r-a) = 0, \quad \operatorname{Re} f(-r-a) = 0.$$

Thus

$$f(\partial U(a; r)) = \{z \in \mathbb{C} : \operatorname{Re} z = 0\}.$$

Combining the above results, we have $f(\partial D) = \partial\Omega$. The image of a point from D will be

$$-a \in D \Rightarrow f(-a) = \frac{1}{3} \in \Omega,$$

hence the function $f(z) = \frac{z}{z-2a}$ conformally maps the domain D onto the domain Ω.

Solution of Exercise 6.5.15

As we have previously mentioned, the required function will be determined by using the images of some known functions. Using the result of Exercise 6.5.14, consider the special case

$$f_1 : D \to \left\{z \in \mathbb{C} : 0 < \operatorname{Re} z < \frac{1}{2}\right\}, \quad \text{with } f_1(z) = \frac{z}{z-2i},$$

obtained for $r = 1$ and $a = i$.

Considering the special case of Exercise 6.5.13,

$$f_2 : \left\{z \in \mathbb{C} : 0 < \operatorname{Re} z < \frac{1}{2}\right\} \to \Omega, \quad \text{with } f_2(z) = e^{2\pi i z},$$

obtained for $a = 0$ and $b = \frac{1}{2}$, from these two results we conclude that the function will be $f = f_2 \circ f_1$, i. e.,

$$f(z) = e^{\frac{2\pi i z}{z-2i}}.$$

Solution of Exercise 6.5.16

The function will be determined by using the images of some known functions. Using the result of Exercise 6.5.14, we have

$$f_1(D) = \left\{z \in \mathbb{C} : 0 < \operatorname{Re} z < \frac{1}{2}, \operatorname{Im} z > 0\right\} = \Delta_1, \quad f_1(z) = \frac{z}{z+r},$$

where $a = -\frac{r}{2}$.

Multiplying the modules of the numbers of the domain Δ_1 by 2π, and translating by $-\frac{\pi}{2}$, i. e.,

$$f_2 : \Delta_1 \to \left\{z \in \mathbb{C} : -\frac{\pi}{2} < \operatorname{Re} z < \frac{\pi}{2}, \operatorname{Im} z > 0\right\} = \Delta_2, \quad f_2(z) = 2\pi z - \frac{\pi}{2},$$

we get $f_2(\Delta_1) = \Delta_2$.

The function

$$f_3 : \Delta_2 \to \Omega, \quad f_3(z) = \sin z$$

has the property that $f_3(\Delta_2) = \Omega$, and the result will be given by $f = f_3 \circ f_2 \circ f_1$, i. e.,

$$f(z) = \sin\left(\frac{2\pi z}{z+r} - \frac{\pi}{2}\right).$$

Solution of Exercise 6.5.17

Similar to the previous problem, according to Exercise 6.5.14 we have

$$f_1(D) = \left\{z \in \mathbb{C} : 0 < \operatorname{Re} z < \frac{1}{2}\right\} = \Delta_1, \quad f_1(z) = \frac{z}{z-r},$$

where $a = \frac{r}{2}$.

Multiplying the modules of the numbers of the domain Δ_1 by 2π, and rotating with the angle π, we get

$$f_2 : \Delta_1 \to \{z \in \mathbb{C} : 0 < \operatorname{Im} z < \pi\} = \Delta_2, \quad f_2(z) = 2\pi i z,$$

hence $f_2(\Delta_1) = \Delta_2$.

Similar to Exercise 6.5.13, the function

$$f_3 : \Delta_2 \to \Omega, \quad f_3(z) = e^z$$

has the property $f_3(\Delta_2) = \Omega$. Thus, the required function will be $f = f_3 \circ f_2 \circ f_1$, i. e.,

$$f(z) = e^{\frac{2\pi i z}{z-r}}.$$

Solution of Exercise 6.5.18

The function

$$f_1 : D \to \{z \in \mathbb{C} : |z| < 1, 0 < \arg z < \pi\} = \Delta_1, \quad f_1(z) = z^2$$

doubles the arguments, hence $f_1(D) = \Delta_1$. Using now the result obtained to Exercise 6.5.11, we have that

$$f_2 : \Delta_1 \to \Omega, \quad f_2(z) = \frac{z^2 + 2iz + 1}{z^2 - 2iz + 1}$$

satisfies $f_2(\Delta_1) = \Omega$. From the above results, we conclude that $f = f_2 \circ f_1$, i. e., $f(z) = \frac{z^4+2iz^2+1}{z^4-2iz^2+1}$.

Solution of Exercise 6.5.19

We will prove that the function

$$f_1 : D \to \mathbb{C} \setminus \{z \in \mathbb{C} : \operatorname{Im} z = 0, \operatorname{Re} z \geq 0\} = \Delta_1, \quad f_1(z) = \frac{z-a}{z-b},$$

conformally maps the domain D onto the domain Ω. The boundary of the domain D is $\partial D = (-\infty, a] \cup [b, +\infty)$.

The function

$$f_1^* : (-\infty, a] \cup (b, +\infty) \to \mathbb{R}, \quad f_1^*(x) = \frac{x-a}{x-b}$$

is decreasing on $(-\infty, a]$ and $(b, +\infty)$. We also have

$$\lim_{x\to-\infty} f_1^*(x) = 1, \quad \lim_{x\to a} f_1^*(x) = 0,$$

and

$$\lim_{x\downarrow b} f_1^*(x) = +\infty, \qquad \lim_{x\to+\infty} f_1^*(x) = 1,$$

because $b > a$. From here, it follows that $f_1^*((-\infty, a] \cup (b, +\infty)) = [0, +\infty)$, thus we proved that $f_1(D) = \mathbb{C} \setminus [0, +\infty) = \Delta_1$.

Now we will return to our problem, where we divide the argument by $\frac{1}{2}$. If

$$f_2 : \Delta_1 \to \Omega, \quad f_2(z) = \sqrt{z},$$

where $f_2(-1) = \sqrt{-1} = i$, then $f_2(\Delta_1) = \Omega$. Composing both of these functions, we have $f = f_2 \circ f_1$, i. e.,

$$f(z) = \sqrt{\frac{z-a}{z-b}}, \quad \text{where } f\left(\frac{a+b}{2}\right) = \sqrt{-1} = i.$$

Solution of Exercise 6.5.20

First, we will make a translation with 1, then we will divide the argument by $\frac{1}{2}$, i. e.,

$$f_1(D) = \{z \in \mathbb{C} : \operatorname{Im} z > 0\} = \Delta_1, \quad f_1(z) = \sqrt{z-1},$$

where $f_1(0) = \sqrt{-1} = i$.

Using now the result of Exercise 6.5.3, we get

$$f_2(\Delta_1) = \Omega, \quad f_2(z) = \frac{z-i}{z+i}.$$

The required function will be $f = f_2 \circ f_1$, i. e.,

$$f(z) = \frac{\sqrt{z-1} - i}{\sqrt{z-1} + i}, \quad \text{where } f(0) = 0.$$

Solution of Exercise 6.5.21

The proof of the first point will be used for all the points of the problem.

1. We will start similarly as to the solution of Exercise 6.5.19. Consider the circular transform

$$f_1 : D \to \Delta_1, \quad f_1(z) = \frac{z-a}{z-b},$$

where

$$\Delta_1 = \mathbb{C} \setminus (\{z \in \mathbb{C} : \operatorname{Im} z = 0, \operatorname{Re} z \le 0\} \cup \{1\}).$$

The boundary of the domain D is

$$\partial D = [a, b] = \{z \in \mathbb{C} : z = z(t) = (1 - t)a + tb, t \in [0, 1]\},$$

and the values of f_1 in these boundary points are

$$f_1(z(t)) = f_1((1 - t)a + tb) = \frac{t}{t - 1}, \quad t \in [0, 1).$$

Let

$$f_1^* : [0, 1) \to \mathbb{R}, \quad f_1^*(t) = \frac{t}{t - 1},$$

a decreasing function on the domain of definition, and

$$\lim_{t \downarrow 0} f_1^*(t) = 0, \quad \lim_{t \uparrow 1} f_1^*(t) = -\infty.$$

From here, we get $f_1^*([0, 1)) = (-\infty, 0]$, and since $\lim_{z \to \infty} f_1(z) = 1$ it follows that $f_1(D) = \Delta_1$. Thus, the function f_1 conformally maps the domain D onto the domain Δ_1.

Now we will divide the arguments of the numbers of Δ_1 by 2, and we will rotate these with the angle $\frac{\pi}{2}$, i. e.,

$$f_2 : \Delta_1 \to \Omega, \quad f_2(z) = i\sqrt{z},$$

where $f_2(2) = i\sqrt{2}$. Since $f_2(\Delta_1) = \Omega$, composing both of these functions we deduce that $f = f_2 \circ f_1$, i. e.,

$$f(z) = i\sqrt{\frac{z - a}{z - b}}, \quad \text{where } f(2b - a) = i\sqrt{2}.$$

2. We will use the result of the point 1, where $T = [1 + i, -3 - i]$, i. e., $a = 1 + i$ and $b = -3 - i$. We obtain

$$f(z) = i\sqrt{\frac{z - 1 - i}{z + 3 + i}}, \quad \text{where } f(-7 - 3i) = i\sqrt{2}.$$

3. In this case, the function will be

$$f(z) = i\sqrt{\frac{z + i}{z - i}}, \quad \text{where } f(3i) = i\sqrt{2}.$$

Here, we used the result obtained in the point 1, for the special case $a = -i$ and $b = i$.

4. Considering in the point 1 the set $T = [-1, 1]$, i. e., $a = -1$ and $b = 1$, then we deduce that

$$f(z) = i\sqrt{\frac{z + 1}{z - 1}}, \quad \text{where } f(3) = i\sqrt{2}.$$

Solution of Exercise 6.5.22

The function will be determined by using the images of some known functions. Let

$$f_1 : D \to \{z \in \mathbb{C} : -1 < \operatorname{Im} z < 0\} = \Delta_1, \quad f_1(z) = \frac{2}{z}.$$

The boundary of the domain D is

$$\partial D = \partial U(i;1) \cup \{z \in \mathbb{C} : \operatorname{Im} z = 0\}.$$

Case 1. If $z \in \{z \in \mathbb{C} : \operatorname{Im} z = 0\}$, then

$$\forall z \in \{z \in \mathbb{C} : \operatorname{Im} z = 0\} \setminus \{0\} \Rightarrow f_1(z) = \frac{2}{z} \in \mathbb{R},$$

hence $f_1(\{z \in \mathbb{C} : \operatorname{Im} z = 0\} \setminus \{0\}) = \mathbb{R}$.

Case 2. If $z \in \partial U(i;1)$, then $z = i + e^{i\theta}$, $\theta \in [0, 2\pi]$. The images of these points by the function f_1 are

$$f_1(z) = f_1(i + e^{i\theta}) = \frac{2}{i + e^{i\theta}} = \frac{\cos\theta}{1 + \sin\theta} - i, \quad \theta \in [0, 2\pi] \setminus \left\{\frac{3\pi}{2}\right\},$$

hence $f_1(\partial U(i;1)) = \{z \in \mathbb{C} : \operatorname{Im} z = -1\}$. The image of a point that lies in D is

$$3i \in D \Rightarrow f(3i) = -\frac{2i}{3} \in \Delta_1,$$

thus the function f_1 conformally maps the domain D onto the domain Δ_1.

Now we will multiply the modules of the numbers of Δ_1 by π, and then we will make a translation with πi, i. e.,

$$f_2 : \Delta_1 \to \{z \in \mathbb{C} : 0 < \operatorname{Im} z < \pi\} = \Delta_2, \quad f_2(z) = \pi z + \pi i,$$

hence $f_2(\Delta_1) = \Delta_2$.

Finally, the function

$$f_3 : \Delta_2 \to \Omega, \quad f_3(z) = e^z$$

satisfies $f_3(\Delta_2) = \Omega$, and composing all the above functions we conclude that $f = f_3 \circ f_2 \circ f_1$, i. e.,

$$f(z) = e^{\frac{\pi(iz+2)}{z}}.$$

Solution of Exercise 6.5.23

We will multiply by 3 the arguments of all the complex numbers of D, i. e.,

$$f_1 : D \to \{z \in \mathbb{C} : \operatorname{Im} z > 0\} = \Delta_1, \quad f_1(z) = z^3,$$

hence $f_1(D) = \Delta_1$.

If

$$f_2 : \Delta_1 \to \{z \in \mathbb{C} : 0 < \operatorname{Im} z < \pi\} = \Delta_2, \quad f_2(z) = \log z,$$

where $\log i = \frac{i\pi}{2}$, then $f_2(\Delta_1) = \Delta_2$.

Letting

$$f_3 : \Delta_2 \to \Omega, \quad f_3(z) = \frac{z}{2},$$

we have $f_3(\Delta_2) = \Omega$. Composing both of the previous functions, we get $f = f_2 \circ f_1$, i. e.,

$$f(z) = \frac{1}{2}\log z^3, \quad \text{where } f(1+i) = \frac{3}{4}\ln 2 + \frac{3\pi i}{8}.$$

Solution of Exercise 6.5.24

We will determine the images of the boundaries of the definition domains, and then we will calculate the image of an interior point.

We have $\partial D = \partial U(a;r) = \{a + re^{i\theta} : \theta \in [0, 2\pi]\}$. Since the given function is a circular transform, and it is well known that it maps any circles of $\mathbb{C}_\infty$ onto the circles of $\mathbb{C}_\infty$, it is sufficient to determine the images of three points of the boundary. Thus, let

$$\{a + r, a - r, a + ir\} \subset \partial D.$$

Since α^* is the inverse of the point α with respect to the circle $\partial U(a;r)$, it follows that

$$(\alpha^* - a)\overline{(\alpha - a)} = r^2 \Leftrightarrow \alpha^* = a + \frac{r^2}{\overline{\alpha - a}}.$$

From here, we get

$$f(a+r) = -\frac{\overline{\alpha - a}}{r}\frac{a - \alpha + r}{\overline{a - \alpha + r}} \Rightarrow |f(a+r)| = \frac{|\alpha - a|}{r},$$

$$f(a-r) = \frac{\overline{\alpha - a}}{r}\frac{a - \alpha - r}{\overline{a - \alpha - r}} \Rightarrow |f(a-r)| = \frac{|\alpha - a|}{r}$$

and

$$f(a - ir) = i\frac{\overline{\alpha - a}}{r}\frac{a - \alpha + ir}{\overline{a - \alpha - ir}} \Rightarrow |f(a + ir)| = \frac{|\alpha - a|}{r},$$

hence

$$\{f(a+r), f(a-r), f(a+ir)\} \subset \partial\Omega.$$

Now we will study both of the points of the problem.

1. The image of the domain of the definition will be given by

$$z = a \in D \Rightarrow |f(a)| = \left|\frac{a-\alpha}{a-\alpha^*}\right| = \left(\frac{|a-\alpha|}{r}\right)^2 < \frac{|a-\alpha|}{r}, \quad \text{because } |a-\alpha| < r,$$

i. e., $f(a) \in \Omega$. From here, it follows that $f(D) = \Omega$, which completes our proof.

2. If α and α^* are two inverse points with respect to the circle $\partial U(a; r)$, then $\alpha = a + r_1 e^{i\varphi}$, $\alpha^* = a + r_2 e^{i\varphi}$, where $r_1 r_2 = r^2$, hence

$$\alpha^* = a + \frac{r^2}{r_1} e^{i\varphi}.$$

The image of the domain of the definition will be given by

$$z = a + \lambda r e^{i\varphi} \in D, \quad \lambda \in (1, +\infty) \setminus \left\{\frac{r}{r_1}\right\} \Rightarrow f(a + \lambda r e^{i\varphi}) = \frac{r_1}{r}\frac{\lambda r - r_1}{\lambda r_1 - r}$$

$$\Rightarrow |f(a + \lambda r e^{i\varphi})| = \frac{r_1}{r}\frac{\lambda r - r_1}{|\lambda r_1 - r|} > 1, \quad \forall \lambda \in (1, +\infty) \setminus \left\{\frac{r}{r_1}\right\} \Rightarrow f(a + \lambda r e^{i\varphi}) \in \Omega,$$

and $\lim_{z\to\infty} f(z) = 1$. We obtained that the function $f(z) = \frac{z-\alpha}{z-\alpha^*}$ conformally maps the set D onto the set Ω.

Solution of Exercise 6.5.25

We will use the result obtained in Exercise 6.5.9, i. e.,

$$z \in \partial D \Rightarrow f(z) = f(e^{i\theta}) = \cos\theta, \quad \theta \in [0, 2\pi],$$

hence $f(\partial D) = [-1, 1]$.

The image of an interior point will be

$$\frac{1}{2} \in D \Rightarrow f\left(\frac{1}{2}\right) = \frac{5}{4} \in \Omega,$$

hence the function f conformally maps the domain D onto the domain Ω.

Solution of Exercise 6.5.26

Considering the Joukowski function

$$f(z) = \frac{1}{2}\left(z + \frac{1}{z}\right),$$

we have $f \in H(D)$. To Exercise 6.5.25, we proved that $f(\partial U(0;1)) = [-1,1]$, hence it is sufficient to show that

$$f\left(\left[\frac{1}{2},1\right)\right) = \left(1,\frac{5}{4}\right].$$

This last result holds, since

$$z = x \in \left[\frac{1}{2},1\right) \Rightarrow f(z) = f(x) = \frac{1}{2}\left(x+\frac{1}{x}\right),$$

and from

$$f'(x) = \frac{1}{2}\left(1-\frac{1}{x^2}\right) < 0, \quad x \in \left[\frac{1}{2},1\right) \Rightarrow f \text{ is a decreasing function,}$$

with

$$f\left(\frac{1}{2}\right) = \frac{5}{4}, \quad \lim_{x\uparrow 1} f(x) = 1.$$

It follows that the Joukowski function conformally maps the domain D onto the domain Ω.

Solution of Exercise 6.5.27

Similar to the previous problem, we define the Joukowski function

$$f_1 : D \to \Delta_1, \quad f_1(z) = \frac{1}{2}\left(z+\frac{1}{z}\right),$$

where the boundary of the domain D is $\partial D = \partial U(0;1) \cup (-1,0] \cup [a,1)$.

To Exercise 6.5.25, we proved that

$$f_1(\partial U(0;1)) = [-1,1]. \tag{7.56}$$

Now we will determine the images of the intervals $(-1,0]$ and $[a,1)$ by the function f_1, i. e.,

$$z = x \in (-1,0] \cup [a,1) \Rightarrow f_1(z) = f_1(x) = \frac{1}{2}\left(x+\frac{1}{x}\right),$$

where we know that f_1 is a decreasing function (see Exercise 6.5.26), and

$$\lim_{x\downarrow -1} f_1(x) = -1, \quad \lim_{x\uparrow 0} f_1(x) = -\infty,$$
$$f_1(a) = \frac{1}{2}\left(a+\frac{1}{a}\right), \quad \lim_{x\uparrow 1} f_1(x) = 1.$$

Thus

$$f_1((-1,0] \cup [a,1)) = (-\infty,-1) \cup \left(1, \frac{1}{2}\left(a + \frac{1}{a}\right)\right]. \tag{7.57}$$

From the relations (7.56) and (7.57), we get

$$f_1(D) = \Delta_1 = \mathbb{C} \setminus \left(-\infty, \frac{1}{2}\left(a + \frac{1}{a}\right)\right].$$

Translating all the numbers of the domain Δ_1 with $\frac{1}{2}(a + \frac{1}{a})$, then we will divide the arguments by 2, and finally we will rotate by the angle $\frac{\pi}{2}$, i. e.,

$$f_2 : \Delta_1 \to \Omega, \quad f_2(z) = i\sqrt{z - \frac{1}{2}\left(a + \frac{1}{a}\right)},$$

where $f_2(a + \frac{1}{2a}) = i\sqrt{\frac{a}{2}}$. Since $f_2(\Delta_1) = \Omega$, composing both of these functions we get $f = f_2 \circ f_1$, i. e.,

$$f(z) = i\sqrt{\frac{1}{2}\left(z + \frac{1}{z}\right) - \frac{1}{2}\left(a + \frac{1}{a}\right)}, \quad \text{where } f\left(\frac{a}{2}\right) = \frac{i\sqrt{a}}{2}.$$

Solution of Exercise 6.5.28

The function will be determined by using the images of some known functions. Using the result of Exercise 6.5.26, we have

$$f_1 : D \to \mathbb{C} \setminus \left[-1, \frac{5}{4}\right] = \Delta_1, \quad f_1(z) = \frac{1}{2}\left(z + \frac{1}{z}\right),$$

and moreover, $f_1(D) = \Delta_1$.

Now we will use the result of the point 1 of Exercise 6.5.21, for the special case $a = -1$ and $b = \frac{5}{4}$:

$$f_2 : \Delta_1 \to \Omega, \quad f_2(z) = i\sqrt{\frac{z+1}{z - \frac{5}{4}}},$$

where $f_2(\frac{7}{2}) = i\sqrt{2}$.

The required function will be $f = f_2 \circ f_1$, i. e.,

$$f(z) = i\sqrt{\frac{2(z+1)^2}{2z^2 - 5z + 2}}, \quad \text{where } f\left(\frac{1}{4}\right) = \frac{5i\sqrt{7}}{7}.$$

Solution of Exercise 6.5.29

We will determine a circular function with the above property. The points $w_1 = 0$ and $w_2 = \infty$ are inverse points with respect to the circles $|w| = \frac{2}{3}$ and $|w| = 1$. Thus $\exists z_1, z_2 \in \mathbb{C} : w(z_1) = 0, w(z_2) = \infty$ and z_1, z_2 are inverse points with respect to the circles $|z-3| = 9$ and $|z-8| = 16$. Then

$$|z_1 - 3||z_2 - 3| = 81 \tag{7.58}$$

$$\arg(z_1 - 3) = \arg(z_2 - 3), \tag{7.59}$$

and respectively,

$$|z_1 - 8||z_2 - 8| = 256 \tag{7.60}$$

$$\arg(z_1 - 8) = \arg(z_2 - 8). \tag{7.61}$$

From (7.59) and (7.61), it follows that $\arg z_1 = \arg z_2 \in \{0, \pi\}$, i. e., $z_1, z_2 \in \mathbb{R}$. Now, from (7.58) and (7.60) we deduce that $z_1 = 0$ and $z_2 = -24$ or conversely, $z_1 = -24$ and $z_2 = 0$.

Case 1. If $z_1 = 0$, $z_2 = -24$, then

$$\begin{cases} w(0) = 0 \\ w(-24) = \infty \end{cases} \Leftrightarrow w(z) = k\frac{z}{z+24}, \quad k \in \mathbb{C}.$$

Let denote by $C_1 = \{z \in \mathbb{C} : |z-3| > 9\}$, $C_2 = \{z \in \mathbb{C} : |z-8| < 16\}$ and $C_1' = \{w \in \mathbb{C} : |w| = \frac{3}{2}\}$, $C_2' = \{w \in \mathbb{C} : |w| = 1\}$. From

$$0 \in U(3;9) \cap U(8;16) \Rightarrow 0 = w(0) \in U\left(0; \frac{2}{3}\right) \cap U(0;1)$$
$$\Rightarrow C_1' = w(C_1), \quad C_2' = w(C_2),$$

we have

$$24 \in C_2 \Rightarrow w(24) \in C_2' \Leftrightarrow |w(24)| = 1 \Leftrightarrow \left|\frac{k}{2}\right| = 1 \Leftrightarrow k = 2e^{i\theta}, \quad \theta \in \mathbb{R},$$

and thus

$$w(z) = 2e^{i\theta}\frac{z}{z+24}, \quad \theta \in \mathbb{R}.$$

Case 2. If $z_1 = -24$, $z_2 = 0$, then

$$\begin{cases} w(-24) = 0 \\ w(0) = \infty \end{cases} \Leftrightarrow w(z) = k\frac{z+24}{z}, \quad k \in \mathbb{C}.$$

From

$$0 \in U(3;9) \cap U(8;16) \Rightarrow \infty = w(0) \in \left(\mathbb{C} \setminus U\left(0; \frac{2}{3}\right)\right) \cap (\mathbb{C} \setminus U(0;1)),$$
$$\Rightarrow C_1' = w(C_2), \quad C_2' = w(C_1),$$

we get

$$24 \in \mathcal{C}_2 \Rightarrow w(24) \in \mathcal{C}_1' \Leftrightarrow |w(24)| = \frac{2}{3} \Leftrightarrow |2k| = \frac{2}{3} \Leftrightarrow k = \frac{e^{i\theta}}{3}, \quad \theta \in \mathbb{R},$$

and thus

$$w(z) = \frac{e^{i\theta}}{3}\frac{z+24}{z}, \quad \theta \in \mathbb{R}.$$

Solution of Exercise 6.5.30

Similar to the previous problem, we will determine a circular function with the above property. The points $w_1 = 0$ and $w_2 = \infty$ are inverse points with respect to the circles $|w| = 1$ and $|w| = \frac{4}{5}\sqrt{\frac{5}{2}}$. Then $\exists z_1, z_2 \in \mathbb{C} : w(z_1) = 0$, $w(z_2) = \infty$ and z_1, z_2 are inverse points with respect to the circles $|z| = 1$ and $|z - \frac{1}{2}| = \sqrt{\frac{5}{2}}$. Then

$$|z_1||z_2| = 1 \tag{7.62}$$

$$\arg z_1 = \arg z_2, \tag{7.63}$$

and respectively,

$$\left|z_1 + \frac{1}{2}\right|\left|z_2 + \frac{1}{2}\right| = \frac{5}{2} \tag{7.64}$$

$$\arg\left(z_1 + \frac{1}{2}\right) = \arg\left(z_2 + \frac{1}{2}\right). \tag{7.65}$$

From (7.63) and (7.65), it follows that $\arg z_1 = \arg z_2 \in \{0, \pi\}$, i. e., $z_1, z_2 \in \mathbb{R}$. Now, from (7.62) and (7.64) we deduce that $z_1 = -\frac{1}{2}$ and $z_2 = -2$ or conversely, $z_1 = -2$ and $z_2 = -\frac{1}{2}$.

Case 1. if $z_1 = \frac{1}{2}, z_2 = 2$, then

$$\begin{cases} w(\frac{1}{2}) = 0 \\ w(2) = \infty \end{cases} \Leftrightarrow w(z) = k\frac{z - \frac{1}{2}}{z - 2}, \quad k \in \mathbb{C}.$$

Denote by $\mathcal{C}_1 = \{z \in \mathbb{C} : |z| > 1\}$, $\mathcal{C}_2 = \{z \in \mathbb{C} : |z + \frac{1}{2}| < \sqrt{\frac{5}{2}}\}$ and $\mathcal{C}_1' = \{w \in \mathbb{C} : |w| = 1\}$, $\mathcal{C}_2' = \{w \in \mathbb{C} : |w| = \frac{4}{5}\sqrt{\frac{5}{2}}\}$. From

$$\frac{1}{2} \in U(0;1) \cap U\left(\frac{1}{2}; \sqrt{\frac{5}{2}}\right) \Rightarrow 0 = w\left(\frac{1}{2}\right) \in U(0;1) \cap U\left(0; \frac{4}{5}\sqrt{\frac{5}{2}}\right)$$
$$\Rightarrow \mathcal{C}_1' = w(\mathcal{C}_1), \quad \mathcal{C}_2' = w(\mathcal{C}_2),$$

we have

$$1 \in \mathcal{C}_1 \Rightarrow w(1) \in \mathcal{C}_1' \Leftrightarrow |w(1)| = 1 \Leftrightarrow \left|\frac{k}{2}\right| = 1 \Leftrightarrow k = 2e^{i\theta}, \quad \theta \in \mathbb{R},$$

and thus

$$w(z) = 2e^{i\theta}\frac{z-\frac{1}{2}}{z-2}, \quad \theta \in \mathbb{R}.$$

Case 2. If $z_1 = 2$, $z_2 = \frac{1}{2}$, then

$$\begin{cases} w(2) = 0 \\ w(\frac{1}{2}) = \infty \end{cases} \Leftrightarrow w(z) = k\frac{z-2}{z-\frac{1}{2}}, \quad k \in \mathbb{C}.$$

From

$$\frac{1}{2} \in U(0;1) \cap U\left(\frac{1}{2}; \sqrt{\frac{5}{2}}\right) \Rightarrow \infty = w\left(\frac{1}{2}\right) \in (\mathbb{C} \setminus U(0;1)) \cap \left(\mathbb{C} \setminus U\left(0; \frac{4}{5}\sqrt{\frac{5}{2}}\right)\right)$$
$$\Rightarrow \mathcal{C}_1' = w(\mathcal{C}_2), \quad \mathcal{C}_2' = w(\mathcal{C}_1),$$

we get

$$1 \in \mathcal{C}_1 \Rightarrow w(1) \in \mathcal{C}_2' \Leftrightarrow |w(1)| = \frac{4}{5}\sqrt{\frac{5}{2}} \Leftrightarrow |2k| = \frac{4}{5}\sqrt{\frac{5}{2}} \Leftrightarrow k = \frac{2}{5}\sqrt{\frac{5}{2}}e^{i\theta}, \quad \theta \in \mathbb{R},$$

and thus

$$w(z) = \frac{4}{5}\sqrt{\frac{5}{2}}e^{i\theta}\frac{z-2}{z-\frac{1}{2}}, \quad \theta \in \mathbb{R}.$$

Solution of Exercise 6.5.31

First, we will rotate the points of the domain D by the angle $-\frac{\pi}{2}$, using the function

$$f_1(D) = \{z \in \mathbb{C} : \operatorname{Re} z > 0\} \setminus (0,1] = \Delta_1, \quad f_1(z) = -iz.$$

Then we will multiply the arguments of the points of the domain Δ_1 by 2, i. e.,

$$f_2(\Delta_1) = \mathbb{C} \setminus (-\infty, 1] = \Delta_2, \quad f_2(z) = z^2.$$

We will translate the image by -1, with the function

$$f_3 : (\Delta_2) = \mathbb{C} \setminus (-\infty, 0] = \Delta_3, \quad f_3(z) = z - 1.$$

We will divide by 2 the arguments of the points of the domain Δ_3, i. e.,

$$f_4(\Delta_3) = \{z \in \mathbb{C} : \operatorname{Re} z > 0\} = \Delta_4, \quad f_4(z) = \sqrt{z},$$

where $\sqrt{1} = 1$. Rotating the domain Δ_4 with the angle $\frac{\pi}{2}$ we get

$$f_5(\Delta_4) = \Omega, \quad f_5(z) = e^{i\frac{\pi}{2}}z = iz.$$

Composing all the above functions, we conclude that $f = f_5 \circ f_4 \circ f_3 \circ f_2 \circ f_1$, and

$$f(z) = i\sqrt{-z^2 - 1}, \quad \text{where } f(i\sqrt{2}) = i.$$

Solution of Exercise 6.5.32

Starting from the Joukowski function,

$$f_1 : D \to \Delta_1, \quad f_1(z) = \frac{1}{2}\left(z + \frac{1}{z}\right),$$

then the domain D has the boundary

$$\partial D = \{e^{i\theta} \in \mathbb{C} : \theta \in [0, \pi]\} \cup [-1, 1] \cup (0, \alpha i].$$

Using the result of Exercise 6.5.6, we get

$$f_1(\{e^{i\theta} \in \mathbb{C} : \theta \in [0, \pi]\} \cup [-1, 1]) = \{z \in \mathbb{C} : \operatorname{Im} z = 0\}.$$

Now we will determine the image of the segment $(0, \alpha i]$ via the function f_1, i. e.,

$$z \in (0, \alpha i] \Rightarrow f_1(z) = f_1(iy) = \frac{1}{2}\left(y - \frac{1}{y}\right)i, \quad y \in (0, \alpha].$$

Letting

$$\varphi_1 : (0, \alpha] \to \mathbb{R}, \quad \varphi_1(y) = \frac{1}{2}\left(y - \frac{1}{y}\right),$$

then φ_1 is an increasing function because $\varphi_1'(y) > 0$, $\forall y \in (0, \alpha]$, and

$$\lim_{y \downarrow 0} \varphi_1(y) = -\infty, \quad \lim_{y \uparrow \alpha} \varphi_1(y) = \frac{1}{2}\left(\alpha - \frac{1}{\alpha}\right).$$

From here, it follows that

$$f_1((0, \alpha]) = \left\{z \in \mathbb{C} : \operatorname{Re} z = 0, \operatorname{Im} z \le \frac{1}{2}\left(\alpha - \frac{1}{\alpha}\right)\right\}.$$

Since

$$z = \frac{(1+\alpha)i}{2} \in D \Rightarrow \operatorname{Im} f_1\left(\frac{(1+\alpha)i}{2}\right) = \operatorname{Im} \frac{(1+\alpha)^2 - 4}{4(1+\alpha)} < 0, \quad \forall \alpha \in (0, 1),$$

we get

$$\Delta_1 = \{z \in \mathbb{C} : \operatorname{Im} z < 0\} \setminus \left\{z \in \mathbb{C} : \operatorname{Re} z = 0, \operatorname{Im} z \le \frac{1}{2}\left(\alpha - \frac{1}{\alpha}\right)\right\},$$

and $f_1(D) = \Delta_1$.

Letting $f_2(z) = \frac{1}{z}$, then

$$f_2(\Delta_1) = \{z \in \mathbb{C} : \operatorname{Im} z > 0\} \setminus \left\{z \in \mathbb{C} : 0 < \operatorname{Im} z \le \frac{2\alpha}{1 - \alpha^2}\right\} = \Delta_2.$$

The new problem that we have obtained using this way is similar to Exercise 6.5.31. Similarly, the function f_3 will be translated not by -1 but with $-\frac{2\alpha}{1-\alpha^2}$, and thus

$$f_3 : \{z \in \mathbb{C} : \operatorname{Im} z > 0\} \setminus \left(0, \frac{2\alpha}{1-\alpha^2} i\right] = \Delta_2 \to \Omega,$$

$$f_3(z) = i\sqrt{-z^2 - \frac{2\alpha}{1-\alpha^2}},$$

where

$$f_3\left(\left(1 + \frac{2\alpha}{1-\alpha^2}\right)i\right) = i\sqrt{1 + \frac{2\alpha}{1-\alpha^2} + \frac{4\alpha^2}{(1-\alpha^2)^2}}.$$

The required function will be $f = f_3 \circ f_2 \circ f_1$, i. e.,

$$f(z) = i\sqrt{-\frac{4z^2}{(z^2+1)^2} - \frac{2\alpha}{1-\alpha^2}},$$

where the branch of the root function is the principal one.

Bibliography

L. Ahlfors: *Complex Analysis*. McGraw-Hill Book Comp., New York, 1966.

N. Boboc: *Complex Functions* (in Roumanian). Didactical and Pedagogical Publ., Bucharest, 1969.

T. Bulboacă, A. B. Németh: *Complex Analysis* (in Hungarian). Ábel Publ., Cluj-Napoca, 2010. Fourth edition (revised and completed).

T. Bulboacă, J. Salamon: *Complex Analysis. Problems and Solutions* (in Hungarian). Ábel Publ., Cluj-Napoca, 2012. Third edition (revised and completed).

H. Cartan: *La théorie élementaire des fonctions analytiques d'une ou plusieurs variables complexes*. Herman, Paris, 1961.

C. Andreian Cazacu (ed.): *Complex Analysis. Clasic and Modern Aspects* (in Roumanian). Scientific and Encyclopedical Publ., Bucharest, 1988.

G. Călugăreanu: *Elements of the Theory of One Complex Variable Functions* (in Roumanian). Didactical and Pedagogical Publ., Bucharest, 1963.

J. B. Conway: *Functions of One Complex Variable*. Springer-Verlag, Berlin, 1973.

J. B. Conway: *Functions of One Complex Variable II*. Graduate Texts in Mathematics 159, Springer-Verlag, New York, 1996.

M. Evgrafov, K. Bejanov, Y. Sidorov, M. Fedoruk, M. Chabounine: *Recueil de problemès sur la théorie des fonctions analytiques*. Mir Publ., Moscow, 1974.

D. Gaşpar, N. Suciu: *Complex Analysis* (in Roumanian). Romanian Academy Publ., Bucharest, 1999.

P. Hamburg, P. Mocanu, N. Negoescu: *Mathematical Analysis (Complex Functions)* (in Roumanian). Didactical and Pedagogical Publ., Bucharest, 1982.

P. Henrici: *Applied and Computational Complex Analysis, vol. 3*. Wiley Classics Library, J. Wiley & Sons, New York, 1993.

A. Kolmogorov, S. Fomine: *Éléments de la théorie des fonctions et de l'analyse fonctionelle* (in French). Mir Publ., Moscow, 1974.

M. A. Lavrentiev, B. Shabat: *Methods of the Theory of Complex Variable Functions* (in Russian). Fizmatgiz, Moscow, 1958.

N. Lebedev: *Special Functions and Their Applications* (in Roumanian). Technical Publ., Bucharest, 1957.

A. I. Markushevich: *The Theory of Analytic Functions* (in Russian). Technical Publ., Moscow, 1950.

O. Mayer: *Function Theory of One Complex Variable* (in Roumanian). Romanian Academy Publ., Bucharest, 1981.

O. Mayer: *Special Problems of the Theory of One Complex Variables* (in Roumanian). Romanian Academy Publ., Bucharest, 1990.

I. M. Milin: *Univalent Functions and Orthonormal Systems*. Nauka Publ., Moscow, 1971.

Gh. Mocanu: *Introduction in the Theory of Complex Functions, vols. I–II* (in Roumanian). Bucharest University Publ., Bucharest, 1996.

Gh. Mocanu, Gh. Stoian, E. Vişinescu: *Function Theory of One Complex Variables – Problems Collections* (in Roumanian). Didactical and Pedagogical Publ., Bucharest, 1970.

P. Mocanu: *Complex Functions* (in Roumanian). Cluj University Publ., Cluj-Napoca, 1972.

G. Mocică: *Problems of Special Functions* (in Roumanian). Didactical and Pedagogical Publ., Bucharest, 1988.

I. P. Natanson: *Function Theory of Real Variable* (in Roumanian). Technical Publ., Bucharest, 1957 (translation from Russian).

W. Rudin: *Real and Complex Analysis*, 3rd Edition. McGraw-Hill, Inc., New-York, 1987.

V. Rudner: *Problems of Special Mathematics* (in Roumanian). Didactical and Pedagogical Publ., Bucharest, 1970.

https://doi.org/10.1515/9783110657869-008

V. Smirnov: *Cours de mathématiques supérieurs. tome III, part 2*, Mir Publ., Moscow, 1972.
M. R. Spiegel, S. Lipschutz, J. J. Schiller, and D. Spellman: *Complex variables with an introduction to conformal mapping and its applications*, 2nd Edition. Schaum's Outline Series, McGraw-Hill, New York, 2009.
S. Stoilov: *Function Theory of Complex Variable* (in Roumanian). vol. I, Romanian Academy Publ., Bucharest, 1954, vol. II, Romanian Academy Publ., Bucharest, 1958.
S. Stoilov: *Function Theory of Complex Variable* (in Roumanian). Didactical and Pedagogical Publ., Bucharest, 1962.
M. Stoka: *Problems of Complex Analysis* (in Roumanian). Technical Publ., Bucharest, 1965.
G. Valiron: *Théorie des fonctions*. Masson et Cie, Paris, 1948.
L. I. Volkovyskii, G. L. Lunts, I. G. Aramanovich: *A Collection of Problems on Complex Analysis*. Mir Publ., Moscow, 1977.

Index